Nanopatterning and Nanoscale Devices for Biological Applications

Devices, Circuits, and Systems

Series Editor
Krzysztof Iniewski
CMOS Emerging Technologies Research Inc.,
Vancouver, British Columbia, Canada

PUBLISHED TITLES:

Atomic Nanoscale Technology in the Nuclear Industry
Taeho Woo

Biological and Medical Sensor Technologies
Krzysztof Iniewski

Building Sensor Networks: From Design to Applications
Ioanis Nikolaidis and Krzysztof Iniewski

Circuits at the Nanoscale: Communications, Imaging, and Sensing
Krzysztof Iniewski

Design of 3D Integrated Circuits and Systems
Rohit Sharma

Electrical Solitons: Theory, Design, and Applications
David Ricketts and Donhee Ham

Electronics for Radiation Detection
Krzysztof Iniewski

Embedded and Networking Systems: Design, Software, and Implementation
Gul N. Khan and Krzysztof Iniewski

Energy Harvesting with Functional Materials and Microsystems
Madhu Bhaskaran, Sharath Sriram, and Krzysztof Iniewski

Graphene, Carbon Nanotubes, and Nanostuctures: Techniques and Applications
James E. Morris and Krzysztof Iniewski

High-Speed Devices and Circuits with THz Applications
Jung Han Choi

High-Speed Photonics Interconnects
Lukas Chrostowski and Krzysztof Iniewski

Integrated Microsystems: Electronics, Photonics, and Biotechnology
Krzysztof Iniewski

Integrated Power Devices and TCAD Simulation
Yue Fu, Zhanming Li, Wai Tung Ng, and Johnny K.O. Sin

Internet Networks: Wired, Wireless, and Optical Technologies
Krzysztof Iniewski

PUBLISHED TITLES:

Labs-on-Chip: Physics, Design and Technology
Eugenio Iannone

Low Power Emerging Wireless Technologies
Reza Mahmoudi and Krzysztof Iniewski

Medical Imaging: Technology and Applications
Troy Farncombe and Krzysztof Iniewski

Metallic Spintronic Devices
Xiaobin Wang

MEMS: Fundamental Technology and Applications
Vikas Choudhary and Krzysztof Iniewski

Micro and Nanoelectronics: Emerging Device Challenges and Solutions
Tomasz Brozek

Microfluidics and Nanotechnology: Biosensing to the Single Molecule Limit
Eric Lagally

MIMO Power Line Communications: Narrow and Broadband Standards, EMC, and Advanced Processing
Lars Torsten Berger, Andreas Schwager, Pascal Pagani, and Daniel Schneider

Mobile Point-of-Care Monitors and Diagnostic Device Design
Walter Karlen

Nano-Semiconductors: Devices and Technology
Krzysztof Iniewski

Nanoelectronic Device Applications Handbook
James E. Morris and Krzysztof Iniewski

Nanopatterning and Nanoscale Devices for Biological Applications
Šeila Selimović

Nanoplasmonics: Advanced Device Applications
James W. M. Chon and Krzysztof Iniewski

Nanoscale Semiconductor Memories: Technology and Applications
Santosh K. Kurinec and Krzysztof Iniewski

Novel Advances in Microsystems Technologies and Their Applications
Laurent A. Francis and Krzysztof Iniewski

Optical, Acoustic, Magnetic, and Mechanical Sensor Technologies
Krzysztof Iniewski

Radiation Effects in Semiconductors
Krzysztof Iniewski

Semiconductor Radiation Detection Systems
Krzysztof Iniewski

Smart Grids: Clouds, Communications, Open Source, and Automation
David Bakken and Krzysztof Iniewski

PUBLISHED TITLES:

Smart Sensors for Industrial Applications
Krzysztof Iniewski

Technologies for Smart Sensors and Sensor Fusion
Kevin Yallup and Krzysztof Iniewski

Telecommunication Networks
Eugenio Iannone

Testing for Small-Delay Defects in Nanoscale CMOS Integrated Circuits
Sandeep K. Goel and Krishnendu Chakrabarty

VLSI: Circuits for Emerging Applications
Tomasz Wojcicki

Wireless Technologies: Circuits, Systems, and Devices
Krzysztof Iniewski

FORTHCOMING TITLES:

Cell and Material Interface: Advances in Tissue Engineering, Biosensor, Implant, and Imaging Technologies
Nihal Engin Vrana

Circuits and Systems for Security and Privacy
Farhana Sheikh and Leonel Sousa

CMOS: Front-End Electronics for Radiation Sensors
Angelo Rivetti

CMOS Time-Mode Circuits and Systems: Fundamentals and Applications
Fei Yuan

Electrostatic Discharge Protection of Semiconductor Devices and Integrated Circuits
Juin J. Liou

Gallium Nitride (GaN): Physics, Devices, and Technology
Farid Medjdoub and Krzysztof Iniewski

High Frequency Communication and Sensing: Traveling-Wave Techniques
Ahmet Tekin and Ahmed Emira

Implantable Wireless Medical Devices: Design and Applications
Pietro Salvo

Laser-Based Optical Detection of Explosives
Paul M. Pellegrino, Ellen L. Holthoff, and Mikella E. Farrell

Mixed-Signal Circuits
Thomas Noulis and Mani Soma

FORTHCOMING TITLES:

MRI: Physics, Image Reconstruction, and Analysis
Angshul Majumdar and Rabab Ward

Multisensor Data Fusion: From Algorithm and Architecture Design to Applications
Hassen Fourati

Nanoelectronics: Devices, Circuits, and Systems
Nikos Konofaos

Nanomaterials: A Guide to Fabrication and Applications
Gordon Harling, Krzysztof Iniewski, and Sivashankar Krishnamoorthy

Optical Fiber Sensors and Applications
Ginu Rajan and Krzysztof Iniewski

Organic Solar Cells: Materials, Devices, Interfaces, and Modeling
Qiquan Qiao and Krzysztof Iniewski

Power Management Integrated Circuits and Technologies
Mona M. Hella and Patrick Mercier

Reconfigurable Logic: Architecture, Tools, and Applications
Pierre-Emmanuel Gaillardon

Radio Frequency Integrated Circuit Design
Sebastian Magierowski

Soft Errors: From Particles to Circuits
Jean-Luc Autran and Daniela Munteanu

Solid-State Radiation Detectors: Technology and Applications
Salah Awadalla

Wireless Transceiver Circuits: System Perspectives and Design Aspects
Woogeun Rhee and Krzysztof Iniewski

Nanopatterning and Nanoscale Devices for Biological Applications

EDITED BY
Šeila Selimović
Harvard Medical School, Cambridge, Massachusetts, USA

MANAGING EDITOR
Krzysztof Iniewski
CMOS Emerging Technologies Research Inc., Vancouver, British Columbia, Canada

CRC Press
Taylor & Francis Group
Boca Raton London New York

CRC Press is an imprint of the
Taylor & Francis Group, an **informa** business

CRC Press
Taylor & Francis Group
6000 Broken Sound Parkway NW, Suite 300
Boca Raton, FL 33487-2742

First issued in paperback 2017

No claim to original U.S. Government works
Version Date: 20140508

ISBN 13: 978-1-138-07262-6 (pbk)
ISBN 13: 978-1-4665-8631-4 (hbk)

Visit the Taylor & Francis Web site at
http://www.taylorandfrancis.com

and the CRC Press Web site at
http://www.crcpress.com

Contents

Preface

Devices utilizing microscale fluidic components (microfluidics) and microelectromechanical systems have become an important focus in a variety of disciplines, including biology and biological engineering, and have revolutionized multiple technological fields, from medical electronic devices to diagnostics and therapeutics. In the last decade, however, there has been additional multidisciplinary effort to gain control over biological systems on the nanoscale, by reducing the size of key elements in such devices down to the submicron scale or by controlling certain substrate properties such as topography and chemistry. These nanoscale approaches are extending the current capabilities of biological research by allowing the experimenter excellent control over multiple system parameters on a molecular scale—while at the same time offering increased performance in terms of data accuracy, precision, and collection time.

This book provides valuable insight into the latest developments in nanoscale technologies for the study of biological systems, in three parts. The first part focuses on device fabrication methods targeting the substrate on the nanoscale through surface modification, for example, by tailoring wettability. This section of the book also explores the generation of nanostructured biointerfaces and bioelectronics elements. This includes a discussion of the physics and modeling of DNA structures in biosensors as a perfect example of nanoscale biological engineering. We will also take a look at microfluidically generated droplets as reactors enabling nanoscale sample preparation and analysis. The second part is devoted to biosensors and integrated devices with nanoscale functionalities. From sensors that monitor the success of tissue implants to nanofluidic devices for nucleic acid analysis, this section of the book gives an overview of the most common and, therefore, perhaps the most relevant biosensor technologies that are based on nanoscale principles. The third part is devoted to a general discussion of the biological applications of nanoscale devices, including a review of nanotechnology in tissue engineering and regenerative medicine. Here, we are particularly interested in utilizing engineering approaches on the nanoscale to generate, control, and monitor tissues, and we discuss bone tissue and vascular structures as typical application areas.

The target audience is academic researchers primarily in biological and biomedical engineering, but the book is also accessible to researchers in related disciplines such as electrical engineering, biophysics, and biochemistry. We thank all the contributors for their insights into the newest technologies and for sharing their knowledge with our audience. We are hopeful that this book will stimulate the reader to pose new questions and develop new solutions and applications in the effort to advance nanoscale engineering in biological fields.

Krzysztof (Kris) Iniewski
Vancouver, British Columbia, Canada

Šeila Selimović
Washington, DC

Editors

Šeila Selimović is currently an AAAS Science and Technology Policy Fellow in Washington, DC, working on science diplomacy issues relating to energy security and scientific cooperation. Previously, she was a postdoctoral research fellow at Harvard Medical School and Brigham & Women's Hospital in Boston, Massachusetts. An author of over 60 research articles, book chapters, and editorials, Dr. Selimović developed an interest in the development of microfluidic and microelectromechanical systems platforms for applications in biophysics and biological engineering. A special emphasis of her work is on biosensors and organ-on-a-chip platforms and has been funded by the US Army. Her research interests include the physics of microscale flows, protein crystallization, and colloidal suspensions, as well as rheology and microrheology. Dr. Selimović earned her PhD and MSc in physics from Brandeis University, where she was a recipient of a 2-year National Science Foundation traineeship (IGERT), and her Bachelor of Arts, also in physics, from Wellesley College. She is a member of Sigma Xi. Outside her work, she enjoys playing the piano and running in long-distance races. She can be contacted at sselimov@gmail.com.

Krzysztof (Kris) Iniewski manages R&D at Redlen Technologies, a start-up company in Vancouver, Canada. Redlen's revolutionary production process for advanced semiconductor materials enables a new generation of more accurate, all-digital, radiation-based imaging solutions. Kris is also president of CMOS Emerging Technologies Research Inc. (www.cmosetr.com), an organization of high-tech events covering communications, microsystems, optoelectronics, and sensors. In his career, Dr. Iniewski has held numerous faculty and management positions at the University of Toronto, the University of Alberta, Simon Fraser University, and PMC-Sierra Inc. He has published over 100 research papers in international journals and conferences. He holds 18 international patents granted in the United States, Canada, France, Germany, and Japan. He is a frequent invited speaker and has consulted for multiple organizations internationally. He has written and edited several books for CRC Press, Cambridge University Press, IEEE Press, Wiley, McGraw-Hill, Artech House, and Springer. His personal goal is to contribute to healthy living and sustainability through innovative engineering solutions. In his leisure time, Kris can be found hiking, sailing, skiing, or biking in beautiful British Columbia. He can be reached at kris.iniewski@gmail.com.

Contributors

Maan M. Alkaisi
Department of Electrical and Computer Engineering
University of Canterbury
Christchurch, New Zealand

Jean Paul Allain
Department of Bioengineering
Micro and Nanotechnology Laboratory
University of Illinois at Urbana-Champaign
Urbana, Illinois

Sandra L. Arias
Department of Bioengineering
Micro and Nanotechnology Laboratory
University of Illinois at Urbana-Champaign
Urbana, Illinois

Y. Emre Arslan
Faculty of Science Tissue Engineering, Biomaterials and Nanobiotechnology Laboratory
Ankara University
Ankara, Turkey

Richard J. Blaikie
Department of Physics
University of Otago
Dunedin, New Zealand

Alexei Bykhovski
North Carolina State University
Raleigh, North Carolina

Yue Cui
Department of Biological Engineering
Utah State University
Logan, Utah

Stephen R. Diegelmann
Department of Materials Science and Engineering
Institute for NanoBioTechnology
Johns Hopkins University
Baltimore, Maryland

Serap Durkut
Faculty of Science Tissue, Engineering Biomaterials and Nanobiotechnology Laboratory
Ankara University
Ankara, Turkey

Monica Echeverry-Rendón
Department of Bioengineering
Micro and Nanotechnology Laboratory
University of Illinois at Urbana-Champaign
Urbana, Illinois

A. Eser Elçin
Stem Cell Institute and Faculty of Science Tissue Engineering, Biomaterials and Nanobiotechnology Laboratory
Ankara University
Ankara, Turkey

Y. Murat Elçin
Stem Cell Institute and Faculty of Science Tissue Engineering, Biomaterials and Nanobiotechnology Laboratory
Ankara University
Ankara, Turkey

Esmaiel Jabbari
Department of Chemical Engineering
University of South Carolina
Columbia, South Carolina

Ryan T. Kelly
Pacific Northwest National Laboratory
Richland, Washington

Paul C.H. Li
Department of Chemistry
Simon Fraser University
Burnaby, British Columbia, Canada

Yurong Liu
Centre for Bioengineering
Trinity College Dublin
Dublin, Ireland

Albana Ndreu-Halili
Department of Computer Engineering
Epoka University
Tirana, Albania

Volker Nock
Department of Electrical and Computer Engineering
University of Canterbury
Christchurch, New Zealand

Hayriye Ozcelik
INSERM Biomaterials and Tissue Engineering Unit
University of Strasbourg
Strasbourg, France

Irina Pascu
Mechanical and Manufacturing Engineering Department
Dublin City University
Dublin, Ireland

Juan Jose Pavón
Department of Bioengineering
Micro and Nanotechnology Laboratory
University of Illinois at Urbana-Champaign
Urbana, Illinois

Craig Priest
Ian Wark Research Institute
University of South Australia
Mawson Lakes, Australia

Rossen Sedev
Ian Wark Research Institute
University of South Australia
Mawson Lakes, Australia

Abootaleb Sedighi
Department of Chemistry
Simon Fraser University
Burnaby, British Columbia, Canada

Şükran Şeker
Stem Cell Institute Tissue Engineering, Biomaterials and Nanobiotechnology Laboratory
Ankara University
Ankara, Turkey

Sirinrath Sirivisoot
Faculty of Engineering
King Mongkut's University of Technology Thonburi
Bangkok, Thailand

Steven S. Smith
Division of Urology
City of Hope
Duarte, California

Xuefei Sun
Biological Sciences Division
Pacific Northwest National Laboratory
Richland, Washington

John D. Tovar
Department of Materials Science and Engineering
Johns Hopkins University
Baltimore, Maryland

Nihal Engin Vrana
INSERM Biomaterials and Tissue Engineering Unit
University of Strasbourg
and
Protip SAS
Strasbourg, France

Brian D. Wall
Department of Materials Science and Engineering
Johns Hopkins University
Baltimore, Maryland

Lin Wang
Department of Chemistry
Simon Fraser University
Burnaby, British Columbia, Canada

Thomas J. Webster
Department of Chemical Engineering
Northeastern University
Boston, Massachusetts

Dwight Woolard
US Army Research Office
Durham, North Carolina

Part 1

Device Fabrication and Operation

1 Interfacial Control of Multiphase Fluids in Miniaturized Devices

Craig Priest and Rossen Sedev

CONTENTS

1.1 INTRODUCTION

Interactions of solid surfaces with droplets, streams, and films of liquid occur in a wide variety of natural processes and are exploited in countless industrial processes and commercial devices. These surfaces, however, are generally heterogeneous, rough, or designed to be structured, and exhibit a diversity of wetting behaviors. At the microscale, these wetting interactions may dominate the other forces acting on the liquid phase, making them central to many microfluidic applications. The focus of this chapter is the interplay of geometry and chemistry in determining wetting behavior and the implications for passive control of fluids in microfluidic systems where immiscible fluids meet.

When a liquid comes in contact with a solid surface in the presence of a vapor or an immiscible liquid, the competition between the two fluids for the solid surface causes one to spread and the other to retreat. The spontaneity of the process is a key factor in many wetting applications, as only the initial contact between the solid and fluid phases is necessary to trigger the wetting response. The application of spontaneous wetting behavior in microfluidic devices is often termed *passive*, due to the lack of moving parts or active switching, and it is exploited in autonomous, capillary-driven microfluidic devices.[1–4]

Wettability is a nontrivial phenomenon, with detailed information about the solid surface being a prerequisite for predicting wetting behavior.[5–10] Depending on the combination of surface geometry, micro- or nanoscale roughness, and chemistry (homogeneous or heterogeneous), one can observe very different wetting phenomena ranging from superhydrophobicity and superhydrophilicity,[11] wetting hysteresis[5,8,9,12–16] (including so-called asymmetric hysteresis[17–21]), and the velocity dependence of the contact angle.[22–25] The small length scales found in microfluidic devices invariably lead to large surface-to-volume ratios, pressures, and velocity ranges, which can result in very different wetting behaviors compared with the wetting of planar (open) surfaces. Even for a microchannel with a square profile and a relatively "large" width, $w = 100\ \mu m$, the surface-to-volume ratio ($4w/w^2$) is $40{,}000\ m^{-1}$. This raises the importance of the surface tensions (and consequently wettability) above body forces, for example, gravity, that may act on the multiphase flow. The spontaneous capillary rise of a liquid against gravity in a porous solid, particle bed, or capillary is a classic example of this dominant interfacial behavior.[26] While the surface wettability of a channel does not affect all microfluidic systems beyond the initial filling of the device with liquid, the proliferation of multiphase microfluidics[27] and the potential for the autonomous operation of microchips[1–4] have brought wettability to the fore in the design and operation of many microfluidic devices.

In this chapter, the fundamentals of surface wettability are revisited (Section 1.2) with respect to ideal and nonideal surfaces, metastable wetting behavior, and wetting dynamics. In Section 1.3, several approaches to modifying microchannel wettability are given to provide context (not a review) for Section 1.4, which is dedicated to a discussion of several key applications of wetting in microfluidic devices and structures. These include wetting-controlled spontaneous filling, valving, flow stability, phase separation, and the role of wettability in droplet (or bubble)-based microfluidics.

1.2 THEORY OF WETTING

1.2.1 Thermodynamic Equilibrium

When a droplet of liquid is placed on a solid surface in the presence of a second fluid (either liquid or vapor), the liquid will temporarily spread over the surface until the liquid front comes to rest. The final state of the liquid may be a thin film (complete wetting) or a partially wetting droplet, depending on the relative magnitude of the three interfacial tensions involved.[28] In the absence or insignificance of gravity, a partially wetting droplet will form a spherical cap on the solid surface bounded by the so-called contact line, where the three phases meet (Figure 1.1a). The characteristic angle measured through the droplet phase between the solid–liquid interface

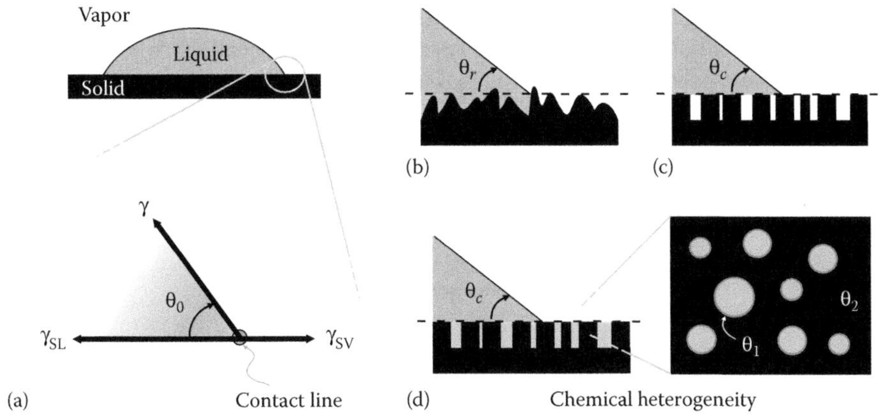

FIGURE 1.1 (a) Illustration of a droplet resting on a solid surface in vapor, showing the contact line and the interfacial tensions acting on it to yield Young's (equilibrium) contact angle. (b, c) Two wetting scenarios for a rough surface: (b) A droplet wetting the full surface area of the rough surface, i.e., the droplet is in the "Wenzel state." (c) A droplet with the ambient fluid remaining underneath the droplet in the rough topography of the surface, i.e., the droplet is in the "Cassie state." (d) A droplet in contact with a flat, chemically heterogeneous surface.

and the plane of the liquid–vapor interface at the contact line is referred to as the contact angle and is the most common measure of wettability. For a simple liquid on a flat, homogeneous, rigid, and chemically inert solid surface (i.e., an ideal surface), the contact line will come to rest only when the three interfacial tensions are perfectly balanced in the plane of the solid surface, according to the well-known Young equation[28]:

$$\gamma \cos\theta_0 + \gamma_{SL} - \gamma_{SV} = 0 \tag{1.1}$$

where:

γ, γ_{SL}, and γ_{SV} are the liquid–vapor, solid–liquid, and solid–vapor interfacial tensions, respectively

θ_0 is the *equilibrium* contact angle

While the above discussion is based on a force balance at the contact line, Young's equation can also be derived via minimization of the surface free energy.

In practice, microfluidic channels, microelectromechanical devices, porous media, and a wide range of natural surfaces exhibit remarkable differences to the ideal surface, which is the basis of Equation 1.1. In particular, microfluidic devices are increasingly moving away from the traditional chemically homogeneous channel with a simple geometry to more complex surface designs that are tailored to specific microfluidic applications. The nonideal nature of these highly functional surfaces can be accounted for by derivation of modified Young equations that take into account roughness or multiple surface components. For a rough surface, the increase in the surface area relative to the projected surface area will enhance the contributions from γ_{SL} and γ_{SV} by a factor r, which is equal to the ratio of the actual to the projected surface area, without impacting γ. The result is Wenzel's equation,[7] which predicts the equilibrium contact angle on a rough (but otherwise ideal) surface, θ_r:

$$\gamma \cos\theta_r + r\left(\gamma_{SL} - \gamma_{SV}\right) = 0 \tag{1.2}$$

or

$$\cos\theta_r = r\cos\theta_0 \tag{1.3}$$

At thermodynamic equilibrium, Wenzel's equation predicts that roughness will increase the contact angle of water on a hydrophobic material ($\theta_r > \theta_0 > 90°$) and decrease the contact angle on a hydrophilic material ($\theta_r < \theta_0 < 90°$). Wenzel's equation assumes that both fluid phases perfectly fill the cavities of the rough surface (Figure 1.1b), so that the second fluid (vapor or liquid) is not trapped beneath the droplet, as shown in Figure 1.1c. This assumption does not always hold in practice and may result in very different behavior.

For a flat and chemically heterogeneous surface, Cassie and Baxter modified Young's equation to account for the different solid–liquid and liquid–vapor interfacial tensions present, weighted by their respective surface area fractions (Figure 1.1d).[6] For a two-component solid surface, free energy minimization gives the Cassie equation:

$$\gamma \cos\theta_c = \phi_1\left(\gamma_{S_1V} - \gamma_{S_1L}\right) + \phi_2\left(\gamma_{S_2V} - \gamma_{S_2L}\right) \tag{1.4}$$

or

$$\cos\theta_c = \phi_1\cos\theta_1 + \phi_2\cos\theta_2 \tag{1.5}$$

where:

θ_c is the equilibrium contact angle on the composite solid surface
ϕ_1 and ϕ_2 are the area fractions of components 1 and 2, respectively ($\phi_1 + \phi_2 = 1$)

The two solid components are indicated in Equation 1.4 by subscripts 1 and 2. Cassie's equation has been applied to a vast number of composite surfaces with variable success.[12,20,29–37]

An important wetting behavior is when the hydrophobicity of a rough or structured surface greatly exceeds the intrinsic hydrophobicity of the material, that is, superhydrophobicity.[38,39] Perhaps the most prominent example is the very high contact angle and low hysteresis observed on the lotus leaf,[40] but the same effect is found on many synthetic and biological surfaces.[41] In all these cases, the liquid rests on a surface composed of solid and vapor domains, as shown in Figure 1.1c. As air is completely hydrophobic ($\theta = 180°$, i.e., the liquid does not come into contact with the solid), Equation 1.5 becomes

$$\cos\theta_c = \phi_1\left(1+\cos\theta_1\right)-1 \tag{1.6}$$

If the area fraction of the solid in direct contact with the liquid, ϕ, is small enough, then the contact angle can be very large (140°–170°). Such behavior is termed *superhydrophobic* because the largest water contact angles on smooth Teflon-like surfaces do not exceed 120°–130°.

In physical terms, the roughness of the hydrophobic material enhances its hydrophobicity by creating a composite surface. Superhydrophobic surfaces effectively contain hydrophobic pores, which the liquid cannot penetrate because capillary pressure holds the liquid back (Figure 1.2a). The maximum pore radius at which the capillary pressure (for water) is of the order of atmospheric pressure is about 1 μm and this is often the magnitude of the roughness needed to induce superhydrophobic behavior. It is possible that the solid protrusions supporting the liquid are themselves rough or structured, as shown in Figure 1.2d and e. This further diminishes the area fraction of the real solid–liquid contacts. Consequently, double- and multiple-scale roughness strongly enhances superhydrophobicity.[42]

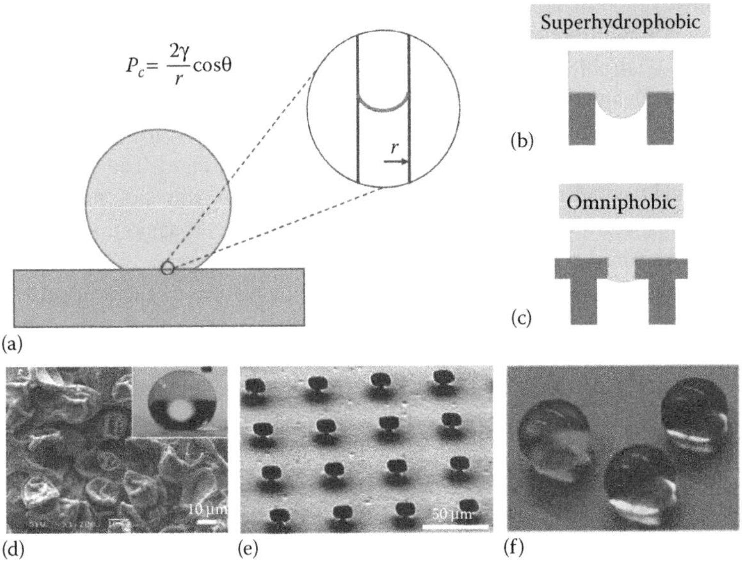

FIGURE 1.2 Superhydrophobicity and omniphobicity: (a) Physics of superhydrophobicity: the capillary pressure, P_C, prevents liquid penetration into the surface pores (radius r). (b) Superhydrophobicity with water (intrinsic contact angle greater than 90°) in a pore with vertical walls. (c) Superhydrophobicity with various liquids (intrinsic contact angle less than 90°) in a pore with overhanging edges. (d) Morphology of an electrospun fabric made of PMMA and fluorodecyl POSS and (inset) a droplet of hexadecane in the Cassie wetting state. (e) A silicon wafer surface with undercut silicon pillars capped with a thin layer of silica. (f) Droplets of heptane (left), methanol (center), and water (right) on a microfabricated surface from (e) hydrophobized with a fluoroalkyl silane. (Figure parts (d), (e), and (f) reprinted from Tuteja, A. et al., Robust omniphobic surfaces, *Proc. Natl. Acad. Sci. USA*, 105, 18200–18205. Copyright 2008 National Academy of Sciences, USA. With permission.)

Note that large contact angles can be obtained in both Wenzel and Cassie wetting states. However, the hysteresis is large in the first case and minimal in the second one. The Wenzel and Cassie states are sometimes designated as "sticky" and "slippery" as liquid droplets on superhydrophobic surfaces are highly mobile (the run-off tilt angle is only a few degrees).

Water has an unusually high surface tension and shows larger contact angles on solid surfaces than most organic liquids. It is technologically relevant to develop a surface that is omniphobic, that is, superhydrophobic with respect to any liquid, not just water.[43,44] A capillary pressure opposing liquid penetration can be achieved for liquids showing acute contact angles on a smooth solid surface but only if the surface features have a reentrant geometry, for example, Figure 1.2c. Surface features with overhangs can be obtained by different methods, including the electrospinning of polymers (Figure 1.2d) or lithographic/etching techniques (Figure 1.2e).[43] Figure 1.2f shows how the wetting behavior is apparently indifferent to the surface tension of the liquid, as droplets of heptane, methanol, and water show superhydrophobic behavior on the surface shown in Figure 1.2e.

The fascination with superhydrophobic surfaces arises from the fact that such surfaces have very low adhesion, the liquid droplets are very mobile, and the friction of liquids flowing past such surfaces is very low.[45] Superhydrophobic surfaces are therefore desirable in a variety of applications including microfluidics (see Section 1.4.2). Despite the research efforts, most superhydrophobic surfaces are mechanically fragile and are easily contaminated. A significant improvement was achieved by locking a lubricating fluid among the surface nano- and microstructure. Various liquids (including complex fluids) show very low adhesion and low contact angle hysteresis, restore after physical damage, resist ice adhesion, and function at up to several hundred atmospheres of pressure.[46]

For sufficiently small droplets or bubbles, the curvature of the contact line may become significant. In much the same way that a curved interface is associated with a three-dimensional (3-D) capillary pressure, a curved contact line is associated with a two-dimensional (2-D) pressure (Figure 1.3). Under such conditions, an additional term must be considered in the so-called augmented Young's equation[47]:

$$\gamma\cos\theta(r) + \gamma_{SL} - \gamma_{SV} + \frac{\tau}{r} = 0 \tag{1.7}$$

where:

τ is the line tension

r is the radius of curvature of the contact line

Note that the contact angle is a function of the curvature. The line tension can be either positive or negative[48] (unlike surface tension, which is always positive). The value of the line tension is estimated as the product of the surface tension and the thickness of the contact line, that is, $|\tau| \cong \gamma\lambda \approx 10^{-11}$ J/m. Experimentalists have struggled to obtain reliable values and the reported numbers vary over orders of magnitude. The effect of the line tension is intimately related to the roughness, heterogeneity, and

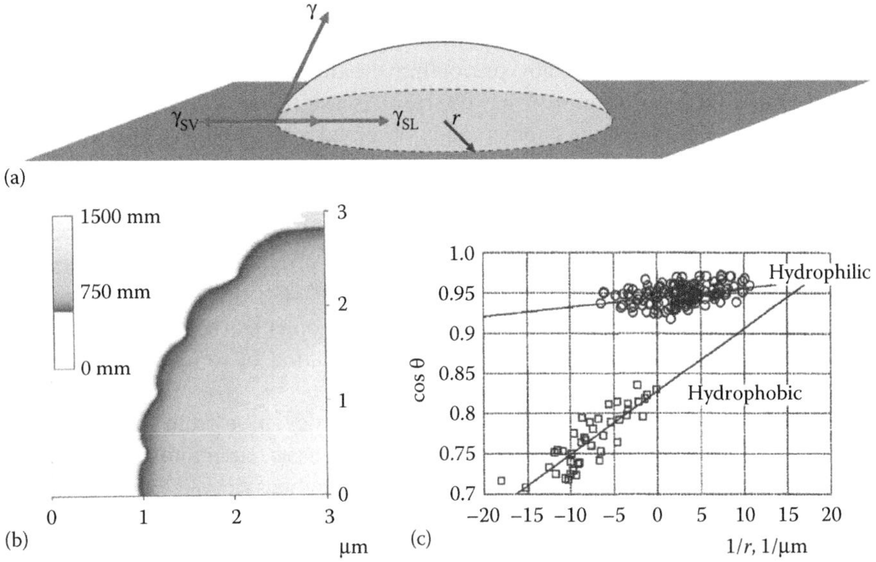

FIGURE 1.3 Line tension: (a) The line tension term, τ/r, indicated by the short arrow drawn in the direction of γ_{SL}, enters the Young equation when the curvature of the contact line, r^{-1}, is large enough. (b) Corrugated contact line of a hexaethylene glycol droplet on a silicon wafer patterned with hydrophobic stripes of a perfluorinated alkylsilane. (c) Local contact angle plotted versus the local curvature of the contact line according to the augmented Young equation. The line tension values are -6×10^{-11} J/m on the hydrophilic domains and -3.5×10^{-10} J/m on the hydrophobic domains.[51] (Figure parts (b) and (c), reprinted from Herminghaus, S. et al., *J. Adhes. Sci. Technol.*, 14, 1767–1782 (2000) by permission of Taylor & Francis Group Ltd.)

stability of the three-phase system.[49] Careful atomic force microscope (AFM) measurements of the shape of microscopic droplets on patterned surfaces[50] have revealed that when the curvature of the contact line is correctly accounted for, τ is indeed of the order of 10^{-11}–10^{-10} J/m. The importance of the last term in Equation 1.7 is amplified by the curvature of the contact line; therefore, the line tension is most relevant when a three-phase contact line is just formed, for example, when a particle pierces a bubble, in heterogeneous nucleation,[47] or when nanoscale curvature of the contact line is induced by surface asperities or chemical heterogeneity.

1.2.2 Wetting Hysteresis

1.2.2.1 Advancing and Receding Contact Angles

The Wenzel and Cassie equations are derived by minimizing the surface free energy of the three-phase system, with no consideration for the local energy barriers associated with the scale or design of the surface features. In other words, the equations consider a droplet that is free to explore the whole surface energy landscape without concern for the path to that free energy minimum. Energy barriers arise when the assumption of an ideal solid surface breaks down; the surface is rough, heterogeneous, elastic, or reactive. In practice, these energy barriers can "pin" the contact line locally in a metastable state, preventing the droplet from achieving the free energy minimum (equilibrium).[5,8–10,52] The result is a significant (and sometimes very large) difference between the contact angle observed after the liquid has advanced over the surface (*static advancing contact angle*) compared with that observed after the liquid has receded (*static receding contact angle*). Contact angle hysteresis is ubiquitous to wetting measurements and no method is available to reliably access the equilibrium contact angle, despite methods being proposed (e.g., the application of mechanical energy to overcome pinning effects[53,54]). For this reason, reporting a static contact angle without specifying whether the droplet has advanced or receded over the surface is less meaningful for the interpretation of wetting behavior and should be avoided.

Contact angle hysteresis by definition involves a deviation from the thermodynamic equilibrium predicted by the Wenzel and Cassie equations (and Young's equation), despite hysteresis also originating from roughness[8,16,17,55–57] and heterogeneity.[15,18,20,58,59] Figure 1.4 shows droplet profiles for water that has partially spread over (advanced) and partially dewetted from (receded) a flat (planar) surface and a structured (pillars) surface.[56] According to Wenzel's equation, the contact angles should be reduced by the presence of roughness because both contact angles on the flat surface are less than 90° (static advancing and receding contact angles are

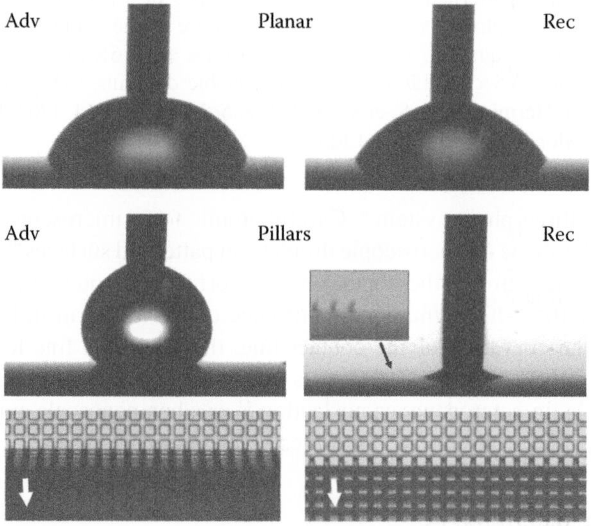

FIGURE 1.4 Static advancing and receding droplet profiles for water droplets on flat (planar) and structured (pillars) surfaces. The contact angle hysteresis is very large on the structured surface due to contact line pinning on the surface features, and cannot be explained using the equilibrium wetting theory. Optical microscopy of the contact line pinned on an individual row of pillars is shown below for the static advancing and receding cases. (Reprinted with permission from Forsberg et al., *Langmuir*, 26, 860. Copyright 2010 American Chemical Society.)

72° and 59°, respectively). In practice, the advancing contact angle dramatically increases and the receding contact angle goes to zero due to pinning of the contact line on the array of pillars. This behavior is inconsistent with the Cassie and Wenzel equations, as well as the qualitative expectation of low hysteresis for Cassie state wetting. These results[56] and similar results from Dorrer et al.[55] were shown to be qualitatively consistent with the effect of contact line pinning on the pillar arrays for liquids resting in the Wenzel state.

This nonideal behavior, however, has not limited the application of these equations to real systems. It is common practice, although not strictly correct, to use the static advancing or receding contact angles in Equations 1.3 and 1.5 to estimate the observed wettability on a rough or heterogeneous surface. While this is a helpful approach, one should keep in mind that the degree and type of roughness or heterogeneity are also responsible for the magnitude of the deviation from these equations, and that the latter does not have a quantitative model. Furthermore, these deviations may be contrary to even the qualitative trends predicted by these equations. In many cases, this non-Cassie and non-Wenzel wetting behavior has significant potential for exploitation in applications including microfluidics.

1.2.2.2 Quantifying Hysteresis Behavior

Researchers have pursued a detailed understanding of wetting hysteresis by setting about explaining the departure from equilibrium on a chemically heterogeneous surface. The early work of Pease addressed the importance of more and less wettable regions on a surface in relation to the motion of the advancing and receding contact line.[52] In essence, Pease suggested that an advancing contact line will be locally "pinned" on less wettable regions, which raises the observed static advancing contact angle. The pinning of the receding contact line on more wettable regions lowers the observed static receding contact angle. This concept has been elaborated on over the years and has been debated at length in the literature[60–62]; however, Pease's simple explanation remains conceptually relevant. In practice, this view requires the replacement of the area fraction used in Cassie's equation with a "line fraction" (i.e., the local surface coverage along the contact line), which has proved to be an effective approximation in some instances.[20,37,56,58] For the simplest case of a droplet resting on a single circular region, this approach is intuitively simple because the line fraction is 0 or 1, depending on whether the contact line is within or outside the region's boundary.[58] For more complex surfaces containing microscopic chemical heterogeneity (well-defined domains), the use of the line fraction in Cassie's equation may approximate the experimental results.[20,37] A similar approach to rough and structured surfaces can be applied, although the complexity of these systems is even greater due to the 3-D geometries involved that may or may not pin the contact line.[17,55–57,63–66] For these cases, theoretical approaches offer greater insight regarding the liquid behavior on a given surface geometry or chemical heterogeneity, revealing detailed information about the meniscus morphology and contact angle hysteresis.[55,65,67–73] The freely available software, Surface Evolver,[74] is able to incrementally modify a meniscus shape until the interfacial free energy of the three-phase system is minimized. This method has proven very effective in replicating wetting behavior and meniscus morphologies (i.e., including the contact angle) on complex solid surfaces, including liquid menisci between particles, on pillar arrays, and in channels.[55,65,67–69] Despite the power of these techniques, it remains prudent to use experimental approaches to verify theoretical predictions, particularly where theoretical models do not capture all the details of the experimental system, for example, a finite curvature at step edges.[55,56]

1.2.3 Dynamic Wetting

The theory discussed thus far relates exclusively to static wetting, where the contact line is at rest—in a thermodynamic equilibrium or in a metastable state. In many applications of wetting behavior, the contact line is moving over the solid at a finite velocity; therefore, dynamic wetting must also be understood. Examples relevant to microfluidics include the spontaneous filling of capillaries and porous materials via the Laplace pressure generated at the liquid–vapor or liquid–liquid interface (consider "paper microfluidics"[1] and other capillary-driven microfluidics[2–4]) or the displacement of one fluid by another in a channel using an external pressure. In either case, the dynamic contact angle, θ_d, can be related to the velocity of the contact line, U, according to a hydrodynamic model or a molecular kinetic model. The most complete description of a hydrodynamic model was presented by Cox[75]; however, simplification by Voinov[76] (Equation 1.8) has proven useful and is correct to better than 1% error for contact angles less than $\theta \cong 135°$:

$$\theta_d^3 = \theta_0^3 + \frac{9\mu U}{\gamma} \ln\left(\frac{L}{l}\right) \tag{1.8}$$

Known as the Cox–Voinov equation, Equation 1.8 includes the viscosity of the liquid, μ, a macroscopic-length scale, L, and a small-length scale, l, which accounts for the violation of the no-slip boundary condition near to the contact line (see Figure 1.5a). While Equation 1.8 accounts for viscous dissipation near the contact line, the molecular kinetic model considers discrete molecular-scale displacements at the contact line, where K_0 and λ are the net displacement frequency (at $U=0$)

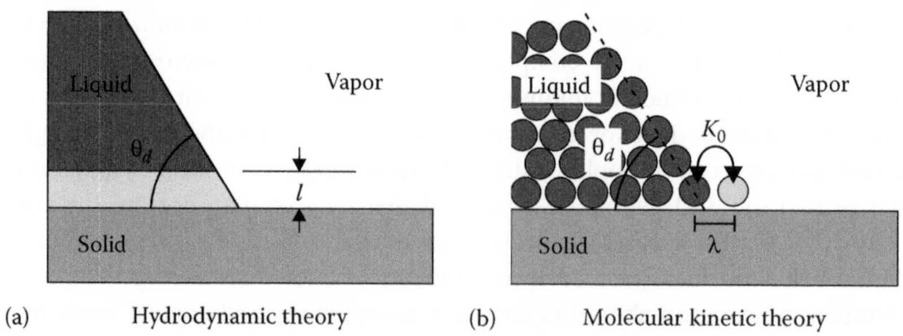

FIGURE 1.5 (a) Illustration of the hydrodynamic model's small length scale, l, accounting for the violation of the no-slip boundary condition at the contact line, and (b) the molecular kinetic model showing molecular-scale liquid displacements with length scale, λ, and frequency, K_0.

and the length of these molecular displacements, respectively (see Figure 1.5b).[77] According to this model, the variation of the dynamic contact angle as a function of the contact line velocity is given by

$$\cos\theta_d = \cos\theta_0 - \frac{2k_B T}{\gamma\lambda^2}\sinh^{-1}\left(\frac{U}{2K_0\lambda}\right) \tag{1.9}$$

where:

k_B is Boltzmann's constant
T is the absolute temperature

Viscous and molecular effects do not act exclusively and, consequently, combined models for dynamic wetting have been proposed.[78,79] For a more thorough discussion of the dynamic wetting theory, see Reference 80.

Dynamic wetting is rarely considered in the design and operation of microfluidic chips, despite the possible consequences being quite remarkable. In the vast majority of microfluidic applications, contact lines are mobile and travel along channels (e.g., transport of droplets[81,82]), cycle their position locally (e.g., drop formation at a channel junction[83–86]), or are created and expanded rapidly (e.g., digital microfluidics[87]). The velocity in these microfluidic systems is dictated by a wettability gradient, capillary pressure, or an externally applied pressure drop. In the case where wetting is forced by an external pressure, there are limits to the maximum contact line velocities achievable, beyond which entrainment of the dewetting phase will occur.[88] While these limits are important in the success of coating technologies,[89] their role in limiting the performance of high-speed microfluidic processes has rarely been considered in the literature, as discussed in Section 1.4.

1.3 TAILORING MICROCHANNEL WETTABILITY

For a given liquid, tuning the wetting behavior is carried out via modification of the surface roughness (i.e., the roughness factor, r, in Wenzel's equation) and the solid–fluid interfacial tensions, γ_{SV} and γ_{SL}, or both. There is a large body of literature that deals with the modification of planar surfaces to influence wetting behavior, which covers thin-film deposition (metal or polymer), self-assembled monolayers (SAMs), gas-phase plasma treatments, and very well-defined fabrication of structures at the micro- and nanoscales using advanced techniques. It is not the intention to present a comprehensive review of these methods here, but rather to highlight several examples of suitable fabrication methods for the modification of microchannel wettability. Additional attention is given to microplasmas as an emerging and potentially very powerful technique for channel modification.

SAMs are well-ordered molecular films that spontaneously adsorb on a surface.[90] The adsorbed molecules typically behave as a surfactant, with the SAM formation driven by a strong interaction between the head-group of the molecule and the surface. Simply changing the outermost (terminal) functional group on the molecule, which contacts the fluid phase, can alter the wetting behavior dramatically. Common examples are alkanethiols on gold substrates and silanes on silica.[90,91] SAMs are also inherently simple to prepare, involving the immersion of the sample in a solution of the SAM molecule for a suitable time. The efficacy of SAMs in wetting studies on planar surfaces has easily transferred to the surface modification of microchannels.[92] The wettability of SAMs can span the full range of accessible contact angles on planar surfaces; from complete wetting (OH or CO_2H-terminated SAMs, $\theta = 0°$)[93] to very hydrophobic surfaces (CH_3 or CF_3-terminated SAMs, $\theta \cong 120°$),[93] and tunable contact angles using "mixed" multicomponent SAMs.[35]

The photosensitivity of some SAMs and other thin films makes them ideal candidates for photoinitiated modification of surface wettability.[92,94] The layers themselves may be removed or replaced completely or partially to generate a pattern,[92,94] or their conformation may be manipulated to induce a switch in wettability.[95] The widespread use of optically transparent chip materials permits the local modification of "closed" microchannels (i.e., after sealing the microchip).[96] Several examples of controlled wetting, described in Section 1.4, rely on photoinitiated surface modification and related patterning techniques.

While SAMs and photopatterning have proven very effective in modifying the wettability of microchannels, these methods typically rely on solution-based chemistry and several processing steps, for example, rinsing and development. In contrast, plasma processing offers a dry (gas-phase) and often single-step alternative. Surface modification is achieved by the exposure of the surface to an excited, ionized gas, which is generated by an electric field. Plasmas contain a variety of highly reactive species, including ions, photons, and radicals, which are able to either treat (e.g., oxidize) the surface or, where a monomer is added to the plasma, deposit polymers.[97–100] It is well known that plasma treatment is widely used in the fabrication of polydimethylsiloxane (PDMS) chips, where the hydrophobic surface is activated by an oxygen plasma prior to bonding. During this process, the surface is rendered hydrophilic, due to the formation of silanol (Si–OH) surface groups. After some time, however, the polymer becomes hydrophobic again.[101] While the majority of reported surface modifications of microchannels have been homogeneous, there has been a growing interest in plasma patterning of microfluidic channels.[102–104] Dixon and Takayama[102] guided a plasma along one side of a linear PDMS channel by offsetting the electrodes positioned at the inlet and outlet. After 5 s of treatment (at high potential; up to 50 kV), parallel regions of treated (hydrophilic, 50 μm wide) and untreated (hydrophobic) surfaces were generated. This method is quite limited in terms of channel geometry and the wettability pattern generated. In earlier work by Klages et al., localized plasma patterning of channels was demonstrated using electrodes that were positioned partway along a channel.[104,105] In this case, plasmas at the millimeter scale could be generated containing hexamethyldisiloxane, which selectively hydrophobized the channel. More recently, Priest et al. demonstrated highly localized helium plasma treatment of regions less than 100 μm long in a 50 μm wide and deep microchannel between embedded gallium electrodes ("injected electrodes") (Figure 1.6).[103] Using this approach, a regular array of hydrophilic regions could be generated along the length of the hydrophobic PDMS channel, shown in Figure 1.6c for 300–800 μm-long channel regions. The authors later showed that the technique

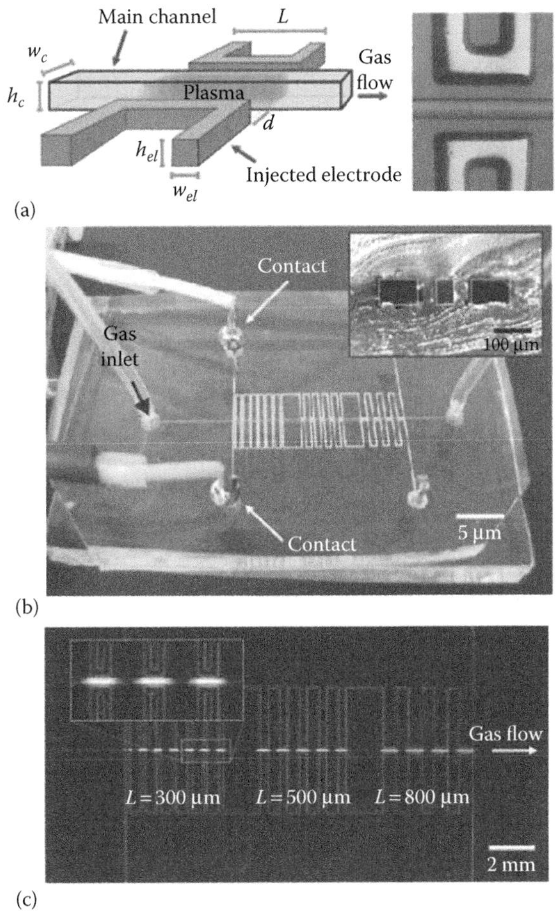

FIGURE 1.6 Plasma patterning of microchannel wettability. (a) Microchannel configuration including "injected" gallium electrodes, with an image of gallium-filled electrodes in polydimethylsiloxane (right). (b) Image of the microchip and, shown inset, a cross section of the gas channel (center) and electrode channels (left and right). (c) Localized helium microplasms with the length of the plasma regions, *L*, noted on the figure. The gas microchannel is 50 μm wide. (From Priest, C. et al., *Lab Chip*, 11, 541–544, 2011. Adapted by permission of The Royal Society of Chemistry.)

is also effective in glass microchannels, where the surface modification is more robust than in PDMS channels.[106]

Despite the availability of many surface treatments, few have the ability to locally modify the chemistry and wettability of an already bonded (closed) microfluidic channel. These techniques generally rely on solution-based adsorption/reaction, oxidation (e.g., by plasma techniques), photoinduced surface modification, and electrochemical deposition (provided electrodes can be prepared during chip fabrication). For a dedicated discussion focused on surface modification techniques for already bonded microchannels, see Reference 96.

The methods discussed above predominantly alter the chemistry, and thus the wettability of the surface, without changing the surface roughness significantly. However, as discussed in Section 1.2, the physical landscape of the solid surface is particularly important in determining the wettability observed in rough or structured microfluidic channels. Generating roughness and designed structures in microchannels can be achieved using particle deposition, wet or dry (e.g., plasma or laser) etching, polymer molding/hot embossing, and surface stress.

Particle deposition is perhaps the most straightforward due to the simplicity of its method. Where the particles are chosen to have an affinity for the microchannel walls, for example, electrostatic attraction, a particle dispersion can be flowed through the microchannel until a sufficient surface coverage is achieved, followed by rinsing and a curing (baking) step.[107] The result is a random array of adsorbed particles that introduce nano- or microscale roughness to the channel, depending on the particle size and the nature of the adsorbed layer (sub-monolayer, monolayer, or multilayer). The method is limited to random roughness at a small scale with respect to the channel dimensions; however, it is very effective in modifying microchannel wettability.[107]

Contact line pinning on microstructures plays a major role in determining the wettability of a surface, as discussed in Section 1.2.2; consequently, the channel geometry and the structures therein must be fabricated for optimal wettability control over a multiphase microfluidic flow. A variety of microscale surface features have been employed, ranging from wet-etched "guide structures" in microsolvent extraction[108] to pillar arrays to comb-like structures[109] and channel constrictions.[110] Most structures of this kind are generated during the fabrication of the microchip using standard photolithography, etching, micromilling, or embossing techniques, due to the difficulty of accessing the microchannel after bonding (in contrast to particle deposition).

The following section will discuss how the theory examined in Section 1.2 and channel structuring and surface modification can be applied to control the flow of multiple fluids in microchannels. Our focus will be on passive applications of wetting in microfluidics, including capillary-driven flows, valves, flow guides, and multiphase flow stability.

1.4 FLUID CONTROL IN MICROCHANNELS

1.4.1 Capillary Flow

The introduction of fluid into microfluidic channels can be achieved through the application of a positive pressure on the fluid phase or, alternatively, through spontaneous penetration of the fluid driven by the capillary pressure. The latter is particularly important for autonomous microfluidic devices, which do not require external pumping to be operated and can therefore be applied in remote areas, at the point-of-care, and using inexpensive equipment. A capillary-driven flow is simply the result of an imbalance of the Laplace pressure (the driving force) against the pressure drop along the length of the liquid filament (the sum of the hydrostatic and hydrodynamic pressures).[26] The Laplace pressure, P_L, is inversely proportional to the radius of the capillary, $1/R$, and proportional to the liquid–vapor interfacial tension, γ, and the wettability of the solid-liquid-fluid system, $\cos\theta_0$, which is given by Equation 1.10 for a cylindrical capillary.[111]

$$P_L = 2\gamma\frac{\cos\theta_0}{R} \tag{1.10}$$

The Laplace pressure may be positive or negative, depending on the wettability of the capillary wall. Where the Laplace pressure opposes the filling of a microchannel, positive pressure must be applied at the device inlet to force the liquid into the channel. However, irrespective of whether the Laplace pressure induces or opposes the flow, the effect is magnified by the small dimensions encountered in microfluidic channels.

Where the capillary pressure is positive (spontaneous filling), no external pressure is required to induce flow. In this case, and assuming gravity is not important (e.g., the flow is perpendicular to the gravitational force), the resistance to the fluid flow is the hydrodynamic pressure drop along the filament of liquid, P_μ, which, for a Poiseuille flow in a cylindrical capillary, is given by the Hagen–Poiseuille equation:

$$P_\mu = \frac{8\mu l}{R^2}U \tag{1.11}$$

where:

μ is the dynamic viscosity
U is the average velocity of the liquid through the capillary
l is the length of the liquid filament

Equations 1.10 and 1.11 can be combined to describe the dynamics of capillary-driven filling of a capillary by a liquid[26]:

$$l^2 = \frac{\gamma R\cos\theta_0}{2\mu}t \tag{1.12}$$

Equation 1.12, known as Washburn's equation, has become a reliable basis for any study of liquid penetration in porous media and capillaries, and has proved to be relevant down to the nanoscale.[112,113] It is clear from the Hagen–Poiseuille equation that the wettability of the capillary or microchannel dictates the direction of flow: either filling or emptying. Washburn's equation is fundamentally important for describing the dynamics of a spontaneous capillary-driven flow in microfluidic devices, whether in open microchannels, bonded (closed) microfluidic and nanofluidic channels, or in so-called paper microfluidics.[114] The following discussion will focus on several features of paper microfluidics due to its recent emergence in the literature; however, for a comprehensive review, see Reference 114.

While capillary-driven transport of liquids through porous media is not a new concept, the coupling of its low cost, autonomy, and vast potential for commercialization and social benefits has popularized paper as a microfluidic tool.[114] The basic concept consists of the capillary-driven transport of liquid samples to reaction sites along hydrophilic channels that are bounded by a hydrophobic material, for example, wax, photoresist, or ink, which is embedded in the paper (Figure 1.7b). Two wettability effects are at work in these devices. The first is capillary-driven flow, which is well described by the Washburn equation (Equation 1.12) and is highly dependent on the physical and chemical properties of the paper used. The second is the guiding of liquid streams using wettability boundaries of nonwetting material. It is the flexibility of these boundary designs that is novel, as the merging and branching of streams are possible[115,116] and, consequently, the delivery of samples to multiple detection sites.[1,117] The paper approach can also embed valves[117] and separate samples[118] in the device, and can include 3-D channel networks using stacked[119] or folded[120] paper. Figure 1.7c shows the simultaneous detection of glucose and bovine serum albumin (BSA) using a branched channel design, as demonstrated by Martinez et al.[1] In their device, the concentration of glucose and BSA was successfully correlated with the intensity of a color change (colorimetric detection),[1] although a variety of other detection methods may be used.[114] Whichever detection method is chosen, it is widely accepted that unambiguous interpretation of the readout is vitally

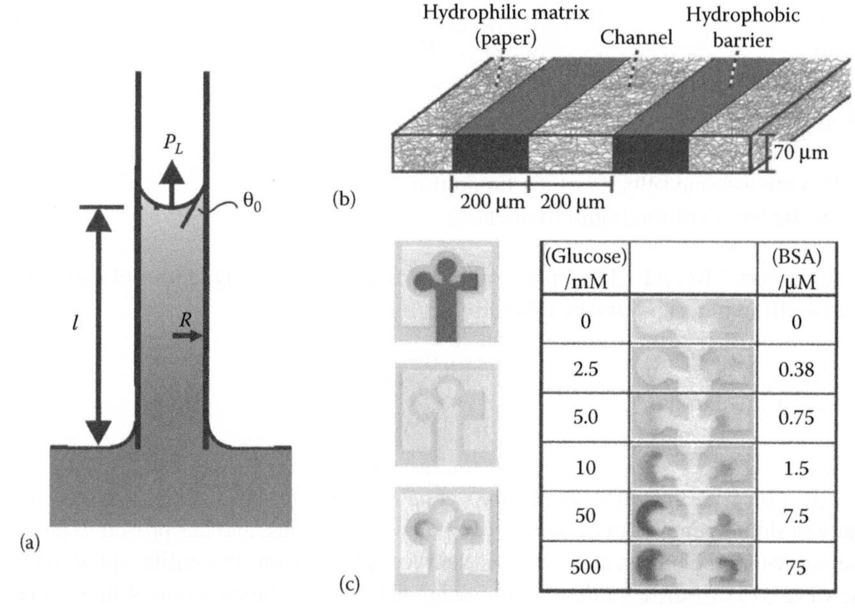

FIGURE 1.7 **(See color insert)** (a) Illustration of the spontaneous capillary rise of water in a hydrophilic capillary. (b, c) An example of "paper microfluidics," where liquid is guided via patterned wettability to reaction sites (circle and square terminus regions) by a capillary-driven flow in the porous paper: (b) Construction of the paper microfluidic device. The hydrophobic barriers consist of printed, then melted wax. (c) Detection of glucose and bovine serum albumin (BSA) at different concentrations (indicated by the table) using two branches of the same paper microfluidic device. (Adapted with permission from Martinez, A.W. et al., *Anal. Chem.*, 82, 3–10. Copyright 2009 American Chemical Society.)

important for the envisaged applications of the technology. Smart phones offer one solution,[121] due to their ability to capture optical images and either send image data for analysis or interpret the data using onboard software. Li et al. proposed a very convenient method for blood-type analysis in which the result appears as text on the paper itself, enabling unambiguous interpretation of the results by any user, without additional technology.[122] Paper microfluidics remains a developing area of research and new approaches to fabrication, functionality, and detection will continue to emerge. However, optimization of capillary flow will remain important to the development of the technology. In a recent review by Li et al., several limitations of the technique were raised, including the inability of some hydrophobic surface treatments to successfully guide samples with a low surface tension.[114] Therefore, precisely tailoring the wettability of the paper (channels and guides) will be central to the ongoing development and ultimate performance of these devices.

1.4.2 Capillary (Laplace) Valves

The ability to drive liquid penetration in microchannels using capillarity is not, however, restricted to the spontaneous filling of microfluidic devices. Local changes in wettability or geometry can be used to manipulate fluids for the precise loading and sampling of nanoliter and picoliter volumes,[107,123] the timing and sequencing of reactions,[110] and other triggered processes.[81,124] The so-called capillary (or Laplace) valves rely on a spatially abrupt change in channel wettability or geometry to control the capillary pressure.[125]

A simple, yet effective, approach is to couple a large and small hydrophobic microchannel in series. Using PDMS microchannels, Yamada and Seki demonstrated the dispensation of 3.5 nL droplets using this technique.[126] The smallest channel dimension (i.e., of the valve channel) was 5 μm, providing almost 20 kPa of capillary pressure for water against its vapor ($\theta_0 = 110°$ for PDMS). In a similar approach, by Lai et al., many small hydrophobic "channels" in between micropillars were fabricated at the sides of a large (millimeter-scale) channel.[123] The pressure required to drive the liquid into the smaller channels was several kilopascals and could be used as a capillary valve in bioanalytical applications.

The most commonly studied type of capillary valve is shown in Figure 1.8, where an expansion in the channel dimensions pins the contact line at a sharp edge. As the contact line cannot advance until the contact angle on the flat surface beyond the edge reaches a value of θ_0, which corresponds to an increase in the interface curvature (shown in Figure 1.8a), the liquid must overcome an additional Laplace pressure to release the valve. The additional pressure can be provided by the centrifugal force on a rotating disk[127–131] or by applying a positive pressure at the liquid inlet.[107,123,132–134] The strength of the capillary valve can be tuned by modifying the expansion angle, β, and changing the wettability of the microchannel. Studies using a combination of theory and experiment have shown that a 3-D model gives the most accurate predictions of the valve strength (maximum pressure).[128,130] The maximum pressure is often called the burst pressure, as it is the pressure that triggers the release of the fluid from the capillary valve. For the simplest case of a cylindrical channel opening into a conical channel (Figure 1.8a), the burst pressure, P_b, is given by

$$P_b = -2\gamma \frac{\cos(\theta_0 + \beta)}{R} \tag{1.13}$$

which can be obtained from Equation 1.10 by including an apparent (or effective) contact angle, $\theta_0+\beta$, observed at the boundary between the cylindrical and conical segments of the channel. The negative sign accounts for the burst pressure opposing the Laplace pressure. The magnitude of the burst pressure can be several kilopascals, depending on the chosen geometry and contact angle.[129,130,135] For the most common channel geometries (e.g., a rectangular cross section), Equation 1.13 should be modified to account for the 3-D geometry of the meniscus that is formed, to avoid considerable deviation from experimental results.[128,130] For example, the expansion of the meniscus may not be possible in all dimensions due to the planar cover that is used to seal the microchip. As Glière and Delattre[134] have shown, this can lead to a dramatic reduction in the burst pressure by more than a factor of two ($\beta=90°$, $\theta_0=60°$, channel height = 15 μm, channel width = 30–115 μm) when compared with the symmetrical case. Thus, in practice, the geometry of capillary valves may be complex and care should be taken when applying predictive models.

While the pursuit of high burst pressures has been the target of many studies, the magnitude of the pressure change required to release the valve has a particularly important role and is therefore addressed here. The transition from the "closed" to the "open" state for a capillary valve may be small or large, depending on the pressure required to drive the liquid to the valve (which may be negative for spontaneous filling) and the burst pressure defined in Equation 1.13. Consider the two examples depicted in Figure 1.8a, where the geometry of the valve is fixed ($\beta=45°$, $R=50$ μm) and water is displacing its vapor in the channel. In the first case, the main channel is hydrophobic ($\theta=100°$) and therefore a positive pressure of 0.5 kPa is required to fill up to the capillary valve. The magnitude of the burst pressure is 2.4 kPa, resulting in 2 kPa more pressure being required to burst the valve above that required for the flow to arrive at the valve. The second microchannel is superhydrophobic ($\theta=150°$), and therefore a positive pressure of 2.5 kPa is required to fill up to the capillary valve. The magnitude of the burst pressure is 2.8 kPa, resulting in only 0.3 kPa more pressure being required to burst the valve. Thus, despite the higher burst pressure, a modest (>0.3 kPa) overpressure during filling of the main channel could result in premature release of the valve, which is unlikely in the former case due to the more than six times greater pressure difference between the closed and open states (2 kPa).

When a pressure-controlled release of the valve is undesirable, for example, where precise pressure control is not available or the combination of two liquid streams should be precisely timed, the liquid-triggered approach reported by Melin et al.[110] offers an elegant solution. Figure 1.8b shows their multichannel design. The first liquid to arrive at the junction stops at the geometry-based capillary valve and remains pinned until the second liquid arrives. The release of the first liquid is initiated by contact with the second liquid (and so on, potentially for several different streams of liquid), breaking the Laplace pressure barrier to release both liquids simultaneously and avoiding trapped air bubbles in the channels.

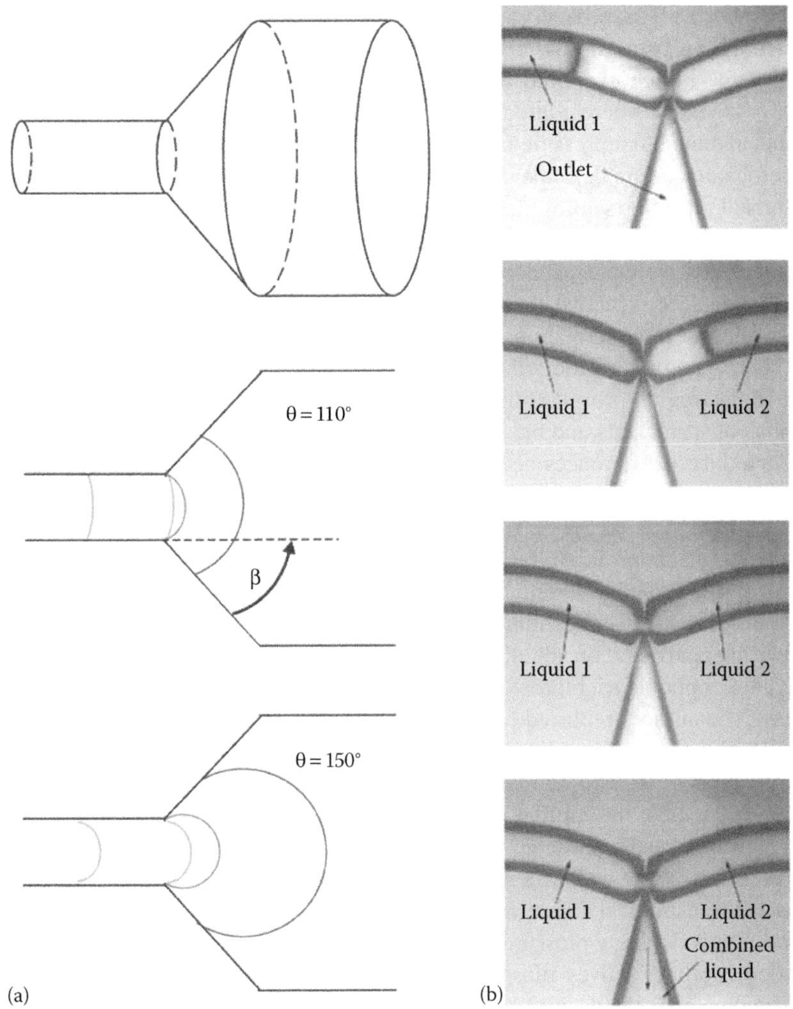

FIGURE 1.8 Geometry-based capillary valves: (a) Capillary burst valve for the symmetrical case of a conical expansion at angle β for a hydrophobic, $\theta = 110°$, and superhydrophobic, $\theta = 150°$, channel. The meniscus is shown to pin at the edge separating the cylindrical and conical regions of the valve, before "bursting" into the conical region. (b) Liquid-initiated release sequence (top to bottom) from a capillary valve as demonstrated by Melin et al. Liquid 1 arrives at the capillary valve and is stopped until Liquid 2 meets Liquid 1, which releases the valve. In the last frame, the liquids are shown to flow together in the outlet channel. (Reprinted from *Sensor Actuat. B Chem.*, Melin, J. et al., A liquid-triggered liquid microvalve for on-chip flow control, 100, 463–468, Copyright (2004), with permission from Elsevier.)

Thus far, only capillary valves that are based on a local change in channel geometry have been discussed. Capillary valves that rely on inhomogeneous surface treatments, however, may have significant advantages over the geometry-based valves, especially where a large difference between the pressure required to fill the channel and the burst pressure (to release the valve) is required. Wettability patterns in microchannels can be achieved using a variety of techniques, including some that are carried out after sealing the microchip.[96] One of the most versatile methods is photolithography, which has been extensively used for making discrete regions of a channel less wettable. Andersson et al. used a photoresist to expose a segment of a linear, deep, reactive, ion-etched silicon microchannel for subsequent coating via plasma polymerization (C_4F_8 monomer).[132,133] The fluorinated region had a final contact angle for water of 105° and a burst pressure magnitude of 760 Pa. This hydrophobic region proved sufficient for valving a wide range of liquids, including solutions of surfactants and biorelevant molecules.

Where larger differences between closed and open states are required, chemical patterning can be coupled with roughness for enhanced wetting transitions at the valve location. The theoretical basis for altering wetting behavior using roughness was given in Section 1.2.1. Takei et al. achieved a superhydrophobic ($\theta > 150°$) to superhydrophilic ($\theta < 9°$) wettability contrast based on nanoroughness and chemical modification.[107] The technique relied on the electrostatic adsorption of titanium dioxide nanoparticles on the silica (Pyrex™) microchannel, surface modification using a hydrophobic trichlorosilane, and ultraviolet (UV) irradiation through a photomask. The authors reported burst pressures up to ~12 kPa and the ability to dispense picoliter volumes of liquid using discrete regions of different wettability in a stepwise process. A well-defined nanoscale roughness can also be used to enhance the performance of the capillary values used in nanofluidics, as shown by Mawatari et al.[136] In this case, nanoscale (tens of nanometers in diameter and height) pillars were etched into one wall of a 200 nm glass nanochannel and used to precisely dispense and transport a 1.7 fL droplet.

The relatively high pressures achievable using simple geometry and surface-modified capillary valves illustrate the importance of tailoring the wettability of microchannels for the desired application. The precise control demonstrated above, including the dispensation and combination of picoliter volumes on demand, is only achievable with well-designed wetting behavior. Furthermore, the phenomenon of contact angle hysteresis, discussed in Section 1.2.2, has the potential to add uncertainty or greater control to capillary valve actuation. It is worth noting here that very few of these examples specifically characterized the static advancing and receding contact angles or considered the influence, if any, of the velocity dependence of the contact angle.

1.4.3 Wettability Flow Guides

The spatially controlled wettability of microchannels is not only useful in valving flow, as described in Section 1.4.2, but it can also be used to guide multiphase flows along microchannels. The physical principles are unchanged from our discussion on capillary valves—the Laplace pressure acts to prevent flow in a particular direction;

however, the flow is now parallel to the wettability boundary (which may be a combination of surface chemistry and geometry).[92,137–142] The boundaries in this configuration have been referred to as "virtual walls" due to the absence of a solid barrier to the flow in this design.[92] The strength of these guides is determined by the Laplace pressure. For the configuration shown in Figure 1.9a, there is a general expression for the Laplace pressure

$$P_L = \gamma\left(\frac{1}{r_1} + \frac{1}{r_2}\right) \tag{1.14}$$

which accounts for the different radii of curvature present at the nonspherical meniscus, r_1 and r_2. For the parallel streams shown in Figure 1.9a, r_2 approaches infinity and $P_L = \gamma/r_1$. This value is half the Laplace pressure of a spherical meniscus ($r_1 = r_2 = R$), for example, as for the capillary valves discussed in Section 1.4.2, for which $P_L = 2\gamma/R$. Despite its diminished value, P_L can be sufficient to guide a liquid along a channel.[92] In practice, the Laplace pressure can be related to the contact angle of the bounding wettability, θ, and the height of the channel, h, according to

$$P_L = \frac{2\gamma\cos\theta}{h} \tag{1.15}$$

The contact angle used in Equation 1.15 is not the equilibrium contact angle but rather the static advancing contact angle on the surface beyond the wettability boundary (as this is the condition for entry to the adjacent stream).

Several examples of guiding multiphase streams using wettability have been reported by Zhao et al.[92,139] The surface modification of a microchannel was carried out using laminar streams of solvent with and without a reagent for hydrophobizing the walls of the microchannel (e.g., octadecyltrichlorosilane) or photolithography. Using laminar flow, the hydrophobic regions corresponded to the path of the stream containing the reagent (which forms a SAM) and were used as boundaries to guide the flow of water (containing dye) (Figure 1.9b). Using photolithography, the whole channel was first modified by a photosensitive SAM, before irradiation with UV through a photomask to render the exposed regions hydrophilic. By choosing different laminar flow regimes or photomasks for the surface modification step, Zhao et al. demonstrated that the wettability boundaries, or *virtual walls*, could be used to control a stream of liquid (or multiple streams) within a single channel. The wettability boundaries could withstand critical pressures of more than 300 Pa—much less than the burst pressures of the valves described earlier, but sufficient to guide the aqueous flow. While the use of laminar flow to pattern wettability in a microchannel is rather restrictive in terms of the pattern design, Kenis et al.[143] demonstrated that discontinuous patterns can be achieved via the partial removal of a preexisting surface coating, that is, a thin gold electrode, by laminar flow of an etch solution. However, photolithography is much more versatile where discontinuous and complex designs are desired.[92,139]

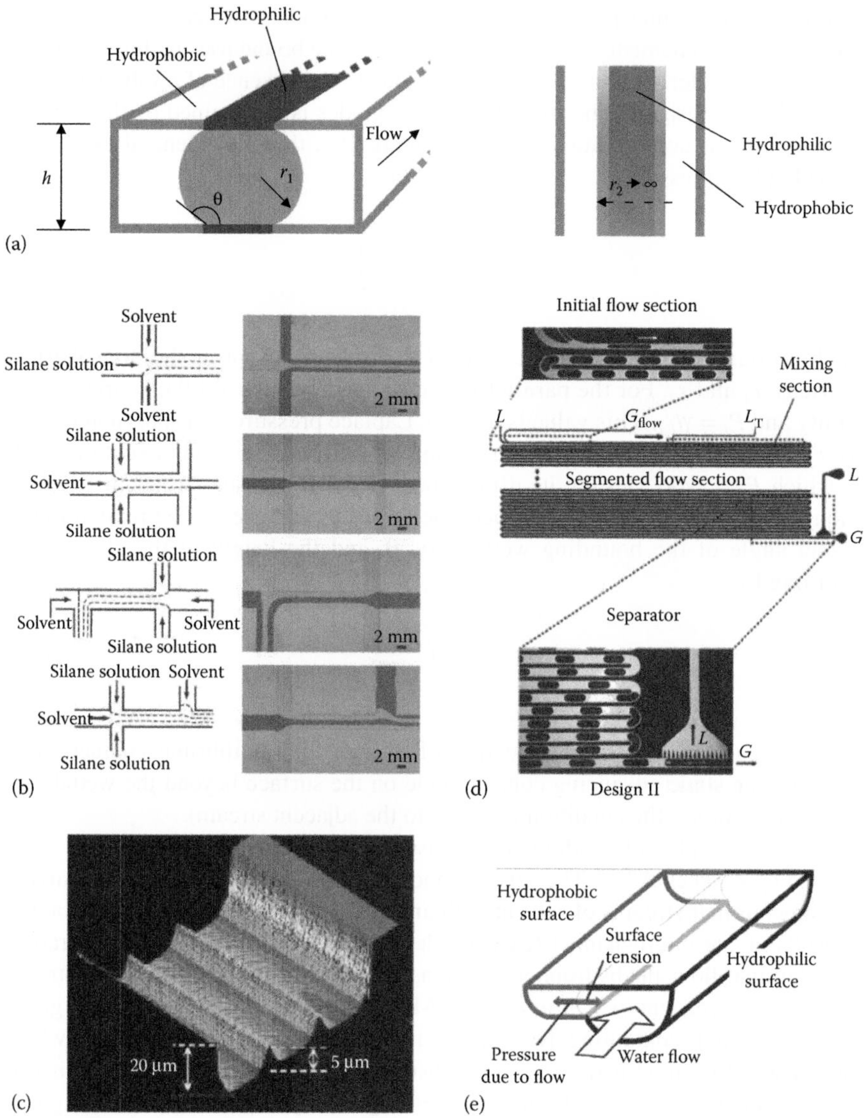

FIGURE 1.9 Guiding fluids in microchannels using "virtual" walls. (a) Side (left) and top (right) views of the "virtual wall" concept showing a liquid following a hydrophilic region along the center of the channel. The radii of curvature, contact angle, and height of the channel used in Equations 1.14 and 1.15 are noted on the figure. (b) Illustrations of several flow regimes for silane solution (octadecyltrichlorosilane in hexadecane; yielding $\theta = 112°$) and solvent (hexadecane) during surface modification of the microchannel (sketches) and the corresponding flow of an aqueous solution of rhodamine B dye, which is confined to the unmodified regions of the channel (images). (From Zhao, B. et al., *Science*, 291, 1023, 2001. Reprinted with permission from AAAS.) (c) "Guide structures" used to pin contact lines, thereby guiding parallel streams of immiscible liquids (two 5 µm high guide structures

In the above examples, the surface wettability (resulting from chemical modification) is the guide; however, as discussed previously in this chapter, surface geometry can play an important, sometimes leading, role in guiding fluids. An example is the so-called guide structure, which has been used to stabilize the parallel flow of multiple streams of liquid for a variety of microfluidic solvent extractions (Figure 1.9c).[108,144–146] While the guide structure provides sufficient contact line pinning to stabilize flow without surface modification,[147,148] the stability window provided by the Laplace pressure is significantly larger when, in addition to channel geometry, one channel is hydrophobized.[140–142,149] The guide structure profile is readily achieved using isotropic etching (e.g., hydrofluoric acid etching of silica) and relies on the pinning of the contact line at the sharp ridge of the guide structure (similar to the geometric capillary valves described in Section 1.4.2). The pinning effect magnifies the local contact angle hysteresis at the guide (and θ in Equation 1.15), which provides greater stability to the interfaces than a planar channel wall. In terms of the contact angle hysteresis discussion in Section 1.2.2, this amounts to a geometrically induced energy barrier to contact line motion (into the adjacent fluid stream).

While the principles are similar to the capillary valve, guiding and stabilizing parallel streams of fluid rely on balancing the pressure difference between the two fluid streams over the full length of contact. This may be several hundred millimeters or more, depending on the application, and therefore the hydrodynamic pressure drops along the length of the two (or more) streams may differ greatly. Assuming that the outlets are at ambient pressure and the two streams are cylindrical, the positive pressure above the ambient pressure required to flow each of the streams is given by Equation 1.11. Consequently, the pressure drop along a given length of channel for dissimilar liquids may be very different and it is the maximum difference at any single point along the streams, ΔP, that determines whether the flow is guided successfully. The condition for guiding and stabilizing a parallel flow in this configuration is therefore $\Delta P < P_L$, where both ΔP and P_L may be positive or negative depending on the geometry, surface wettability, and fluid properties.

These "virtual" walls, however, are not impenetrable to any second phase of liquid. While one liquid may not be able to pass through due to the Laplace pressure barrier involved, the other may pass through freely. In addition, a dispersed phase may pass

FIGURE 1.9 **(Continued)** designed to accommodate three parallel streams are shown). (Reprinted from *Adv. Drug Deliv. Rev.*, 55, Sato, K. et al., Microchip-based chemical and biochemical analysis systems, 379–391, Copyright (2003), with permission from Elsevier.) (d) Phase separation of ethanol or water from gas through a comb-like structure (a series of small side channels) in a polydimethylsiloxane chip. (Reprinted with permission from Günther, A. et al., *Langmuir*, 21, 1547–1555. Copyright (2005) American Chemical Society.) (e) An illustration of the membrane-free phase separation microchip of Hibara et al. The shallow channel in this glass chip is hydrophobized with octadecyltrichlorosilane to maximize the Laplace pressure that opposes entry by the liquid phase (from the larger channel). (Reprinted with permission from Hibara, A. et al., *Anal. Chem.*, 77, 943. Copyright (2005) American Chemical Society.)

through the wettability "wall" via coalescence (i.e., coalescence eliminates the interface responsible for the Laplace pressure), leaving immiscible phases behind. Thus, the two fluids in an emulsion phase can be readily separated under continuous flow in a well-designed microchannel (geometry and contact angle). While an alternative—and very elegant—phase separation of droplets may be triggered by "active" methods, for example, electrocoalescence,[150] the spontaneous coalescence of droplets with a flowing stream segregated by wettability alone reduces device complexity. Several examples of this microfluidic phase separation mechanism have been reported, based on patterned wettability[137] or a channel structure with appropriate surface wettability (including channels separated by membranes).[109,141,151] Logtenberg et al. used the physisorption of two polymer solutions under laminar flow to generate parallel regions of different wettability along a PDMS microchannel.[137] The pattern was terminated partway along the channel via the introduction of a gas bubble. Using the side-by-side wettability difference generated by the two polymers, the authors demonstrated the phase separation of slug flows of 1-octanol and water.

In general, though, a combination of channel (or membrane) geometry and an appropriate contact angle provides the most robust method for microphase separations. For example, Günther et al. fabricated a comb-like structure (a series of small channels) through which a gas phase could selectively escape, leaving a liquid-phase reaction mixture in the main channel (Figure 1.9d).[109] The capillary pressure opposing the entry of the liquid phase into the gas outlet was more than 7 and 2 kPa for water and ethanol phases, respectively. In a similar approach using a fluoropolymer membrane with submicron pore dimensions, Kralj et al. estimated a capillary pressure as high as 20 kPa opposing entry of the aqueous phase in a liquid–liquid separation.[151] While effective, porous membranes and fine channel structures may suffer from fouling in particular applications; therefore, the membrane-free phase separation chip presented by Hibara et al. is an attractive solution (Figure 1.9e).[141] In this chip, a hydrophobic shallow channel (8.6–39 μm deep) and a hydrophilic deep channel (100 μm deep) meet in parallel for a contact length of tens of millimeters. The Laplace pressure opposing entry of the aqueous phase into the shallow channel (against a gas phase) is up to ~8 kPa (depending on the depth of the shallow channel), in agreement with theoretical estimates and sufficient to achieve rapid phase separation of a gas phase from water.

1.4.4 Dispersed-Phase Microfluidics

Dispersed-phase microfluidics refers to the generation, manipulation, and combination (coalescence) of droplets or bubbles within microfluidic channels to carry out chemical, physical, and biological processes in discrete fluid volumes.[152–154] The various manipulations possible may include capture and storage,[155] mixing,[156,157] sorting,[153] spatial reorganization,[158,159] solidification (particle formation),[160,161] and encapsulation.[162] While the dispersed-phase approach (sometimes termed *droplet based*) is a powerful technique to increase sample throughput and screening, the stability and flow behavior of these multiphase systems is very susceptible to microchannel wettability. This section first discusses the influence of wetting phenomena as a passive approach to achieving these processes.

The use of droplets in microfluidic applications has grown rapidly since the first demonstrations of extremely monodispersed bubble/droplet generation at microfluidic junctions (e.g., T-junction,[160] flow-focusing junction,[84,163] and abrupt changes in microchannel geometry[85]). While several droplet formation mechanisms exist, the wettability of a microchannel surface remains a critical factor for success. In the vast majority of cases, the dispersed phase should travel along a microchannel without strong interaction with the walls to avoid cross contamination between droplets, to minimize liquid loss through entrainment (note that the flow velocities may be very high), and to reduce the pressure drop through the channel. It is therefore desirable for the continuous phase to completely wet the solid wall, that is, a dispersed-phase contact angle equal to 180°, which is usually achieved by adding a surfactant to the continuous phase. While the addition of a surfactant is simple and effective, applications where the interfacial properties of the dispersed phase are to be studied may require surfactant-free systems,[164] a surfactant in both phases,[83,165] or additional stability above that provided by a surfactant. These applications may require the microchannel wettability to be modified in a well-controlled and robust manner, which, for example, is straightforward for glass (using silane chemistry) but more challenging for PDMS.[101] Perhaps one of the best illustrations of the need to control wettability in microfluidic emulsification is a study by Nisisako et al.[83,165] The authors used two junctions situated in series to prepare double emulsions, where one or more droplets was contained within a larger droplet, which was dispersed in a continuous phase. By modifying the surface chemistry so that one of the junctions was hydrophilic and the other hydrophobic, water-in-oil and oil-in-water double emulsions could be generated with a high degree of control (Figure 1.10a–c). In contrast, wettability patterns can also be used to fuse droplets, where the role of the channel wall is now to trap one droplet until a subsequent droplet arrives and the two droplets fuse together.[166] Figure 1.10d shows a time sequence for the fusion of droplets with and without dye at a hydrophilic region of the microchannel, as demonstrated by Fidalgo et al.[166] The authors showed that after fusion, the combined droplet is released from the hydrophilic site via the viscous drag on the immobilized liquid. This method, however, may be limited in applications by the presence and concentration of surfactants in the liquid phases, which may prevent the wetting and/or fusion events.

The importance of microchannel wettability extends beyond the ability to generate droplets (or bubbles) at the microscale to the downstream flow behavior in these systems. The downstream flow behavior of droplet- and bubble-based systems is important in practical applications such as heat exchangers and oil recovery, where the type of flow directly affects their performance.[167,168] Several different flow regimes are possible, depending on the wettability of the microchannel and the volume fraction of the dispersed phase.[82,167–172] These regimes are shown for water-air systems flowing in a hydrophilic and hydrophobic channel in Figure 1.11. For hydrophilic channels, shown in Figure 1.11a, Cubaud et al. showed several flow regimes from "bubbly flow," "wedging flow," "slug flow," and "annular flow," to "dry flow" for increasing the volumetric flow fraction of the gas phase.[82] Bubbly flow refers to bubbles that are smaller than the channel dimensions and therefore move freely to temporarily interact with one another and the walls. Wedging flow and slug flow are

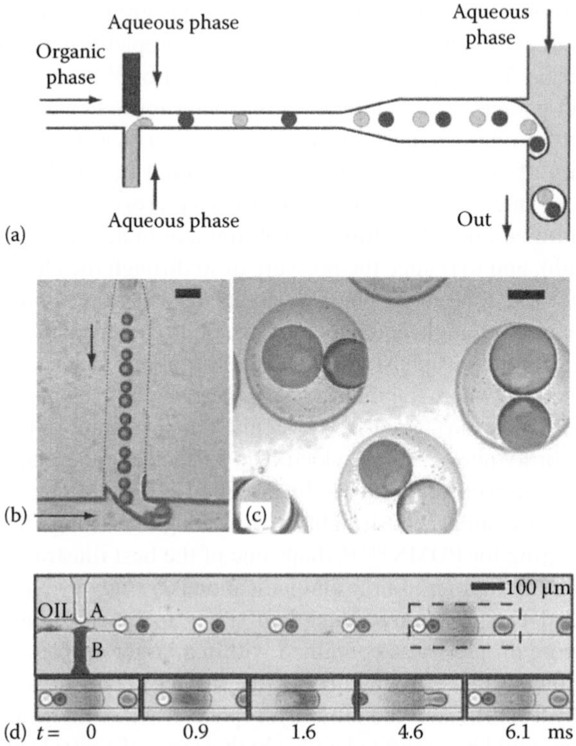

FIGURE 1.10 (a–c) Generation of double emulsions at sequential junctions in a microfluidic chip. The first junction is hydrophobic to force the aqueous phases to disperse, while the second is hydrophilic to disperse the organic phase (containing the aqueous phase droplets). (c) The authors demonstrated the precise loading of the double emulsion with pairs of droplets (labeled with two different dyes). (Figure parts (a), (b), and (c), from Nisisako, T. et al., *Soft Matter*, 1, 23–27, 2005. Reproduced by permission of The Royal Society of Chemistry.) (d) Droplet fusion: Time sequence for surface-induced droplet fusion in a 50 μm wide and 25 μm deep channel. An approximately 100 μm long hydrophilic region of the channel (within the dashed rectangle) was generated to trap (time = 0.9 ms) and fuse (time = 1.6 ms) droplets at the microchannel wall. The fused droplet is released from the interface between 4.6 and 6.1 ms. (From Fidalgo, L.M. et al., *Lab Chip*, 7, 984, 2007. Reproduced by permission of The Royal Society of Chemistry.)

collectively referred to as "segmented flow" and exist when bubbles are elongated due to the confinement of the channel dimensions.[171] Fewer flow regimes and less regularity are observed in hydrophobic channels due to wetting interactions with microchannel walls (Figure 1.11b). The gas bubbles spread on hydrophobic Teflon-coated channels ($\theta \approx 120°$) and the flow becomes unsteady, with gas temporarily (or even permanently) held up in corners and at surface defects, consistent with the discussion in Section 1.2.2. Similar results were observed by Fang et al., where condensate flow was studied for hydrophobic ($\theta = 123°$) and hydrophilic ($\theta = 25°$) microchannels; however, for an intermediate contact angle ($\theta = 91°$), dropwise condensation

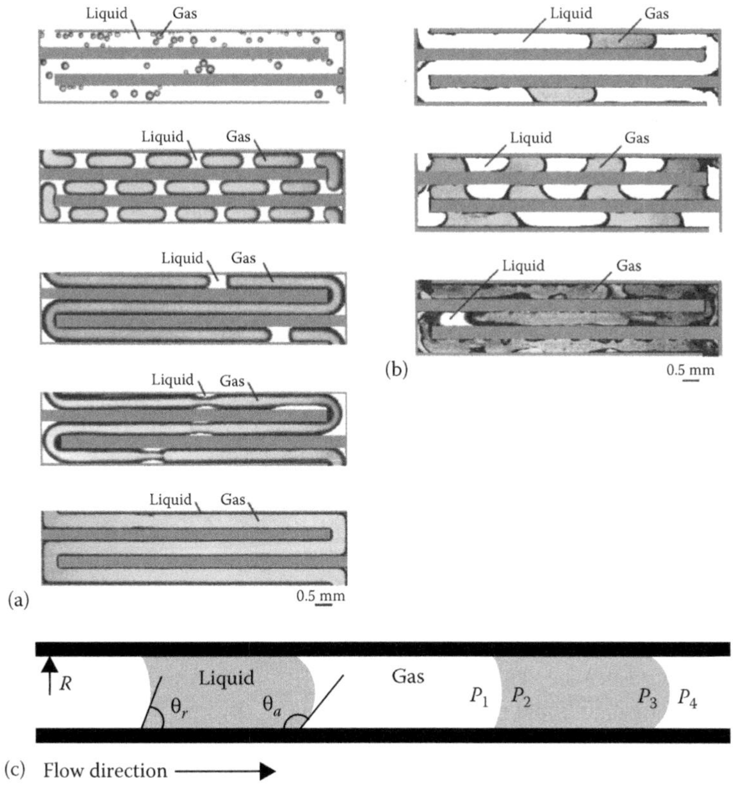

FIGURE 1.11 Flow regimes for the gas–water multiphase system, as reported by Cubaud et al. for (a) hydrophilic channels and (b) hydrophobic channels. From top to bottom, with the increasing flow ratio of the gas phase: (a) "bubbly flow," "wedging flow," "slug flow," "annular flow," and "dry flow," and (b) irregular flow patterns and entrainment of fluid at the walls and corners of the channel. (Figure parts (a) and (b), reprinted with permission from Cubaud, T. et al., Two-phase flow in microchannels with surface modifications, *Fluid Dyn. Res.*, 38, 772–786. Copyright 2006 IOP Publishing.) (c) Illustration of droplets under conditions where the liquid partially wets the microchannel walls. Contact angle hysteresis causes the profile of the lead and trailing interfaces to differ in curvature, leading to a wetting-induced pressure drop opposing flow. (Figure redrawn from Lee, C.Y. and Lee, S.Y., *Exp. Therm. Fluid Sci.*, 34, 1–9, 2010.)

coexisted upstream with stratified flow on the side walls.[168] While the differences in wettability are rather large in these cases, experiments by Salim et al.[167] revealed that changing from a rounded Pyrex™ to a near-rectangular quartz channel, which exhibit similar wettability, can change the flow behavior from parallel streams to a slug flow for a range of flow rate ratios. Thus, both the contact angle and the channel geometry should be measured precisely to ensure that the desired flow regime will be achieved.

A precise characterization of wettability in multiphase microfluidic systems must include an evaluation of contact angle hysteresis. As discussed earlier, static and dynamic advancing and receding contact angles may be vastly different, depending on the nature of the surface and the geometry of the microfluidic environment. Nonetheless, contact angle hysteresis is seldom considered in microfluidic studies, with many authors quoting a single measured value for the contact angle (usually termed the *static* or, incorrectly, "equilibrium" contact angle). Contact angle hysteresis is an important factor in determining the pressure drop along channels containing droplets, due to the different contact angles (and therefore meniscus curvature) at the leading and trailing interfaces (Figure 1.11c).[173–175] Figure 1.11c illustrates a characteristic droplet shape in a segmented flow with a finite contact angle hysteresis. The capillary pressures at the lead and trailing interfaces of the droplet are therefore distinct and related to the dynamic (or static, when there is no droplet motion) advancing, θ_a, and receding, θ_r, contact angles according to Equation 1.10 (for a cylindrical geometry). Thus, the pressure drop across a single droplet due to wetting alone (i.e., excluding hydrodynamic flow resistance, i.e., P_2-P_3), ΔP_θ, is

$$\Delta P_\theta = P_1 - P_4 = \frac{2\gamma}{R}\left(\cos\theta_a - \cos\theta_r\right) \quad (1.16)$$

For the condition $\theta_a=\theta_r$, ΔP_θ is zero and, excluding any influence of wettability on the hydrodynamic slip at the microchannel walls, wetting effects do not contribute to the flow resistance. This condition, however, is never practically observed, as the dynamic advancing and receding contact angles diverge with increasing velocity (see Equations 1.8 and 1.9). Measured pressure drops due to the wetting behavior of segmented flows are significant, with values ranging from several 100 Pa to 2 kPa.[173–175] The magnitude of the overall pressure drop increases, however, with the number of droplets[173] and is therefore not limited to a maximal value. As a consequence, these effects are likely to be significant where channels are partially wet and very long (e.g., in heat transfer applications), while in other applications, the effect may be negligible.

1.5 SUMMARY AND OUTLOOK

Wetting behavior is a fundamental consideration in the design of any microfluidic device where multiple fluid phases meet. The complexity of wetting behavior, including the effects of chemical heterogeneity and roughness, coupled with different channel geometries, can lead to enhanced wetting or dewetting behavior and substantial energy barriers to the reversibility of contact line motion (hysteresis). While much of the theoretical considerations of wettability reflect thermodynamic equilibrium, the measured wettability of real, rough, and heterogeneous solid surfaces often reflects large energy barriers that prevent the strict application of wetting theory, for example, the Wenzel or Cassie equation. Nonetheless, these theories may be used as approximations in microfluidics for nonideal planar surfaces, provided their limitations are fully understood, accounted for, and deficiencies supplemented by wettability measurements. This chapter has shown how the wettability of a solid can be employed as a major driver for the microhandling of fluids. Tailored wettability via controlled chemistry and well-designed geometries can be used to generate a capillary (Laplace) pressure that drives, regulates (as a valve), guides, and phase separates fluids in a passive manner.

The so-called paper microfluidics and other autonomous point-of-care microfluidic devices rely on capillary pressure for their function, and, in many cases, they have a patterned surface wettability to guide the liquid. In valve applications, the capillary pressure may be several kilopascals, depending on the design of the chip, and it may be used to dispense nano- and picoliter volumes of liquid in a controlled manner. Where this valving action is applied perpendicular to the flow direction, the capillary pressure acts to maintain the stability of the laminar streams against mixing. This is particularly useful for reactions and separations that occur at the liquid–liquid (or fluid) interface. Fluid dispersions (droplets and bubbles) that partially wet the microchannel wall generate a wetting-induced flow resistance, which can be directly related to wetting hysteresis. Finally, droplets and bubbles can be phase separated from streams using the capillary pressure generated using surface chemistry and a particular geometry. In this case, the capillary pressure is different for the dispersed and continuous phases (due to preferential wetting by the dispersed phase) and selectively allows the dispersed phase to pass through a particular channel geometry.

The use of wettability as a passive tool in microfluidic devices is already well established; however, there remains significant scope to expand beyond the current applications. In particular, contact angle hysteresis is neglected in most microfluidics studies, yet there is a potential richness in exploiting hysteresis for advanced applications. The same assertion applies to wetting dynamics (the velocity-dependent contact angle and the onset of fluid entrainment) in microchannels, which may lead to a greater understanding of both the potential and the limitations of high-velocity fluid processing. Nonetheless, even on planar surfaces, studying wetting hysteresis and dynamics has proven challenging to date and harnessing these wetting phenomena in the confinement and complexity of microfluidic channels is unlikely to be straightforward.

ACKNOWLEDGMENT

The authors acknowledge financial support from the Australian Research Council.

REFERENCES

1. Martinez, A. W., Phillips, S. T., Whitesides, G. M., Carrilho, E., Diagnostics for the developing world: Microfluidic paper-based analytical devices. *Anal. Chem.* 82(1), 3–10, 2009.
2. Gervais, L., Delamarche, E., Toward one-step point-of-care immunodiagnostics using capillary-driven microfluidics and PDMS substrates. *Lab Chip* 9(23), 3330–3337, 2009.
3. Juncker, D., Schmid, H., Drechsler, U., Wolf, H., Wolf, M., Michel, B., de Rooij, N., Delamarche, E., Autonomous microfluidic capillary system. *Anal. Chem.* 74(24), 6139–6144, 2002.
4. Zimmermann, M., Hunziker, P., Delamarche, E., Autonomous capillary system for one-step immunoassays. *Biomed. Microdevices* 11(1), 1–8, 2009.

5. Good, R. J., Thermodynamic derivation of Wenzel's modification of Young's equation for contact angles: Together with a theory of hysteresis. *J. Am. Chem. Soc.* 74, 5041–5042, 1952.
6. Cassie, A. B. D., Baxter, S., Wettability of porous surfaces. *Trans. Faraday Soc.* 40, 547–551, 1944.
7. Wenzel, R. N., Resistance of solid surfaces to wetting by water. *Ind. Eng. Chem.* 28(8), 988, 1936.
8. Johnson, R. E. Jr., Dettre, R. H., Contact angle hysteresis I. Study of an idealized rough surface. *Adv. Chem. Ser.* 43, 112–135, 1964.
9. Johnson, R. E. Jr., Dettre, R. H., Contact angle hysteresis III. Study of an idealized heterogeneous surface. *J. Phys. Chem.* 68, 1744, 1964.
10. Neumann, A. W., Good, R. J., Thermodynamics of contact angles I. Heterogeneous solid surfaces. *J. Colloid Interf. Sci.* 38, 341, 1972.
11. Bico, J., Thiele, U., Quéré, D., Wetting of textured surfaces. *Colloids Surf. A* 206, 41–46, 2002.
12. Dettre, R. H., Johnson, R. E. J., Contact angle hysteresis – Porous surfaces. *Soc. Chem. Ind. Monogr.* 25, 144–163, 1967.
13. Dettre, R. H., Johnson, R. E. J., Contact angle hysteresis. IV. Contact angle measurements on heterogeneous surfaces. *J. Phys. Chem.* 69(5), 1507–1515, 1965.
14. Shanahan, M. E. R., Di Meglio, J. M., Wetting hysteresis: Effects due to shadowing. *J. Adhes. Sci. Technol.* 8, 1371, 1994.
15. Joanny, J. F., de Gennes, P. G., A model for contact angle hysteresis. *J. Chem. Phys.* 81, 552–562, 1984.
16. Huh, C., Mason, S. G., Effects of surface roughness on wetting (theoretical). *J. Colloid Interf. Sci.* 60, 11, 1977.
17. Priest, C., Albrecht, T. W. J., Sedev, R., Ralston, J., Asymmetric wetting hysteresis on hydrophobic microstructured surfaces. *Langmuir* 25(10), 5655, 2009.
18. Priest, C., Sedev, R., Ralston, J., Asymmetric wetting hysteresis on chemical defects. *Phys. Rev. Lett.* 99, 026103, 2007.
19. De Jonghe, V., Chatain, D., Experimental study of wetting hysteresis on surfaces with controlled geometrical and/or chemical defects. *Acta Metall. Mater.* 43, 1505, 1995.
20. Naidich, Y. V., Voitovich, R. P., Zabuga, V. V., Wetting and spreading in heterogeneous solid surface-metal melt systems. *J. Colloid Interf. Sci.* 174, 104, 1995.
21. Anantharaju, N., Panchagnula, M. V., Vedantam, S., Asymmetric wetting of patterned surfaces composed of intrinsically hysteretic materials. *Langmuir* 25(13), 7410–7415, 2009.
22. McHale, G., Newton, M. I., Shirtcliffe, N. J., Dynamic wetting and spreading and the role of topography. *J. Phys. Condens. Matter* 21(46), 464122, 2009.
23. Semal, S., Voué, M., de Ruijter, M. J., Dehuit, J., De Coninck, J., Dynamics of spontaneous spreading on heterogeneous surfaces in a partial wetting regime. *J. Phys. Chem. B* 103(23), 4854–4861, 1999.
24. Fetzer, R., Ralston, J., Influence of nanoroughness on contact line motion. *J. Phys. Chem. C* 114(29), 12675–12680, 2010.
25. Fetzer, R., Ralston, J., Dynamic dewetting regimes explored. *J. Phys. Chem. C* 113(20), 8888–8894, 2009.
26. Washburn, E. W., The dynamics of capillary flow. *Phys. Rev.* 17, 273–283, 1921.
27. Gunther, A., Jensen, K. F., Multiphase microfluidics: From flow characteristics to chemical and materials synthesis. *Lab Chip* 6(12), 1487–1503, 2006.
28. Young, T., An essay on the cohesion of fluids. *Phil. Trans. R. Soc. Lond.* 95(1), 65, 1805.
29. Kitaev, V., Seo, M., McGovern, M. E., Huang, Y.-J., Kumacheva, E., Mixed monolayers self-assembled on mica surface. *Langmuir* 17(14), 4274–4281, 2001.

30. Dettre, R. H., Johnson, R. E. J., Contact angle hysteresis IV. Contact angle measurements on heterogeneous surfaces. *J. Phys. Chem.* 69, 1507, 1965.
31. Crawford, R., Koopal, L. K., Ralston, J., Contact angles on particles and plates. *Colloids Surf.* 27, 57–64, 1987.
32. Diggins, D., Fokkink, L. G. J., Ralston, J., The wetting of angular quartz particles: Capillary pressure and contact angles. *Colloids Surf.* 44, 299–313, 1990.
33. Woodward, J. T., Gwin, H., Schwartz, D. K., Contact angles on surfaces with mesoscopic chemical heterogeneity. *Langmuir* 16(6), 2957–2961, 2000.
34. Folkers, J. P., Laibinis, P. E., Whitesides, G. M., Self-assembled monolayers of alkanethiols on gold: Comparisons of monolayers containing mixtures of short- and long chain constituents with CH_3 and CH_2OH terminal groups. *Langmuir* 8(5), 1330–1341, 1992.
35. Imabayashi, S.-I., Gon, N., Sasaki, T., Hobara, D., Kakiuchi, T., Effect of nanometer-scale phase separation on wetting of binary self-assembled thiol monolayers on Au(111). *Langmuir* 14(9), 2348–2351, 1998.
36. Rousset, E., Baudin, G., Cugnet, P., Viallet, A., Screened offset plates: A contact angle study. *J. Imaging Sci. Technol.* 45, 517, 2001.
37. Cubaud, T., Fermigier, M., Advancing contact lines on chemically patterned surfaces. *J. Colloid Interf. Sci.* 269(1), 171–177, 2004.
38. Quéré, D., Wetting and roughness. *Annu. Rev. Mater. Res.* 38, 71–99, 2008.
39. Shirtcliffe, N. J., McHale, G., Atherton, S., Newton, M. I., An introduction to superhydrophobicity. *Adv. Colloid Interf. Sci.* 161(1–2), 124–138, 2010.
40. Barthlott, W., Neinhuis, C., The purity of sacred lotus or escape from contamination in biological surfaces. *Planta* 202, 1–8, 1997.
41. Liu, K., Jiang, L., Bio-inspired self-cleaning surfaces. *Annu. Rev. Mater. Res.* 42, 231–263, 2012.
42. Gao, L., McCarthy, T. J., The "lotus effect" explained: Two reasons why two length scales of topography are important. *Langmuir* 22(7), 2966–2967, 2006.
43. Tuteja, A., Choi, W., Mabry Joseph, M., McKinley Gareth, H., Cohen Robert, E., Robust omniphobic surfaces. *Proc. Natl. Acad. Sci. USA* 105(47), 18200–18205, 2008.
44. Ahuja, A., Taylor, J. A., Lifton, V., Sidorenko, A. A., Salamon, T. R., Lobaton, E. J., Kolodner, P., Krupenkin, T. N., Nanonails: A simple geometrical approach to electrically tunable superlyophobic surfaces. *Langmuir* 24(1), 9–14, 2008.
45. Bhushan, B., Jung, Y. C., Natural and biomimetic artificial surfaces for superhydrophobicity, self-cleaning, low adhesion, and drag reduction. *Prog. Mater. Sci.* 56(1), 1–108, 2010.
46. Wong, T.-S., Kang, S. H., Tang, S. K. Y., Smythe, E. J., Hatton, B. D., Grinthal, A., Aizenberg, J., Bioinspired self-repairing slippery surfaces with pressure-stable omniphobicity. *Nature* 477(7365), 443–447, 2011.
47. Toshev, B. V., Platikanov, D., Scheludko, A., Line tension in three-phase equilibrium systems. *Langmuir* 4(3), 489–499, 1988.
48. Indekeu, J. O., Line tension at wetting. *Int. J. Mod. Phys. B* 8(3), 309–345, 1994.
49. Drelich, J., The effect of drop (bubble) size on contact angle at solid surfaces. *J. Adhes.* 63(1–3), 31–51, 1997.
50. Pompe, T., Herminghaus, S., Three-phase contact line energetics from nanoscale liquid surface topographies. *Phys. Rev. Lett.* 85(9), 1930–1933, 2000.
51. Herminghaus, S., Pompe, T., Fery, A., Scanning force microscopy investigation of liquid structures and its application to fundamental wetting research. *J. Adhes. Sci. Technol.* 14(14), 1767–1782, 2000.
52. Pease, D. C., The significance of the contact angle in relation to the solid surface. *J. Phys. Chem.* 49, 107–110, 1945.
53. Sedev, R., Fabretto, M., Ralston, J., Wettability and surface energetics of rough fluoropolymer surfaces. *J. Adhes.* 80, 497–520, 2004.

54. Decker, E. L., Garoff, S., Using vibrational noise to probe energy barriers producing contact angle hysteresis. *Langmuir* 12(8), 2100–2110, 1996.
55. Dorrer, C., Rühe, J., Drops on microstructured surfaces coated with hydrophilic polymers: Wenzel's model and beyond. *Langmuir* 24, 1959, 2007.
56. Forsberg, P. S. H., Priest, C., Brinkmann, M., Sedev, R., Ralston, J., Contact line pinning on microstructured surfaces for liquids in the Wenzel state. *Langmuir* 26, 860, 2010.
57. Dorrer, C., Rühe, J., Advancing and receding motion of droplets on ultrahydrophobic post surfaces. *Langmuir* 22, 7652, 2006.
58. Extrand, C. W., Contact angles and hysteresis on surfaces with chemically heterogeneous islands. *Langmuir* 19, 3793, 2003.
59. Marmur, A., Contact angle hysteresis on heterogeneous smooth surfaces. *J. Colloid Interf. Sci.* 168, 40–46, 1994.
60. Gao, L., McCarthy, T. J., How Wenzel and Cassie were wrong. *Langmuir* 23(7), 3762–3765, 2007.
61. Marmur, A., Bittoun, E., When Wenzel and Cassie are right: Reconciling local and global considerations. *Langmuir* 25(3), 1277–1281, 2009.
62. McHale, G., Cassie and Wenzel: Were they really so wrong? *Langmuir* 23(15), 8200–8205, 2007.
63. Spori, D. M., Drobek, T., Zürcher, S., Spencer, N. D., Cassie-state wetting investigated by mean of a hole-to-pillar denisty gradient. *Langmuir* 26, 9465–9473, 2010.
64. Priest, C., Forsberg, P. S. H., Sedev, R., Ralston, J., Structure-induced wetting of liquid in micropillar arrays. *Microsyst. Technol.* 18, 167, 2012.
65. Semprebon, C., Herminghaus, S., Brinkmann, M., Advancing modes on regularly patterned substrates. *Soft Matter* 8(23), 6301–6309, 2012.
66. Spori, D. M., Drobek, T., Zürcher, S., Ochsner, M., Sprecher, C., Mühlebach, A., Spencer, N. D., Beyond the lotus effect: Roughness influences on wetting over a wide surface energy range. *Langmuir* 24, 5411–5417, 2008.
67. Choi, W., Tuteja, A., Mabry, J. M., Cohen, R. E., McKinley, G. H., A modified Cassie–Baxter relationship to explain contact angle hysteresis and anisotropy on non-wetting textured surfaces. *J. Colloid Interf. Sci.* 339(1), 208–216, 2009.
68. Gögelein, C., Brinkmann, M., Schröter, M., Herminghaus, S., Controlling the formation of capillary bridges in binary liquid mixtures. *Langmuir* 26(22), 17184–17189, 2010.
69. Seemann, S., Brinkmann, M., Kramer, E. J., Lange, F. F., Lipowsky, R., Wetting morphologies at microstructured surfaces. *Proc. Natl. Acad. Sci. USA* 102, 1848, 2005.
70. Kusumaatmaja, H., Pooley, C. M., Girardo, S., Pisignano, D., Yeomans, J. M., Capillary filling in patterned channels. *Phys. Rev. E.* 77, 067301, 2008.
71. Kusumaatmaja, H., Yeomans, J. M., Modeling contact angle hysteresis on chemically patterned and superhydrophobic surfaces. *Langmuir* 23, 6019, 2007.
72. Mognetti, B. M., Kusumaatmaja, H., Yeomans, J. M., Drop dynamics on hydrophobic and superhydrophobic surfaces. *Faraday Discuss.* 146, 153–165, 2010.
73. Lundgren, M., Allan, N. L., Cosgrove, T., George, N., Molecular dynamics study of wetting of a pillar surface. *Langmuir* 19(17), 7127–7129, 2003.
74. Brakke, K., The surface evolver. *Exp. Math.* 1, 141, 1992.
75. Cox, R. G., The dynamics of the spreading of liquids on a solid surface. Part 1. Viscous flow. *J. Fluid Mech.* 168, 169–194, 1986.
76. Voinov, O. V., Hydrodynamics of wetting. *Mekhanika Zhidkosti i Gaza* 5, 714–721, 1976.
77. Blake, T. D., Haynes, J. M., Kinetics of liquid/liquid displacement. *J. Colloid Interf. Sci.* 30(3), 421–423, 1969.
78. Petrov, P. G., Petrov, J. G., A combined molecular-hydrodynamic approach to wetting kinetics. *Langmuir* 8(7), 1762–1767, 1992.

79. Brochard-Wyart, F., de Gennes, P. G., Dynamics of partial wetting. *Adv. Colloid Interf. Sci.* 39, 1–11, 1992.
80. Blake, T. D., The physics of moving wetting lines. *J. Colloid Interf. Sci.* 299(1), 1–13, 2006.
81. Takahashi, K., Mawatari, K., Sugii, Y., Hibara, A., Kitamori, T., Development of a micro droplet collider; the liquid–liquid system utilizing the spatial–temporal localized energy. *Microfluid. Nanofluid.* 9(4), 945–953, 2010.
82. Cubaud, T., Ulmanella, U., Ho, C.-M., Two-phase flow in microchannels with surface modifications. *Fluid Dyn. Res.* 38(11), 772–786, 2006.
83. Nisisako, T., Okushima, S., Torii, T., Controlled formulation of monodispere double emulsions in a multiple-phase microfluidic system. *Soft Matter* 1, 23–27, 2005.
84. Anna, S. L., Bontoux, N., Stone, H. A., Formation of dispersions using "flow focusing" in microchannels. *Appl. Phys. Lett.* 82, 364, 2003.
85. Priest, C., Herminghaus, S., Seemann, R., Generation of monodisperse gel emulsions in a microfluidic device. *Appl. Phys. Lett.* 88, 024106, 2006.
86. Sugiura, S., Nakajima, M., Seki, M., Prediction of droplet diameter for microchannel emulsification. *Langmuir* 18, 3854, 2002.
87. Choi, K., Ng, A. H. C., Fobel, R., Wheeler, A. R., Digital microfluidics. *Annu. Rev. Anal. Chem.* 5(1), 413–440, 2012.
88. Bertrand, E., Blake, T. D., De Coninck, J., Dynamics of dewetting. *Colloids Surf. A Physicochem. Eng. Asp.* 369(1–3), 141–147, 2010.
89. Blake, T. D., Clarke, A., Ruschak, K. J., Hydrodynamic assist of dynamic wetting. *AIChE J.* 40(2), 229–242, 1994.
90. Bain, C. D., Evans, S. D., Laying it on thin. *Chem. Br.* 31(1), 46–48, 1995.
91. Bain, C. D., Troughton, E. B., Tao, Y.-T., Evall, J., Whitesides, G. M., Nuzzo, R. G., Formation of monolayer films by the spontaneous assembly of organic thiols from solution onto gold. *J. Am. Chem. Soc.* 111(1), 321–335, 1989.
92. Zhao, B., Moore, J., Beebe, D. J., Surface directed liquid flow inside microchannels. *Science* 291, 1023, 2001.
93. Chidsey, C. E., Loiacono, D. N., Chemical functionality in self-assembled monolayers: Structural and electrochemical properties. *Langmuir* 6(3), 682–691, 1990.
94. Tarlov, M. J., Burgess, D. R. F. Jr., Gillen, G., UV Photopatterning of alkanethiolate monolayers self-assembled on gold and silver. *J. Am. Chem. Soc.* 115(12), 5305–5306, 1993.
95. Lake, N., Ralston, J., Reynolds, G., Light-induced surface wettability of a tethered DNA base. *Langmuir* 21, 11922, 2005.
96. Priest, C., Surface patterning of bonded microfluidic channels. *Biomicrofluidics* 4(3), 032206–032213, 2010.
97. Kreitz, S., Penache, C., Thomas, M., Klages, C.-P., Patterned DBD treatment for area-selective metallization of polymers-plasma printing. *Surf. Coat. Technol.* 200, 676, 2005.
98. Siow, K. S., Britcher, L., Kumar, S., Griesser, H. J., Plasma methods for the generation of chemically reactive surfaces for biomolecule immobilization and cell colonization: A review. *Plasma Process. Polym.* 3, 392, 2006.
99. Dai, L., Griesser, H. J., Mau, A. W. H., Surface modification by plasma etching and plasma patterning. *J. Phys. Chem. B* 101, 9548, 1997.
100. Aizawa, H., Makisako, T., Reddy, S. M., Terashima, K., Kurosawa, S., Yoshimoto, M., On-demand fabrication of microplasma-polymerized styrene films using automatic motion controller. *J. Photopolym. Sci. Technol.* 20(2), 215, 2007.
101. Zhou, J., Ellis, A. V., Voelcker, N. H., Recent developments in PDMS surface modification for microfluidic devices. *Electrophoresis* 31, 2, 2010.
102. Dixon, A., Takayama, S., Guided corona generates wettability patterns that selectively direct cell attachment inside closed microchannels. *Biomed. Microdevices* 12, 769, 2010.

103. Priest, C., Gruner, P. J., Szili, E. J., Al-Bataineh, S. A., Bradley, J. W., Ralston, J., Steele, D. A., Short, R. D., Microplasma patterning of bonded microchannels using high-precision "injected" electrodes. *Lab Chip* 11, 541–544, 2011.
104. Klages, C.-P., Hinze, A., Lachmann, K., Berger, C., Borris, J., Eichler, M., von Hausen, M., Zänker, A., Thomas, M., Surface technology with cold microplasmas. *Plasma Process. Polym.* 4, 208, 2007.
105. Klages, C.-P., Berger, C., Eichler, M., Thomas, M., Microplasma-based treatment of inner surfaces in microfluidic devices. *Contrib. Plasma Phys.* 47(1–2), 49, 2007.
106. Szili, E. J., Al-Bataineh, S. A., Priest, C., Gruner, P. J., Ruschitzka, P., Bradley, J. W., Ralston, J., Steele, D. A., Short, R. D., Integration of microplasma and microfluidic technologies for localised microchannel surface modification. In: Juodkazis, S., Gu, M., (eds). *Smart Nano-Micro Materials and Devices*, vol. 8204, p. 82042J. SPIE, Bellingham, WA, 2011.
107. Takei, G., Nonogi, M., Hibara, A., Kitamori, T., Kim, H.-B., Tuning microchannel wettability and fabrication of multiple-step Laplace valves. *Lab Chip* 7, 596, 2007.
108. Sato, K., Hibara, A., Tokeshi, M., Hisamoto, H., Kitamori, T., Microchip-based chemical and biochemical analysis systems. *Adv. Drug Deliv. Rev.* 55(3), 379–391, 2003.
109. Günther, A., Jhunjhunwala, M., Thalmann, M., Schmidt, M. A., Jensen, K. F., Micromixing of miscible liquids in segmented gas–liquid flow. *Langmuir* 21(4), 1547–1555, 2005.
110. Melin, J., Roxhed, N., Gimenez, G., Griss, P., van der Wijngaart, W., Stemme, G., A liquid-triggered liquid microvalve for on-chip flow control. *Sensor Actuat. B Chem.* 100(3), 463–468, 2004.
111. Adamson, A. W., Gast, A. P., *Physical Chemistry of Surfaces*, 6th edn. Wiley, New York, 1997.
112. Haneveld, J., Tas, N. R., Brunets, N., Jansen, H. V., Elwenspoek, M., Capillary filling of sub-10 nm nanochannels. *J. Appl. Phys.* 104(1), 014309, 2008.
113. Han, A., Mondin, G., Hegelbach, N. G., de Rooij, N. F., Staufer, U., Filling kinetics of liquids in nanochannels as narrow as 27 nm by capillary force. *J. Colloid Interf. Sci.* 293(1), 151–157, 2006.
114. Li, X., Ballerini, D. R., Shen, W., A perspective on paper-based microfluidics: Current status and future trends. *Biomicrofluidics* 6(1), 011301–011313, 2012.
115. Osborn, J. L., Lutz, B., Fu, E., Kauffman, P., Stevens, D. Y., Yager, P., Microfluidics without pumps: Reinventing the T-sensor and H-filter in paper networks. *Lab Chip* 10(20), 2659–2665, 2010.
116. Rezk, A. R., Qi, A., Friend, J. R., Li, W. H., Yeo, L. Y., Uniform mixing in paper-based microfluidic systems using surface acoustic waves. *Lab Chip* 12(4), 773–779, 2012.
117. Li, X., Tian, J., Shen, W., Progress in patterned paper sizing for fabrication of paper-based microfluidic sensors. *Cellulose* 17(3), 649–659, 2010.
118. Yang, X., Forouzan, O., Brown, T. P., Shevkoplyas, S. S., Integrated separation of blood plasma from whole blood for microfluidic paper-based analytical devices. *Lab Chip* 12(2), 274–280, 2012.
119. Martinez, A. W., Phillips, S. T., Whitesides, G. M., Three-dimensional microfluidic devices fabricated in layered paper and tape. *Proc. Natl. Acad. Sci. USA* 105(50), 19606–19611, 2008.
120. Liu, H., Crooks, R. M., Three-dimensional paper microfluidic devices assembled using the principles of origami. *J. Am. Chem. Soc.* 133(44), 17564–17566, 2011.
121. Delaney, J. L., Hogan, C. F., Tian, J., Shen, W., Electrogenerated chemiluminescence detection in paper-based microfluidic sensors. *Anal. Chem.* 83(4), 1300–1306, 2011.
122. Li, M., Tian, J., Al-Tamimi, M., Shen, W., Paper-based blood typing device that reports patient's blood type "in writing". *Angew. Chem. Int. Ed.* 51(22), 5497–5501, 2012.

123. Lai, H.-H., Xu, W., Allbritton, N. L., Use of a virtual wall valve in polydimethylsiloxane microfluidic devices for bioanalytical applications. *Biomicrofluidics* 5(2), 024105–024113, 2011.
124. Takahashi, K., Sugii, Y., Mawatari, K., Kitamori, T., Experimental investigation of droplet acceleration and collision in the gas phase in a microchannel. *Lab Chip* 11(18), 3098–3105, 2011.
125. Oh, K. W., Ahn, C. H., A review of microvalves. *J. Micromech. Microeng.* 16(5), R13, 2006.
126. Yamada, M., Seki, M., Nanoliter-sized liquid dispenser array for multiple biochemical analysis in microfluidic devices. *Anal. Chem.* 76(4), 895–899, 2004.
127. Moore, J., McCuiston, A., Mittendorf, I., Ottway, R., Johnson, R., Behavior of capillary valves in centrifugal microfluidic devices prepared by three-dimensional printing. *Microfluid. Nanofluid.* 10(4), 877–888, 2011.
128. Chen, J., Huang, P.-C., Lin, M.-G., Analysis and experiment of capillary valves for microfluidics on a rotating disk. *Microfluid. Nanofluid.* 4(5), 427–437, 2008.
129. Duffy, D. C., Gillis, H. L., Lin, J., Sheppard, N. F., Kellogg, G. J., Microfabricated centrifugal microfluidic systems: Characterization and multiple enzymatic assays. *Anal. Chem.* 71(20), 4669–4678, 1999.
130. Leu, T.-S., Chang, P.-Y., Pressure barrier of capillary stop valves in micro sample separators. *Sensor Actuat. A Phys.* 115(2–3), 508–515, 2004.
131. Madou, M. J., Lee, L. J., Daunert, S., Lai, S., Shih, C.-H., Design and fabrication of CD-like microfluidic platforms for diagnostics: Microfluidic functions. *Biomed. Microdevices* 3(3), 245–254, 2001.
132. Andersson, H., van der Wijngaart, W., Griss, P., Niklaus, F., Stemme, G., Hydrophobic valves of plasma deposited octafluorocyclobutane in DRIE channels. *Sensor Actuat. B Chem.* 75(1–2), 136–141, 2001.
133. Andersson, H., van der Wijngaart, W., Stemme, G., Micromachined filter-chamber array with passive valves for biochemical assays on beads. *Electrophoresis* 22(2), 249–257, 2001.
134. Glière, A., Delattre, C., Modeling and fabrication of capillary stop valves for planar microfluidic systems. *Sensor Actuat. A Phys.* 130–131, 601–608, 2006.
135. Cho, H., Kim, H.-Y., Kang, J. Y., Kim, T. S., Capillary passive valve in microfluidic systems. In: *NSTI-Nanotech 2004*, vol. 1, pp. 263–266, 2004.
136. Mawatari, K., Kubota, S., Xu, Y., Priest, C., Sedev, R., Ralston, J., Kitamori, T., Femtoliter droplet handling in nanofluidic channels: A Laplace nanovalve. *Anal. Chem.* 84(24), 10812–10816, 2012.
137. Logtenberg, H., Lopez-Martinez, M. J., Feringa, B. L., Browne, W. R., Verpoorte, E., Multiple flow profiles for two-phase flow in single microfluidic channels through site-selective channel coating. *Lab Chip* 11(12), 2030–2034, 2011.
138. Watanabe, M., Microchannels constructed on rough hydrophobic surfaces. *Chem. Eng. Technol.* 31(8), 1196–1200, 2008.
139. Zhao, B., Moore, J. S., Beebe, D. J., Pressure-sensitive microfluidic gates fabricated by patterning surface free energies inside microchannels. *Langmuir* 19, 1873, 2003.
140. Aota, A., Nonaka, M., Hibara, A., Kitamori, T., Countercurrent laminar microflow for highly efficient solvent extraction. *Angew. Chem. Int. Ed.* 45, 1, 2006.
141. Hibara, A., Iwayama, S., Matsuoka, S., Ueno, M., Kikutani, Y., Tokeshi, M., Kitamori, T., Surface modification method of microchannels for gas-liquid two-phase flow in microchips. *Anal. Chem.* 77, 943, 2005.
142. Hibara, A., Nonaka, M., Hisamoto, H., Uchiyama, K., Kikutani, Y., Tokeshi, M., Kitamori, T., Stabilization of liquid interface and control of two-phase confluence and separation in glass microchips by utilizing octadecylsilane modification of microchannels. *Anal. Chem.* 74, 1724, 2002.

143. Kenis, P. J. A., Ismagilov, R. F., Whitesides, G. M., Microfabrication inside capillaries using multiphase laminar flow patterning. *Science* 285, 83, 1999.
144. Minagawa, T., Tokeshi, M., Kitamori, T., Integration of a wet analysis system on a glass chip: Determination of Co(II) as 2-nitroso-1-naphthol chelates by solvent extraction and thermal lens microscopy. *Lab Chip* 1, 72, 2001.
145. Tokeshi, M., Minagawa, T., Kitamori, T., Integration of a microextraction system on a glass chip: Ion-pair solvent extraction of Fe(II) with 4,7-diphenyl-1,10-phenanthrolinedisulfonic acid and tri-n-octylmethylammonium chloride. *Anal. Chem.* 72(7), 1711–1714, 2000.
146. Tokeshi, M., Minagawa, T., Uchiyama, K., Hibara, A., Sato, K., Hisamoto, H., Kitamori, T., Continuous-flow chemical processing on a microchip by combining microunit operations and a multiphase flow network. *Anal. Chem.* 74, 1565–1571, 2002.
147. Priest, C., Zhou, J., Klink, S., Sedev, R., Ralston, J., Microfluidic solvent extraction of metal ions and complexes from leach solutions containing nanoparticles. *Chem. Eng. Technol.* 35(7), 1312–1319, 2012.
148. Priest, C., Zhou, J., Sedev, R., Ralston, J., Aota, A., Mawatari, K., Kitamori, T., Microfluidic extraction of copper from particle-laden solutions. *Int. J. Miner. Process.* 98(3–4), 168–173, 2011.
149. Aota, A., Mawatari, K., Kitamori, T., Parallel multiphase microflows: Fundamental physics, stabilization methods and applications. *Lab Chip* 9(17), 2470–2476, 2009.
150. Kralj, J. G., Schmidt, M. A., Jensen, K. F., Surfactant-enhanced liquid–liquid extraction in microfluidic channels with inline electric-field enhanced coalescence. *Lab Chip* 5, 531, 2005.
151. Kralj, J. G., Sahoo, H. R., Jensen, K. F., Integrated continuous microfluidic liquid–liquid extraction. *Lab Chip* 7, 256, 2007.
152. Song, H., Chen, D. L., Ismagilov, R. F., Reactions in droplets in microfluidic channels. *Angew. Chem. Int. Ed.* 45, 7336, 2006.
153. Tan, Y.-C., Fisher, J. S., Lee, A. I., Christini, V., Lee, A. P., Design of microfluidic channel geometries for the control of droplet volume, chemical concentration, and sorting. *Lab Chip* 4, 292, 2004.
154. Teh, S.-Y., Lin, R., Hung, L.-H., Lee, A. P., Droplet microfluidics. *Lab Chip* 8, 198, 2008.
155. Boukellal, H., Selimovic, S., Jia, Y., Cristobal, G., Fraden, S., Simple, robust storage of drops and fluids in a microfluidic device. *Lab Chip* 9(2), 331–338, 2009.
156. Song, H., Bringer, M. R., Tice, J. D., Gerdts, C. J., Ismagilov, R. F., Experimental test of scaling of mixing by chaotic advection in droplets moving through microfluidic channels. *Appl. Phys. Lett.* 83, 4664, 2003.
157. Tice, J. D., Lyon, A. D., Ismagilov, R. F., Effects of viscosity on droplet formation and mixing in microfluidic channels. *Anal. Chim. Acta* 507, 73, 2004.
158. Evans, H. M., Surenjav, E., Priest, C., Herminghaus, S., Seemann, R., In situ formation, manipulation, and imaging of droplet-encapsulated fibrin networks. *Lab Chip* 9, 1933, 2009.
159. Surenjav, E., Priest, C., Herminghaus, S., Seemann, R., Manipulation of gel emulsions by variable microchannel geometry. *Lab Chip* 9, 325–330, 2009.
160. Nisisako, T., Torii, T., Microfluidic large-scale integration on a chip for mass production of monodisperse droplets and particles. *Lab Chip* 8, 287, 2008.
161. Nisisako, T., Torii, T., Takahashi, T., Takizawa, Y., Synthesis of monodisperse bicoloured Janus particles with electrical anisotropy using a microfluidic co-flow system. *Adv. Mater.* 18, 1152, 2006.
162. Priest, C., Quinn, A., Postma, A., Zelikin, A. N., Ralston, J., Caruso, F., Microfluidic polymer multilayer adsorption on liquid crystal droplets for microcapsule synthesis. *Lab Chip* 8, 2182, 2008.

163. Gañán-Calvo, A. M., Gordillo, J. M., Perfectly monodisperse microbubbling by capillary flow focusing. *Phys. Rev. Lett.* 87(27), 274501, 2001.
164. Priest, C., Reid, M. D., Whitby, C. P., Formation and stability of nanoparticle-stabilised oil-in-water emulsions in a microfluidic chip. *J. Colloid Interf. Sci.* 363(1), 301–306, 2011.
165. Okushima, S., Nisisako, T., Torii, T., Higuchi, T., Controlled production of monodisperse double emulsions by two-step droplet breakup in microfluidic devices. *Langmuir* 20(23), 9905–9908, 2004.
166. Fidalgo, L. M., Abell, C., Huck, W. T. S., Surface-induced droplet fusion in microfluidic devices. *Lab Chip* 7, 984, 2007.
167. Salim, A., Fourar, M., Pironon, J., Sausse, J., Oil–water two-phase flow in microchannels: Flow patterns and pressure drop measurements. *Can. J. Chem. Eng.* 86(6), 978–988, 2008.
168. Fang, C., Steinbrenner, J. E., Wang, F.-M., Goodson, K. E., Impact of wall hydrophobicity on condensation flow and heat transfer in silicon microchannels. *J. Micromech. Microeng.* 20(4), 045018, 2010.
169. Dreyfus, R., Tabeling, P., Willaime, H., Ordered and disordered patterns in two-phase flows in microchannels. *Phys. Rev. Lett.* 90(14), 144505, 2003.
170. Ody, C., Capillary contributions to the dynamics of discrete slugs in microchannels. *Microfluid. Nanofluid.* 9(2), 397–410, 2010.
171. Kim, N., Evans, E., Park, D., Soper, S., Murphy, M., Nikitopoulos, D., Gas–liquid two-phase flows in rectangular polymer micro-channels. *Exp. Fluids* 51(2), 373–393, 2011.
172. Huh, D., Kuo, C. H., Grotberg, J. B., Takayama, S., Gas–liquid two-phase flow patterns in rectangular polymeric microchannels: Effect of surface wetting properties. *New J. Phys.* 11(7), 075034, 2009.
173. Lee, C. Y., Lee, S. Y., Pressure drop of two-phase dry-plug flow in round mini-channels: Effect of moving contact line. *Exp. Therm. Fluid Sci.* 34(1), 1–9, 2010.
174. Yu, D., Choi, C., Kim, M., The pressure drop and dynamic contact angle of motion of triple-lines in hydrophobic microchannels. *ASME Conf. Proc.* 2010(54501), 1453–1458, 2010.
175. Rapolu, P., Son, S., Characterization of wettability effects on pressure drop of two-phase flow in microchannel. *Exp. Fluids* 51(4), 1101–1108, 2011.

2 Nanostructured Biointerfaces

Jean Paul Allain, Monica Echeverry-Rendón, Juan Jose Pavón, and Sandra L. Arias

CONTENTS

2.1 INTRODUCTION

The structural diversity of material surfaces in contact with biological organisms such as cells, tissue, and the extracellular matrix (ECM) can dictate the proliferation, differentiation, and overall behavior of cell function. The ability to control the surface chemistry, topology, and elastomechanical strength of a biointerface can have important implications for the multifunctionality of modern biomaterials. In this chapter, we present a generalized summary of nanostructured biointerfaces and their synthesis. We focus primarily on the properties of nanostructured biointerfaces and the process–structure–property (PSP) relationship strategies to design them. The chapter summarizes key biofunctional properties that dictate both their function and behavior. To conclude the chapter, three emergent biotechnological applications are summarized with respect to the design of nanostructured biointerfaces.

2.2 BIOMATERIAL INTERFACE: SURFACE PROPERTIES

The biomaterial interface is a complex interchange between inorganic, organic, and bioactive systems coupled via complex molecular and chemical interactions. The interface between living cells and materials, for example, relies on interdependent signaling between proteins on the surfaces/interfaces of the ECM or inorganic material and cell-adhering receptors or molecules (e.g., integrins and proteins). The hierarchy of surface properties shown in Figure 2.1 illustrates the underlying mechanisms involved in the interaction between cells and biomaterials. The interactions can be generally described by four main mechanisms in order of hierarchy: (1) the surface energy can influence the hydrophilicity and the hydrophobicity of a biointerface and

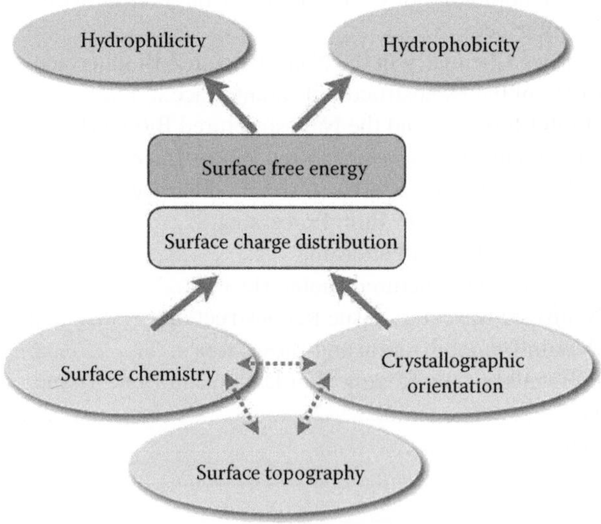

FIGURE 2.1 A schematic illustrating the properties responsible for controlling the hydrophobicity and hydrophilicity of a material biointerface.

is correlated by (2) the electrical charge distribution, which is dependent on (3) the surface chemistry and crystallographic orientation, which are codependent with (4) the surface topography.

2.2.1 Cell–Surface Interactions

Biomedical applications frequently require the use of biomaterials in the fabrication of devices and implants in order to improve the behavior of an organ or tissue, or, sometimes, for its replacement. Consequently, it is important to have multiple alternatives in terms of the design and function of a biomaterial to guarantee an appropriate interaction with the host tissue. There are a variety of biomaterials according to their characteristic nature and class of material (e.g., metals, polymers ceramics, or a combination thereof), which also depend on their application and expected function. Likewise, the surface of the material requires detailed studies since it is in intimate contact with the tissue; defined as the biointerface, it is where the most important reactions occur. The biointerface mechanisms responsible for cell–surface interactions consist of two primary effects: (1) protein adsorption and (2) cell response. Proteins in the appropriate conformation and orientation can stimulate a constructive cell response that could include, for example, favoring wound repair and tissue reconstruction. The cell response depends on specific proteins to achieve anchorage and receive extracellular instructions that can guide its differentiation and proliferation [1].

When a biomaterial is implanted in the body, a chain of reactions is activated and this interaction is magnified at the surface of the material. Factors at the interface of the biomaterial, such as physical, chemical, physicochemical, and mechanical properties, can dictate the biological response defined by biocompatibility, adhesion and cell differentiation and proliferation. Therefore, one can determine the body's localized or general response to the implant in terms of integration, function, and stability [2]. On the other hand, key aspects at the biointerface can influence the stability of a material. For example, in metals, corrosion resistance plays an important role in achieving a high level of biocompatibility. Typically, the top surface layer that cover a metal are oxides that act like a protection barrier between the bulk material and tissue, preventing chemical reactions between the material surface and the environment [3]. Later in the chapter, more detail is provided on the importance of chemistry at the biointerface. Therefore, in this section, some of the most important physical characteristics involved in the interaction between tissue and the material surface will be addressed.

In terms of physical properties, the topography is considered one of the most crucial parameters in the biological response after the implantation of a biomaterial. In several studies, different morphology configurations at micro- and nanolevels demonstrated that the roughness profile and surface distribution have a strong and direct influence on the cellular behavior, showing that it is possible to appreciate that cells prefer texturized surfaces in comparison with smooth ones [4]. Similarly, isotropic and aligned fibers can guide cells to grow in the same direction and orientation following the pattern. Moreover, texturized surfaces with pores, canals, or tunnels can promote interactions between different kinds of molecules and even cells, due to an increase in their surface area, facilitating the deposition of proteins [5]. The role of surface topography has been extensively studied in the last few decades due to the well-known fact that modifications at the nano- and microlevel can affect cellular activity [6]. This effect has been used as a strategy to improve biomechanical fixation, thereby guaranteeing the stability of an implant. In other cases, surface roughness has been used to immobilize proteins or peptides that can promote cellular adhesion and proliferation [7]. In other instances, antimicrobial activity through the deposition of antibiotics or antimitotic drugs has been used to control the growth of bacteria and fungus in implants [8].

The surface of an implant can be modified to transform the bulk material or for coating deposition. In the latter case, it can be an inorganic material, such as metal, ceramic, or polymer, or an organic compound. Cells might be able to sense different surface conditions such as texture orientation, morphology, isotropy, and anisotropy, as well as mechanical, physical, and chemical signals that translate them into stimuli to promote cell motility, alignment, and other functions. Cell shape and phenotypic responses seem most pronounced when the scale of the surface feature is some fraction of the cell size, (e.g., from tens to hundreds of nanometers) [9].

In addition, the physical properties that can interfere with surface modifications are strength and flexibility. Therefore, Young's moduli, which are close in magnitude, can introduce an adequate transition between materials with different features, thus avoiding stress shielding [3]. Furthermore, signals can mediate the histological response in the surrounding structures. In this way, it is possible to identify many key factors that have definite implications on the success or failure of an implantation.

2.2.2 Surface Free Energy and Biointerface

The surface free energy can be defined as the change or variation in interatomic or intermolecular bonds to form a surface. The surface energy can also be defined as the excess energy at the material surface compared with its material bulk. Vogler, for example, defines surface energy as "an intensive thermodynamic property of a material that arises from the loss of nearest-neighbor interactions among atoms or molecules at the boundary. This excess energy most prominently manifests itself in adhesion and adsorption reactions at the surface" [9]. The surface in a material is a two-dimensional (2-D) defect and thus atomic and defect mobilities can be orders of magnitude higher than in the material bulk. Biomaterials are intrinsically dependent on surface free energy given that at this interface living organisms can interact via complex biochemical bonding channels, for example, in the case of cell adhesion. The biointerface of any biomaterial introduced *in vivo* will interact primarily with water given that it is a major component of biological fluids. Consequently, the role of the biointerface and its interaction with a water molecule can dictate the design of a particular biomaterial interface and the strategy used in developing biomaterials around that particular dynamic medium. The surface energy of a biomaterial is also closely associated with the charge distribution on a surface. The combination of the charge state and the surface energy of a particular biomaterial surface (biointerface) can therefore dictate the surface wettability that ultimately influences protein adsorption. However, it is not intuitively obvious how wettability can control protein adsorption. Generally, hydrophilic surfaces can improve adhesion (e.g., water, protein, cells, etc.); however, in the case of proteins, the probability of adhesion is largely determined by the local immunological response at the biointerface. For example, conformational changes at a biointerface by processes such as inflammation, coagulation, and foreign-body response can initiate reactions that lead to tight adherence to hydrophobic biomaterial surfaces [10]. In addition to the characteristics of the surface and its effect on the biocompatibility of a biomaterial, there is also its complex contact with body fluids in a region defined by Vogler as a "pseudo two-dimensional zone of water directly adjacent to the surface, referred to as *interfacial* or sometimes *vicinal* water" [9]. This region of water at the biomaterial surface can dictate how the surface chemistry influences the biomaterial effects on the body. The strong coupling at this interface can also dictate the strategy selected for surface modification depending on the desired functional outcome of the biomaterial. Moreover, the PSP relationships dictating biomaterials design must also assess the dynamic environments where the biomaterial is exposed.

2.2.3 Process–Structure–Property Relationships in Nanostructured Biointerfaces

The PSP relationships in the materials design of nanostructured biointerfaces are intimately connected to their performance characteristics. The fundamental paradigm in material science enables us to find the most convenient combination of processing parameters, in search of a desired structure that will ensure the required properties for a specific application. Therefore, with the aim of designing optimum biointerfaces for biomedical applications (biocompatibility, tissue repair, tissue growth, or tissue regeneration), we need first to address the global biomaterial/tissue system, as well as the specific roles and interactions of the interfaces associated with given design requirements.

2.2.3.1 Biointerface and Its Properties

Before addressing the details of the actual role of biointerfaces in a system where the new implant and regenerative tissue interact, one must place in context the assumptions made about a biointerface and its interaction with the complex *in vivo* human environment. An interface between two physical entities that are chemically or physically dissimilar or both, is assumed, in principle, as that measurable border or limit between material species that forms an intimate contact, either chemical or physical or both, between the atoms or molecules or both that make up the outermost surface of any material system [11]. One of the most distinctive features of any interface is that every surface associated with it has a special structural organization and reactivity, which will dictate any biochemical and immunological interactions with the body and ultimately any new entities that could be generated from those interactions. Because of these potential complex interactions around any interface (i.e., whether in the body or other applications), advanced characterization techniques are always necessary and, in particular, those that can elucidate complex interfacial properties (e.g., material surfaces exposed to the corrosive environment of the body). Another important factor always associated with any surface is the assumption that it has unique chemical reactivity characteristics [11]; this is the rule of thumb for most materials design approaches for interface analysis. It implies important restrictions and sometimes when it is overlooked, outcomes are unreliable. The unique chemical reactivity of any surface is the main factor responsible for its fundamental mechanisms such as surface oxidation in metals and other chemical or biochemical reactions at the biointerface.

2.2.3.2 Broad Perspective of the Interface Between Biomaterials and Tissue

A traditional approach found in the literature regarding factors that affect biomaterials/tissue systems normally addresses a few central issues, including the effects of the implant on the host tissue and the host tissue on the implant surface, mediating complications, and device failure [12]. To a great extent, factors arise from alterations in the physiological (normal) processes (e.g., immunity, inflammation, blood coagulation) comprising host defense mechanisms that function to protect an organism from deleterious external threats, such as bacteria, other microbiological organisms, injury, and foreign materials.

In addition to the conventional approach toward a fundamental understanding of the biomaterial/vicinal region interface summarized earlier, here the authors introduce a new multidimensional biomaterials design framework, which includes, for the first time, the intrinsic properties of materials and the *in situ* mechanical stimulation of the system as critical factors for an optimum materials design. In other words, not only is the material characterized, but ultimately the performance and the function of the biomaterial are assessed with *in situ* techniques that are able to capture its physical and biophysical properties (e.g., chemical, electronic, thermal) *during* exposure to a simulated environment in which it will perform.

This new approach can be better appreciated through Figures 2.2 and 2.3. First, this holistic approach enables the classification of biomaterials design in four parts: (1) Material factors are represented by the intrinsic properties of the biomaterial or biomaterials (biocomposites) in contact with the tissue, as well as by the geometry and shape parameters of the component. (2) The role of the host tissue will depend on the available quantity of tissue in contact with the biomaterial, as well as on the healthy state of the tissue, which is directly related to its chemistry, biochemistry, and mechanical properties. (3) The biointerface role is arguably the most critical factor, being the biophysical border between the physical and biochemical entities. Its importance in dictating the performance of the whole system depends on the topography, chemistry, biochemistry, surface energy, and elastomechanical properties of the biointerface. Using this basis for design, one can move from options that are provided by conventional biointerfaces (i.e., first- and second-generation biomaterials; the effects at the microscale level [13]), up to advanced biointerfaces (third-generation biomaterials; the effects at the molecular and nanoscale levels [13]). Finally, (4) environmental stimulation is a collection of complex factors that are also

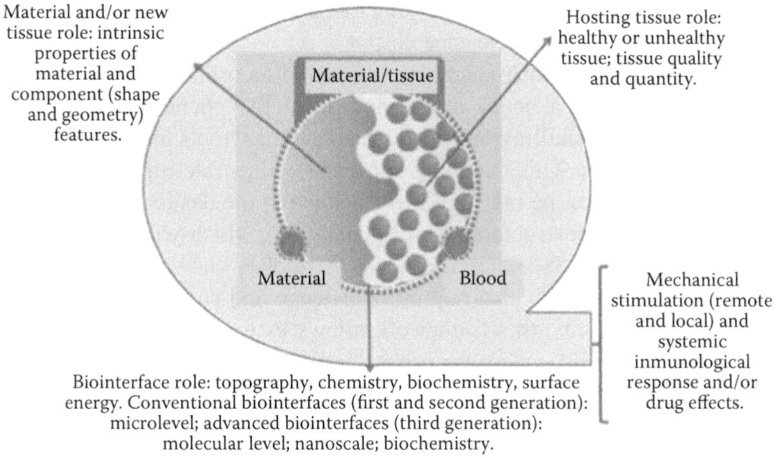

FIGURE 2.2 A general multidimensional framework of the factors involved in a biomaterial/tissue interface system.

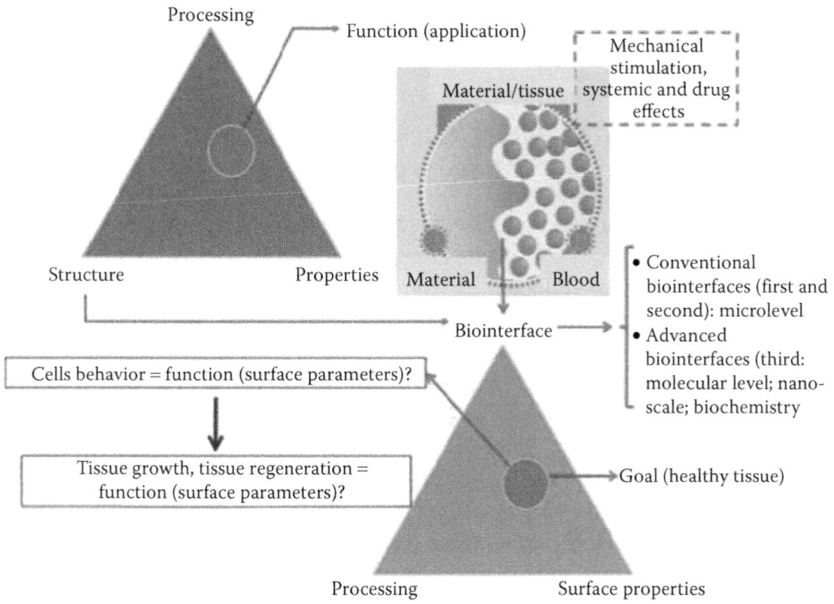

FIGURE 2.3 A multidimensional approach for the process–structure–property design of a biointerface in a biomaterial/tissue biological system.

difficult to elucidate and handle. They consist of both the remote and local mechanical stimulations of the above three factors (i.e., conventional physical therapies and, more recently, ultrasound and magnetic external protocols), drugs, medicine, and systemic effects. Natural and modulated immunological responses are also included in this group. The emphasis on the role of the biointerface, within the context of all the above factors, can be addressed by trying to match the above scheme (Figure 2.2) with the conventional materials science design.

This new paradigm, which matches the multidimensional framework shown above, not only enables a better understanding of the phenomenon, but it also improves the favorable handling of those elements as new tools for new biomaterials development and design. This matching approach is illustrated in Figure 2.3, from which several aspects can be inferred. First, achieving the desired or optimum bio-interface properties (nanostructured, third generation), which will ensure a healthy tissue in contact with the biomaterial properties, implies knowledge and control of the surface properties, which also require knowledge and control of the processing conditions. In other words, this framework allows us to identify the relationships between cell behavior and surface properties and, simultaneously, identify the relationships correlated to the surface processing conditions. Even more ambitious is the idea that this approach will allow us to establish desired relationships between directed tissue behavior and both surface properties and surface processing conditions. Finally, this approach also opens routes for synergistic contingencies obtained once the optimum relationships are known, simultaneously applying the suitable mechanical and general environmental stimulation. Of course, this approach is more than challenging from the aspect of optimizing the multifactorial problem, on which the authors are currently working from experimental, mathematical, and computational points of views.

2.3 BIOMECHANICAL PROPERTIES OF NANOSTRUCTURED BIOINTERFACES

As was previously discussed and described, the physicochemical and topographical properties of surfaces are important for their thorough assessment. These properties are normally reflected in the elastic-plastic behavior of surfaces, depending on the kind of loading that they are subjected to. Therefore, the study of the mechanical effects of biointerfaces, closely correlated to their working environment, specifically on cells, subcellular components, and biomolecules, demands an understanding of the relationships between physicochemical, topographical, and surface mechanical properties, as well as the available measurement techniques for these purposes. Furthermore, as alluded to in the previous section, *dynamic* and *in situ* measurements to correlate surface properties with their ultimate function and performance are desirable in an overall biointerface design strategy.

2.3.1 Elastic-Plastic Properties of Surfaces and Their Measurement

A general description of the elastic-plastic properties of surfaces implies two different approaches: the elastic-plastic micromechanical properties of conventional microstructured engineering surfaces and the elastic-plastic nanomechanical properties of advanced nanostructured biointerfaces. In that context, a mechanical analysis of a nanostructured biointerface, in which an engineered surface is in contact with a biological environment, can result in the movement from a bio-micromechanical to a bio-nanomechanical framework. The elastic properties of materials engineering surfaces are usually different from those of the same material in bulk. The basic elastic constants, such as Young's modulus, E, Poisson ratio, v, and shear elasticity, G, inherent to any surface, as well as surface plastic properties such as hardness, H, yield strength, σ_y, mechanical strength, σ_0, and residual stress, σ_{res}, are usually difficult to estimate. This implies the necessity of using advanced techniques for estimating surface mechanical properties. Figure 2.4 shows a summary of the conventional and advanced techniques for measuring the elastic-plastic micro- and nanomechanical properties of surfaces.

2.3.2 Importance of Elastic Properties in Conventional and Advanced Biointerfaces

The elastic properties of surfaces are relevant for any interface system, when they are analyzed from interactions with any mechanical surrounding environment, such as applied stresses and/or layers and substances in contact. Detailed knowledge of the surface elastic behavior can provide, in principle, an indicator of the surface resistance to undergo a nonpermanent deformation due to applied stress. On the other hand, the plastic properties of a surface allow us to know the surface resistance to undergo permanent damage, which is mostly related to focused or contact stresses, instead of uniform or distributed applied stresses. Within this context, any mechanical interactions between a soft material in contact with a stiffer and harder material, as in many biomedical systems, will be focused on the influence of the surface elastic properties of the stiffer material in the final mechanical response of the softer material; good examples of this situation are all components of any biological environment in contact with artificial biomaterials (cells, proteins, ECMs, etc.).

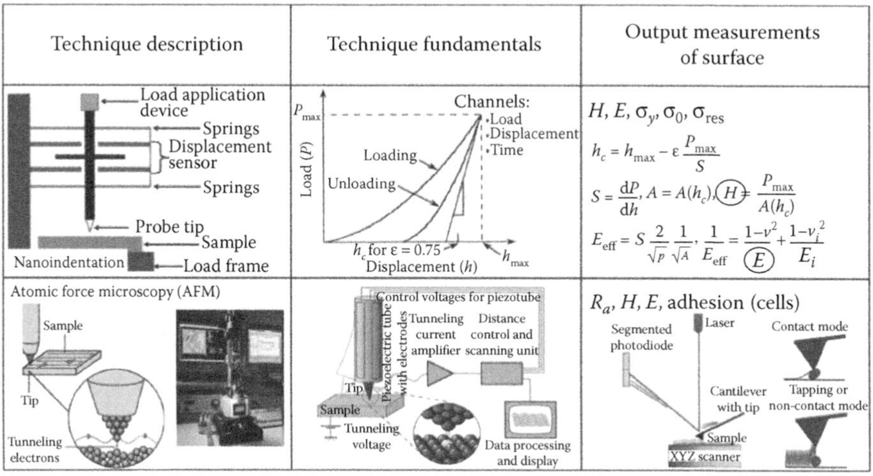

FIGURE 2.4 A summary schematic of the characterization of a biomaterial surface. Nanoindentation and atomic force microscopy are shown.

Before the emergence of nanotechnology in materials and biomaterials science, most of the interest in functional surface properties relied on plastic and topographical properties, meaning hardness, *H*, yield strength, σ_y, mechanical strength, σ_0, and residual stress, σ_{res}, due to their direct relationship with overloading, wear, and fatigue life resistance. In addition, with respect to topographical features (first- and second-generation biomaterials), it was assumed that the most important parameters were the microroughness parameters [13]. The relevance of nanoscale features has emerged during the last decade in the area of biomaterials. The development of third-generation biomaterials has opened up enormous potential for applications of nanostructured surfaces, due to the required manipulation of the biological environment at the molecular and subcellular levels. In the context of the evolution of functional biomaterials, remarkable work can be found in the literature from the last decade, showing the feasibility for nanostructuring many biomaterials. Similar studies have also explored key nanostructured features in order to manipulate cell response in contact with these surfaces. Almost all possible nanostructures from the available fabrication techniques have been successfully produced and suitably characterized, including complete evaluations of biological behavior in contact with a broad spectrum of cells, from stem cells up to all possible differentiated cell lines.

The relationship between elasticity and length scale can be best described by the work of Bao and Suresh [14]. They suggested a strategic approach to deciphering the importance of the elastic properties of surfaces at the nanoscale level with respect to cell response. From this approach, it is understandable how living cells can sense mechanical forces and convert them into biological responses, as long as the nanostructured features can function at the molecular level, including surface elastic features at that scale.

Studies on the mechanics of single cells, subcellular components, and biological molecules have rapidly evolved during the past decade with significant implications for biotechnology and human health. Details of the mechanical deformation of proteins and nucleic acids may provide key insights for understanding the changes in cellular structure, response, and function under external stimuli, and offer new opportunities for the diagnosis and treatment of diseases [14]. Also, due to the unique effect of surface nanostructured features on the cellular mechanical response, the controlled design of biointerface nanostructures will not only enable cell behavior modulation, but it will also result in an estimate of the elastic response of cells with their medium when the elastic properties of the associated biointerface are known.

2.3.3 Mechanobiology of Cells and Their Interaction with Biointerfaces

Most biological cells are between 1 and 100 μm in size, and they comprise many constituents. The cell is covered by a phospholipid bilayer membrane reinforced with protein molecules, and the interior of the cell contains a liquid phase (cytosol), a nucleus, the cytoskeleton consisting of networks of microtubules, actin, and intermediate filaments, organelles of different sizes and shapes, and other proteins. The deformability of cells is basically determined by their cytoskeleton, whose rigidity is influenced by the mechanical and chemical environments, including cell–cell and cell–ECM interactions [13,14]. All these small-scale features of cells and their high sensitivity to mechanical stimulation help explain a cell's sensitivity to the nanostructured features of biointerfaces. A good example of the important role of mechanical loading in cell response is the well-known bone mechanobiology. In this process, osteoblasts and osteoclasts modulate their functions in terms of both the level and the direction of applied stress to bone [13,14]. In the same sense, it can be stated in general that the mechanical loading of cells induces deformation and remodeling, which influence many aspects of human health and disease.

The mechanical scenario in which cells are well adhered to any biointerface is even more complex than it appears. Cells are not only able to mechanically sense the applied load, but they are also able to generate support forces. This means that any intended mechanical stimulation that is induced by any designed structured surface has to take into account these complex interactions. The goal of generating some specific mechanical stimulation in a cell remains a big challenge, and how cells sense mechanical forces and how deformation converts such signals into biological responses are not well understood [15]. The forces associated with the Brownian motion, as well as stochastic processes and nonequilibrium thermodynamics, should be accounted for in an overall interpretation of the mechanical response of cells [14].

One of the first attempts to elucidate the role of surface features on the elasticity of cell adhesion was addressed by the work of Ward and Hammer [16]. They developed a model of adhesion strengthening, which predicts large increases in adhesion strength following increased receptor clustering and adhesion size, marked by an elongation of the adhesion plaque. This process is understood to be due to an increase in tension at the adhesion site, because the focal adhesion size has been shown to be proportional to the force applied to it by the cell [17]. In the same sense, other authors have discovered integrin clustering [18], which has a discrete lateral spacing that lies in the realm of 15–30 nm and, as will be discussed, is a key indicator of the mechanisms involved in the nanofeature-mediated perturbation of focal adhesion formation. Biointerface topography and, in particular, nanoscale features can affect cell behavior and integrin-mediated cell adhesion, which is now evident from studies with fabricated topographical features [19]. The extent to which nanotopography influences cell behavior *in vitro* remains unclear, and investigations into this phenomenon are ongoing. The processes that mediate the cellular reaction to nanoscale surface structures are also not well understood and may be direct [20] (a direct result of the influence of the surface topography) or indirect (where the surface structure has affected the composition, orientation, or conformation of the adsorbed ECM components) [21,22].

Cellular mechanotransduction relies on the ability of the focal adhesion proteins to change their chemical activity state when physically distorted, converting mechanical energy into biochemical energy by modulating the kinetics of protein–protein or protein–ligand interactions within the cell [19]. Thus, nanotopography can be considered an important mediator of both cellular adhesion and differential function, acting to impart changes in cellular behavior through the modulation of focal adhesion reinforcement and protein interaction kinetics. Within this context, it can be stated that the important role of the elasticity of a nanostructured biointerface relies on the cell's mechanotransduction mechanisms. It is in the area of mechanotransduction that one likely finds the connection between chemical signals, topographic signals, and the life and behavior of the cell [23]. Many cell types generate mechanical tension in their surroundings, including other nearby cells. These forces are of the order of 1 nN per cell [17,24].

Thus, mechanical forces applied to a cell may be linked to the apparatus for gene expression. The reaction of cells to the mechanical properties of the substratum has been described and named durotaxis [24]. In particular, it refers to the cell migration directed by a gradient in the stiffness along a particular surface (e.g., gradient in surface morphology). Presumably, the cells can detect these properties by applying mechanical forces to the underlying layers and measuring their movement. That is why the elasticity of a nanostructured biointerface becomes so important. Thus, cells are sensing and reacting to an appreciable thickness of the substratum. In this way, cells may be able to detect "buried" features. Thus, the elasticity of a nanostructured surface should affect the mechanical properties of any overlying layers to some extent by altering the compliance of the substratum, and the cell will detect this as the strains coming back to its sensing systems. Since nanoimprinting of the cell can occur, this is likely to be a mechanical process and can add itself onto any chemical interactions.

2.4 ROLE OF SURFACE CHEMISTRY IN THE NANOSTRUCTURED BIOINTERFACE

As is well known, chemical interactions mediate many reactions in the body, and, in this way, the reactions at the interfaces between materials and tissues are no exception. At this level, some of the most important parameters that play a major role in the biological responses following an implantation are the energy of surfaces in terms of charges, mobility, and wettability [3]. Interfacial chemical reactions depend on the class of the material (metal, polymer, and ceramic respond differently) and the state of the host tissue. Moreover, surfaces may react in different ways depending on the specific characteristics of the environment in which they are situated.

The surface free energy is an important factor in the initial response of the body to an interface, because it promotes the formation of new bonds and the production of new compounds in the interface, allowing the absorption of proteins that can trigger a cell-mediated response. Surface charges in the interface may define the velocity and efficiency in recruiting proteins in the area of interest and then in the cellular attachment [3]. Surface segregation is related to the redistribution of solute atoms between the bulk of a material and its surface. Consequently, the total energy of the material can be affected. The surface of a material may promote the migration of functionality domains in the surface-to-bulk direction or in the opposite direction through diffusion mechanisms. In general, the surface chemistry of a material can affect its surface segregation in response to reducing the interfacial tension of the external environment [25]. This phenomenon is commonly present in metals and is sometimes accompanied by intragranular corrosion or stress corrosion cracking. In the same way in polymers, these phenomena can produce different responses depending on the phase formed, which could lead to a uniform distribution of molecules and atoms. Segregation may be considered a favorable or unfavorable event for the interaction of materials with cells, depending on the application [25].

The ability of a liquid to stay in contact with a surface is known as wettability and this phenomenon occurs as a result of several molecular interactions. The degree of wetting for each material is determined by a force balance between adhesive and cohesive forces and can be determined by measuring the contact angle. A surface is considered hydrophilic if the contact angle is $<65°$ and hydrophobic if the contact angle is $>65°$ [9]. Wettability has a great effect on the biological response because the first step following the introduction of a material into the body is the contact between the blood and the surface, which permits the hydration of the interfacial area, promoting migration of proteins to the site of interest [9].

The primary component of blood is water, making up approximately 45% of its composition and promoting the redistribution of ions near to the surface. The first process that occurs on the surface is the hydration of the surface and both the structure and reactivity of water have an important and controversial responsibility in the biological response of the interaction with biomaterials. Depending on the application, it is better to have some kind of configuration: hydrophilic or hydrophobic. For example, in a cardiovascular application, hydrophilic surfaces can be used to help in the activation of platelets, resulting in blood coagulation that helps promote the formation of a thrombus [9].

Materials can be exposed to different thermal and chemical techniques in order to change some of their surface properties. These can range from exposing the material to high temperatures or to harsh chemicals. As a result of these treatments, the surface chemistry can be altered and thus the subsequent biological response [26,27]. In some cases, pretreatment can decrease the biocompatibility of the biointerface; in others, the addition of new compounds can promote the formation of new reactive species, increasing the velocity of precipitation and the deposition of proteins on the interface and, consequently, the migration and adhesion of cells [28]. Several studies show that the interaction between proteins and the biointerface can improve with the addition at the surface of some chemical groups, such as carboxyl and hydroxyl, as well as amines, fatty acids, salts, and thiol groups [9].

2.5 BIOCHEMISTRY OF THE NANOSTRUCTURED BIOINTERFACE

As mentioned in the previous section, adhesion and cell proliferation depend primarily on the interaction of proteins with the materials–tissue interface, and in turn, the adsorption of proteins is determined by the chemistry and surface physicochemical characteristics. The hydration of the surface is the first activity that occurs when a material is implanted. Surfaces are exposed to a corrosive biological environment that degrades or modifies resorbable materials or may help corrode unstable metallic surfaces by means of propagated reactions outward from the surface of the material. The biological response in the dynamic interphase involves the triggering of linked enzymes reactions that can amplify the local response in the biomaterial. The biointerface physiochemical reactions that catalyze, mediate, and control the biological responses to the material start to occur. The biointerface mechanisms that are responsible for cell–surface interactions consist of two primary effects: (1) protein adsorption and (2) cell response. Proteins in the appropriate conformation and orientation can stimulate a constructive cell response that could include, for example, favoring wound repair and tissue integration. The cell response depends on specific proteins being correctly presented to achieve anchorage and receive extracellular instructions that can guide the cell's migration, proliferation, and apoptosis [1].

Protein absorption is the principal precursor of cellular migration near the implant zone. This process begins when different proteins migrate to the area and are absorbed to the previously hydrated surface, creating a complex called interphase. The interphase is a dynamic region where proteins start their interaction, absorbing and desorbing, causing a flux of water, proteins, and associated ions into and out of the interphase. The biological response is a consequence of this chemical response [9]. The adsorption process of proteins consists of the following steps: (1) the hydration of an adsorbent surface brought into contact with an aqueous protein solution; (2) the formation of a thin pseudo-2-D interface between the adsorbent and the protein solution; (3) the diffusion of protein molecules into this interface; and (4) the formation of a three-dimensional (3-D) interphase composed of arriving proteins. Finally, this interphase undergoes a time-dependent decrease in volume by expulsion of either or both the interphase water and the initially adsorbed proteins. Displacement of interphase water requires an amount of energy that depends on the adsorbent surface chemistry and energy. Subsequently, interaction between antibodies, enzymes, hormones, and other signals can activate the cell response and their migration to specific locations [9].

Adsorbed proteins such as fibronectin (FN), vitronectin, and osteopontin play a major role in regulating the initial cell adhesion and spreading on biointerfaces, through which cells bind via diverse molecular mechanisms including integrins [29]. The integrin is a heterodimeric cell surface receptor that is able to link specific proteins of the ECM and actin filaments. FN can also be secreted by cells as a disulfide-bonded dimer, which self-assembles forming an FN fibrillar matrix. This self-association of FN units is encouraged by integrins, which display their self-assembly sites [30]. The FN serves as a core assembly for multiple components of the ECM (e.g., collagen I and II, fibrinogen); it gives structural support for cell adhesion via integrins; and it controls the availability of some growth factors, such as transforming growth factor β, by regulating their activation from latent complexes [31]. The binding between integrins and FN molecules generates mechanical strain through the actin cytoskeleton, which is able to stimulate signal transduction pathways that result in a switch between cell growth and differentiation.

The interaction between proteins at different velocities, the control of the pH, and the ionic strength also play important roles in the interaction between biomaterials and tissue [3]. Through biomimetic processes, it is possible to change the configuration of the surface of materials by incorporating or immobilizing biological molecules with specific functions, such as glycoproteins, peptides, phospholipids, proteins, and saccharides among others in this case nonspecific adsorption of macromolecules and covalent bonding there are used [9].

2.6 FOREIGN-BODY RESPONSE AND THE NANOSTRUCTURED BIOINTERFACE

The reaction of the body in response to the presence of a foreign agent implies the participation of the body's immune system, where macrophages and other cells populate the tissue–material interfaces in order to detect and recognize the new material and recruit more cells into the area. This phenomenon at the tissue–material interface of medical devices, prostheses, and biomaterials is important to mediate the process of rejection or wound healing. Host reactions after implantation of biomaterials include production of injury, interaction between fluids (blood) and the surface, formation of a provisional matrix, local inflammation, and formation of granulated tissue around the biomaterial interface [32].

In the process of recognizing the presence of a biomaterial, the body activates different molecules such as mitogens, chemoattractants, cytokines, growth factors, and other bioactive molecules that can modulate the activity of macrophages, along with the proliferation and activation of other cell populations in the inflammatory and wound-healing responses, which are identified by the presence of mononuclear cells such as monocytes and lymphocytes, at the biomaterial implant site. Chemokines are cytokines that are chemoattractants. They coordinate cellular migration in inflammation to produce an efficient and effective process of wound healing and are involved in other processes such as hematopoiesis, angiogenesis, and cellular differentiation. The production of cytoquines such as interleukin-4 and interleukin-13 plays a significant role in the reaction of the body after exposure to a foreign agent [32,33].

Despite the fact that the FN fibrillar matrix plays a pivotal role in supporting cell adhesion and subsequent signaling cascades, its presence *per se* on the surface does not indicate that the cell–surface interface is functional and is therefore able to trigger cell biological activity. In this case, no functional FN fibrillar matrix can provide cell adhesion, but not cell spreading and traction as has been depicted by Girós et al. [34]. Upon coating titanium substrates with FN, Pegueroles et al. found that osteoblasts (MG63) tended to agglomerate on the microstructure peaks where the FN was mostly accumulated [35].

Luthen et al. showed that integrins $\alpha 5$ and $\beta 1$ did not form focal adhesions on the surfaces of microstructures with a high roughness profile [36]. The FN bound to those surfaces did not display its active sites such as III_{10}RGD, which is an issue for cell migration. These results indicate that a cellular response can be triggered not only by the presence of the key proteins for cell adhesion, but also by the distribution of those proteins on the surface. Following the initial interface interactions and provisional matrix formation, acute and chronic inflammation occurs in a sequential fashion as expected [32].

Other cells that play an important role in the body's recognition of biomaterials include neutrophils and mast cells that mediate inflammatory responses to implanted biomaterials, modulated by histamine-mediated phagocyte recruitment and phagocyte adhesion to implanted surfaces facilitated by the adsorbed host fibrinogen [32]. Following the resolution of acute and chronic inflammatory responses, granulation tissue is identified by the presence of macrophages, the infiltration of fibroblasts, and neovascularization in the new healing tissue. Granulation tissue is the precursor to fibrous capsule formation, and it is separated from the implant or biomaterial by the cellular components of the foreign-body reaction [37,38].

2.7 CELL REGULATION AND THE NANOSTRUCTURED BIOINTERFACE

Since cells in their natural environment are surrounded by nanoscale features linked to their ECM, nanotopographical traits become an essential part of the design of biomaterials for tissue formation and repair. The biointerfaces designed for tissue engineering are thought to mimic the microenvironment surrounding cells, giving them the ability to regulate and control the cells' fate. Different mechanisms affect the cellular function on interfaces, many of them depending on the surface nanotopographical cues.

The initial attachment of cells to biomaterial surfaces determines subsequent signal transduction processes such as adhesion, morphology, migration, orientation, proliferation, and, ultimately, differentiation [23,39,40]. The nanoscale features can direct the way as the adhesive peptides are linked to the surface, as well as their spatial arrangement. The FN correctly presented can support the initial cell interaction with the surface, and facilitates the formation of more complex structures such as the FN fibrillar matrix. This matrix serves as (1) an assembly site for other components of the ECM; (2) it controls the availability of growth factors; and (3) it provides structural support for cell adhesion while the cell membrane receptors sense and transduce signals from the ECM. This initial integrin–ligand binding is followed by the formation of focal adhesions and actin stress fibers that enhance and strengthen cell adhesion to the surface by the recruitment of additional proteins such as the focal adhesion kinase (FAK) [40].

Focal adhesions are the first route by which nanotopographical traits are sensed by the cells, and they play a pivotal role in the subsequent downstream signaling in response to nanotopography, such as gene transcription. This phenomenon is known as mechanotransduction [41]. The possible mechanisms whereby the tension from the focal adhesions may be converted into a gene or protein response are understood to be (i) the increased access to DNA by transcriptional factors; (ii) the formation of transcriptional factor complexes derived from the focal adhesion tension strain; and (iii) the alteration in mRNA transport, inducing protein production. Studies suggest that a remarkably small modification in surface nanotopography could allow mesenchymal stem cell growth and development, indicating that changes in such nanotopographical features have a direct influence on the adhesion/tension balance, permitting self-renewal or targeted differentiation [42].

2.8 EMERGENT PROCESSING OF NANOSTRUCTURED BIOINTERFACES

The processing of nanostructured biointerfaces is intimately connected with the function and properties of a biomaterial, as discussed in Section 2.1. In this section, we summarize the emergent processing technologies of nanostructured biointerfaces.

One key aspect of processing nanostructured biointerfaces is the understanding of the fundamental mechanisms responsible for the modification of material surfaces. Conventional materials processing has been dominated by the development of advanced top-down fabrication techniques that include lithography-based techniques (e.g., ultraviolet and extreme ultraviolet lithography), which are capable of fabricating surface structures on the order of 50–100 nm. Additional techniques include focused-beam lithographies using electron or ion energetic particles and scanning probe lithographies [43–45]. The limitations of these conventional techniques have mainly been attributed to their physical limitations in fabricating structures smaller than approximately 50 nm in size and their high-volume manufacturing scalability. Therefore, bottom-up techniques that rely on self-assembly, self-organization, and local patterning have become emergent technologies that are capable of patterning biocompatible surface nanostructures.

In this context, the ability to process materials over multiple scales (e.g., from nanoscale structures coupled to micron- and mesoscale structures) has also become increasingly important in enabling materials to acquire multiple functions as the demands for so-called third-generation materials [46] require not only biocompatibility but also other functional properties, including sensing, surface adsorption, magnetic attraction, drug encapsulation and delivery, rigidity and toughness, and antifouling, and many other biomaterial requirements. The desire to design biointerfaces with multiple functions has ushered in a new class of biomaterials that requires higher-fidelity processing and fabrication control, and takes advantage of emergent processing techniques that include bottom-up and hybrid top-down/bottom-up approaches. Furthermore, the growth in biocompatible and multifunctional thin films coupled with the design of advanced nano- and microscale composites has also inspired thinking about combining bioinspired and bioderived materials coupled with conventional materials scaffolding for applications in tissue regeneration and wound-healing *in vivo* protocols. This section summarizes some recent work in the processing of biocompatible thin films with multifunctional properties followed by emergent techniques for multiscale surface patterning of biomaterials. This is followed by a thorough discussion of irradiation-driven fabrication techniques that induce nanopatterning and enable multiscale composite fabrication. This section concludes with a brief outlook on the use of computational tools for nanostructured biointerface design.

2.8.1 Biocompatible Thin-Film Processing

The design of biomaterials with multifunctionality has taken advantage of the use of advanced thin-film technology. One primary advantage of using thin films in biomaterial interfaces is their inherent low-dimensional state that allows for multifunctional properties in their application. For example, using thin films, one can introduce a biocompatible property into the first 100–1000 nm of a biomaterial surface while keeping the desired macroscopic bulk material properties of a substrate, which include the desired mechanical and thermal properties. However, the deposition of thin films can become a complex process given that their interface on the substrate (e.g., metal, semimetal, ceramic, etc.) can have indirect and direct effects on their inherent properties.

Furthermore, challenges exist with thin-film integrity (e.g., adhesion to a substrate, stability) in extreme environments of the body, which include the corrosive conditions of body fluid and stringent elastomechanical constraints (i.e., in the application of prosthesis devices or biomedical implants). The paradigm in the materials design approach using thin films expands the traditional PSP relationships in traditional "bulk" materials design. Figure 2.5 illustrates the methodology in thin-film processing design. In particular, the deposition process must account for low-dimensional state mechanisms such as the rate of film growth and the residual stress in the film due to interfacial forces between thin-film atoms and substrate surface atoms. Lattice matching effects can influence the formation of a polycrystalline film, and a defect density at the interface. The real structure of the system can also induce constraints on the orientation of thin-film polycrystalline grains and the ultimate "texture" or preferential crystallographic orientation of the thin-film grains during growth.

These interfacial parameters in turn will influence properties such as surface charge and chemistry, which can dictate biocompatible functions in the thin-film biointerface. There are numerous techniques to deposit thin films and here we list a few that are typically used in the processing of thin-film biomaterials. Numerous techniques are used with different classes of materials depending on the technique employed for modification, fabrication, and chemical functionalization. For example, physical vapor deposition techniques can include plasma-enhanced techniques, magnetron sputtering, and ion-beam-assisted deposition. For oxide-based materials, chemical vapor deposition or atomic layer deposition techniques can be used. For polymers or other organic-based materials, liquid-precursor-based techniques can be used, including electrophoretic deposition and spray deposition methods or spin- or dip-coating techniques, or both. Each technique has its unique limitations that are dictated by the type of composition/chemical structure desired.

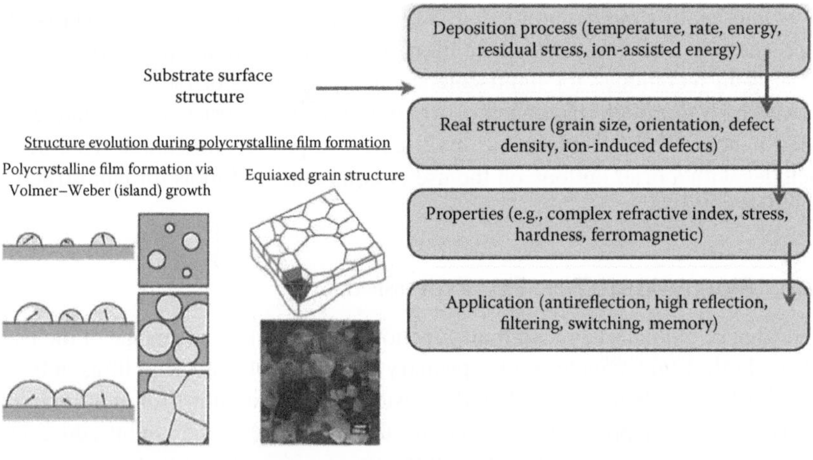

FIGURE 2.5 A materials design with thin films elucidating the role at the thin-film/substrate interface. The structural evolution of polycrystalline thin films is dictated by the kinetics of the deposited material and island/coalescence during the growth period. (Adapted from Kaiser, N., *Appl. Optics*, 41, 3053, 2002; Thompson, C.V., *Ann. Rev. Mater. Sci.*, 2000.)

2.8.2 Irradiation-Driven Patterning

Ion beams can be used to induce patterned structures and a unique topography at the nanoscale by means of sputtering and other surface-related processes [47–55]. This approach can have significant implications for the design of nanostructured surfaces used in cell engineering and cell biology [20,23,56,57] as an example. Irradiation-driven systems have been explored in similar moderate energy regimes dominated by knock-on atom displacement regimes for semiconductor metallization microstructure control [58], artificial texturing of ceramics (Harper and others with the ion-beam-assisted deposition [IBAD]), engineering of nanostructured carbon [59], and compositional patterning of immiscible alloys [60]. Such findings suggest that irradiation-driven self-organized structures can have critical effects on the surface properties of low-dimensional state structures.

Irradiation-driven systems have been limited to the study of dissipative systems, which lead to permanent damage by operating in energy and flux regimes that limit the self-organization of low-dimensional state structures. Despite the use of low-energy ion beams, difficulties remain with the control of short- and long-range ordering, surface chemistry, and topographical control. In addition, despite numerous experiments and models that have investigated ion-induced surface nanopatterning, the synthesis design for the functionalization of ion-induced biomaterials remains absent. Ion-induced strategies are only used for biocompatibility enhancement. One important limitation of current nanomanufacturing is its dependence on naturally self-ordered processes that balance kinetic and thermodynamic dissipative forces in the absence of irradiation techniques [61]. This dependence leads to limited control of surface chemistry and functionalization. Furthermore, top-down and bottom-up surface patterning is predominantly dependent on focused beam technology. One example is the use of electron and ion-beam nanolithography, which have order-of-magnitude limitations toward high-volume manufacturing. Directed irradiation synthesis (DIS) addresses this limitation by introducing a synthesis process that is scalable to high-volume manufacturing by virtue of its intrinsic large-area simultaneous exposure of materials surfaces and interfaces. Broad-beam ions combined with rastered focused ions and gradient ion-beam profiles are sequenced and/or combined with reactive and/or nonreactive thermal beams that control the surface topography, chemistry, and structure at the micro- and nanoscale. DIS exploits the self-organized regime in irradiated surfaces dominated by ion-induced erosion and surface diffusion. The novelty of DIS is significant because it will transform the synthesis and design of nanostructured systems by identifying the composition-dependent mechanisms that drive self-organization on irradiation-driven micro- and nanostructures, enabling tunability and control of their chemical and electronic properties. In particular, this invention has important ramifications for biomaterials. This is critical to introducing novel design pathways that tune the bioactive properties of the thin-film coatings used in multiple applications for biocompatibility and biosurface material adaptability.

An example of an irradiation-driven surface modification is the deposition of nanostructured cobalt (Co) films for biomedical applications. In this application, the desire to functionalize magnetic biointerfaces is driven by applications in endovascular reconstruction strategies. Irradiation-driven approaches can lead to novel surface nanostructures of Co-deposited particles that can induce significant changes in the material's properties, such as wettability. Figure 2.6 illustrates contact angle measurements that clearly show the irradiation-driven modification inducing significant variation in the contact angle of the Co-based nanostructures on the surface. Figure 2.6 shows a high-resolution micrograph of the DIS-designed nanostructures. Another example of irradiation-driven nanostructures is the synthesis of Sn-oxide and Sn-metal nanoparticles on various substrates including Au, Ru, Rh, and Si. Figure 2.7 shows

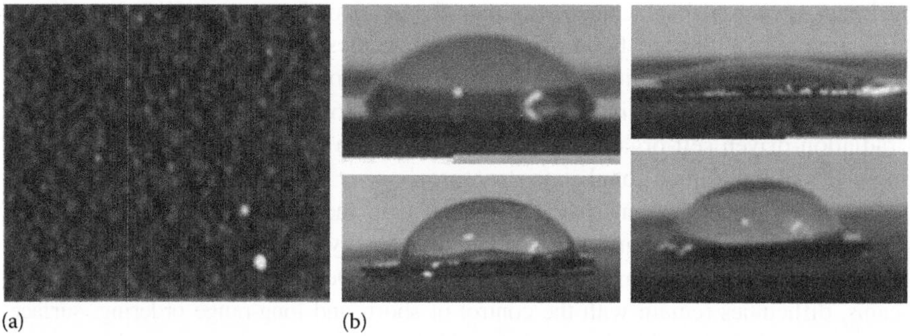

FIGURE 2.6 Contact angle measurements of Au- and Co-deposited and irradiated surfaces with DIS. Note the dramatic decrease in contact angle for Co nanostructures. (a) An AFM micrograph of a Co-deposited and irradiated surface morphology. (b) Before (top) and after (bottom) irradiation contact angle measurements for Au-deposited nanostructured films (left) compared with Co-deposited nanostructured films (right).

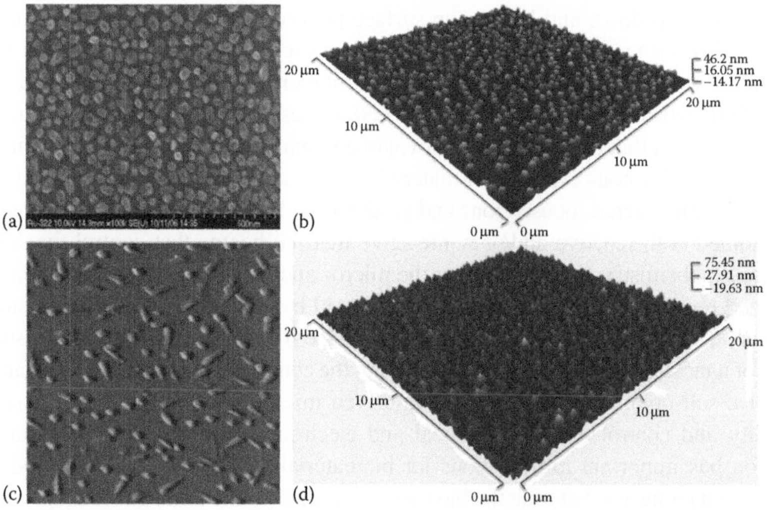

FIGURE 2.7 (a) High-resolution scanning electron microscope (SEM) micrograph of Sn nano- and microstructures synthesized with directed irradiation synthesis for a given thermal Sn to energetic Xe+ ion flux ratio. (b) 3-D atomic force micrograph illustrating evidence of Sn microstructure separation variability with a change in irradiation conditions at room temperature. (c) 3-D modification of the surface morphology of Sn structures as templates for biofunctionality and (d) more dense Sn microstructures compared with (b) by varying incident particle energy.

the surface nanostructures of Sn-oxide on Rh synthesized from a combined thermal and ion energetic source. Sn-oxide can be used as nanotubes for sensing, thin films as biocompatible interfaces, indium-tin-oxide (ITO) as thin film for biosensing, and Sn-oxide as semiconductor applications in biomaterials. In this particular example, SnO thin films and nanostructures can be grown as ultrathin films that can be integrated with biocompatible materials such as Si membranes or polydimethylsiloxane (PDMS).

2.9 APPLICATIONS OF NANOSTRUCTURED BIOINTERFACES

2.9.1 Neuroendovascular Tissue Reconstruction

The role of nanotechnology in the endovascular treatment of cerebral aneurysms is becoming increasingly critical for localized, asymmetric treatments (e.g., the localized treatment of cerebral aneurysms) as articulated in this chapter. One example where nanotechnology can have a significant impact on the treatment of aneurysms and vascular disease is in the synthesis and manipulation of the surfaces and interfaces of devices deployed for reconstruction and tissue repair (e.g., 3-D scaffolds with commercial stents). The role of surface modification is critical due to the unique relationship between protein adsorption processes and how they are affected by a biomaterial's surface properties at the nanoscale [62].

Advanced coatings can be designed to induce functionality on 3-D scaffolds. Multifunctional properties could be tailored with the use of directed irradiation synthesis, which was pioneered by Allain et al. [63]. This method is a low-energy (below threshold damage) ion- and atom-beam modification technique that aids in the growth of thin films while inducing structure and topography at the nanoscale. Thus, one could envision coupling various functionalities (i.e., biocompatibility, hydrophilicity, magnetic attraction, surface chemistry, surface topography, and cell regulation) in one multicoating system that would be deposited, for example, on stent surfaces. Ion-beam-induced nanopatterns can couple organic/inorganic systems as well as immiscible materials at the nanoscale. Recently, Allain et al. have developed advanced *in situ* synthesis methods to elucidate on ion-irradiation mechanisms that can manipulate surface chemistry and surface morphology to ultimately synthesize functional coatings for 3-D scaffold systems. Figure 2.8 shows the current approach to designing irradiation-driven coatings, which characterizes these materials during synthesis and couples morphology measurements with surface chemistry in a new facility known as particle and radiation interaction with hard and soft matter (PRIHSM).

Manipulating nanoscale systems and nanotopography can be critical to cell and subcellular regulation and manipulation, ultimately influencing the growth and viability of tissue [64]. Surface-modified systems can be influential in cell growth on desired regions of a stent scaffold. Nanostructures increase cell adhesion because of their influence on surface wettability, increased surface area, and cell shape [65,66]. Cells are often directed through the ridges/grooves of nanostructures and will align themselves in the direction of these ridges. The multiple ridges/grooves increase the cell contact inhibition effect as well as making it difficult for cells to grow over

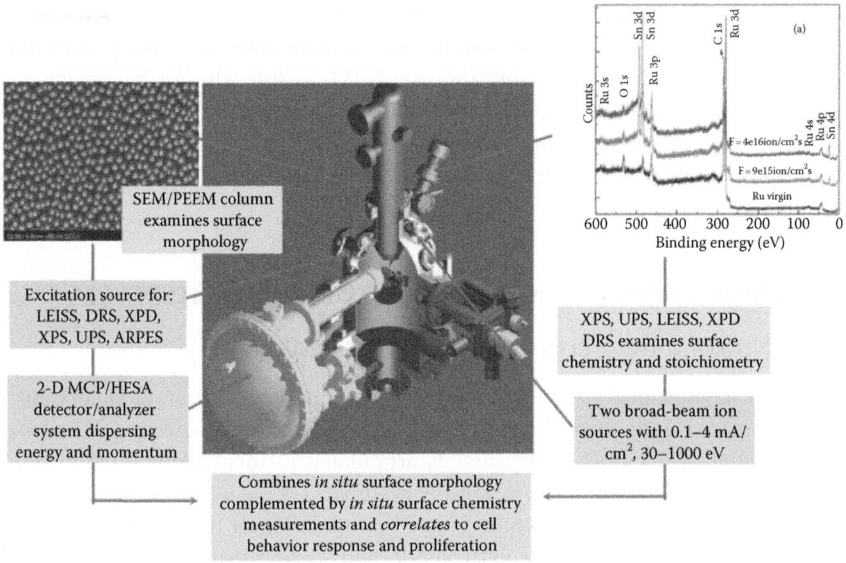

FIGURE 2.8 The PRIHSM experimental facility that couples *in situ* surface morphology evolution with surface chemistry. Coupling *in situ*, these surface properties are critical to the correlation with the cellular response to surface nanotopography.

the ridges. This "contact guidance" phenomenon generates cell orientation whose arrangement is orderly over a large surface area [67,68]. An ECM provides support and anchorage for cells, as well as directional cues; nanostructures provide these same directional cues [69]. Allain et al. have recently devised a design for nanostructured biointerfaces that induces three main functionalities: (1) renders an asymmetric region of the stent surface magnetic; (2) induces nanotopography for cell adhesion, migration, and proliferation; and (3) induces biocompatibility. Figure 2.9 illustrates the design of these nanostructured biointerfaces coupled to neuroendovascular stents that are locally introduced within arterial vessels in the brain against cerebral aneurysms with the ultimate outcome to reconstruct the tunica media and seal the aneurysm shut.

A cell's shape is a major determinant of its growth because as a cell spreads, it elongates. In response to nanostructures, endothelial cells exhibit the same arcuate, or curved, morphology that is typical of the vascular endothelium [70]. This well-spread morphology and a well-developed cytoskeleton have been observed for a nano-island height of 13 nm [71,72]. Once the cells have adhered and formed a monolayer, a second specifically aligned monolayer can form on top of it without having direct contact with the nanostructure substrate [73,74]. This can allow for multiple aligned layers of cells in the desired region on the stent. The goal of nanostructures is to mimic the *in vivo* environment for cell growth and proliferation. Realizing the scale and size of the stent sections involved is the daunting challenge faced by Allain et al. in their recent work on bioactive neuroendovascular stents, foretold in a statement by Hopkins on assessing the frontiers

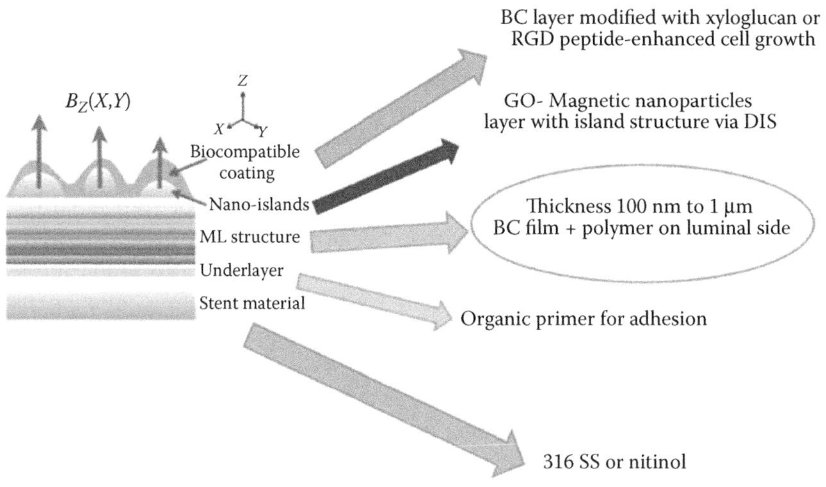

FIGURE 2.9 Illustration of an advanced nanostructured bioactive biointerface design that couples dissimilar materials to conformal shapes for neuroendovascular vessel reconstruction.

he saw for the future treatment of cerebral aneurysms: "The potential for nanotechnology in this field and the future endovascular suite are envisioned" [75].

2.9.2 Protein Immobilization and Biosensors

The role of the biointerface and nanostructured biomaterial surface on the functionality of biosensors and their ability to immobilize proteins has become another emergent area in biotechnology. Biosensors are devices currently under extensive research due to their applications for fast and precise detection of viral agents. Several configurations of biosensors devices are currently being studied, including those based on optical, thermal, and acoustical processes [76]. In particular, biosensors based on surface acoustic waves (SAWs) enable biomolecule detection in real time with a high degree of sensitivity and at a very low cost; they have recently been used in the detection of bacteria and viruses [77,78]. An important aspect of research in SAW advanced biosensors is the role of sustainable materials (e.g., non-Pb materials) that may be used in their fabrication, in particular oxide-based materials [79,80]. Furthermore, considering advanced SAW devices, the piezoelectric properties of bioinspired and bioderived materials remain marginally unexplored. Given the broad applicability of SAW biosensors in the areas of regenerative medicine, oncology, virology, and many other biomolecular tracing methods, the need for novel synthesis approaches to integrate sustainable materials in SAW biosensors is critical. Novel synthesis approaches at attaining both attractive biosensing properties and achieving them with sustainable materials have become an emergent area of research to establish a platform of advanced biosensors fabricated from sustainable materials with a benign environmental impact.

A SAW device mainly consists of a planar structure with electrodes on a piezoelectric material including interdigital transducers (IDTs) [81]. Of all the known piezoelectric materials, Pb-based piezoelectric materials, such as lead zirconate titanate (PZT), are widely used given their excellent piezoelectric properties. However, lead toxicity has motivated studies of materials without lead for environmentally friendly and biocompatible attributes. IDTs based on nonlead materials can consist of lead-free piezoelectric materials such as $LiNbO_3$, ZnO, or compound III–V semiconductors. Unexplored materials that can usher in a new class of advanced sustainable materials for SAW devices may include surface-doped grapheme [82] and bioderived materials such as cellulose [83,84]. One of the most commonly used substrates that matches many of the required conditions for the proper generation of SAWs is lithium tantalate ($LiTaO_3$) [81]. Applying a high-frequency (generally between 80 and 100 MHz) alternative voltage by means of the IDTs, acoustic waves are generated on the substrate, which, in turn, results in a specific resonance frequency that is characteristic of the acoustic wave speed propagation on such a substrate. This frequency is sensitive to measurable changes on the substrate surface, for instance, those produced by the interaction of the surface with specific biological materials.

SAW devices have been used, for example, in detecting viruses such as the Coxsackei virus, a member of the enterovirus family, which also includes poliovirus and hepatitis A virus [81,85]. Recently, fast chromatography combined with SAW quartz sensors (GC/SAW, zNose) has been presented as one of the most achievable new methods. GC/SAW sensors [86–88] are mainly vapor analyzers and their performance principles mimic those of the human sensorial system. The advantages of these devices include their overall high degree of simplicity, fast analysis, portability, and low cost in comparison with chromatographic systems combined with mass spectrometry. As a consequence, the GC/SAW devices exhibit the ability to recognize and quantify several complex fragrances. This measurement is manifested as a fragrance pattern, known as a "vaporprint," obtained from the frequency of a SAW sensor. Furthermore, it is capable of reaching sensitivity up to parts per billion (ppb) levels for some volatile compounds and up to parts per trillion (ppt) for semivolatile compounds. The SAW device methods have demonstrated their adaptability in several applications, including the monitoring of environmental pollution, geographic and botanic discrimination of organisms, bad flavor detection in food preparation, disease diagnosis through the air exhaled by patients, drugs and explosive tracking, and measuring the fragrance changes of living organisms in real time [87]. However, as the SAW sensor response critically depends on the interaction between an analyte and a substrate surface in several environmental conditions, extensive research is necessary in this field not only to improve the efficiency of such devices, but ultimately to understand how nanoscale surface modification can lead to transformational designs leading to advanced SAW biosensor devices. For example, the adhesion properties between a substrate and an analyte have to be optimized and the damage induced on the substrate surface after lapping and polishing processes should be eliminated. All these facts require effective surface modification techniques aimed, apart from elucidating on the fundamental mechanisms behind these processes, at enhancing the performance of SAW sensors.

Additional critical areas where nanoscale surface modification can have a significant impact include the development of microfluidic acoustically driven devices and tissue manipulation-on-chip platforms, and high-resolution sensing of tissue/cell behavior. Furthermore, one key limitation in achieving high-performance piezoelectric materials has been the need for the synthesis of complex high-quality epitaxial films requiring expensive, time-consuming, and environmentally unfriendly processing techniques. Recent work has identified novel properties of oxide-based heterostructures for the discovery of sustainable electronic materials, and with the advent of techniques such as pulsed laser deposition (PLD), many advanced heterostructures of complex oxide-based materials have been achieved. However, PLD techniques lack the ability to control both thin-film deposition and nanostructure formation necessary to induce the attractive enhanced SAW properties described above.

One of the most effective methods for surface modification is ion irradiation. It is well known that ion beam and plasma processing of materials has found prolific applications in the areas of nanoelectronics, biomaterials, nuclear materials, and energy materials. Even with these advances, the mechanisms by which ion-induced processes modify and functionalize surfaces and interfaces (in particular, low-dimensional state systems; e.g., nanostructures and ultrathin films) at particle energies near the damage threshold remain largely unknown. This is particularly true for complex multicomponent surfaces and interfaces where preferential sputtering and intermixing effects may introduce a new facet of driven self-organization. Therefore, the structure and properties of these surfaces/interfaces depend strongly on the conditions during ion-beam modification.

2.9.3 Biointerfaces for Treatment of Damage in Bone Tissue

Bone is one of the tissues that has had a substantial impact on biomaterials. Commonly, bone problems result from trauma events or from degenerative diseases. The most important cells involved in the formation processes of bone are the osteoblasts, which regulate the maturation and mineralization of the bone ECM. After some process of differentiation, osteoblasts may become preosteoblasts to finally mature into osteocytes or lining cells. Osteoblasts can synthesize specific proteins such as alkaline phosphatase, collagen I, osteonectin, osteocalcin, osteopontin, and bone sialoproteins, among others [89]. On the other hand, bone is an anisotropic, heterogeneous, nonlinear, thermorheologically complex, viscoelastic material at the biomechanical level. Implants that guarantee the optimal performance of their mechanical, biological, and structural behavior are the best option in the treatment of bone problems. Some metals such as titanium and its alloys, chromium, cobalt, and molybdenum and some steels such as 316L have shown good responses to loads and physical exertion and, additionally, are biocompatible and exhibit good corrosion resistance; however, their "almost inert" nature hinders their integration with the surrounding tissue [90]. This problem can be solved by physicochemically modifying the material surface or by introducing a new bioactive material such as hydroxyapatite, tricalcium phosphate, or bioglass, which may attach bioactive properties by promoting the growth of bone tissue around the implant. When using a bioactive ceramic, is important to know if it is resorbable or nonresorbable. If the bioactive ceramic is resorbable, the rate of its solubility should be considered, because if this is higher compared with bone formation, then the stability of the implant can be compromised due to the early loss of coating [91]. The bioactivity index (BI) of a material is defined according to the time it takes for more than 50% of the surface to attach to the bone. There are two types of bioactivity: Class A, or osseoproductive, is the process by which a bioactive surface is colonized by osseogenic mesenchymal cells near the implant site [92] and Class B for the case of an osseoconductive interface that only allows cellular migration through it [93]. In either case, the depositions of bioactive materials on the surface of metallic materials are used as coatings in order to improve the processes of osseointegration.

The clinical success of an implant will depend on various conditions, such as the status of the host tissue, the physical and chemical characteristics of the implant, the quality and durability of the bonding interface between the prosthesis and the host tissue, and the resistance to mechanical loads that it will be subjected to [94,95]. Once the material is implanted, several events occur that favor binding to the tissue. Depending on the nature of the material, two types of interactions can occur: extracellular and intracellular. An extracellular interaction is determined by the characteristics of the material surface, which facilitates the absorption of proteins, growth factors, and collagen, as well as the activation of enzymes. Intracellular interaction exists in the communication and exchange between materials and cells, stimulating cell mitotic activity, which is a precursor for proliferation processes and cell differentiation. In the case of osteoblasts, they have been found to increase DNA synthesis and the production of alkaline phosphatase and osteocalcin [96]. A bioactive material has an attachment in which the interfacial bond between tissue and material is strong both chemically and mechanically; this type of attachment is time dependent [97] and is one of the reasons why it has been found that the topographical configuration is also a factor that greatly influences the tissue binding to the implant [5,98,99]. In conclusion, an implant should be similar to the host tissue at the interface level and regarding the material response to physical stimuli [100]. The nature, composition, and mechanical properties of materials make biological responses at the time of implantation differ in aspects such as integration time with tissues and the attachment mechanism, among others [97].

2.10 SUMMARY

In summary, nanostructured biointerfaces can help mediate the cellular functionality and behavior in the body with a higher degree of fidelity than conventional biomaterials, which are limited to aspects of biocompatibility (e.g., so-called first- and second-generation biomaterials). Although numerous efforts in past decades have focused on understanding the response of cells to their interacting medium, there is still only limited knowledge about the fine-tuning of cellular behavior on the scale of several nanometers (or molecular scales) to subtissue spatial scales. Nanostructured biointerfaces have also been tailored to provide a limited number of functions that enable the integration of materials into the body. These functions include scaffolding, wound healing, drug delivery, and sensoring. The synthesis of these nanostructured biointerfaces is critical for materials with favorable properties. The PSP relationships in traditional materials science are complemented with additional relationships related to the *in vivo* function and performance of the proposed biomaterial. This approach leads to the design and synthesis of new biomaterials, the *in situ* performance characteristics that enable a new class of materials. DIS might be an alternative approach, using scalable modification and fabrication techniques to design novel nanostructured biointerfaces with multiple functions.

ACKNOWLEDGMENTS

The authors would like to acknowledge the support from grant DOD # W81XWH-11-2-0067 and the support for J.P. Allain from the Paul Zmola Young Scholar Award. The authors would like to also thank Emily Gordon for initial literature research on biosensors and Eric Yang for computational simulation figures.

REFERENCES

1. Parhi, P., A. Golas, and E. A. Vogler, 2010, Role of proteins and water in the initial attachment of mammalian cells to biomedical surfaces. A review. *Journal of Adhesion Science and Technology*, 24, 853–888.
2. Koegler, P., A. Clayton, H. Thissen, G. N. C. Santos, and P. Kingshott, 2012, The influence of nanostructured materials on biointerfacial interactions. *Advanced Drug Delivery Reviews*, 64, 1820–1839.
3. Bauer, S., P. Schmuki, K. von der Mark, and J. Park, 2012, Engineering biocompatible implant surfaces. *Progress in Materials Science*, 58, 261–326.
4. Deligianni, D., N. Katsala, S. Ladas, D. Sotiropoulou, J. Amedee, and Y. Missirlis, 2001, Effect of surface roughness of the titanium alloy Ti-6Al-4V on human bone marrow cell response and on protein adsorption. *Biomaterials*, 22, 1241–1251.
5. Huang, H., C. Ho, T. Lee, T. Lee, K. Liao, and F. Chen, 2004, Effect of surface roughness of ground titanium on initial cell adhesion. *Biomolecular Engineering*, 21, 93–97.
6. Liu, X., P. K. Chu, and C. Ding, 2010, Surface nano-functionalization of biomaterials. *Materials Science and Engineering: R: Reports*, 70, 275–302.
7. Senta, H., H. Park, E. Bergeron, O. Drevelle, D. Fong, E. Leblanc, F. Cabana, S. Roux, G. Grenier, and N. Faucheux, 2009, Cell responses to bone morphogenetic proteins and peptides derived from them: Biomedical applications and limitations. *Cytokine and Growth Factor Reviews*, 20, 213–222.
8. Kazemzadeh-Narbat, M., J. Kindrachuk, K. Duan, H. Jenssen, R. E. W. Hancock, and R. Wang, 2010, Antimicrobial peptides on calcium phosphate-coated titanium for the prevention of implant-associated infections. *Biomaterials*, 31, 9519–9526.
9. Vogler, E. A., 2013, Surface modification for biocompatibility, in A. Lakhtakia and R. J. Martín-Palma (eds), *Engineered Biomimicry*. Boston, MA, Elsevier, pp. 189–220.
10. Thevenot, P., W. Hu, and L. Tang, 2008, Surface chemistry influence implant biocompatibility. *Current Topics in Medicinal Chemistry*, 8, 270.
11. Ratner, B., A. S. Hoffman, F. Schoen, and J. E. Lemons (eds), 2004, Classes of materials used in medicine, in *Biomaterials Science: An Introduction to Materials in Medicine*. San Diego, CA, Elsevier, pp. 162–164.
12. Anderson, J. M., A. G. Gristina, S. R. Hanson, L. A. Harker, R. J. Johnson, K. Merritt, P. T. Naylor, et al., 1996, Host reactions to biomaterials and their evaluation, in B. D. Ratner, A. S. Hoffman, F. J. Schoen, and J. E. Lemons, *Biomaterials Science*. San Diego, CA, Academic Press, pp. 165–214.

13. Hench, L. L. and I. Thompson, 2010, Twenty-first century challenges for biomaterials. *Journal of the Royal Society Interface*, 7, S379–S391.
14. Bao, G. and S. Suresh, 2003, Cell and molecular mechanics of biological materials. *Nature Materials*, 2, 715–725.
15. Zhu, C., G. Bao, and N. Wang, 2000, Cell mechanics: Mechanical response, cell adhesion, and molecular deformation. *Annual Review of Biomedical Engineering*, 2, 189–226.
16. Ward, M. D. and D. A. Hammer, 1993, A theoretical analysis for the effect of focal contact formation on cell-substrate attachment strength. *Biophysical Journal*, 64, 936–959.
17. Balaban, N. Q., U. S. Schwarz, D. Riveline, P. Goichberg, G. Tzur, I. Sabanay, D. Mahalu, S. Safran, A. Bershadsky, and L. Addadi, 2001, Force and focal adhesion assembly: A close relationship studied using elastic micropatterned substrates. *Nature Cell Biology*, 3, 466–472.
18. Takagi, J., B. M. Petre, T. Walz, and T. A. Springer, 2002, Global conformational rearrangements in integrin extracellular domains in outside-in and inside-out signaling. *Cell*, 110, 599–611.
19. Biggs, M. J. P., R. G. Richards, and M. J. Dalby, 2010, Nanotopographical modification: A regulator of cellular function through focal adhesions. *Nanomedicine: Nanotechnology, Biology and Medicine*, 6, 619–633.
20. Dalby, M. J., N. Gadegaard, R. Tare, A. Andar, M. O. Riehle, P. Herzyk, C. D. Wilkinson, and R. O. Oreffo, 2007, The control of human mesenchymal cell differentiation using nanoscale symmetry and disorder. *Nature Materials*, 6, 997–1003.
21. Andersson, A.-S., J. Brink, U. Lidberg, and D. S. Sutherland, 2003, Influence of systematically varied nanoscale topography on the morphology of epithelial cells. *IEEE Transactions on NanoBioscience*, 2, 49–57.
22. Martines, E., K. Seunarine, H. Morgan, N. Gadegaard, C. D. Wilkinson, and M. O. Riehle, 2005, Superhydrophobicity and superhydrophilicity of regular nanopatterns. *Nano Letters*, 5, 2097–2103.
23. Curtis, A. S., M. Dalby, and N. Gadegaard, 2006, Cell signaling arising from nanotopography: Implications for nanomedical devices. *Nanomedicine (London)*, 1, 67–72.
24. Lo, C.-M., H.-B. Wang, M. Dembo, and Y.-l. Wang, 2000, Cell movement is guided by the rigidity of the substrate. *Biophysical Journal*, 79, 144–152.
25. Yang, S. Y., E.-S. Kim, G. Jeon, K. Y. Choi, and J. K. Kim, 2013, Enhanced adhesion of osteoblastic cells on polystyrene films by independent control of surface topography and wettability. *Materials Science and Engineering: C*, 33, 1689–1695.
26. Chu, P. K., 2012, Surface engineering and modification of biomaterials. *Thin Solid Films*, 528, 93–105.
27. Zreiqat, H., S. Valenzuela, B. Nissan, R. Roest, C. Knabe, R. Radlanski, H. Renz, and P. Evans, 2005, The effect of surface chemistry modification of titanium alloy on signalling pathways in human osteoblasts. *Biomaterials*, 26, 7579–7586.
28. Diener, A., B. Nebe, F. Lüthen, P. Becker, U. Beck, H. Neumann, and J. Rychly, 2005, Control of focal adhesion dynamics by material surface characteristics. *Biomaterials*, 26, 383–392.
29. Larsen, M., V. V. Artym, J. A. Green, and K. M. Yamada, 2006, The matrix reorganized: Extracellular matrix remodeling and integrin signaling. *Current Opinion in Cell Biology*, 18, 463–471.
30. Leiss, M., K. Beckmann, A. Girós, M. Costell, and R. Fässler, 2008, The role of integrin binding sites in fibronectin matrix assembly in vivo. *Current Opinion in Cell Biology*, 20, 502–507.
31. Dallas, S. L., P. Sivakumar, C. J. Jones, Q. Chen, D. M. Peters, D. F. Mosher, M. J. Humphries, and C. M. Kielty, 2005, Fibronectin regulates latent transforming growth factor-β (TGFβ) by controlling matrix assembly of latent TGFβ-binding protein-1. *Journal of Biological Chemistry*, 280, 18871–18880.

32. Anderson, J. M., A. Rodriguez, and D. T. Chang, 2008, Foreign body reaction to biomaterials. *Seminars in Immunology*, 20, 86–100.
33. Ekdahl, K. N., J. D. Lambris, H. Elwing, D. Ricklin, P. H. Nilsson, Y. Teramura, I. A. Nicholls, and B. Nilsson, 2011, Innate immunity activation on biomaterial surfaces: A mechanistic model and coping strategies. *Advanced Drug Delivery Reviews*, 63, 1042–1050.
34. Girós, A., K. Grgur, A. Gossler, and M. Costell, 2011, α5β1 integrin-mediated adhesion to fibronectin is required for axis elongation and somitogenesis in mice. *PLoS ONE*, 6, e22002.
35. Pegueroles, M., C. Aparicio, M. Bosio, E. Engel, F. Gil, J. Planell, and G. Altankov, 2010, Spatial organization of osteoblast fibronectin matrix on titanium surfaces: Effects of roughness, chemical heterogeneity and surface energy. *Acta Biomaterialia*, 6, 291–301.
36. Lüthen, F., R. Lange, P. Becker, J. Rychly, U. Beck, and J. Nebe, 2005, The influence of surface roughness of titanium on β1-and β3-integrin adhesion and the organization of fibronectin in human osteoblastic cells. *Biomaterials*, 26, 2423–2440.
37. Chellat, F., Y. Merhi, A. Moreau, and L. H. Yahia, 2005, Therapeutic potential of nanoparticulate systems for macrophage targeting. *Biomaterials*, 26, 7260–7275.
38. Thomsen, P. and C. Gretzer, 2001, Macrophage interactions with modified material surfaces. *Current Opinion in Solid State and Materials Science*, 5, 163–176.
39. Guduru, D., M. Niepel, J. Vogel, and T. Groth, 2011, Nanostructured material surfaces—Preparation, effect on cellular behavior, and potential biomedical applications: A review. *The International Journal of Artificial Organs*, 34, 963–985.
40. McNamara, L. E., R. J. McMurray, M. J. Biggs, F. Kantawong, R. O. Oreffo, and M. J. Dalby, 2010, Nanotopographical control of stem cell differentiation. *Journal of Tissue Engineering*, 1, 120623.
41. Park, J., P. Kim, W. Helen, A. J. Engler, A. Levchenko, and D.-H. Kim, 2012, Control of stem cell fate and function by engineering physical microenvironments. *Integrative Biology*, 4, 1008–1018.
42. McMurray, R. J., N. Gadegaard, P. M. Tsimbouri, K. V. Burgess, L. E. McNamara, R. Tare, K. Murawski, E. Kingham, R. O. Oreffo, and M. J. Dalby, 2011, Nanoscale surfaces for the long-term maintenance of mesenchymal stem cell phenotype and multipotency. *Nature Materials*, 10, 637–644.
43. Piner, R. D., J. Zhu, F. Xu, S. Hong, and C. A. Mirkin, 1999, "Dip-pen" Nanolithography. *Science*, 283(5402), 661–663.
44. Marrian, C. R. K. and D. M. Tennant, 2003, Nanofabrication. *Journal of Vacuum Science and Technology*, A21, S207.
45. Fuhrer, A., S. Lüscher, T. Ihn, T. Heinzel, K. Ensslin, W. Wegscheider, and M. Bichler, 2001, Energy spectra of quantum rings. *Nature*, 413, 822.
46. Hench, L. L. and J. M. Polak, 2002, Third-generation biomedical materials. *Science*, 295(5557), 1014–1017.
47. Chan, W. L. and E. Chason, 2007, Making waves: Kinetic processes controlling surface evolution during low energy ion sputtering. *Journal of Applied Physics*, 101, 121301.
48. Erlebacher, J., M. J. Aziz, E. Chason, M. B. Sinclair, and J. A. Floro, 1999, Spontaneous pattern formation on ion bombarded Si (001). *Physical Review Letters*, 82, 2330.
49. Facsko, S., T. Dekorsy, C. Koerdt, C. Trappe, H. Kurz, A. Vogt, and H. L. Hartnagel, 1999, Formation of ordered nanoscale semiconductor dots by ion sputtering. *Science*, 285, 1551–1553.
50. Frost, F. and B. Rauschenbach, 2003, Nanostructuring of solid surfaces by ion-beam erosion. *Applied Physics A*, 77, 1–9.
51. Frost, F., A. Schindler, and F. Bigl, 2000, Roughness evolution of ion sputtered rotating InP surfaces: Pattern formation and scaling laws. *Physical Review Letters*, 85, 4116.

52. Kim, T., M. Jo, Y. Kim, D. Noh, B. Kahng, and J.-S. Kim, 2006, Morphological evolution of ion-sputtered Pd (001): Temperature effects. *Physical Review B*, 73, 125425.
53. Makeev, M. A. and A.-L. Barabási, 1997, Ion-induced effective surface diffusion in ion sputtering. *Applied Physics Letters*, 71, 2800–2802.
54. Qian, J., Y. Kang, Z. Wei, and W. Zhang, 2009, Fabrication and characterization of biomorphic 45S5 bioglass scaffold from sugarcane. *Materials Science and Engineering: C*, 29, 1361–1364.
55. Ziberi, B., F. Frost, B. Rauschenbach, and T. Hoche, 2005, Highly ordered self-organized dot patterns on Si surfaces by low-energy ion-beam erosion. *Applied Physics Letters*, 87, 033113.
56. Dalby, M. J., 2009, Nanostructured surfaces: Cell engineering and cell biology. *Nanomedicine (London)*, 4(3), 247–248.
57. Stevens, M., 2008, Biomaterials for bone tissue engineering. *Materials Today*, 11, 18–25.
58. Harper, J. and K. Rodbell, 1997, Microstructure control in semiconductor metallization. *Journal of Vacuum Science and Technology B: Microelectronics and Nanometer Structures*, 15, 763–779.
59. Krasheninnikov, A. and F. Banhart, 2007, Engineering of nanostructured carbon materials with electron or ion beams. *Nature Materials*, 6, 723–733.
60. Vo, N., R. Averback, P. Bellon, S. Odunuga, and A. Caro, 2008, Quantitative description of plastic deformation in nanocrystalline Cu: Dislocation glide versus grain boundary sliding. *Physical Review B*, 77, 134108.
61. Barth, J. V., G. Costantini, and K. Kern, 2005, Engineering atomic and molecular nanostructures at surfaces. *Nature*, 437, 671–679.
62. Berry, C. C., S. Wells, S. Charles, and A. S. Curtis, 2003, Dextran and albumin derivatised iron oxide nanoparticles: Influence on fibroblasts in vitro. *Biomaterials*, 24, 4551–4557.
63. Allain, J. P., T. Tigno, and R. Armonda, 2013, Nanotechnology for cerebral aneurysm treatment, in B. Kateb and J. D. Heiss (eds), *The Textbook of Nanoneuroscience and Nanoneurosurgery*. Boca Raton, FL, CRC Press, pp. 259–282.
64. Planell, J. A., M. Navarro, G. Altankov, C. Aparicio, E. Engel, J. Gil, M. P. Ginebra, and D. Lacroix, 2010, Materials surface effects on biological interactions, *Advances in Regenerative Medicine: Role of Nanotechnology, and Engineering Principles*. Netherlands, Springer, pp. 233–252.
65. Allain, J., M. Nieto, M. Hendricks, P. Plotkin, S. Harilal, and A. Hassanein, 2007, IMPACT: A facility to study the interaction of low-energy intense particle beams with dynamic heterogeneous surfaces. *Review of Scientific Instruments*, 78, 113105.
66. Moroni, L., P. Habibovic, D. J. Mooney, and C. A. van Blitterswijk, 2010, Functional tissue engineering through biofunctional macromolecules and surface design. *MRS Bulletin*, 35, 584–590.
67. Grinnell, F. and M. Feld, 1982, Fibronectin adsorption on hydrophilic and hydrophobic surfaces detected by antibody binding and analyzed during cell adhesion in serum-containing medium. *Journal of Biological Chemistry*, 257, 4888–4893.
68. Norman, J. J. and T. A. Desai, 2006, Methods for fabrication of nanoscale topography for tissue engineering scaffolds. *Annals of Biomedical Engineering*, 34, 89–101.
69. Kim, P., D. Kim, B. Kim, S. Choi, S. Lee, A. Khademhosseini, R. Langer, and K. Suh, 2005, Fabrication of nanostructures of polyethylene glycol for applications to protein adsorption and cell adhesion. *Nanotechnology*, 16, 2420.
70. Zhu, B., Q. Zhang, Q. Lu, Y. Xu, J. Yin, J. Hu, and Z. Wang, 2004, Nanotopographical guidance of C6 glioma cell alignment and oriented growth. *Biomaterials*, 25, 4215–4223.
71. Dalby, M., M. Riehle, H. Johnstone, S. Affrossman, and A. Curtis, 2002, In vitro reaction of endothelial cells to polymer demixed nanotopography. *Biomaterials*, 23, 2945–2954.

72. Dalby, M. J., G. E. Marshall, H. J. Johnstone, S. Affrossman, and M. O. Riehle, 2002, Interactions of human blood and tissue cell types with 95-nm-high nanotopography. *IEEE Transactions on NanoBioscience*, 1, 18–23.
73. Dalby, M. J., S. J. Yarwood, M. O. Riehle, H. J. Johnstone, S. Affrossman, and A. S. Curtis, 2002, Increasing fibroblast response to materials using nanotopography: Morphological and genetic measurements of cell response to 13-nm-high polymer demixed islands. *Experimental Cell Research*, 276, 1–9.
74. Gonsalves, K., C. Halberstadt, C. T. Laurencin, and L. Nair, 2007, *Biomedical Nanostructures*. Hoboken, NJ, Wiley.
75. Zhu, B., Q. Lu, J. Yin, J. Hu, and Z. Wang, 2004, Effects of laser-modified polystyrene substrate on CHO cell growth and alignment. *Journal of Biomedical Materials Research Part B: Applied Biomaterials*, 70, 43–48.
76. Deisingh, A. K. and M. Thompson, 2004, Biosensors for the detection of bacteria. *Canadian Journal of Microbiology*, 50, 69–77.
77. Berkenpas, E., P. Millard, and M. Pereira da Cunha, 2006, Detection of *Escherichia coli* O157: H7 with langasite pure shear horizontal surface acoustic wave sensors. *Biosensors and Bioelectronics*, 21, 2255–2262.
78. Branch, D. W. and S. M. Brozik, 2004, Low-level detection of a *Bacillus anthracis* simulant using Love-wave biosensors on 36° YX $LiTaO_3$. *Biosensors and Bioelectronics*, 19, 849–859.
79. Cloft, H. J. and D. F. Kallmes, 2002, Cerebral aneurysm perforations complicating therapy with Guglielmi detachable coils: A meta-analysis. *American Journal of Neuroradiology*, 23, 1706–1709.
80. Heber, J., 2009, Materials science: Enter the oxides. *Nature*, 459, 28.
81. Moll, N., E. Pascal, D. H. Dinh, J.-P. Pillot, B. Bennetau, D. Rebière, D. Moynet, Y. Mas, D. Mossalayi, and J. Pistré, 2007, A Love wave immunosensor for whole *E. coli* bacteria detection using an innovative two-step immobilisation approach. *Biosensors and Bioelectronics*, 22, 2145–2150.
82. Ong, M. T. and E. J. Reed, 2012, Engineered piezoelectricity in graphene. *ACS Nano*, 6, 1387–1394.
83. Ding, B., M. Wang, X. Wang, J. Yu, and G. Sun, 2010, Electrospun nanomaterials for ultrasensitive sensors. *Materials Today*, 13, 16–27.
84. Moon, R. J., A. Martini, J. Nairn, J. Simonsen, and J. Youngblood, 2011, Cellulose nanomaterials review: Structure, properties and nanocomposites. *Chemical Society Reviews*, 40, 3941–3994.
85. Hanel, R. A., A. S. Boulos, E. G. Sauvageau, E. I. Levy, L. R. Guterman, and L. N. Hopkins, 2005, Stent placement for the treatment of nonsaccular aneurysms of the vertebrobasilar system. *Neurosurgical Focus*, 18, 1–9.
86. Kunert, M., A. Biedermann, T. Koch, and W. Boland, 2002, Ultrafast sampling and analysis of plant volatiles by a hand-held miniaturised GC with pre-concentration unit: Kinetic and quantitative aspects of plant volatile production. *Journal of Separation Science*, 25, 677–684.
87. Lammertyn, J., E. A. Veraverbeke, and J. Irudayaraj, 2004, zNose™ technology for the classification of honey based on rapid aroma profiling. *Sensors and Actuators B: Chemical*, 98, 54–62.
88. Staples, E. J., 2000, The zNose, a new electronic nose using acoustic technology. *Journal of the Acoustical Society of America*, 108, 2495.
89. Aubin, J. E. and J. N. M. Heersche, 2003, Bone cell biology: Osteoblasts, osteocytes, and osteoclasts, in H. Jüppner (ed.), *Pediatric Bone, Biology and Diseases*. San Diego, CA, Academic Press, pp. 43–75.
90. Fathi, M. and A. Doostmohammadi, 2009, Bioactive glass nanopowder and bioglass coating for biocompatibility improvement of metallic implant. *Journal of Materials Processing Technology*, 209, 1385–1391.

91. Dorozhkin, S., 2010, Bioceramics of calcium orthophosphates. *Biomaterials*, 31, 1465–1485.
92. Wilson, J. and E. Chaikof, 2008, Challenges and emerging technologies in the immunoisolation of cells and tissues. *Advanced Drug Delivery Reviews*, 60, 124–145.
93. LeGeros, R., 2002, Properties of osteoconductive biomaterials: Calcium phosphates. *Clinical Orthopaedics and Related Research*, 395, 81.
94. Albrektsson, T., G. Zarb, P. Worthington, and A. Eriksson, 1986, The long-term efficacy of currently used dental implants: A review and proposed criteria of success. *The International Journal of Oral and Maxillofacial Implants*, 1, 11–25.
95. Izquierdo-Barba, I., F. Conde, N. Olmo, M. Lizarbe, M. Garcia, and M. Vallet-Regi, 2006, Vitreous SiO_2-CaO coatings on Ti6Al4V alloys: Reactivity in simulated body fluid versus osteoblast cell culture. *Acta Biomaterialia*, 2, 445–455.
96. Sikavitsas, V. I., J. S. Temenoff, and A. G. Mikos, 2001, Biomaterials and bone mechanotransduction. *Biomaterials*, 22, 2581–2593.
97. Cao, W. and L. Hench, 1996, Bioactive materials. *Ceramics International*, 22, 493–507.
98. Ku, C., D. Pioletti, M. Browne, and P. Gregson, 2002, Effect of different Ti-6Al-4V surface treatments on osteoblasts behaviour. *Biomaterials*, 23, 1447–1454.
99. Wall, I., N. Donos, K. Carlqvist, F. Jones, and P. Brett, 2009, Modified titanium surfaces promote accelerated osteogenic differentiation of mesenchymal stromal cells in vitro. *Bone*, 45, 17–26.
100. Hench, L. and E. Ethridge, 1982, *Biomaterials: An Interfacial Approach*. New York, Academic Press.
101. Ionita, C. N., A. M. Paciorek, A. Dohatcu, K. R. Hoffmann, D. R. Bednarek, J. Kolega, E. I. Levy, L. N. Hopkins, S. Rudin, and J. D. Mocco, 2009, The asymmetric vascular stent efficacy in a rabbit aneurysm model. *Stroke*, 40, 959–965.

3 Biological Sample Preparation and Analysis Using Droplet-Based Microfluidics

Xuefei Sun and Ryan T. Kelly

CONTENTS

3.1 INTRODUCTION

Modern biological research often requires massively parallel experiments to analyze a large number of samples in order to find biomarkers, screen drugs, or elucidate complex cellular pathways. These processes frequently involve time-consuming sample preparation and expensive biochemical measurements. Another constraint frequently encountered in bioanalysis is limited amounts of available sample. Microfluidics or lab-on-a-chip platforms offer promise for addressing the challenges encountered in biological research because a large number of small samples can be handled and processed with different functional elements in an automated fashion.

Droplet-based microfluidics, in which reagents of interest are compartmentalized within femto- to nanoliter-sized aqueous droplets or plugs that are encapsulated and dispersed in an immiscible oil phase, has emerged as an attractive platform for small-volume bioanalysis [1–9]. This new platform elegantly addresses the challenges encountered with conventional continuous flow systems by, for example, limiting reagent dilution caused by diffusion and Taylor dispersion and minimizing cross-contamination and surface-related adsorptive losses [10]. The microdroplets isolated by the immiscible liquid can serve as microreactors, allowing for high-throughput chemical reaction screening and extensive biological research [1]. Droplet-based microfluidics also offers great promise for reliable quantitative analysis because monodisperse microdroplets can be generated with controlled sizes and preserve temporal information that is easily lost to dispersion in continuous flow systems [11,12].

Biological analysis begins with sample selection and preparation. The initial sampling can comprise cell sorting, tissue dissection, or extraction of protein or other analytes of interest from cells or tissues [13]. The biological samples are then prepared by, for example, combining reagents, mixing, incubating, purifying, and/or enriching. Depending on the complexity of the sample, the subsequent analytical measurements can be very simple, employing, for example, laser-induced fluorescence (LIF) to detect a single-labeled analyte. With more complex samples having multiple analytes of interest, chemical separations including capillary electrophoresis (CE) and liquid chromatography (LC) and information-rich detection methods such as mass spectrometry (MS) become necessary. To date, many operational components for microdroplets have been well developed to perform most of these basic operations. For example, stable aqueous droplets dispersed in an oil phase can be generated using various droplet generator designs for sampling in a confined small volume, the most common of which are the T-junction [14,15] and flow-focusing [16,17] geometries. The addition of reagents to existing droplets can be realized by fusion with other droplets enabling the initiation and termination of the compartmentalized reactions confined in the microdroplets [18,19]. The rapid mixing of fluids within droplets enables a homogeneous reactive environment to be achieved and can be enhanced by means of chaotic advection [20]. In addition, droplets can be incubated in delay lines [21] or stored in reservoirs [22,23] or traps [24,25] for extended periods of time to complete reactions or facilitate the biological processes.

Droplet-based microfluidic platforms have been successfully applied in a variety of chemical and biological research areas. For example, a droplet-based platform for polymerase chain reaction (PCR) amplification has proven able to significantly improve amplification efficiency over conventional microfluidic formats [26], which is mainly due to the elimination of both reagent dilution and adsorption on the channel surfaces. Droplets have also been employed to encapsulate, sort, and assay single cells [12,27] or microorganisms [24], study enzyme kinetics [11] and protein crystallization [28], and synthesize small molecules and polymeric micro- and nanoparticles.

Although droplet-based microfluidic technology has developed to a degree where droplets can be generated and manipulated with speed, precision, and control, some real challenges still exist that limit the widespread use of these systems. One challenge of concern is how to extract and acquire the enormous amount of chemical information that can be contained in picoliter-sized droplets. The detection of droplet contents has historically been limited to optical methods such as LIF, while coupling with chemical separations and nonoptical detection has proven difficult. Combining the advantages of droplet-based platforms with more information-rich analytical techniques including LC, CE, and MS can greatly extend their reach. This requires that the droplets be extracted from the oil phase for downstream analysis and detection.

This chapter focuses primarily on the integrated droplet-based microsystems having the ability to couple with chemical separations and nonoptical detection, allowing for *ex situ* analysis and the identification of the biochemical components contained in the microdroplets. Some unit operations for microdroplets will be briefly introduced, including droplet generation, fusion, and incubation. All approaches and techniques that have been developed for droplet detection, droplet extraction, coupling to CE separation, and electrospray ionization (ESI)-MS detection will be reviewed. An example of integrated droplet-based microfluidics, including on-demand droplet generation and fusion, robust and efficient droplet extraction, and a monolithically integrated nano-ESI emitter, will be given to demonstrate its potential for chemical and biological research.

3.2 DROPLET-BASED OPERATIONS

3.2.1 Droplet Generation

Currently, most planar microfluidic droplet generators are designed using T-junction [14,15] and flow-focusing [16,17] geometries, in which small droplets are spontaneously formed at an intersection taking advantage of the interface instability between oil and aqueous streams. Using these approaches, droplets can be generated over a broad range of frequencies ranging from ~0.1 Hz to 10 kHz and using flow rates on the order of 0.1–100 μL/min [29]. Droplet volume and generation frequency depend on several factors, including the physical properties of the immiscible phases, flow rates, intersection geometry, and so on. For a given geometry and solvent composition, flow-focusing and T-junction interfaces exhibit interdependence between the flow rate and droplet generation frequency and cannot be easily modulated over short timescales.

For lower frequencies and applications for which the ability to rapidly change droplet size and generation frequency is desirable, on-demand droplet generation strategies become more favorable, as they ensure precise control and fine manipulation of individual droplets. Various approaches have been developed to generate droplets on demand, for example, by carefully balancing the pressure and flow in the system [27], as well as electrical [30] or laser pulsing [31] and piezoelectric actuation [32]. Pneumatic valving has also been explored and has been found to provide facile, independent control over both droplet size and generation frequency [33–36]. Galas et al. utilized a single pneumatic valve that was embedded in an active connector and assembled close to a T-junction to regulate the flow of the dispersed phase [33]. Constant pressures were applied on the inlets of two immiscible liquids to drive the flow in the microchannel. Individual droplets were created by briefly opening the valve. The aqueous droplet size depended on the valve actuation time and frequency, as well as the pressure applied at the oil inlet. Therefore, the droplet volume, spacing, and speed could be controlled accurately and independently. This device not only generated periodic sequences of identical droplets, but it also enabled the production of nonperiodic droplet trains with different droplet sizes or spacing. Lin and coworkers also reported a similar platform for pneumatic valve–assisted on-demand droplet generation [34]. Negative pressure was applied at the outlet of the device to drive the flow of the two immiscible liquids through the microchannel. The dependence of the droplet size on the valve actuation time and applied pressure was investigated. In addition, the authors utilized several aqueous flow channels, each with independently controlled microvalves, to generate arrays of droplets containing different compositions by alternately actuating the valves.

We have also investigated valve-controlled, on-demand droplet generation. To minimize the dead volume and control the droplet volume precisely, the pneumatic valve was placed over the side channel exactly at the T-junction (Figure 3.1a). Carrier oil flow was driven by a syringe pump, and the dispersed aqueous phase was injected by finely controlled air pressure. Figure 3.1b shows the generation process of an individual droplet. The valve is initially closed and the aqueous fluorescein solution is confined in the side channel. On opening the valve briefly, a small volume of aqueous solution is dispensed into the oil channel to form a droplet, which is then flushed downstream by the carrier oil flow. Compared with conventional microfluidic droplet generation techniques based on a T-junction or flow focusing, the valve-integrated system can generate droplets with precise control

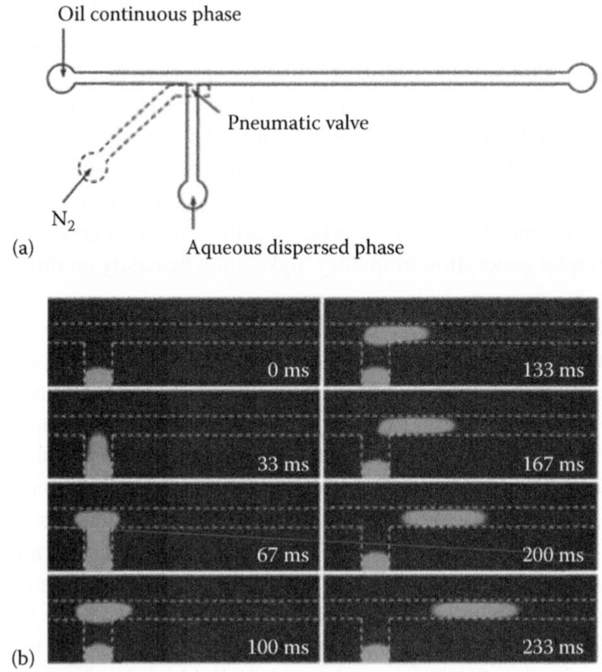

FIGURE 3.1 (a) Schematic depiction of a T-junction droplet generator controlled by a pneumatic valve. (b) Micrograph sequences depicting pneumatic valve-controlled generation of an individual fluorescein droplet. The width of the oil flow channel is 100 μm, the oil flow rate is 0.5 μL/min, and the sample injection pressure is 8 psi. The valve actuation time and pressure are 33 ms and 25 psi, respectively.

over the droplet volume, generation frequency, and velocity. The droplet velocity is determined by the syringe pump driving the oil stream, while the droplet generation rate is controlled by the valve actuation frequency as defined in the software. The droplet spacing is determined by the interval between the valve openings and the oil flow velocity. The droplet volume depends on several parameters including the valve actuation time, the pressure of the aqueous solution, the oil-phase flow rate, and the valve control pressure (Figure 3.2).

3.2.2 In-Droplet Reagent Combination and Mixing

Besides controlled droplet generation, droplet fusion is of crucial importance in relation to the development of microreactors because it allows precise and reproducible mixing of reagents at well-defined points to initiate, modify, and terminate reactions [6]. Ismagilov and coworkers carried out pioneering work to combine different reagents into individual droplets by flowing two reagent solutions in a microchannel as two laminar streams [37]. To prevent prior contact between the two reagents before droplet generation, an inert center stream was used to separate them. Thus, three streams were continuously injected into an immiscible carrier oil phase to form droplets. The gradient droplets in the reagent concentrations were achieved by varying the relative flow rates of the three streams [11,38]. A subsequent winding channel

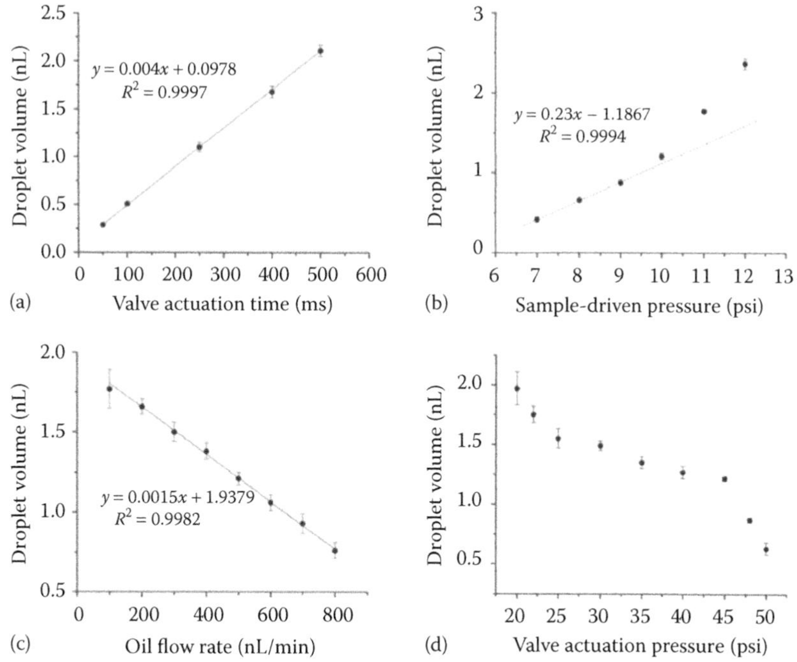

FIGURE 3.2 Plots of droplet volume dependence on (a) valve actuation time, (b) sample-driven pressure, (c) oil flow rate, and (d) valve actuation pressure. The channel dimensions are the same as for Figure 3.1.

was designed to accelerate mixing by chaotic advection [20,37]. This approach has been widely employed to control networks of chemical reactions [37], study reaction kinetics [11], screen protein crystallization conditions [38], and investigate single-cell-based enzyme assays [39] and protein expression [12].

Recently, Weitz and colleagues presented a robust picoinjector to add reagents to droplets in microfluidic systems [40]. The picoinjector was controlled by an electric field to trigger the injection of a controlled volume of reagents into each droplet. The injection volume was precisely controlled by adjusting the droplet velocity and injection pressure. Selective injection was realized by switching the electric field on and off, at kilohertz frequencies.

In-channel droplet fusion is another attractive approach to combine different reagents in individual droplets to initiate or terminate the confined reactions. The process of droplet merging introduces convective flows into the system, resulting in far more rapid mixing than relying on diffusion alone [41]. In-channel droplet fusion is readily achieved by bringing two or more surfactant-free droplets into contact. Both passive and active methods have been developed to control droplet fusion. For passive fusion devices, droplet coalescence is usually initiated by utilizing specially designed fusion elements in the channel. For example, Bremond et al. incorporated an expanded coalescence chamber in the channel network in which two droplets were brought into close proximity and merged together before entering a narrow channel [42]. Fidalgo et al. reported a method for droplet fusion based on a surface energy pattern inside a microfluidic channel where the segmented flow was disrupted and the droplets were trapped and fused together [43]. In this case, full control of droplet fusion could be achieved by varying the channel and pattern dimensions, as well as the fluid flow. This surface-induced droplet fusion method enabled the merging of multiple droplets containing different reagents to form a large droplet. However, this approach could potentially cause cross contamination between droplets from the patterned surface. Niu et al. developed a pillar-induced droplet merging device, in which rows of pillars were constructed in the channel network serving as passive fusion elements or chambers [44]. The pillar array trapped droplets and drained the carrier oil phase through the apertures between the pillars. The first trapped droplet was suspended and merged with succeeding droplets until the surface tension was overwhelmed by the hydraulic pressure. The merging process depended on the droplet size, and the number of droplets that could be merged relied on the mass flow rate and the volume ratio between the droplets and the merging chamber.

Active fusion methods that can be controlled externally and selectively have also been developed using, for example, electric fields [45–48] and laser pulses [49] to trigger coalescence. To perform active droplet fusion effectively, the synchronization of droplets is a key factor because fusion efficiency relies on the droplets being in very close proximity [50]. Currently, special designs are often employed to synchronize droplets in two parallel channels, which then merge into a single channel downstream to realize droplet coalescence [34,42,51]. However, this system can potentially be disturbed by a few factors such as the flow rate and back pressure in the channel, which may reduce the fusion efficiency. Recently, Jambovane et al. used valve-based droplet generation for multiple reagents to perform controlled reactions and establish chemical gradients among arrays of droplets [52]. Droplets were generated at a valve-controlled side channel and then different reagents were added to the droplets as they passed by similar side channels downstream.

An efficient method for reagent combination that we have developed recently employed two pneumatic valves integrated at a double-T intersection (Figure 3.3a). Reagents were introduced through the different side channels, each controlled by a separate valve, and the simultaneous opening of the valves resulted in the creation of an aqueous plug containing both reagents. Upon actuation, the oil between the two side channels was quickly displaced, and the two aqueous streams collided and combined. No sample cross contamination was observed because of the applied pressures, the rapid valve actuation, and the offset between the two side channels. The two liquids mixed together by a combination of diffusion and convection caused by an equalization of internal pressures following the combining of the aqueous streams. The linear dependence of the droplet volume on the valve actuation time and the independent valving for the two aqueous streams provide a high degree of control over droplet composition. Figure 3.3b shows arrays of six droplets containing different ratios of two dyes that were created by controlling the operation of two valves. Such control should be useful for optimizing or screening reactions and for studying reaction kinetics.

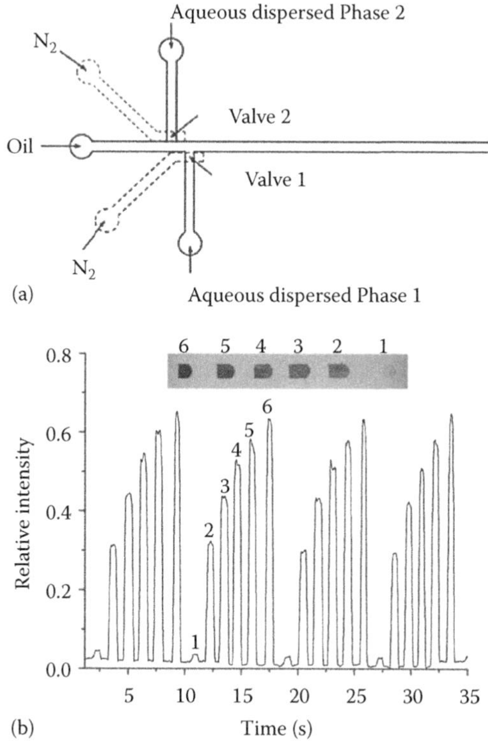

FIGURE 3.3 (a) Schematic depiction of the droplets generation, fusion, and mixing portion of the device. (b) Relative intensity of an array of six droplets containing different volume ratios of colored dyes.

3.2.3 Droplet Incubation

Many biological assays involving, for example, enzymatic reactions have relatively slow kinetics, requiring microdroplets to be incubated for minutes to hours for an efficient reaction. Similarly, studies involving cell incubation or protein expression require extended incubations. A straightforward method for microdroplet incubation is to simply increase the channel length following droplet generation [53,54], but increased back pressure and disruption of droplet formation can quickly become an issue. Frenz et al. incorporated deeper and wider delay lines following the droplet generation section, which enabled reactions in the droplets to increase from 1 min to >1 h [21]. Similarly, Kennedy and coworkers have interfaced capillaries or Teflon tubing with droplet generation devices to collect and store sample plugs for 1–3 h [55]. For longer-term online incubation, droplets can be stored in reservoirs, traps, or dropspot arrays. For example, Courtois et al. fabricated a large reservoir for droplets storage over periods of up to 20 h to study the retention of small molecules in droplets [23]. Huebner et al. designed a droplet trapping array to store and incubate picoliter-sized droplets for extended periods of time to investigate the encapsulated cells and enzymatic reactions [25]. Weitz and colleagues introduced a "dropspots" device to immobilize and store thousands of individual droplets in a round chamber array over a 15 h incubation period [56]. Droplets can also be incubated off-chip from minutes to several days when using appropriate surfactants to stabilize the droplets [57,58]. The incubated droplets can then be reinjected into microfluidic devices for further processing and detection.

3.2.4 Droplet Readout Strategies

To date, in-droplet fluorescence detection remains the most widely used method for analyzing the contents of droplets due to its ability to measure in real time and with high sensitivity. Fluorescence detection has been implemented to study enzyme kinetics within droplets [11,59,60], characterize the behavior of encapsulated single cells [12,27], detect PCR products, [61,62], and investigate the interactions between biological samples [63]. Fluorescence detection is ideally suited to the rapid, sensitive detection of a small number of distinct species. For cases in which a large number of analytes need to be detected and identified (e.g., proteomics and metabolomics) and where fluorescent labeling is not desirable, alternative measurement strategies are needed.

In-droplet Raman spectroscopy has recently been used to detect and analyze droplets contents [64,65]. It is a nondestructive and label-free detection approach with high molecular selectivity, which can track the droplets in real time to provide the fundamental droplet properties and the chemical contents within the droplets, including droplet sizes, encapsulated species, structures, and concentrations. Surface-enhanced Raman spectroscopy (SERS) can offer a higher sensitivity and reproducible quantitative analysis of the droplets due to the enhancement of the Raman signal intensity [66]. Electrochemical detection is an inexpensive and label-free approach to collect information on the physical and chemical properties of droplets and can monitor droplet production and measure droplet length, frequency, and velocity [67]. It can provide the chemical information when the reaction within the droplets involves an electrochemically active reactant or product [68]. Another advantage of electrochemical measurements is their compatibility with alternative chip materials, including opaque substrates, which are difficult to implement for conventional optical detection strategies such as fluorescence. Nuclear magnetic resonance (NMR) has also been used for droplets or segmented flow analysis. Karger and coworkers developed a microcoil NMR probe for high-throughput analysis of sample plugs in dimethyl sulfoxide (DMSO) [69].

While the above detection strategies can be employed *in situ*, others require the contents to be removed from the droplet for subsequent analysis. Once extracted to an aqueous stream, the droplet contents can be analyzed using more information-rich techniques including LC, CE, and MS. MS is an especially attractive technique for an in-depth, label-free biological analysis because of its ability to identify and provide structural information for hundreds of more unique species in a given analysis [70]. Below, we detail the methods used for droplet extraction and subsequent analysis.

Ismagilov and colleagues used a microfluidic system to screen and optimize organic reaction conditions in microdroplets detected using matrix-assisted laser desorption ionization mass spectrometry (MALDI-MS) [71]. The incubated reaction plugs were deposited onto a sample plate for MALDI-MS analysis. Kennedy and coworkers directly pumped nanoliter plugs of sample into a mass spectrometer for analysis through a metal-coated capillary nanospray emitter, separating the analyte from the carrier at the emitter itself [72,73]. Teflon tubing was placed close to the emitter tip to siphon the accumulated oil away from the tip, which could maintain a stable electrospray at flow rates as high as 2000 nL/min. However, it is generally necessary to extract the aqueous droplet from the oil phase for further separation or online MS analysis to avoid contamination of the mass spectra with peaks from the oil and to maintain the electrospray Taylor cone in the most efficient cone-jet mode of operation.

Edgar et al. first reported the extraction of aqueous droplet contents into a channel for CE separation [74]. A femtoliter-volume aqueous droplet was directly delivered to fuse with the aqueous phase in a separation channel for CE separation. Niu et al. employed a similar method to inject the droplets in which the LC eluent was fractionated into a CE channel for comprehensive two-dimensional separations in both time and space [75]. A pillar array was constructed at the interface to evacuate the carrier oil phase prior to loading samples into the separation channel. In these two cases, it was very difficult to maintain a robust extraction because the segmented flow was perpendicular to the CE separation channel.

Kennedy and colleagues exploited a surface modification method to form a stable interface at the junction between two immiscible phases in a microchannel [76–78]. They selectively patterned glass surfaces in the segmented flow channel to be hydrophobic in order to stabilize the oil–water interface and facilitate droplet extraction. But in some cases, only part of each droplet was extracted due to the presence of a "virtual wall," which was not suitable for quantitative analysis because of irreproducibility and loss of information [76]. Fang and coworkers employed a similar surface modification technique to obtain a hydrophilic tongue-based droplet extraction interface, which could control the droplet extraction by regulating the waste reservoir height [79]. The extracted droplet contents were then detected by MS through an integrated ESI emitter. More recently, Filla et al. used a corona treatment to hydrophilize a portion of a polydimethylsiloxane (PDMS) chip to establish an extraction interface [80]. Aqueous droplets were transferred into the hydrophilic channel when the segmented flow encountered the interface. The droplet contents were subsequently analyzed by electrochemistry or microchip-based electrophoresis with electrochemical detection.

Huck and coworkers employed electrocoalescence to control droplet extraction [81,82]. The segmented flow and the continuous aqueous flow met in a rectangular chamber where an interface between two immiscible phases was built up. A pulsed electric field was applied over the chamber to force droplets to coalesce with the continuous aqueous stream, which then delivered the droplet contents to a capillary emitter for ESI-MS detection [82]. This droplet extraction approach required careful adjustments of the flow of the two immiscible phases to maintain a stable interface in the extraction chamber and avoid cross contamination of the aqueous and oil streams. In addition, the severe dilution of the droplet contents resulted in high detection limits (~500 μM bradykinin). Lin and coworkers used an electrical-based method to control droplet breaking and extraction at the stable oil–water interface [83]. One reported issue in this case was the difficulty of achieving complete extraction with high efficiency, which limited its compatibility with quantitative analysis.

Kelly et al. invented a droplet extraction interface, which was constructed with an array of cylindrical posts to separate the segmented flow channel and the continuous aqueous-phase channel [84]. When the aqueous stream and the carrier oil phase flow rates were well controlled to balance the pressure at the junction, a stable oil–aqueous interface based on interfacial tension alone was formed to prevent bulk crossover of the two immiscible streams. The droplets could be transferred through the apertures to the continuous aqueous stream and finally detected by ESI-MS with virtually no dilution, enabling nanomolar detection limits.

Most of the reported methods and techniques for droplet extraction, as mentioned above, need to adjust two immiscible liquid flow rates to stabilize the interface and extract entire droplets. It is desirable to perform effective and complete droplet extraction independent of the flow rates, which would provide added flexibility for device operation. Recently, we have developed a robust interface for reliable and efficient droplet extraction, which was integrated in a droplet-based PDMS microfluidic assembly. The droplet extraction interface consisted of an array of cylindrical posts (Figure 3.4a), the same as was previously reported [84], but the aqueous stream microchannel surface was selectively treated by corona discharge to be hydrophilic. The combination of different surface energies and small flow-through apertures (~3 × 25 μm) enabled a very stable liquid interface between two immiscible steams to be established over a broad range of aqueous and oil flow rates. All aqueous droplets were entirely transferred to the aqueous stream (Figure 3.4b) and detected by MS following ionization at a monolithically integrated nanoelectrospray emitter.

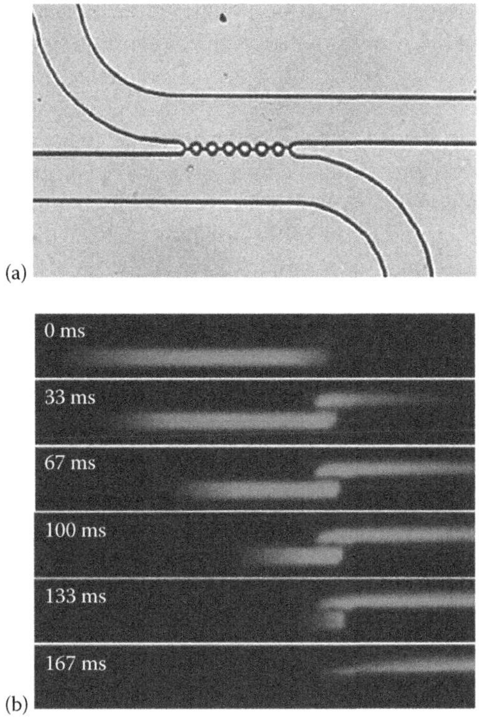

FIGURE 3.4 (a) Photograph of the droplet extraction region of the device. Water and oil fill the top and bottom channels, respectively, and the interface of the two liquids can be seen between the circular posts. (b) Micrograph sequences depicting the extraction of an individual fluorescein droplet. The flow rate in both channels is 400 nL/min.

3.3 PERSPECTIVES FOR DROPLET-BASED MICROFLUIDICS

As mentioned above, droplet-based microfluidics have been employed for a wide range of analyses and due to their unique advantages, their use will undoubtedly grow. Below, we outline a few promising applications that will leverage the strengths of the platform.

3.3.1 Enhanced LC/MS-Based Proteomic Analysis

MS-based proteomics studies are vital for biomarker discovery, identification of drug targets, and fundamental biological research. In a typical "bottom-up" proteomics workflow [85], proteins are extracted from a sample, purified and enzymatically digested into peptides. The peptides are then separated by LC, ionized by ESI, identified by MS, and those identified peptides are then matched to their corresponding proteins based on genomic information. Alternatively, for "top-down" proteomics [86], intact proteins are separated and identified directly by MS, providing potentially more complete sequence information and the ability to characterize posttranslational modifications. However, MS identification of intact proteins is far more challenging and lower in throughput, currently limiting the widespread use of top-down approaches.

As top-down and bottom-up proteomics approaches each have unique and complimentary advantages, it would be especially attractive to obtain both intact protein and peptide-level information from a single analysis. We propose that this could be achieved by encapsulating separated proteins into droplets as they elute from an LC column, thus preserving temporal information and separation resolution while enabling further processing. For example, using our droplet-on-demand and droplet merging technologies, we could encapsulate eluting proteins into droplets, and selectively add reagents for digestion to alternating droplets. The droplets could then be incubated in a delay line to allow sufficient reaction time prior to extracting and ionizing the droplets. The result would be that each droplet containing an unreacted protein would be followed by a droplet containing a digested peptide such that conventional bottom-up MS would be complemented with the intact molecular mass.

To this end, we have begun combining proteins with proteases in droplets to evaluate the conditions needed for digestion. The platform incorporates our integrated droplet-on-demand interface that enables controlled in-droplet reactions, incubation in the oil stream, extraction from the aqueous stream, and ionization of the droplet contents at an integrated nanoelectrospray (nano-ESI) emitter [87] for MS analysis (Figures 3.5 and 3.6). This integrated microfluidic platform has been successfully utilized to combine myoglobin and pepsin from separate aqueous streams into droplets to perform rapid in-droplet digestions that were detected and identified online by nano-ESI-MS following droplet extraction (Figure 3.7). Given the short incubation time (18 s), the digestion did not go to completion such that peaks from the intact protein were still evident in the mass spectrum, but numerous peptides were confidently identified based on their *m/z* ratio as well (Table 3.1). We expect that simply extending the incubation time will dramatically improve digestion efficiency and enable the application of the platform to combined top-down/bottom-up proteomic analyses.

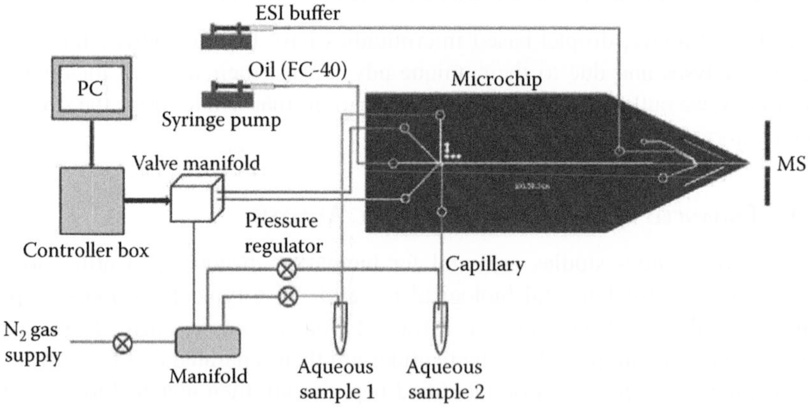

FIGURE 3.5 Schematic of the experimental setup for droplet generation, fusion, mixing, extraction, and MS detection.

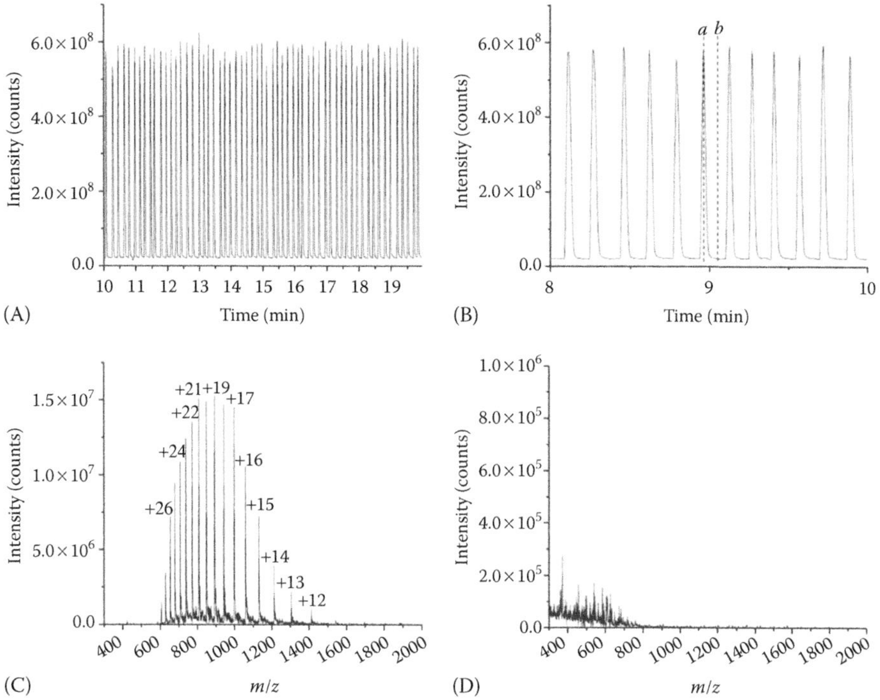

FIGURE 3.6 (A) MS detection of extracted 1 μg/μL apomyoglobin droplets. Oil-phase flow rate is 100 nL/min, ESI buffer flow rate is 400 nL/min, and droplet generation frequency is 0.1 Hz. (B) Detailed view of the MS-detected extracted apomyoglobin droplets. (C, D) Mass spectra obtained from the peak and baseline indicated as (a) and (b) in Figure 3.6B, respectively.

3.3.2 Single-Cell Chemical Analysis

Sensitivity limitations on biochemical measurements typically dictate that large samples are required comprising populations of cells. These ensemble measurements average over important cell-to-cell differences. Direct chemical analysis at the single-cell level will enable the heterogeneity that is currently obscured to be better understood. The sensitivity of the MS instrumentation that is used for proteomic and metabolomic studies has increased to the point that such single-cell measurements are now feasible. For example, while inefficient ionization and transmission of ions generated at atmospheric pressure to the high vacuum region of the mass spectrometer were previously prohibitive, recent improvements have produced combined efficiencies that can exceed 50% in some cases [88]. Indeed, around 50 proteins have been identified from samples containing just 50 pg of protein [89], which is as much protein as is contained in an average eukaryotic cell [90]. However, despite having adequate analytical sensitivity, existing methods for sample preparation, involving manual pipetting and multiple reaction vessels, are incompatible with single cells. This is another area where droplet-based microfluidics should be able to

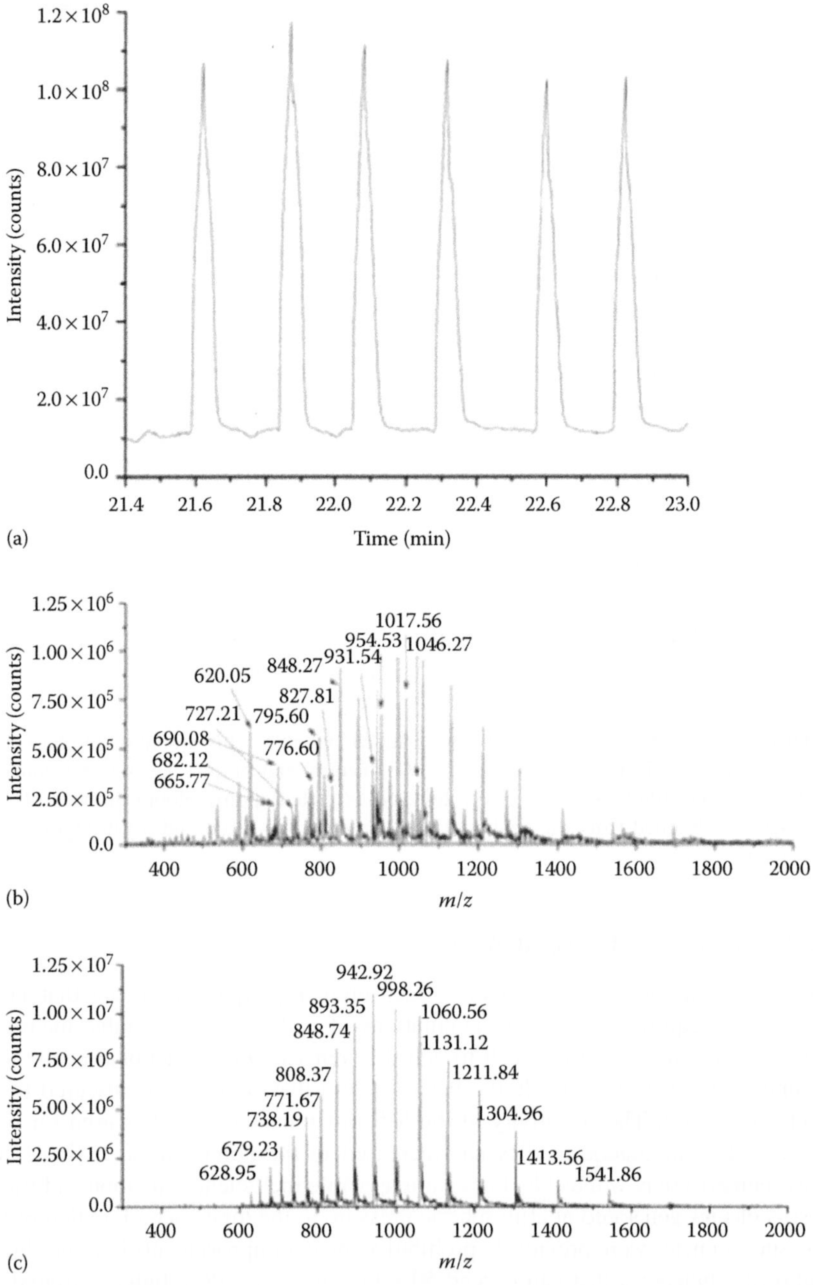

FIGURE 3.7 (a) MS detection of fused droplets mixing 1 μg/μL apomyoglobin with 1 μg/μL pepsin in water containing 0.1% formic acid (pH~3). The flow rates of the oil and ESI buffer streams are 0.1 and 0.4 μL/min, respectively. (b) and (c) MS spectra of the fused droplet and 1 μg/μL apomyoglobin, respectively. The sequences of some digested peptide fragments labeled in (b) are listed in Table 3.1.

TABLE 3.1
Sequence of Apomyoglobin and Identification of Peptide Fragments from In-Droplet Digested Apomyoglobin Shown in Figure 3.7b

m/z	Mass	*z*	Position	Sequence
620.05	1856.0	3+	138–153	FRNDIAAKYKELGFQG
690.08	4133.9	6+	70–106	TALGGILKKKGHHEAELKPLAQSHATKHKIPIKYLEF
827.81		5+		
665.77	4653.4	7+	30–69	IRLFTGHPETLEKFDKFKHLKTEAEMK
776.60		6+		ASEDLKKHGTVVL
931.54		5+		
682.12	4767.5	7+	110–153	AIIHVLHSKHPGDFGADAQGAMTKALE
795.60		6+		LFRNDIAAKYKELGFQG
954.53		5+		
727.21	5082.8	7+	107–153	ISDAIIHVLHSKHPGDFGADAQGAMTK
848.27		6+		ALELFRNDIAAKYKELGFQG
1017.56		5+		
1046.27	3133.6	3+	1–29	GLSDGEWQQVLNVWGKVEADIAGHGQEVL

Note: GLSDGEWQQVLNVWGKVEADIAGHGQEVLIRLFTGHPETLEKFDKFKHLKTEAEMKAS EDLKKHGTVVLTALGGILKKKGHHEAELKPLAQSHATKHKIPIKYLEFISDAIIHVLHSK HPGDFGADAQGAMTKALELFRNDIAAKYKELGFQG (apomyoglobin sequence).

meet the need. Droplets have previously been used for single-cell encapsulation, and cells have also been lysed within droplets, with the surrounding oil preventing further dilution of the contents. Using such technologies for encapsulation and lysis in combination with our approaches for reagent mixing and droplet compatibility with ultrasensitive MS should enable us to dig deeper into the proteome and metabolome of single cells than has been accomplished previously.

3.4 CONCLUSIONS

Droplet-based microfluidics has developed substantially as a technology will likely assume a higher-profile role in biological analyses moving forward. Not only are much smaller amounts of reagents and samples consumed, but also thousands of reactions and screening experiments can be performed within droplets simultaneously. Perhaps more importantly, droplet-based microfluidics is a promising tool to help us understand some fundamental biological questions such as enzymatic reactions in a confined and crowded environment, protein–protein or protein–ligand interactions, interfacial functions in biological systems, and single-cell proteomics and metabolomics. A number of operational units have been well developed for droplet-based microfluidics, including droplet generation, fusion, and incubation. Others, such as droplet extraction for subsequent analysis of the contents have been developed recently and promise to add versatility to the platform. The robust integration of multiple functions to create a true "lab-on-a-chip" continues to be a challenge, but the unique advantages of droplets for sample-limited biological analyses will undoubtedly spawn further development and we anticipate significant growth in the number of applications that rely on this technology in the coming years.

ACKNOWLEDGMENTS

Support was provided by the intramural program of the William R. Wiley Environmental Molecular Sciences Laboratory (EMSL). The EMSL is a national scientific user facility sponsored by the U.S. Department of Environment's Office of Biological and Environmental Research and located at Pacific Northwest National Laboratory in Richland, WA.

REFERENCES

1. Song, H., D.L. Chen, and R.F. Ismagilov, Reactions in droplets in microfluidic channels. *Angew. Chem. Int. Ed.*, **45**: 7336–7356, 2006.
2. Huebner, A., et al., Microdroplets: A sea of applications? *Lab Chip*, **8**: 1244–1254, 2008.
3. Teh, S.-Y., et al., Droplet microfluidics. *Lab Chip*, **8**: 198–220, 2008.
4. Chiu, D.T., R.M. Lorenz, and G.D.M. Jeffries, Droplets for ultrasmall-volume analysis. *Anal. Chem.*, **81**: 5111–5118, 2009.
5. Chiu, D.T. and R.M. Lorenz, Chemistry and biology in femtoliter and picoliter volume droplets. *Acc. Chem. Res.*, **42**(5): 649–658, 2009.
6. Theberge, A.B., et al., Microdroplets in microfluidics: An evolving platform for discoveries in chemistry and biology. *Angew. Chem. Int. Ed.*, **49**(34): 5846–5868, 2010.
7. Yang, C.-G., Z.-R. Xu, and J.-H. Wang, Manipulation of droplets in microfluidic systems. *Trends Anal. Chem.*, **29**: 141–157, 2010.
8. Kintses, B., et al., Microfluidic droplets: New integrated workflows for biological experiments. *Curr. Opin. Chem. Biol.*, **14**: 548–555, 2010.
9. Casadevall i Solvas, X. and A.J. deMello, Droplet microfluidics: Recent developments and future applications. *Chem. Commun.*, **47**: 1936–1942, 2011.
10. Roach, L.S., H. Song, and R.F. Ismagilov, Controlling nonspecific protein adsorption in a plug-based microfluidic system by controlling interfacial chemistry using fluorous-phase surfactants. *Anal. Chem.*, **77**: 785–796, 2005.
11. Song, H. and R.F. Ismagilov, Millisecond kinetics on a microfluidic chip using nanoliters of reagents. *J. Am. Chem. Soc.*, **125**: 14613–14619, 2003.
12. Huebner, A., et al., Quantitative detection of protein expression in single cells using droplet microfluidics. *Chem. Commun.*, (12): 1218–1220, 2007.
13. Aebersold, R. and M. Mann, Mass spectrometry-based proteomics. *Nature*, **422**(6928): 198–207, 2003.
14. Thorsen, T., et al., Dynamic pattern formation in a vesicle-generating microfluidic device. *Phys. Rev. Lett.*, **86**: 4162–4166, 2001.
15. Garstecki, P., et al., Formation of droplets and bubbles in a microfludic T-junction-scaling and mechanism of break-up. *Lab Chip*, **6**: 437–446, 2006.
16. Anna, S.L., N. Bontoux, and H.A. Stone, Formation of dispersions using "flow focusing" in microchannels. *Appl. Phys. Lett.*, **82**: 364–366, 2003.
17. Ward, T., et al., Microfluidic flow focusing: Drop size and scaling in pressure versus flow rate driven pumping. *Electrophoresis*, **26**: 3716–3724, 2005.
18. Baroud, C.N., F. Gallaire, and R. Dangla, Dynamics of microfluidic droplets. *Lab Chip*, **10**(16): 2032–2045, 2010.

19. Gu, H., M.H.G. Duits, and F. Mugele, Droplets formation and merging in two-phase flow microfluidics. *Int. J. Mol. Sci.*, **12**(4): 2572–2597, 2011.
20. Song, H., et al., Experimental test of scaling of mixing by chaotic advection in droplets moving through microfluidic channels. *Appl. Phys. Lett.*, **83**: 4664–4666, 2003.
21. Frenz, L., et al., Reliable microfluidic on-chip incubation of droplets in delay lines. *Lab Chip*, **9**: 1344–1348, 2009.
22. Courtois, F., et al., An integrated device for monitoring time-dependent in vitro expression from single genes in picolitre droplets. *ChemBioChem*, **9**: 439–446, 2008.
23. Courtois, F., et al., Controlling the retention of small molecules in emulsion microdroplets for use in cell-based assays. *Anal. Chem.*, **81**: 3008–3016, 2009.
24. Shi, W., et al., Droplet-based microfluidic system for individual *Caenorhabditis elegans* assay. *Lab Chip*, **8**: 1432–1435, 2008.
25. Huebner, A., et al., Static microdroplet arrays: A microfluidic device for droplet trapping, incubation and release for enzymatic and cell-based assays. *Lab Chip*, **9**: 692–698, 2009.
26. Schaerli, Y., et al., Contunuous flow polymerase chain reaction of single copy DNA in microfluidic microdroplets. *Anal. Chem.*, **81**: 302–306, 2009.
27. He, M., et al., Selective encapsulation of single cells and subcellular organelles into picoliter- and femtoliter-volume droplets. *Anal. Chem.*, **77**: 1539–1544, 2005.
28. Lau, B.T.C., et al., A complete microfluidic screening platform for rational protein crystallization. *J. Am. Chem. Soc.*, **129**: 454–455, 2007.
29. Yobas, L., et al., High performance flow focusing geometry for spontaneous generation of monodispersed droplets. *Lab Chip*, **6**: 1073–1079, 2006.
30. He, M., J.S. Kuo, and D.T. Chiu, Electro-generation of single femtoliter- and picoliter-volume aqueous droplets in microfluidic systems. *Appl. Phys. Lett.*, **87**: 031916, 2005.
31. Park, S.-Y., et al., High-speed droplet generation on demand driven by pulse laser-induced cavitation. *Lab Chip*, **11**: 1010–1012, 2011.
32. Bransky, A., et al., A microfluidic droplet generator based on a piezoelectric actuator. *Lab Chip*, **9**: 516–520, 2009.
33. Galas, J.C., D. Bartolo, and V. Studer, Active connectors for microfluidic drops on demand. *New J. Phys.*, **11**: 075027, 2009.
34. Zeng, S., et al., Microvalve-actuated precise control of individual droplets in microfluidic devices. *Lab Chip*, **9**: 1340–1343, 2009.
35. Choi, J.-H., et al., Designed pneumatic valve actuators for controlled droplet breakup and generation. *Lab Chip*, **10**: 456–461, 2010.
36. Abate, A.R., et al., Valve-based flow focusing for drop formation. *Appl. Phys. Lett.*, **94**: 023503, 2009.
37. Song, H., J.D. Tice, and R.F. Ismagilov, A microfluidic system for controlling reaction networks in time. *Angew. Chem. Int. Ed.*, **42**: 768–772, 2003.
38. Zheng, B., L.S. Roach, and R.F. Ismagilov, Screening of protein crystallization conditions on a microfluidic chip using nanoliter size droplets. *J. Am. Chem. Soc.*, **125**: 11170–11171, 2003.
39. Huebner, A., et al., Development of quantitative cell-based enzyme assays in microdroplets. *Anal. Chem.*, **80**: 3890–3896, 2008.
40. Abate, A.R., et al., High throughput injection with microfluidics using picoinjectors. *Proc. Natl. Acad. Sci. USA*, **107**: 19163–19166, 2010.
41. Rhee, M. and M.A. Burns, Drop mixing in a microchannel for lab on a chip platforms. *Langmuir*, **24**: 590–601, 2008.
42. Bremond, N., A.R. Thiam, and J. Bibette, Decompressing emulsion droplets favors coalescence. *Phys. Rev. Lett.*, **100**: 024501, 2008.
43. Fidalgo, L.M., C. Abell, and W.T.S. Huck, Surface-induced droplet fusion in microfluidic devices. *Lab Chip*, **7**: 984–986, 2007.

44. Niu, X., et al., Pillar-induced droplet merging in microfluidic circuits. *Lab Chip*, **8**: 1837–1841, 2008.
45. Priest, C., S. Herminghaus, and R. Seemann, Controlled electrocoalescence in microfluidics: Targeting a single lamella. *Appl. Phys. Lett.*, **89**: 134101, 2006.
46. Link, D.R., et al., Electric control of droplets in microfluidic devices. *Angew. Chem. Int. Ed.*, **45**: 2556–2560, 2006.
47. Zagnoni, M. and J.M. Cooper, On-chip electrocoalescence of microdroplets as a function of voltage, frequency and droplet size. *Lab Chip*, **9**: 2652–2658, 2009.
48. Niu, X., et al., Electro-coalescence of digitally controlled dropelts. *Anal. Chem.*, **81**: 7321–7325, 2009.
49. Baroud, C.N., M.R. de Saint Vincent, and J.P. Delville, An optical toolbox for total control of droplet microfluidics. *Lab Chip*, **7**: 1029–1033, 2007.
50. Thiam, A.R., N. Bremond, and J. Bibette, Breaking of an emulsion under an ac electric field. *Phys. Rev. Lett.*, **102**: 188304, 2009.
51. Frenz, L., et al., Microfluidic production of droplet pairs. *Langmuir*, **24**: 12073–12076, 2008.
52. Jambovane, S., et al., Creation of stepwise concentration gradient in picoliter droplets for parallel reactions of matrix metalloproteinase II and IX. *Anal. Chem.*, **83**: 3358–3364, 2011.
53. Agresti, J.J., et al., Ultrahigh throughtput screening in drop based microfluidics for directed evolution. *Proc. Natl. Acad. Sci. USA*, **107**: 4004–4009, 2010.
54. Brouzes, E., et al., Droplet microfluidic technology for single-cell high throughput screening. *Proc. Natl. Acad. Sci. USA*, **106**: 14195–14200, 2009.
55. Slaney, T.R., et al., Push-pull perfusion sampling with segmented flow for high temporal and saptial resolution in vivo chemical monitoring. *Anal. Chem.*, **83**: 5207–5213, 2011.
56. Schmitz, C.H.J., et al., Dropspots: A picoliter array in a microfluidic device. *Lab Chip*, **9**: 44–49, 2009.
57. Mazutis, L., et al., Multi-step microfluidic droplet processing: Kinetic analysis of an in vitro translated enzyme. *Lab Chip*, **9**: 2902–2908, 2009.
58. Clausell-Tormos, J., et al., Droplet based microfluidic platforms for the encapsulation and screening of mammalian cells and multicellular organisms. *Chem. Biol.*, **15**: 427–437, 2008.
59. Damean, N., et al., Simultaneous measurements of reactions in microdroplets filled by concentration gradients. *Lab Chip*, **9**: 1707–1713, 2009.
60. Bui, M.P.N., et al., Enzyme kinetic measurements using a droplet based microfluidic system with a concentration gradient. *Anal. Chem.*, **83**: 1603–1608, 2011.
61. Beer, N.R., et al., On chip, real time, single copy polymerase chain reaction in picoliter droplets. *Anal. Chem.*, **79**: 8471–8475, 2007.
62. Beer, N.R., et al., On-chip single-copy real-time reverse-transcription PCR in isolated picoliter droplets. *Anal. Chem.*, **80**: 1854–1858, 2008.
63. Srisa-Art, M., et al., Monitoring of real time streptavidin biotin binding kinetics using droplet microfluidics. *Anal. Chem.*, **80**: 7063–7067, 2008.
64. Marz, A., et al., Droplet formation via flow-through microdevices in Raman and surface enhanced Raman spectroscopy—Concepts and applications. *Lab Chip*, **11**: 3584–3592, 2011.
65. Cristobal, G., et al., On-line laser Raman spectroscopic probing of droplets engineered in microfluidic devices. *Lab Chip*, **6**: 1140–1146, 2006.
66. Strehle, K.R., et al., A reproducible surface enhanced Raman spectroscopy approach. Online SERS measurements in a segmented microfluidic system. *Anal. Chem.*, **79**: 1542–1547, 2007.
67. Liu, S., et al., The electrochemical detection of droplets in microfluidic devices. *Lab Chip*, **8**: 1937–1942, 2008.
68. Han, Z., et al., Measuring rapid enzymatic kinetics by electrochemical method in droplet-based microfluidic devices with pneumatic valves. *Anal. Chem.*, **81**: 5840–5845, 2009.

69. Kautz, R.A., W.K. Goetzinger, and B.L. Karger, High throughput microcoil NMR of compound libraries using zero-dispersion segmented flow analysis. *J. Comb. Chem.*, **7**: 14–20, 2005.
70. Liu, T., et al., Accurate mass measurements in proteomics. *Chem. Rev.*, **107**(8): 3621–3653, 2007.
71. Hatakeyama, T., D.L. Chen, and R.F. Ismagilov, Microgram-scale testing of reaction conditions in solution using nanoliter plugs in microfluidics with detection by MALDI-MS. *J. Am. Chem. Soc.*, **128**: 2518–2519, 2006.
72. Pei, J., et al., Analysis of samples stored as individual plugs in a capillary by electrospray ionization mass spectrometry. *Anal. Chem.*, **81**: 6558–6561, 2009.
73. Li, Q., et al., Fraction collection from capillary liquid chromatography and off-line electrospray ionization mass spectrometry using oil segmented flow. *Anal. Chem.*, **82**: 5260–5267, 2010.
74. Edgar, J.S., et al., Capillary electrophoresis separation in the presence of an immiscible boundary for droplet analysis. *Anal. Chem.*, **78**(19): 6948–6954, 2006.
75. Niu, X.Z., et al., Droplet-based compartmentalization of chemically separated components in two dimensional separations. *Chem. Commun.*, (41): 6159–6161, 2009.
76. Roman, G.T., et al., Sampling and electrophoretic analysis of segmented flow streams using virtual walls in a microfluidic device. *Anal. Chem.*, **80**: 8231–8238, 2008.
77. Wang, M., et al., Microfluidic chip for high efficiency electrophoretic analysis of segmented flow from a microdialysis probe and in vivo chemical monitoring. *Anal. Chem.*, **81**: 9072–9078, 2009.
78. Pei, J., J. Nie, and R.T. Kennedy, Parallel electrophoretic analysis of segmented samples on chip for high-throughput determination of enzyme activities. *Anal. Chem.*, **82**: 9261–9267, 2010.
79. Zhu, Y. and Q. Fang, Integrated droplet analysis system with electrospray ionization-mass spectrometry using a hydrophilic tongue-based dropelt extraction interface. *Anal. Chem.*, **82**: 8361–8366, 2010.
80. Filla, L.A., D.C. Kirkpatrick, and R.S. Martin, Use of a coroma discharge to selectively pattern a hydrophilic/hydrophobic interface for integrating segmented flow with microchip electrophoresis and electrochemical detection. *Anal. Chem.*, **83**: 5996–6003, 2011.
81. Fidalgo, L.M., et al., From microdroplets to microfluidics: Selective emulsion separation in microfluidic devices. *Angew. Chem. Int. Ed.*, **47**: 2042–2045, 2008.
82. Fidalgo, L.M., et al., Coupling microdroplet microreactors with mass spectrometry: Reading the contents of single droplets online. *Angew. Chem. Int. Ed.*, **48**(20): 3665–3668, 2009.
83. Zeng, S., et al., Electric control of individual droplet breaking and droplet contents extraction. *Anal. Chem.*, **83**: 2083–2089, 2011.
84. Kelly, R.T., et al., Dilution-free analysis from picoliter droplets by nano-electrospray ionization mass spectrometry. *Angew. Chem. Int. Ed.*, **48**(37): 6832–6835, 2009.
85. Swanson, S.K. and M.P. Washburn, The continuing evolution of shotgun proteomics. *Drug Discov. Today*, **10**(10): 719–725, 2005.
86. Zhou, H., et al., Advancements in top-down proteomics. *Anal. Chem.*, **84**(2): 720–734, 2012.
87. Sun, X., et al., Ultrasensitive nanoelectrospray ionization-mass spectrometry using poly(dimethylsiloxane) microchips with monolithically integrated emitters. *Analyst*, **135**: 2296–2302, 2010.
88. Marginean, I., et al., Achieving 50% ionization efficiency in subambient pressure ionization with nanoelectrospray. *Anal. Chem.*, **82**(22): 9344–9349, 2010.
89. Shen, Y., et al., Ultrasensitive proteomics using high-efficiency on-line micro-SPE-nanoLC-nanoESI MS and MS/MS. *Anal. Chem.*, **76**(1): 144–154, 2004.
90. Zhang, Z.R., et al., One-dimensional protein analysis of an HT29 human colon adenocarcinoma cell. *Anal. Chem.*, **72**(2): 318–322, 2000.

4 Recent Developments toward the Synthesis of Supramolecular Bioelectronic Nanostructures

John D. Tovar, Stephen R. Diegelmann, and Brian D. Wall

CONTENTS

Supramolecular materials derived from the assembly of molecularly well-defined precursors are being applied in many areas of current technology. Their length scales are in a regime not easily accessed through chemical synthesis, lithography or nanofabrication. Although increasingly small nanostructured components can be fabricated for integrated circuitry, replicating comparable processes in the soft materials domain is very difficult. Self-assembly has emerged as a powerful tool to construct advanced materials from molecular components with pre-programmed interactions that ultimately lead to the formation of defined objects with higher structural order and with lengths in one, two or three dimensions in the 1–100 nm regime. In this chapter, we will discuss briefly some of the enthalpic and entropic currencies (as referred to in energy terms of kcal/mol) that are exchanged to build supramolecular materials, and then highlight how they can be used collectively to construct novel electronic structures. Organic electronic materials in general are becoming increasingly important for many clinical biomaterials efforts, and the ability to scale their dimensionality down to the nanoscale will lead to many potentially innovative new therapies. We discuss some of the current organic bioelectronic state of the art and then discuss emerging strategies from our laboratory and others to render the organic electronic materials into one-dimensional nanoscale fibrils that resemble in part the structural components of the extracellular matrix.

4.1 "SUPRAMOLECULAR SYNTHONS" USED TO CONSTRUCT SELF-ASSEMBLED MATERIALS

The self-assembly of hierarchically structured objects from molecular precursors comes with a severe entropic penalty since particular conformations must be established in the molecular framework, and degrees of freedom in the system as a whole are lost as multiple molecules come together into defined and relatively speaking more ordered objects. However, enthalpic gains in system stability can successfully overcome the entropic costs to provide a net thermodynamic favorability. These enthalpic stabilizations reflect in many cases an electrostatic attraction that can be built into a self-assembling strategy through molecular design. These interactions are individually weak, usually being on the order of 1–10 kcal/mol compared to the ca. 100 kcal/mol strength of a typical hydrocarbon covalent bond. However, when several of these interactions can be set into play, the collective stabilization becomes quite significant. Desiraju classified these general interactions as *supramolecular synthons* after the synthons of retrosynthetic analysis strategy championed by the eminent organic chemist E. J. Corey for the synthesis of complex organic molecules. We will only briefly describe these interactions here, and we refer the reader to more specialized textbooks and reviews that elaborate upon the physics and applications of these interactions in the context of supramolecular assembly and π-conjugated materials.[1–3]

There are several flavors of electrostatic interactions wielded by the supramolecular engineer to promote a particular motif of self-assembly. The strongest of these are classical point charge pairings among positively and negatively charged atoms or functional groups that can provide up to 60 kcal/mol of stabilization. Common examples are formal ionic bonds and coordination complexes. These interactions are tempered by the nature of the solvation sphere or the medium that can screen the extent of the coulombic attraction such as the case for protein salt bridges between ammonium salts and carboxylates at physiological pH. Dipoles within molecules can also exert electrostatic interactions with point charges (worth 15–50 kcal/mol) or even with other dipoles (worth 1–5 kcal/mol), such as water molecules solvating a sodium ion, or liquid crystalline molecules interacting in a particular phase. The lone pairs on oxygen are not formally charged but still exert an attractive force upon a nearby center of positive charge. The classical hydrogen bond is simply a variation upon this ion-dipole theme. The relative charge distributions within more ionizable X-H bonds (such as $R\text{-}NH_2$ and R-OH) lead to a formal δ+ charge on the proton that then can partake in stabilizing interactions with nearby anions or dipoles, typically gaining up to 15 kcal/mol of enthalpic stabilization.

Quadrupoles also wage unique forms of enthaplic stablization (ca. 0–5 kcal/mol in aqueous media). Although subtle in their effects, quadrupolar interactions are important for the energy transporting characteristics of organic electronic materials in an intermolecular sense. The non-spherical charge distribution of an aromatic ring places substantial negative charge above and below the ring, leading to charge deficiency at the edges of the ring. Thus, an aromatic system can stack favorably through an "edge-to-face" T-shaped arrangement, or less so through a "slip stack" where the π systems are planar but offset. It should be stressed that the common depiction of a π-stack involving perfect centered alignment of two aromatics is actually quite repulsive! As optimal π-electron overlap for most aromatic molecules would be desirable for energy transport, a substantial amount of research has been devoted to learning how to overcome quadrupolar repulsions in thin films or even single crystals.[4,5] The quadrupolar charge distribution can also exert a stabilizing influence with ions and dipoles depending on the electronics involved among the two interacting species. One important example is the cation–π interaction,[6] a fairly recently elucidated interaction that has very important implications for protein folding and ligand-receptor interactions.

These examples of enthalpic stabilization have originated from static ions or dipoles existing within the molecular framework. As Nature abhors a vacuum, a molecule in the presence of a solvating field must find some way to interact favorably with the solvent at the molecular level. If specific ion, dipolar or quadrupolar interactions are not present, *induced* influences must be established. This is not unreasonable since the electric field exerted by an ionic charge is on the order of 10^6 V/cm! The polarizability of a covalent bond, even within "inert" hydrocarbons, will allow for a redistribution of charge in the presence of an electric field associated with a nearby charge, dipole, etc. thereby generating a local δ^+ and δ^- within the bond. This induced charge can now act in an electrostatically favorable sense even though the interacting unit is formally neutral. As may be expected, induced dipole interactions are fairly short range and weak in their strength. One might be quick to rule these interactions to be minor players, but they play very important roles in the stabilization of self-assembled monolayers made up of long hydrocarbon chains. With no clear sources of electrostatic stabilization existing among hydrocarbon chains, both interacting partners must mutually induce dipoles that can stabilize each other. These are referred to as London forces, dispersion forces, or van der Waals forces, and can provide less than 1 kcal/mol *per interaction*.

Finally, there is one very important entropic effect that crucial for realizing supramolecular assembly schemes, the *hydrophobic effect*. As mentioned above, a solute is stabilized by the medium via collective electrostatic interactions from the individual solvent molecules. Therefore, as solutes come together to react, to bind or to assemble, they must shed or alter their solvation sphere. If we consider 5 molecules that assemble into 1 larger structure, we see this as an entropic loss. If the enthalpic gains established from new intermolecular interactions within the aggregate cannot compensate for this entropic loss, we would not expect the assembly process to be favorable or thermodynamically spontaneous. However, this analysis neglects the fact that the 5 molecules establishing favorable interactions among themselves in essence acts as a new solvation sphere. The solvents previously needed to solvate the assembling molecule are now "freed" from their interactions thereby increasing the *overall* disorder within the system as a whole. The gained entropy within the solvent plays (plus any enthalpic gains from enhanced solvent-solvent interactions) an incredibly influential role within any type of supramolecular assembly scheme, be it a self-assembling molecule or a biological protein.

Using these enthalpic and entropic considerations as a foundation, an elegant range of self-assembled supramolecular low-dimensional materials have been prepared. These range from zero-dimensional (0-D, where all three dimensions are in the 1–100 nm regime) spherical micelles and vesicles formed from simple surfactants, to 1-D cylindrical structures from polymers or peptide amphiphiles (where two dimensions are in the 1–100 nm range) to 2-D structures such as lipid bilayers (where one dimension is in the 1–100 nm range). Although supramolecular assembly is relevant to the formation of liquid crystalline bulk phases, block copolymer morphologies and other extended materials, we will focus here on the supramolecular synthesis of well-defined and isolable 1-D objects. The geometries of the assembled objects in many instances can be dictated by choice of molecular geometry as articulated by Israelachvili.[7] Although a comprehensive review of molecular self-assembly is well beyond the scope of this chapter, we will highlight a few key examples of discrete 1-D nanostructures with electronically interesting functionality built by employing these supramolecular design principles. The interested reader is referred to the excellent review put out by Schenning and Meijer in 2005 for a broader coverage of supramolecular electronic materials.[2]

4.2 ONE-DIMENSIONAL SUPRAMOLECULAR ASSEMBLIES OF π-ELECTRON MATERIALS

Difficult synthetic challenges must be overcome to prepare complex one-dimensional aromatic architectures capable of entering the realm of devices for bioelectronics. π-conjugated organic polymers can give well-ordered molecular architectures but exact control over polymer aggregation, π-electron backbone planarization and domain morphology leading to desired electronic properties in the bulk solid-state structure is not trivial. For bioelectronic materials, conductive features with sizes on the order of 10 s–100 s of nm are needed for intimate interactions with cellular components. This size regime is relatively untouched in the soft materials arena as it is beyond the limits of most top-down fabrication techniques (such as nanolithography), and beyond the synthetic ability and efficiency for covalently-linked bottom-up designs. Nature's ability to utilize this regime is beginning to be mimicked with fundamental understandings of noncovalent interactions and the emergence of supramolecular biomaterials chemistry. The use of noncovalent interactions to construct supramolecular assemblies of π-electronic materials is maturing from proof-of-principle examples to promising systems showing very elegant control of molecular packing and π-electron interactions at the nanoscale, as discussed in a very brief sampling of instructive examples below.

Schenning, Meijer and coworkers developed functionalized oligo(p-phenylene vinylene)s such as OPV-3 (Figure 4.1) substituted with hydrogen-bonding self-complementary ureidotriazine units. The ureidotriazine undergoes strong

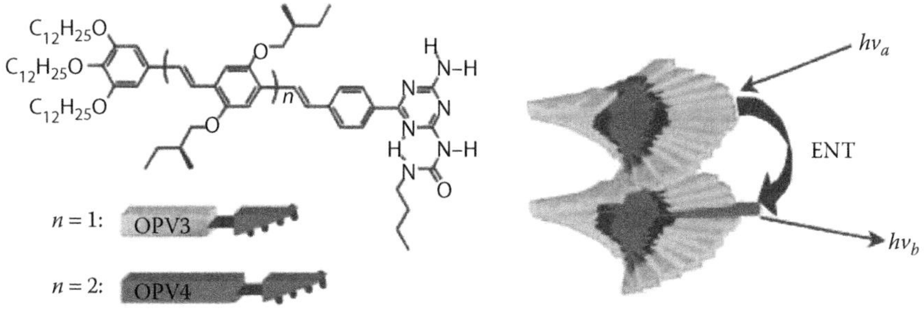

FIGURE 4.1 Oligophenylene vinylenes that undergo dimerization and self-assembly into 1-D helical nanostructures capable of energy transfer to low bandgap dopants. Figure taken from reference 8 with permission. Copyright 2004, Wiley-VCH Verlag GmbH & Co. KGaA.

hydrogen-bonding in organic solvents leading to the formation of dimeric structures. The hydrogen-bonded dimers then aggregate in a one-dimensional fashion leading to the formation of well-defined helical and columnar nanoscopic objects.[8] These supramolecular structures form when dissolved in dodecane, a nonpolar solvent that allows for the ureidotriazene moieties to establish hydrogen bonding interactions within a supramolecular dimer without significant competition from the solvent. The chiral stacks were characterized by UV-Vis, fluorescence and circular dichroism (CD) spectroscopy, all of which were compared to the same spectroscopic measurements taken in chloroform, a solvent that inhibits assembly by more effectively solvating the triazine. The results obtained for the samples in chloroform were shown to be that of the molecularly dissolved species due to the absence of a Cotton effect in the CD spectra and lack of spectral shifts in the UV-Vis. The samples in dodecane showed a blue-shifted absorbance and a red-shifted quenched luminescence, along with a strong bisignated Cotton effect consistent with the exciton coupling model for chromophores locked in a twisted and chiral environment. All of these trends differed from the molecularly dissolved species indicating that they arise from the intermolecular π-electron interactions within the aggregate. Heating the spectroscopic solutions provided enough of a stimulus to disrupt the aggregate into the molecular components, a process that was reversible upon cooling back to the initial temperature. This thermochromism was readily tracked through spectroscopic observation of the optical signatures corresponding to the aggregated and molecularly-dissolved species.

To show the versatility that this complimentary hydrogen-bonding subunit has for creating optoelectronic supramolecular aggregates, several variants were made to explore the potential for organic electronic devices. The first variant involved the inclusion of oligomers of greater π-electron conjugation length (OPV-4, a tetramer with a smaller optical bandgap).[9] The combination of the longer conjugated monomer into ordered aggregates of OPV-3 at a low mole percent (1.2 mol%) created an acceptor unit mixed within the stack of higher-energy OPV-3 chromophores. After excitation of the dominant OPV-3 trimer within the aggregate, the excitation energy was efficiently transferred to the lowest lying electronic state of the more conjugated OPV-4 oligomer dopant as evidenced by very efficient quenching of the less conjugated OPV-3 oligomer fluorescence and enhanced fluorescence corresponding to the longer conjugated OPV-4. This effect was observed only after a mixture of the OPV-3 and OPV-4 molecules were heated above the transition temperature (from assembled aggregate to molecularly dissolved) and then allowed to cool thus reforming a random dispersal of OPV-4 hydrogen-bonded dimers within the 1-D supramolecular stacks. The general dynamic properties of the ensemble allow for the incorporation of different acceptor units while still fostering self-assembly and allow for a way to tune the emission wavelength of the overall system through tuning the electronic properties of the component molecular parts.

Wurthner and coworkers have studied the assembly properties of perylene diimides (PDIs) in the context of light harvesting materials and organogelators.[10] They found that the unsubstituted imide units on the PDI molecules could form hydrogen bonds with alkyl-substituted melamines to self-assemble into mesoscopic superstructures.[11] With the success of the complimentary hydrogen bonding subunits attached to OPVs and the evidence of energy migration through supramolecular aggregates, more complex electronic donor-acceptor variants were developed whereby the PDI with its two pre-existing hydrogen bonding triads acted as a complementary hydrogen bonding unit to the Meijer OPVs (Figure 4.2). The OPVs were substituted with a melamine subunit that would allow for the hydrogen bonding with the PDI unit but would not readily allow for the formation of noncovalent OPV dimers.[12] The supramolecular organization of the donor-acceptor-donor chromophore triads (D-A-D) formed helical columnar stacks as been previously shown with substituted OPVs. AFM confirmed the chiral aggregates formed in methylcyclohexane which due to their D-A-D aggregation of p-type electron donating OPVs and n-type electron accepting PDIs displayed very efficient photoinduced electron transfer and might potentially be used for the development of supramolecular photovoltaics.

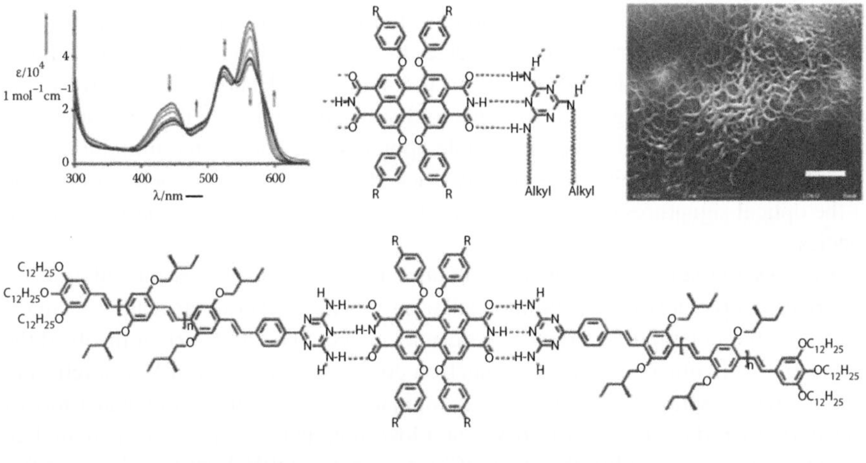

FIGURE 4.2 A perylene diimide (PDI)–melamine self-assembly scheme leading to fibrous materials (top) that was then extended into OPV–PDI triads (bottom). Modified figures taken from references 11 and 12 with permission. Copyrights 1999, Wiley-VCH Verlag GmbH & Co. KGaA, and 2002, American Chemical Society.

Stupp and coworkers established the self-assembling properties of dendron rod-coil molecules (DRCs), triblock molecules containing 3,5 dihydroxy-benzoic ester dendritic segments, biphenyl ester rigid-rod segments, and flexible coil-like polymers.[13] The DRC motif was designed to encourage hydrogen bonding among the dendrons, π–π stacking among the rod segments to create one-dimensional growth, and to discourage crystallization due to the flexible coils. These molecules formed birefringent gels in dichloromethane, a sign of the presence of 1-D nanostructures that were identified as such with transmission electron microscopy (TEM) and atomic force microscopy (AFM). Much like the Meijer OPVs, the DRC nanostructures arise from hydrogen bonding among the dendron segment followed by directional aggregation into ribbon-like structures. Different solvents led to different morphologies: the DRCs also formed self-supporting gels in ethyl methacrylate, but in this solvent they formed defined helical materials that could be use to sequester cadmium ions from the nonpolar medium. Upon exposure of the cadmium templated nanostructures to hydrogen sulfide gas, a helical morphology of CdS semiconductor was obtained.[14] Later work extended this motif into a more general form that enabled the inclusion of different π-conjugated rod segments such as oligo(thiophene), oligo(phenylene-vinylene), and oligo(phenylene), where Figure 4.3 shows a quaterthiophene variant.[15] The gelation and assembly of these structures is the same as seen previously, where a 1 wt% solution of the quaterthiophene DRCs in toluene was molecularly dissolved at high temperature but led to the gelation upon cooling. The assembled quaterthiophene structures showed a similar spectroscopic response discussed previously, with a blue-shifted absorbance and a red-shifted and quenched luminescence. To test the feasibility of these aggregates for devices, a solid-state film was cast from an iodine-doped 1 wt% assembled gel that exhibited a conductivity of 7.9×10^{-5} S/cm. This value was three orders of magnitude larger then the films constructed from the molecularly dissolved monomer units thus revealing the influence and the functional attributes of the supramolecular π-stacking as a conduit for electrical charge migration within nanoscale dimensions.

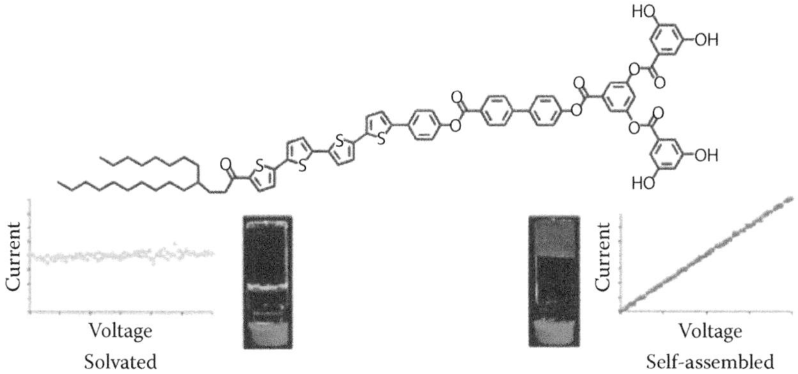

FIGURE 4.3 An oligothiophene-based dendron rod-coil molecule that leads to electrically conductive nanoribbons upon chemical doping. Modified figure taken from reference 15 with permission. Copyright 2004, American Chemical Society.

For electronic materials to be developed to interface with biologically relevant species, the aggregated structure needs to present a biologically benign surface capable of integrin-mediated recognition.[16] One approach is to use oligopeptides to develop a hierarchical structure. Peptides are well known for forming common hydrogen-bonding motifs like the a-helix and the P-sheet, and recent research has used the natural propensities for peptides to adopt these secondary structures as a means of making a new electronic materials. Frauenrath and coworkers constructed self-assembling polymer-oligopeptide conjugates bearing diacetylene units (Figure 4.4).[17] The formation of P-sheet secondary structural motifs in halogenated organic solvents (such as methylene chloride) among the molecular components led to the formation of aggregates that simultaneously aligned the diacetylene subunits favorably for an efficient topochemical polymerization induced by UV irradiation. This created a conjugated polydiacetylene polymer covalently linked with periodically-spaced peptide chains. Several new self-assembling compounds have since been synthesized with varying abilities to assemble and/or to undergo topochemical polymerization.[18] Hydrogen bonding was shown to be an important factor for the formation of robust organogels and for an efficient topochemical polymerization: molecules capable of supporting multiple stabilizing hydrogen-bond interactions about both sides of the diacetylene unit gelled and formed polydiacetylenes upon irradiation, where those with minimal or no hydrogen bonding capabilities on one end of the diacetylene did not polymerize at all into polydiacetylene. This again highlights the stringent requirement for preorganization of monomers through noncovalent interactions that was necessary to form highly conjugated polymers.

The 1-D supramolecular structures discussed thus far have relied to different extents on specific hydrogen-bonding motifs that when coupled with π–π interactions led to clear electronic delocalization within the resulting nanostructured aggregates. These nanomaterials have demonstrated their potential for energy migration, photoinducted electron transfer, and electrical conductivity. Given the influence of

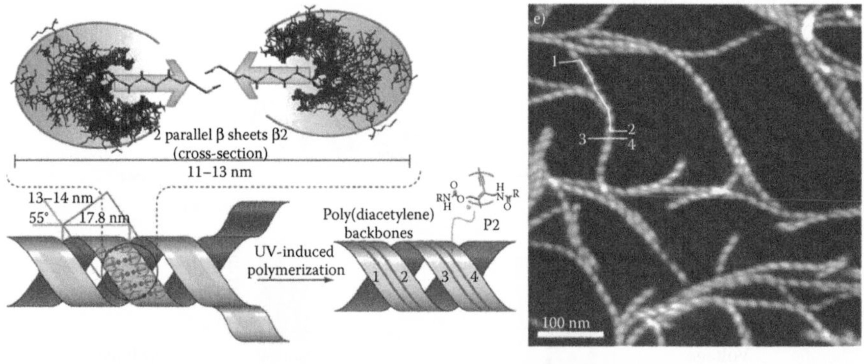

FIGURE 4.4 **(See color insert)** Self-assembly of oligopeptides linked with coil polymer segments and diacetylenes positions the alkynes in favorable geometries to facilitate the topochemical polymerization into polydiacetylene ultimately leading to a helical nanostructured material. Modified figure taken from reference 17 with permission. Copyright 2006, Wiley-VCH Verlag GmbH & Co. KGaA.

the hydrogen bonding to direct molecular self-assembly into nanoscale objects, the *organic* solvents used to dissolve the molecules were relatively nonpolar. From a biological application standpoint, it would be preferable to construct these materials from *aqueous* and *physiologically relevant* solutions. Self-assembly in aqueous conditions becomes more difficult as the solvent (water) now competes with hydrogen bonding and other electrostatic interactions making the desired intermolecular solute-solute interactions less potent. In addition, the π-conjugated moieties themselves tend to be very hydrophobic, making their extension into water even more difficult. The development of new assembly schemes whereby water screening becomes less detrimental to supramolecular interactions will be needed to promote the aqueous self-assembly of artificial electronic materials.

Aida and coworkers reported an amphiphilic hexabenzocoronene (HBC) that underwent assembly in THF solvent as well as in water-THF mixtures (Figure 4.5).[19,20] The HBC motif is well-known to form discotic liquid crystals with strong π-stacking among the flat aromatic units. The Aida molecular design installed hydrophobic alkane chains on one edge of the HBC core and hydrophilic tetraethylene oxide chains on the other. Dissolution was achieved in THF at elevated temperatures, and upon cooling the self-assembly occurred. The proposed molecular assembly involves a bilayer of HBC molecules, whereby the alkyl chains are interdigitated in a crystalline manner, leaving the surfaces of the bilayer decorated with the ethylene oxide chains. These polar groups interact favorably with the THF solvent and prevent extended multilayer formation. In THF, the formation of large nanotubes was observed (20 nm diameter, much larger than typical single-walled carbon nanotubes, CNTs), but in aqueous cosolvents a helical morphology was obtained. Upon chemical oxidation, the generation of radical cation charge carriers was evident in optical absorption and in actual electrical measurements. They have used these materials for photoconductivity[21] and as templates for other supramolecular interactions through the incorporation of positively-charged isothiouronium ion side chains.[22] In the latter example, the cationic isothiouronium surface allowed for the complexation with anionic species such as anthraquinone-2-carboxylate that quenched the aggregate emission through photoinduced electron transfer. This new example has shown

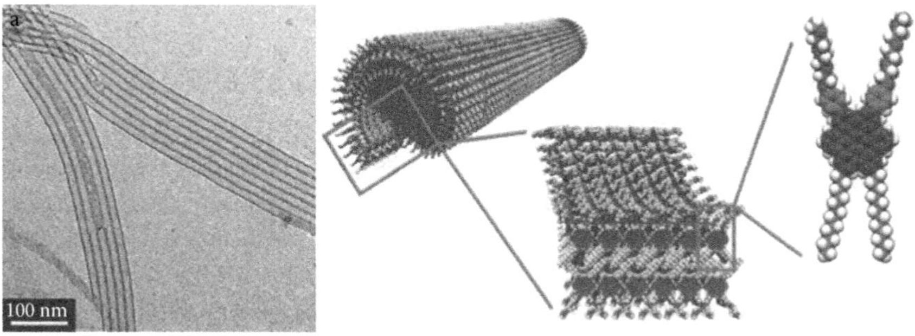

FIGURE 4.5 Amphiphilic hexabenzocoronenes that self-assemble into well-defined nanotubes. Modified figure taken from reference 20 with permission. Copyright 2005, National Academy of Sciences (USA).

that well-defined monomers can construct higher ordered nanostructures through noncovalent interactions that can persist in aqueous environments.

Lee and coworkers have studied several amphiphilic rod-coil oligo(phenylene) self assembling systems leading to micelles, vesicles, helical structures and many other low-dimensional motifs.[23] Their design strategy involves the alteration of the lengths of rod segment (oligophenylene) and coil segment (oligoethylene oxide) in order to control the nature of the resulting aggregate. More complex molecular architectures such as branched or dendritic presentation of the ethylene oxide fragments can also be used to control the shapes of the supramolecular objects. In many cases, these assemblies form in organic solvents, but with sufficient hydrophilic character the molecules are water-soluble and also assemble in aqueous conditions. Although the aqueous assembly of these oligophenylene molecules is indeed an important advance, it is known that the presentation of oligoethylene oxide moieties can be used to resist the adhesion of proteins and might therefore impact the nanostructure's ability to foster protein-based molecular recognition with cells.[24] To overcome this limitation, specific carbohydrates have also been attached covalently to the hydrophilic coil segments thereby providing a possible handle for cellular recognition.[25] Given that the molecular assembly leads to a nanostructure presenting multiple sugar units, this scheme could prove to be a powerful method to encourage the multivalency in ligand-receptor interactions known to enhance the binding between cells and artificial surfaces.

From single monomers, very complex and elegant supramolecular architectures can be constructed with varying properties. These nanoscale objects materials fill the void in accessible size regimes from typical micro- or nano-fabrication techniques used to process soft materials. The use of fundamental intermolecular interactions to construct supramolecular structures from well-characterized monomeric precursors is a powerful route to nanoelectronic materials. Protein-based hydrogen bonding motifs should allow for the ability to construct a biocompatible material and possibly incorporate a secondary function that may promote cell adhesion or other biologically important interactions, an important consideration for compatibility in biologically relevant environments. The examples shown here give a broad and representative view of the many promising and elegant contemporary systems for application as bioelectronic nanomaterials.

4.3 ELECTRICALLY CONDUCTIVE POLYMERS AS BIOMATERIALS

The nanomaterials described in the last section could be considered as exotic designer semiconductors—they are available through sometimes lengthy chemical synthesis on material scales usually on the order of grams or less. Conducting polymers are more readily available on multi-gram scales and above, but the ability to fashion them into isolated 1-D objects is difficult with current lithography. In specialized cases, block copolymers can form nanoscale materials, but the individual components are now intimately if not covalently linked within a large area network or domain rather than being isolable and manipulatable structures. Like most plastics, π-electron conjugated polymers in their pristine neutral state are not electrically conductive. Redox chemistry (most commonly leading to polymer oxidation) induces a change in electronic structure that leads to an electrically conductive material. This property can be switched on and off repeatedly by simple application of an electrical potential, or through exposure to suitable oxidants and reductants. The intricacies of this process are well-established, and we refer curious readers to the recent edition of the *Handbook of Conducting Polymers* for more information.[26] Conductive polymers have been used successfully in many biomedical applications such as for controlled drug delivery because of their switching capabilities and biocompatibility.[27]

Two major conducting polymers are quite prevalent in the biomedical engineering (Figure 4.6): polypyrrole (PPy) and poly(3,4-ethylenedioxythiophene, PEDOT). Both of these polymers are derived from the polymerization of electron rich heteroaromatic rings, and the resulting polymers have favorable materials properties such as aqueous processability, established biocompatibility, ease of deposition via electropolymerization, and electrically responsive switching between insulating and conductive states. PPy unfortunately suffers from long-term instability, performance variation due to coupling defects, and the possibility of reduction (back to the neutral insulating form) by biologically relevant reducing agents. PEDOT has most commonly been used as a surface modification agent for optoelectronic devices, specifically to make conductive transparent oxide surfaces less rough yet still highly electrically conductive. It is now starting to find use as a biological coating as well.[28] Other specialty polymers have been employed in biological settings such as polythiophene electrolytes or oxygenated polypyrroles.[29,30]

4.3.1 Electrodes at Neural and Cardiac Interfaces

Neural and cardiac electrodes are seeing increased usage in a number of diagnostic, therapeutic, and treatment applications; everything from brain implants (Figure 4.7), pacemakers, and cochlear implants, to the treatment of epilepsy, depression, chronic

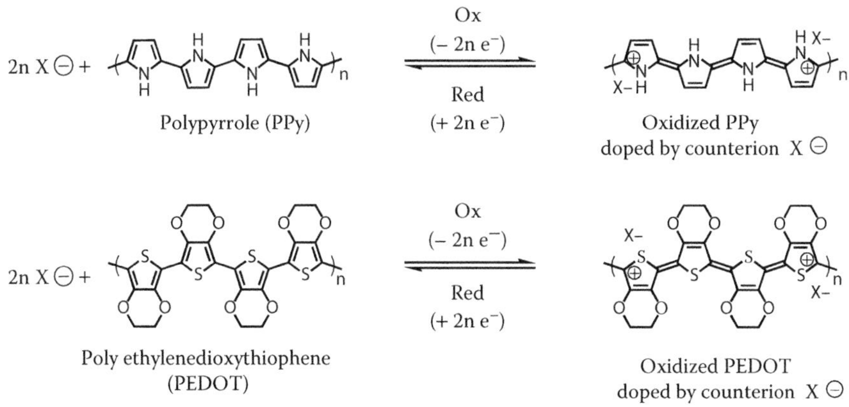

FIGURE 4.6 Neutral (insulating) and oxidized (conductive) chemical structures depicting the important forms of PPy and PEDOT. Note that the actual localization of charges here is only schematic.

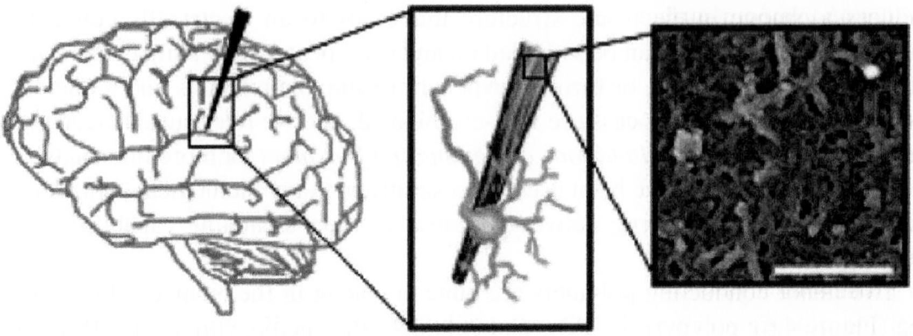

FIGURE 4.7 Biocompatible conductive coatings play a major role in the performance at the neuralelectrode interface by encouraging the adhesion of cells to the electrode surface as schematically depicted on different length scale: Direct insertion into the brain (left) should ideally elicit cellelectrode interactions (center) that are facilitated by proper surface roughness and chemistry (right). Figure taken from reference 43 with permission. Copyright 2009, American Chemical Society.

pain, and Parkinson's disease, to the regulation of breathing, bladder, and bowel control. These implanted electrodes work by sensing or sending electrical pulses and are therefore highly dependent on the contact at the tissue-electrode interface. 'Classical' electrodes, made from materials like gold, platinum, iridium, titanium, or steel are biocompatible or bioinert but suffer the drawback of having poor contact with tissue. This poor contact can provoke an immune response and scar tissue formation around the electrode which drastically diminishes the long term electrode performance. Furthermore, these interfaces may not be mechanically robust which poses a problem for implants subject to muscle contractions during device operation. Ideally, an electrode coating is needed that is conductive, adhesive and biocompatible to address the electrode-tissue interface.

A neurological probe's ability to sense an action potential from a neuron is dependent on both the surface area of the recording site and the impedance of the surface. Recording the activity of a single neuron requires a decreased electrode size down to that of a single cell. Unfortunately, a decreased electrode surface area causes an increase in the impedance, leading to reduced sensitivity and charge-transfer capability.[31,32] Conductive polymers such as PPy have been used for just this purpose, and because of rough surface morphology, electropolymerized PPy leads to an increased overall surface area for a given geometric space. This in effect increases the capacitance at the recording site and decreases the impedance. These films are also porous and allow for effective ion exchange into and out of surrounding tissue which is important for rapid electrochemical response of the polymer. The redox chemistry occurring at the recording site transduces the ionic current associated with ion channel flux into electric current detected with common electronics. PPy, unfortunately, is not well suited for long-term implantation and PEDOT has begun to receive increased interest. In time-based studies, PPy/poly(styrene sulfonate, PSS) retained only 5% of its original activity over 16 hours, whereas PEDOT/PSS retained 89% over the same time period.[33]

Probes using conductive materials like silicon, gold, and iridium are currently being used to both stimulate and record electronic impulses from cells. Unfortunately, the flat and smooth surfaces of these metal probes do not encourage the adhesion of cells and thus limit their overall utility. Conductive polymers have been employed as biomaterials for effective surface modifications to optimize interaction between the stimulating/recording electronics and living cells. Although PPy and PEDOT have been used successfully for this purpose, Martin's group also demonstrated that hydroxymethylated PEDOT (PEDOT-MeOH) provides a more effective probe surface modifier. PEDOT-MeOH has lower impedance that PPy, is more stable to biological reductants for long-term implant potential, and adds more aqueous solubility for processing and biocompatibility. It is also possible to dope PEDOT-MeOH with PSS and short oligopeptides (such as the amino acid sequence CDPGYIGSR, a substructure of laminin known to encourage cell adhesion). These two dopants allow for better processing than dopants used with PPy and provide a rougher surface morphology which in turn leads to enhanced cell adhesion.[34] Recently, Ying, Yu and coworkers have developed a series of functionalized PEDOT materials with a variety of linker groups that are covalently attached to the polymer backbone (Figure 4.8).[28] These monomers could be polymerized from aqueous emulsions and deposited on a wide array of substrates such as metal electrodes or even Tygon tubing. These polymers express established chemical linkers that could be used in the future to conjugate biosignals onto the PEDOT coatings and should lead to better immobilization of expressed signals without dramatically impacting film forming or electronic properties.

4.3.2 Cell Growth and Guidance

Electrically conductive biomaterials are also being used to control the shape and growth of cells. Size, shape and growth of cultured cells can somewhat be controlled

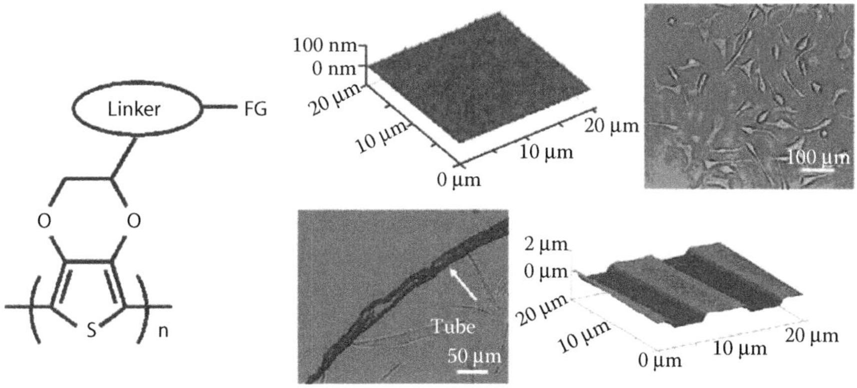

FIGURE 4.8 Functionalized EDOT monomers can be polymerized from aqueous emulsions to provide conductive polymers (left) for coatings on a variety of surfaces [such as macroelectrodes (upper middle AFM) and microelectrode arrays (lower right) that encourage cell adhesion (upper right image). Substrates as goofy as plastic surgical tubing could even be used. Figure taken from reference 28 with permission. Copyright 2008, American Chemical Society.

by environmental conditions including type of cell culture medium, supplements in the medium, inclusion or exclusion of growth factors, or addition of hormones. These control techniques address the environment surrounding the cultured cells but overlooked the attachment surface. Since mammalian and other cells must attach to a surface to survive, the nature of the attachment itself provides another handle for cell morphology control. Many factors play a role to influence what type of surface is presented to an attached cell; surface charge density, wettability, and roughness can all be used to influence cellular attachment, metabolism, and/or function. PPy again has potential as a conductive biomaterial due to its switchable surface characteristics. For example, PPy in its reduced insulating form inhibits proper cellular adhesion, while oxidation of the polymer into the conductive state leads to much greater degrees of cell spreading. Both of these electronic states have been shown to be nontoxic to cells. This material acts as an active and switchable surface to control cell growth and differentiation,[35] a property that has been applied recently for tissue engineering. Likewise, PEDOT films also encourage cell adhesion and spreading on a variety of planar and curved surfaces (Figure 4.8).

In the context of cellular engineering, the polymer surfaces mentioned above are *hundreds* of microns in size and thereby provide isotropic guidance cues to a specific cell with respect to any directionality in the growth or extension of cellular processes. Nevertheless, prospects for engineering at the *tissue* level are quite high, such as for the development of nerve guidance channels. A current clinical approach to repair peripheral nerve damage requires an autologous nerve graft, where the graft is removed from the patient's own nerve tissue at a healthy donor site. Although this has the advantage of tissue compatibility, the graft causes loss of function at the donor site and brings the possibility of mismatched dimensions between donor and acceptor site.[36] Alternatives to autologous nerve grafts are synthetic tubular guidance channels made from materials like collagen, laminin, fibronectin, silicone, or other polymers. The goal of these types of natural and synthetic nerve guidance channels is to direct axon resprouting through physical cues. Electrical stimulation may also play an inherent role in the function of autologous nerve grafts. Electrical charge and external electromagnetic fields can stimulate the extension, proliferation, differentiation and regeneration of many different cell types. Electrically conductive biomaterials offer two key advantages for nerve tissue engineering applications. First, electrical stimulation through the cell adherent surface would localize the electromagnetic field to the polymer surface, providing more spatial control over externally applied fields. Secondly, many conductive polymers present unique surfaces properties that can be varied through the type of dopant and surface morphology in order to tailor the types of cells adhered or even allow for surface patterning. Overall, using PPy as a nerve cell guidance material has been shown to enhance nerve cell differentiation in culture and increase the neurite growth length while invoking little immune response in animal implants.[37] Recent work with PPy-coated electrospun polymer nanofibers has shown preferential cellular elongation on aligned nanomaterials (Figure 4.9).[38]

Other electronic materials making impacts in the biomaterials realm are the many types of carbon nanotubes (CNTs). Unlike the other 1-D materials discussed in this chapter, CNTs can be considered as single molecules since they are composed

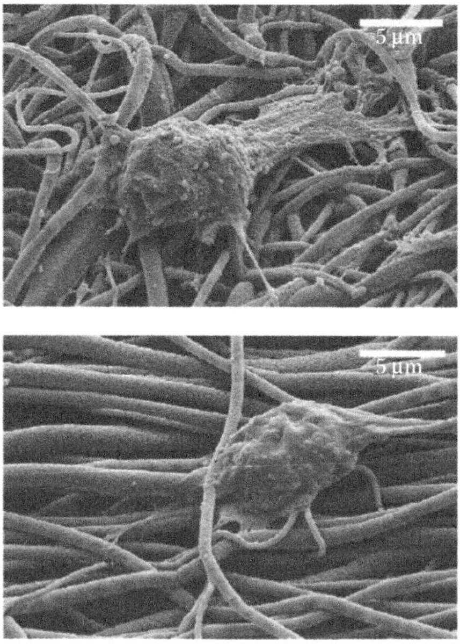

FIGURE 4.9 Rat PC-12 nerve cells adhered to both random and aligned polypyrrole fibers. Figure taken from reference 38 with permission. Copyright 2009, Elsevier.

of covalently linked graphene sheets that have been schematically rolled up and stitched together. CNTs can be produced industrially and are now commercially available in bulk quantities with varying degrees of purity. Their mechanical strength and toughness as well as chemical stability and high conductivity make them an obvious candidate for a conductive biomaterial electrode coating, particularly in applications where the mechanical rigidity of the underlying matrix is important. The performance of CNT-based electrodes is much better than that of current state-of-the-art-electrodes in the ability to reduce impedance, increase charge capacity, and facilitate charge transfer and they provide dual electronic and ionic conductivities. However, CNTs still need to have improvement in their electrochemical properties, long-term durability, and biocompatibility. CNTs are also prepared as mixtures of single and multi-walled nanotubes with different diameters and different electronic properties (some are insulating, some semiconductive, and some metallic!). These as-synthesized nanotubes are incredibly hydrophobic and often difficult to dissolve in aqueous media with out the help of surfactants and other solubilizers. Purification, functionalization and separation on the basis of electronic structure or size remains a substantial challenge in modern CNT research.[39,40]

One strategy to address the shortcomings of CNTs is to co-deposit or layer CNTs with other known conductive biomaterials (such as PEDOT) to take advantage of the individual beneficial properties of each. The surface of CNTs can also be modified directly in order to covalently install functional groups for cell adhesion.[41] It has been shown that CNTs can promote neurite growth to a greater extent by coating them with bioactive molecules. For example, a coating of 4-hydroxynonenal was applied to CNTs to encourage adhesion, and this composite system was able to elicit neurons to extend multiple neurites with extended branching versus CNTs without the coating.[42] CNTs have also been modified in a layer-by-layer approach with polyelectrolytes.[43] Although these CNT surface alterations have helped to make the nanotubes more soluble in aqueous environments and more adherent for cells, these alterations can also alter the inherent electronic properties of the nanotubes. CNTs have also be covalently surface modified with nerve growth factors to successfully regulate the survival and differentiation of neurons.[44] The conductance of CNTs can also be enough to stimulate a neuronal response where the surface modification simply acts as a solubilizing agent. For example, CNTs were functionalized with poly (ethyleneglycol) (PEG) to make them aqueous soluble and increase their bioavailability. These PEG-CNTs were then deposited onto glass cover slips in varying thicknesses in order to control the overall conductivity. By looking neuronal interactions with a range of conductivities it was shown that there is an ideal range for the promotion of neuronal growth and neurite extension and this data is now being applied to other CNT systems as well as other conductive electrode surface materials.[45]

4.4 PEPTIDE-OLIGOTHIOPHENE CONJUGATES FOR BIONANOSTRUCTURES

There is now a broad recognition of the importance of developing biologically relevant nanomaterials with electronic functions. Self-assembly of π-conjugated molecules into low-dimensional objects has usually been executed in organic solvents thus precluding useful transitions into biological applications. Likewise, the current difficulties in fashioning conducting polymers into defined nanomaterials have precluded the development of stand-alone conductive nanomaterials. Carbon nanotubes have been one approach to meet this challenge, but there are several issues of purity, composition, and ability to achieve biological recognition that affect the size and the bioelectronic properties of the nanotube samples. The structural proteins of the extracellular matrix are comprised of 1-D materials, and the ability to bring useful optoelectronic properties with this dimensionality could have major impacts on bridging the biotic-abiotic interface. One class of organic semiconductors that has received much interest in this regard is the *oligothiophene* family of p-channel (electron-donating) materials. Over the past year, new work has been reported with oligothiophene 1-D nanomaterials built from the assembly of peptide-based components (Figure 4.10). The intention with these approaches is to bring the electronic properties of the oligothiophenes into biologically compatible applications.

Bauerle and coworkers prepared a designer oligo(ethyleneoxide)-peptide sequence with sufficient solubility such that they could achieve the solution-phase "click chemistry" functionalization of the peptides onto a centeral quaterthiophene unit (Figure 4.11, left and center).[46] The peptide π-conjugated hybrid was indeed soluble in many common organic solvents but there was still a tendency to have aggregation to some extent. The final molecular architecture presented oligoethylene oxide units at the two termini of the linear molecule, and these in practice minimized the intermolecular π-electron delocalization within

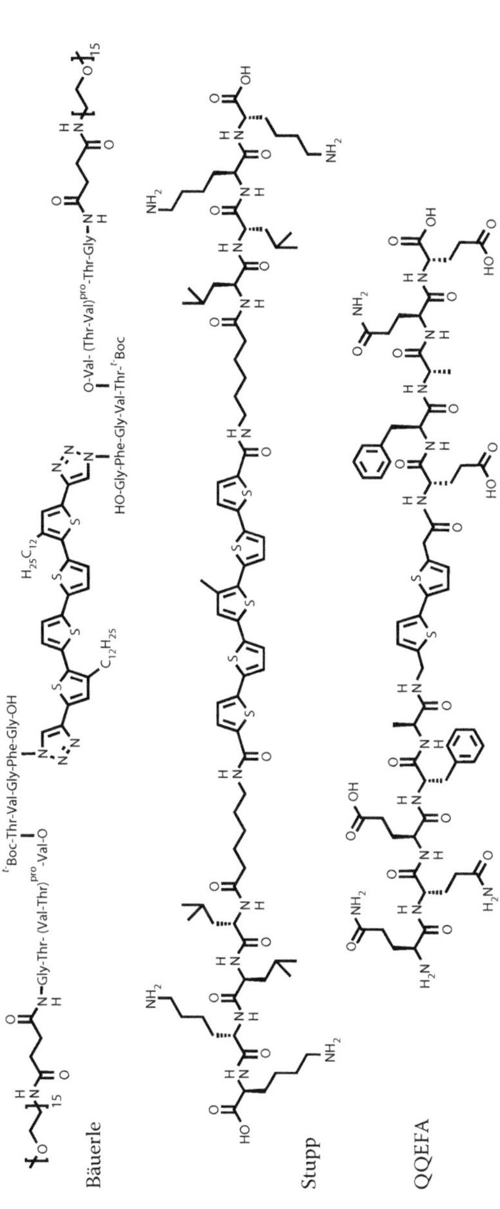

FIGURE 4.10 Oligothiophene peptides recently reported by the Bauerle, Stupp and Tovar groups as reported in references 46, 47 and 48.

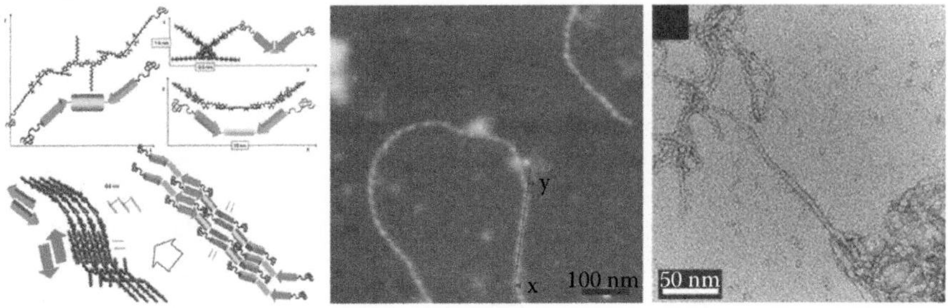

FIGURE 4.11 A model proposed for the assembly of the Bäuerle peptide (left) along with images formed by the Bäuerle peptide (middle, AFM) and the Stupp peptide (right, TEM). Modified figure taken from references 46 and 47 with permission. Copyrights 2009, Wiley-VCH Verlag GmbH & Co. KGaA and American Chemical Society.

the resulting nanomaterials necessary for delocalized electronic structures. These elegant multifunctional molecules offer many design strategies for continued development such as the novel switch in peptide backbone connectivity that could be used for promoting solubility until a late stage of synthesis. A careful switching condition was developed for more controlled self assembly leading to well-defined 1-D nanostructures between 8–12 nm in height with the functional quaterthiophene unit embedded within.

Stupp shortly thereafter reported an even longer quinquethiophene segment covalently attached within the peptide backbone (Figure 4.11, right).[47] In their synthesis approach, a small peptide sequence with multiple ionizable groups (lysine here) was prepared independently using standard solution-phase coupling techniques. The peptide was installed to the central quinquethiophene through solution phase amidation between a quinquethiophene diacid and the N-termini of the peptide sequence. These molecules formed self-assembled structures in water that were observed to be on the order of 5–10 nm in height. Here, methanolic solutions could be used to study the spectroscopic properties of molecularly dissolved material while addition of water led to extended self-assembly. There were pronounced temperature effects whereby heating led to the establishment of β-sheet-rich assemblies. It is clear from the Bäuerle and Stupp work that there is great promise for peptidic materials bearing π-electron function that can form well-defined nanomaterials under organic or even completely aqueous conditions.

In our own research, we sought a synthetic strategy that would allow for the preparation of the electronic peptides directly on the solid-phase resin support commonly used for automated peptide synthesis.[48] This required the independent preparation of π-conjugated "amino acids" that would react analogously to the amino acid building blocks commercially available for peptide solid-phase synthesis. To validate this approach, we prepared one such amino acid around a bithiophene π-electron unit. Indeed, this unit could be built into a peptide sequence with no need for solution-phase synthetic chemistry that can often require laborious purification steps to isolate desired peptidic materials from reaction byproducts. One such π-conjugated peptide prepared through this solid-phase approach is the QQEFA peptide illustrated at the bottom of Figure 4.10.

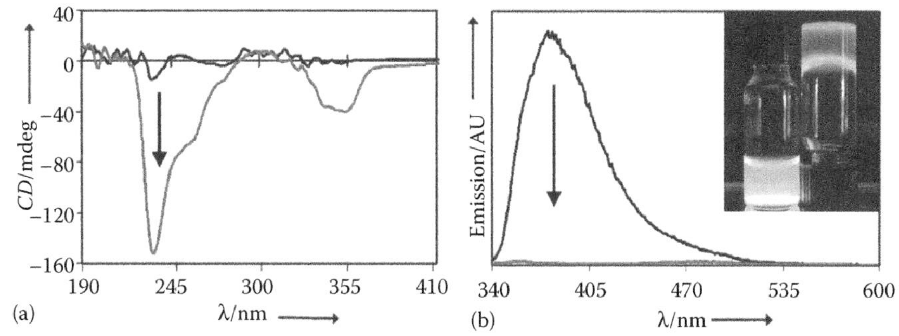

FIGURE 4.12 Spectroscopy of the QQEFA peptide prepared in our labs. (a) Circular dichroism of the QQEFA peptide in basic solution (molecularly dissolved, black line) and in acidic solution (assembled, gray line); (b) fluorescence spectra of the basic solution (black line) and once acidified (gray line). Inset to panel (b) shows a photo of the bulk phenomena, where the basic molecular solution is to the left, and the self-assembled hydrogel is to the right. Modified figure taken from reference 48 with permission. Copyright 2008, American Chemical Society.

Standard spectroscopies were used to characterize the assembly process (Figure 4.12). The infrared (IR) spectrum for QQEFA displayed characteristic absorptions found in other synthetic β-sheet systems. Circular dichroism of freely-soluble QQEFA yielded no meaningful absorption while acidification and assembly led to intense absorptions associated with the π-conjugated unit (Figure 4.12a). The data indicate that the bithiophenes rest in twisted chiral environments as expected from the slight twists found in natural β-sheet motifs. The CD cross-over at 320 nm (coincident with the low-energy UV-vis absorption λ_{max}) is a classic signature for exciton coupling observed routinely in other optoelectronic molecules assembled from organic solvents. Exciton coupling is one observation of a well-defined intermolecular π-electron interaction among two or more chromphore units leading to a direct perturbation of the overall electronic properties. The exciton coupling was further verified by examining the fluorescence of QQEFA in solution and in aggregates (Figure 4.12b). Molecularly dissolved QQEFA absorbs at 320 nm resulting in strong photoluminescence, but the fluorescence is dramatically quenched upon acidification.[49] These photophysical behaviors support the formation of tape-like aggregates with intimate π–π contact, a design element that will play a critical role in bioelectronic applications requiring π-stacking or spectroscopic observables that vary with assembly or solvation.

Atomic force microscopy was used to visualize the components that comprise the self-supporting gel (Figure 4.13). The subtle energy landscape for amyloid assembly offers many thermodynamic sinks on the road from free molecule to large amyloid-like fiber. The predominant features were high aspect-ratio fibrillar 1-D nanostructures typically 2–5 nm in height. In some cases, we could experimentally resolve the appearance of superstructural helicity with fairly regular 76 nm periodicities as is common for many other naturally-occurring amyloid plaques.[50] Molecular modeling of a model peptide revealed an anti-parallel β-sheet assembly motif shown in Figure 4.13 (right) that is consistent with the observed spectroscopy

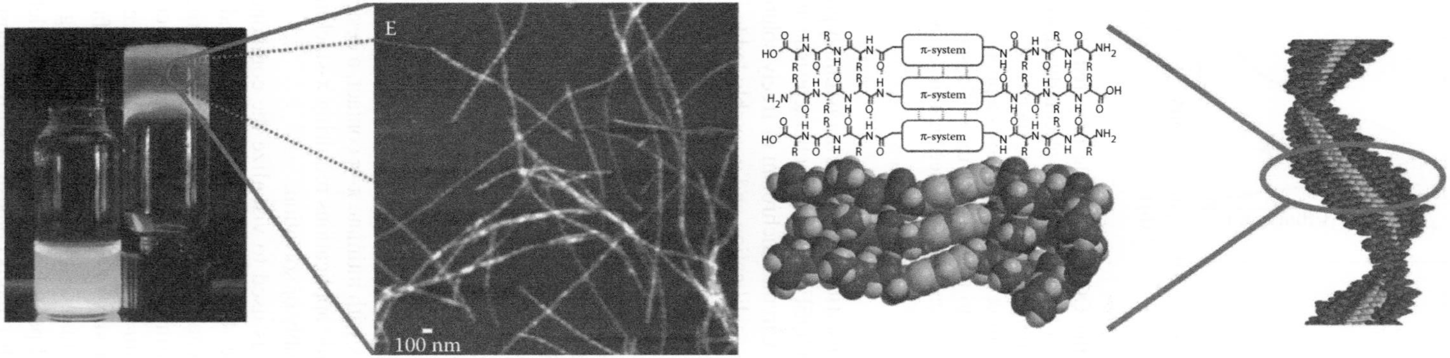

FIGURE 4.13 AFM image of the self-supporting hydrogel that formed upon acidification of a solution of the QQEFA peptide showing the one-dimensional nanomaterials that comprise the gel (left side). A space-filling molecular modeling depiction of the structure of the resulting aggregates showing the helical twist sense of the β-sheet. Modified figure taken from reference 48 with permission. Copyright 2008, American Chemical Society.

and microscopy. This leads to twists similar to those within natural β-sheet structures *(ca.* 9° between individual molecules) and the intermolecular π-stacking distance falls at about 5 Å. This distance is longer than what one typically associates with strong π-stacking, but the perturbation in the present case is sufficient enough to observe exciton-coupling and has been previously exploited for energy transfer among indole side-chains of assembled tryptophan containing peptides as well as among stilbenes covalently attached to a-helical peptides.[51,52]

Our work is continuing to expand the scope of π-electron units that can be incorporated within these structures. We have developed new reaction strategies to diversify these materials even more through the use of easily prepared (or even commercially available) building blocks with varying electronic properties (such as easily oxidized, easily reduced, highly fluorescent, etc.). We have also devised new strategies to control the presentation of the p-sheet peptide polarity, another important consideration in terms of influencing how the peptide P-sheets interact in an intermolecular sense within a given supramolecular aggregate. Finally, since the nanostructure formation seems to be a general approach, we have successfully incorporated bioactive signals on the peptide termini such that the resulting nanostructures present a dense concentration of bioadhesive signals for cell adhesion and proliferation. Our progress along these new fronts will be reported in the near future.

4.5 CONCLUDING REMARKS

Supramolecular assembly is a powerful approach to construct a variety of low-dimensional objects with a variety of unique functions. Conducting polymers are also powerful biomaterials for nerve repair and cardiac regulation. Exciting new developments in the supramolecular assembly of well-defined π-conjugated electronic materials in physiological environments is now underway. It will soon be possible to construct 1-D nanomaterials that can directly interface useful electrical function with living cells, be it for therapeutic or diagnostic applications. The ability to influence cells through external nanostructured biomaterials as well as the ability to detect cellular events or harness biological energy sources will provide many new and exciting scientific inquiries in the coming years, and self-assembled organic electronic materials stand to play large roles in these important pursuits.

ACKNOWLEDGEMENTS

Our work in this area is supported by the Johns Hopkins University. S.R.D. was supported by the Institute for NanoBioTechnology as an NSF-IGERT Predoctoral Fellow (DGE-0549350).

REFERENCES

1. J. W. Steed, J. L. Atwood, *Supramolecular Chemistry,* 1 ed., John Wiley & Sons, Ltd., New York, **2000**.
2. F. J. M. Hoeben, P. Jonkheijm, E. W. Meijer, A. P. H. J. Schenning. *Chem. Rev.* **105,** 1491 (2005).

3. E. V. Anslyn, D. A. Dougherty, *Modern Physical Organic Chemistry,* 1 ed., University Science Books, Sausalito (CA), **2006**.
4. J. Cornil, D. Beljonne, J.-P. Calbert, J.-L. Brédas. *Adv. Mater.* **13,** 1053 (2001).
5. J. E. Anthony, J. S. Brooks, D. L. Eaton, S. R. Parkin. *J. Am. Chem. Soc.* **123,** 9482 (2001).
6. J. C. Ma, D. A. Dougherty. *Chem. Rev.* **97,** 1303 (1997).
7. J. N. Israelachvili, *Intermolecular and Surface Forces,* 2 ed., Academic Press, San Diego, **1992**.
8. A. Schenning, P. Jonkheijm, E. Peeters, E. W. Meijer. *J. Am. Chem. Soc.* **123,** 409 (2001).
9. F. J. M. Hoeben, L. M. Herz, C. Daniel, P. Jonkheijm, A. P. H. J. Schenning, C. Silva, S. C. J. Meskers, D. Beljonne, R. T. Phillips, R. H. Friend, E. W. Meijer. *Angew. Chem. Int. Ed.* **43,** 1976 (2004).
10. X. Q. Li, V. Stepanenko, Z. J. Chen, P. Prins, L. D. A. Siebbeles, F. Würthner. *Chem. Commun.,* 3871 (2006).
11. F. Würthner, C. Thalacker, A. Sautter. *Adv. Mater.* **11,** 754 (1999).
12. A. P. H. J. Schenning, J. von Herrikhuyzen, P. Jonkheijm, Z. Chen, F. Würthner, E. W. Meijer. *J. Am. Chem. Soc.* **124,** 10252 (2002).
13. E. R. Zubarev, M. U. Pralle, E. D. Sone, S. I. Stupp. *J. Am. Chem. Soc.* **123,** 4105 (2001).
14. E. D. Sone, E. R. Zubarev, S. I. Stupp. *Angew. Chem. Int. Ed.* **41,** 1705 (2002).
15. B. W. Messmore, J. F. Hulvat, E. D. Sone, S. I. Stupp. *J. Am. Chem. Soc.* **126,** 14452 (2004).
16. S. E. Sakiyama-Elbert, J. A. Hubbell. *Ann. Rev. Mater. Res.* **31,** 183 (2001).
17. E. Jahnke, I. Lieberwirth, N. Severin, J. P. Rabe, H. Frauenrath. *Angew. Chem. Int. Ed.* **45,** 5383 (2006).
18. E. Jahnke, N. Severin, P. Kreutzkamp, J. P. Rabe, H. Frauenrath. *Adv. Mater.* **20,** 409 (2008).
19. J. P. Hill, W. S. Jin, A. Kosaka, T. Fukushima, H. Ichihara, T. Shimomura, K. Ito, T. Hashizume, N. Ishii, T. Aida. *Science* **304,** 1481 (2004).
20. W. Jin, T. Fukushima, M. Niki, A. Kosaka, N. Ishii, T. Aida. *Proc. Nat. Acad. Sci. USA* **102,** 10801 (2005).
21. Y. Yamamoto, T. Fukushima, Y. Suna, N. Ishii, A. Saeki, S. Seki, S. Tagawa, M. Taniguchi, T. Kawai, T. Aida. *Science* **314,** 1761 (2006).
22. G. Zhang, W. Jin, T. Fukushima, A. Kosaka, N. Ishii, T. Aida. *J. Am. Chem. Soc.* **129,** 719 (2007).
23. J.-H. Ryu, D.-J. Hong, M. Lee. *Chem. Commun.,* 1043 (2008).
24. K. L. Prime, G. M. Whitesides. *J. Am. Chem. Soc.* **115,** 10714 (1993).
25. J. H. Ryu, E. Lee, Y. B. Lim, M. Lee. *J. Am. Chem. Soc.* **129,** 4808 (2007).
26. T. A. Skotheim, J. R. Reynolds (eds.) *Handbook of Conducting Polymers* (CRC Press, New York, 2007).
27. P. M. George, D. A. LaVan, J. A. Burdick, C. Y. Chen, E. Liang, R. Langer. *Adv. Mater.* **18,** 577 (2006).
28. S. C. Luo, E. M. Ali, N. C. Tansil, H. H. Yu, S. Gao, E. A. B. Kantchev, J. Y. Ying. *Langmuir* **24,** 8071 (2008).
29. A. Herland, O. Inganas. *Macromol. Rapid Commun.* **28,** 1703 (2007).
30. C. A. Thomas, K. Zong, P. Schottland, J. R. Reynolds. *Adv. Mater.* **12,** 222 (2000).
31. S. J. Paik, Y. Park, D. I. Cho. *J. Micromech. Microeng.* **13,** 373 (2003).
32. K. A. Ludwig, J. D. Uram, J. Y. Yang, D. C. Martin, D. R. Kipke. *J. Neural Eng.* **3,** 59 (2006).
33. H. Yamato, M. Ohwa, W. Wernet. *J. Electroanal. Chem.* **397,** 163 (1995).
34. Y. H. Xiao, X. Y. Cui, J. M. Hancock, M. B. Bouguettaya, J. R. Reynolds, D. C. Martin. *Sens. Actuators B, Chem.* **99,** 437 (2004).

35. J. Y. Wong, R. Langer, D. E. Ingber. *Proc. Nat. Acad. Sci. USA* **91,** 3201 (1994).
36. L. Dahlin, F. Johansson, C. Lindwall, M. Kanje. *Internat. Rev. Neurobiol.* **87,** 507 (2009).
37. C. E. Schmidt, V. R. Shastri, J. P. Vacanti, R. Langer. *Proc. Natl. Acad. Sci. USA* **94,** 8948 (1997).
38. J. Y. Lee, C. A. Bashur, A. S. Goldstein, C. E. Schmidt. *Biomaterials* **30,** 4325 (2009).
39. M. S. Strano, C. A. Dyke, M. L. Usrey, P. W. Barone, M. J. Allen, H. W. Shan, C. Kittrell, R. H. Hauge, J. M. Tour, R. E. Smalley. *Science* **301,** 1519 (2003).
40. M. S. Arnold, S. I. Stupp, M. C. Hersam. *Nano Lett.* **5,** 713 (2005).
41. N. Saito, Y. Usui, K. Aoki, N. Narita, M. Shimizu, K. Hara, N. Ogiwara, K. Nakamura, N. Ishigaki, H. Kato, S. Taruta, M. Endo. *Chem. Soc. Rev.* **38,** 1897 (2009).
42. M. P. Mattson, R. C. Haddon, A. M. Rao. *J. Mol. Neurosci.* **14,** 175 (2000).
43. E. Jan, J. L. Hendricks, V. Husaini, S. M. Richardson-Burns, A. Sereno, D. C. Martin, N. A. Kotov. *Nano Lett.* **9,** 4012 (2009).
44. K. Matsumoto, C. Sato, Y. Naka, A. Kitazawa, R. L. D. Whitby, N. Shimizu. *J. Biosci. Bioeng.* **103,** 216 (2007).
45. E. B. Malarkey, K. A. Fisher, E. Bekyarova, W. Liu, R. C. Haddon, V. Parpura. *Nano Lett.* **9,** 264 (2009).
46. E. K. Schillinger, E. Mena-Osteritz, J. Hentschel, H. G. Börner, P. Bäuerle. *Adv. Mater.* **21,** 1562 (2009).
47. D. A. Stone, L. Hsu, S. I. Stupp. *Soft Matter* **5,** 1990 (2009).
48. S. R. Diegelmann, J. M. Gorham, J. D. Tovar. *J. Am. Chem. Soc.* **130,** 13840 (2008).
49. M. Kasha, H. R. Rawls, M. Ashraf El-Bayoumi. *Pure Appl. Chem.* **11,** 371 (1965).
50. P. Mesquida, C. K. Riener, C. E. MacPhee, R. A. McKendry. *J. Mater. Sci. Mater. Med.* **18,** 1325 (2007).
51. V. Kayser, D. A. Turton, A. Aggeli, A. Beevers, G. D. Reid, G. S. Beddard. *J. Am. Chem. Soc.* **126,** 336 (2004).
52. O. Y. Kas, M. B. Charati, K. L. Kiick, M. E. Galvin. *Chem. Mater.* **18,** 4238 (2006).

5 Physics and Modeling of DNA-Derivative Architectures for Long-Wavelength Bio-Sensing

Alexei Bykhovski and Dwight Woolard

CONTENTS

5.1 INTRODUCTION

The physics and modeling of biological molecular components is discussed in the context of defining DNA-based biological molecule switches (BMSs) that will be useful for incorporating into larger DNA-based nanoscaffolds for the purposes of defining a novel class of smart materials for terahertz (THz) and/or very far-infrared (far-IR) based biological sensing. Here, theoretical research will be discussed that is being applied to define molecular-level functionality that will be useful for realizing THz/IR-sensitive materials. Synthetic DNA-derivative architectures with sensitivity to THz and/or Far-IR signals are of interest because many biological (and chemical) molecules possess unique spectral fingerprints in the far-IR and THz frequency (~3.0–0.3 millimeters) regimes. However, practical problems (e.g., weak signatures, limited number of discernable features, and sensitivity to environmental factors) often reduce the effectiveness of THz/IR spectroscopy in practice, and this motivates the development of novel smart material paradigms that can be used to extract nanoscale information (e.g., composition, dynamics, conformation) through electronic/photonic transformations to the macroscale. Hence, this research seeks to define DNA-derivative architectures with new spectral-based sensing modalities that will be useful for long-wavelength bio-sensing applications [1]. In addition, since these same smart material paradigms can be used to define antibody (i.e., structures that capture) and receptors (i.e., structures that capture and report) mimics, they also have important relevance to recognition-based detection and future medical applications (e.g., in developing synthetic vaccines) [2].

5.2 BIOLOGICAL-SEMICONDUCTING INTERFACES

The exploration of bio-organic device functionality and sensing in the future will require interfacing to traditional electronic materials and/or structures. An example of one such interface was recently considered in context of the resonant sensing of bio-molecules [3]. Resonant far-IR spectroscopy is a common technique for the characterization of biological (bio) molecules. The lower portion of the THz spectrum of DNA, RNA and proteins is also being actively studied using both experimental and computational methods. To date, some progress has been made in the detection and identification of bio-materials and interest is rapidly increasing across the scientific and technology communities. Most experimental and theoretical work has considered bulk quantities of bio-molecules in solution or as dehydrated powders or thin films. However, reliable and repeatable collection of the spectral signatures of DNA/RNA targets by conventional optical and quasi-optical techniques is hampered by effects (e.g., statistical, geometrical, environmental, etc.), which often alter and/or degrade the available sequence (i.e., structural) information. Hence, this issue motivates the search for alternative methods for extracting long-wavelength spectral information from single (or few) numbers of bio-molecules.

One potential alternative for the detection of ultra-low concentrations of bio-material was recently suggested [4]. In this sensor concept, a known single-strand of DNA is to be bonded to Silicon nano-probes and active THz illumination will be applied to induce phonon vibrations that are sensed by direct electron current measurements. In order to fully realize the potential of any type of ultra-sensitive DNA detection that involves contact probing, a realistic analysis of the DNA bonded to nano-scale substrate features is required. Currently, the bonding of DNA to semiconductor substrates is less understood than the dynamics and spectral characteristics of DNA in solution. In order to fill this gap, a first principle study of deOxyGuanosine (dG) residues chemically bonded via $(CH)_2$ linkers to the surface of hydrogen-terminated Silicon(111) nano-sized clusters was recently performed [3]. These investigations utilize differing types of Silicon surface structure models and quantum mechanical molecular descriptions were generated to accommodate one or two dG residues. In particular, a model for two different dG residues that form a single strand of DNA in the lateral direction (along the surface of a nanodot) was developed. The number of Silicon atoms included in the four-layer nano-size Silicon quantum dot (q-dot) structure that were considered varied from 49 (smallest) to 82 (largest). First principle simulations with valence electron basis and effective core potentials were conducted. Here, for each molecular system defined by a chosen DNA chain bonded to a particular Silicon nanodot, all-atom geometric optimizations without constraints were performed to determine the final conformations and normal mode analyses were applied to derive the spectral absorption information. More specifically, stable dG conformations on Silicon were obtained for varying types of DNA chain length and nanodot size/shape. These simulation studies provided new insights into the collective dynamics of the combined DNA chain and Silicon nanodot system and the resulting optically-active absorption modes within the THz and Far-IR regimes. Furthermore, these studies demonstrated that it is possible to identify THz spectral signatures of DNA chains chemical bonded to nanodots that are essentially insensitive to the specific geometrical characteristics of the nanodots. For example, Figure 5.1 provides the simulation results for two dG residues (dGG) bonded to Silicon q-dots that contain 82 (Si82) and 49 (Si49) atoms, respectively. Figure 5.1a illustrates snapshots of the conformation results (side and above views) for the two compound structures, with the larger dGG/Si82 on the left and the smaller dGG/Si49 on the right. Figure 5.1b compares the spectral absorption arising within the dGG/Si82 and dGG/Si49 structures below 6 THz. Here, one can easily identify absorption lines that are uniquely defined by the size of the q-dot (e.g., dGG/Si49 exhibits unique lines at 80 and 89 cm^{-1}) but there are also lines that persist independent of q-dot size (e.g., at 176 cm^{-1}). These results are particularly important in the context of the DNA-derivative architectures under discussion here for application to long-wavelength based biological sensing. Specifically, these results demonstrate the general feasibility of using highly complex DNA nanoscaffolds for the purpose of organizing target DNA-derivatives to enable an enhanced and extended capability for collecting the THz signatures associated with the target DNA molecules. A novel concept for using biological architectures to enhance and multiply the available THz spectral fingerprints in DNA molecules will be discussed in the next section.

5.3 NOVEL BIOLOGICAL ARCHITECTURAL CONCEPTS BASED UPON LIGHT-INDUCED SWITCHING

While long-wavelength spectroscopic analysis has been shown to have potential for the detection and characterization of biological materials and agents, the practical application of THz fingerprinting to macroscopic bio-samples is very difficult because the already very-weak spectral signatures can be masked and/or altered by structural/geometrical effects and by external environmental influences, and there are a relatively limited number of available spectral signatures (~<100) associated with any particular specimen in its natural (e.g., ground) state [5] These facts have motivated research into new types of biological architectures that can be used to actively control the conformational properties of target bio-molecules at the nanoscale, and to precisely extract information regarding their THz-frequency spectral absorption properties [7]. Here, the first goal is to define man-engineered nanoscaffolds that allow for the use of external stimulus (e.g., light-based excitation) to excite the target molecules into as many alternative metastable states as possible

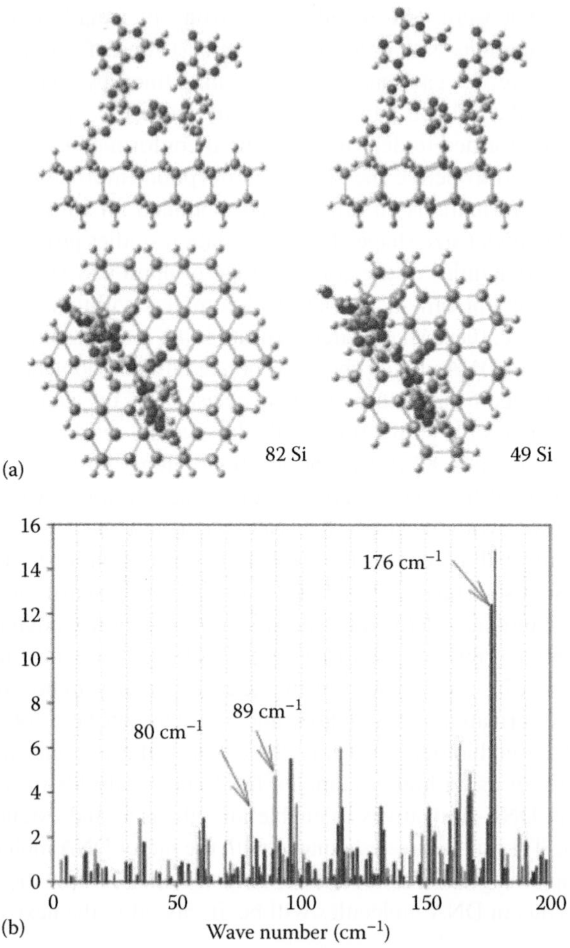

FIGURE 5.1 **(See color insert)** Simulation results for two compound structures dGG/Si82 and dGG/Si49. (a) Snapshots of the conformation for dGG/Si82 (left) and dGG/Si49 (right). (b) Spectral absorption for dGG/Si82 (black lines) and dGG/Si49 (red lines) below 200 cm^{-1} (6 THz).

(i.e., where they will exhibit multiple sets of THz fingerprints) as this will enable a type of multistate spectral sensing (i.e., with increased amounts of information for identification and characterization). The second goal is to use these nanoscaffolds to define highly organized systems (e.g., where the target molecules might be aligned as in a crystal) such that the absorption associated with the target molecules is significantly enhanced, and/or a novel electro-optical based transduction can be defined that allows for efficient extraction of the THz properties (e.g., dielectric response) of the target molecules. Here, it is also desired that these bio-architectures initially offer measurement apertures of sufficient size (~order of the spectral wavelength) to be amenable for traditional spectroscopic characterization, although if novel interfacing to the nanoscale could be defined some time in the future, then the sensitivity and discrimination capability would be increased even for very small nanoscaffold-based systems.

One novel bio-based architectural concept of the general type discussed above currently under consideration by our research group [1] involves the molecular engineering of THz/IR smart materials that can be used for biological (and chemical) threat agent sensing. Here, the specific goal is to define switchable molecular components that when incorporated into larger DNA-based nanoscaffolds lead to THz and/or IR regime electronic and/or photonic material properties that are dictated in a predictable manner by novel functionality paradigms. This research is considering both organic molecular switches (OMSs) and biological molecular switches (BMSs) that can be incorporated into DNA-based nanoscaffolds to define novel transduction of the spectral information related to the particular target molecules that are under consideration. Note that these smart material type paradigms are envisioned both for use in sensing an exposure event to some external molecular agent (i.e., either by binding or bonding interactions) or for use in extracting nanoscale information (e.g., composition, dynamics, conformation) related to select molecules that were used in the construction of the bio-architecture itself (e.g., in the case where the BMS was made up of genetic material).

These smart materials may be constructed through the use of designer DNA origami structures [6,7] and/or general motif DNA nanoscaffolds [8] that will allow for the incorporation of organized molecular functionality. Figure 5.2 provides an illustration of how DNA origami architectures could be applied to realize THz/IR smart material systems. First, Figure 5.2a illustrates the construction of a hypothetical DNA-based unit-cell that spans 3-D coordinate space. Here, each face of this DNA unit-cell is made up of a DNA origami cassette that has been designed with a window for incorporating a switchable molecule. Next, Figure 5.2b illustrates how these DNA unit-cells might be combined (e.g., through predefined bonding points on the origami structures or through the use of additional DNA motif structures) to arrive at large scale 2-D structures with a overall dielectric tensor that is strongly influenced by the switchable molecules that were incorporated into the windows. It is relevant to note that while these structures might be novel, they are definitely not fantasy, as active self-assembly research is being pursued at this time [6–8] that have already produced the real world constructs of this type. Here, the DNA cassettes shown in Figure 5.2c and the chain of DNA cassettes shown in Figure 5.2d represent the first nanoscaffold prototypes for these architectural concepts. These structural designs are of relevance to the discussion presented here because if one were to choose DNA-derivative as the BMSs, then it should be possible to extract information on the spectral absorption of the DNA based upon how this property modifies the electro-optical transmission (amplitude and phase) through the material. Furthermore, if DNA-derivatives are used that allow for being light-induced into many different metastable states, then the large amount of resulting spectral signature information would allow for full identification and characterization of the associated genetic sequence information. The sections that follow below will address the theoretical problem of defining and analyzing DNA-based BMSs (e.g., stilbene-DNA complexes) useful for these types of sensing applications.

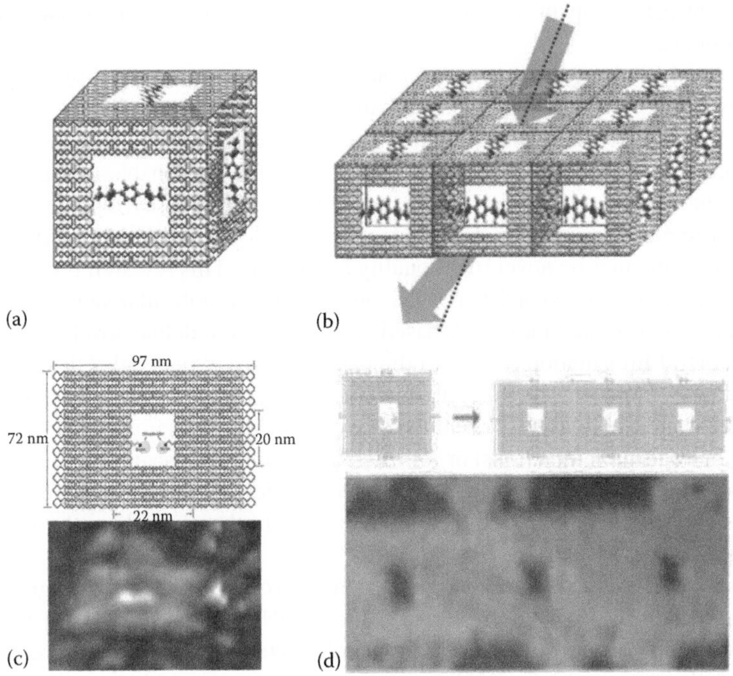

FIGURE 5.2 (See color insert) (a) DNA-based unit cell that spans 3-D coordinate space. (b) Smart material using the unit cell from (a) with measurable absorption/refraction. (c) An actual 97×72 nm DNA origami cassette design with biotinbiotin/streptavidin adapter molecules (upper) and AFM image of assembled structure (lower). (d) Linear array design (upper) using DNA cassette from (c) and AFM image of assembled array (lower).

5.4 HYBRID AB-INITIO/EMPERICAL MODELING OF STILBENE-DERIVATIVE/DNA COMPLEXES

5.4.1 Background

Both molecular mechanical (MM) and first principle (or ab-initio) quantum mechanical (QM) methodologies are actively used in nanoscale phenomenology and nano-structures related research that involve organic and biological (bio) molecules. In particular, simulations on molecular conformation, dynamics and associated interactions (i.e., including far-IR and THz spectral absorption and emission) usually require the application of ab-initio models. However, research into new nanotechnology concepts that seeks to identify novel functionality derived from nanoscale-dictated electronic, photonic and/or phonon processes usually involves modeling and simulation of very large and complex bio-organic molecular systems [3]. Even a single bio-molecule can consist of hundreds of thousands of atoms. Moreover, it is often necessary to consider the effects of local atomic interactions and chemical reactions. Essentially all quantum mechanical methods (i.e., including parameterized methods such as tight-binding) are usually not feasible for such complex problems since too many atoms are involved in the simulations and this same fact makes all first principle approaches prohibitively expensive. On the other hand, empirical molecular mechanics is unable to solve many important problems including those involving: changes in electronic states; chemical reactions; and breaking or formation of chemical bonds. Furthermore, it is not possible to adequately model the structure and dynamics of some bio-organic nano-structures with the required accuracies using predefined molecular mechanical force fields which only depend on the local atoms that are involved and which have effective charges that do not depend on molecular conformations.

Therefore, serious difficulties arise in the context of molecular-scale modeling and simulation when one seeks to explore new types of nanoscale devices and architectures that incorporate organic and/or bio-structures with complex geometries and/or functionality. Specifically, very high quality (accuracy, fidelity, etc.) in physical modeling and simulation is required to achieve a basic understanding of the fundamental mechanisms and for successfully investigating specific bio-electronic applications that rely upon organic/bio based molecular structure and/or functionality [5]. Furthermore, since these investigations will inherently require the interfacing of traditional electronic materials/devices with novel organic-and/or bio-based systems, the progression to practical sensors and sensor systems will present computational problems of ever increasing size and complexity. Hybrid approaches that combine different levels of theory such as QM and MM methodologies may help to bridge the gap in understanding electronic and atomic structure and light-induced interactions in complex bio-organic nano-structures. Therefore identification of novel functionality derived from nanoscale-dictated electronic, photonic and phonon processes requires simulation tools that execute a successful merger of the molecular mechanical and first principle quantum mechanical techniques. To this end, the feasibility of using hybrid theoretical approaches for a proper description of bio-organic nanoscale device structures needs to be explored.

Stilbene-functionalized bio-molecular structures have a potential to become future sensing elements in both single devices and integrated arrays. Previous studies have demonstrated that stilbene derivatives have potential uses in various applications and fields, in particular in biological nanoscale sensors. For example, a chemically reactive blue-fluorescent optical sensor based on stilbene derivatives was demonstrated that exhibits strong fluorescence in specific protein environments [9]. Another example can be found in synthetic DNA structures where the fluorescence quenching from stilbene derivatives was observed in the presence of G bases and modified Z bases [10]. In Ref. [11], hybrid QM/MM modeling and simulation techniques were explored for their utility in describing complex covalently bonded bio-molecular structures such as a double-stranded (ds) DNA fragment capped with trimethoxystilbene carboxamide (TMS), which is a well known stilbene derivative. Also, various single-stranded (ss) DNA fragments capped with TMS were studied. This broad study was conducted because both ss- and ds-based structures play important roles in bio-sensing. Since the stilbene-functionalized DNA molecules are viewed as prime candidates for integration into new types of nanoscale sensing architectures, the dependence of spectral absorption (i.e., in the terahertz (THz), infrared (IR) and ultraviolet (UV) regimes) on conformation and DNA sequence types were investigated to provide information on the potential use of active modalities (i.e., light-induced transitions) in detection and identification applications. In Ref. [11], for each molecular system defined by a chosen DNA chain bonded to TMS, all-atom geometric optimizations without constraints were performed to determine the final conformations and normal mode analyses were applied to derive the spectral absorption information. Time-dependent Hartree-Fock (TDHF) and density functional theory (TDDFT) were applied to derive electronic states and light-induced transitions for studied hybrid systems. For TMS-ss-DNA, hybrid semi-empirical QM/first-principle simulations were performed using the very well known AM1 [12] method (which determines various molecular energy terms from semiempirical expressions with parameters specifically tailored for the atomic components of DNA—i.e., H, C, O, P and N) to describe the DNA part of the molecule and the HF approximation to model the TMS section of the molecule bonded to one of the nucleotides. The simulation studies of Ref. [11] provided detailed insight into the electronic structure and dynamics of the combined DNA chain and TMS system and the resulting electronic transitions and optically-active absorption modes within the THz and Far-IR regimes. These studies demonstrated that it is possible to identify THz spectral signatures of specific DNA chains capped with TMS. This work also suggested that light-induced transitions could be a useful tool for the identification of the DNA sequence type. Furthermore, since the predicted two conformations of the TGCGCA duplex capped with TMS agree well with Nuclear Magnetic Resonance (NMR[1]) data [13], this research demonstrated that hybrid QM/MM methodologies can be a valuable theoretical tool for the study and design of novel sensor functionality based upon electronic, photonic and phonon processes in bio-organic structures.

5.4.2 Molecular-Mechanical Studies

While molecular mechanical (MM) modeling techniques have their obvious limitations (e.g., not able to provide excited state information), they are very useful for estimating the basic molecular conformation and for providing approximate insights into the static and dynamic phenomenology associated with both small and large molecular systems. Indeed, molecular mechanics is widely used in simulations of multi-atom/molecular systems that sometimes include hundreds of thousands of atoms. At the same time, it is often a prerequisite for generating input parameters needed in hybrid QM/MM simulations. MM simulations rely on molecular mechanical force fields that effectively describe covalent/Van-der-Waals/Coulomb interactions that are known to occur between individual bonded and non-bonded atom pairs. Here, specific force characteristics are assigned to the atoms that depend on the expected types of chemical bonds. Individual force types such as long-range electrostatic, van-der-Waals (VDW) forces, covalent bond stretch, bond angles and torsions are possible contributors to the overall effective force field. The force field models offered by the Amber simulation package are proven to perform well in describing biological systems that include nucleic acids and proteins [14–18]. Here, it is important to note that the accurate specification of the effective MM force is highly dependent on the overall molecular system. For example, the use of Amber to model general organic molecules (i.e., other than carbohydrates) requires the user to specify additional force field parameters to obtain accurate results. Also excluded from the Amber library are many "nonstandard" nucleic acid structures, such as nucleotides with the phosphate group protonated. Despite these general limitations, the Amber force field model is highly applicable to the TMS-DNA molecules under discussion here, and Amber-generated results established an important physical foundation for the hybrid QM/MM studies of the next section [11].

5.4.3 Hybrid QM/MM Studies

Results from prior research [11] has clearly illustrated the strong influence of the TMS molecule on the resulting conformations, spectral absorption and stability behavior of compound TMS-DNA type structures. Indeed, the previous MM-based studies demonstrated that when stilbene derivatives are bound to ss-DNA or ds-DNA that there are identifiable structural characteristics of the stilbene derivatives that dictate many properties of the entire molecule. Hence, when one views these molecular mechanisms from a classical point-of-view where certain semi-isolated components (or segments) of the compound structure are known to determine the overall properties, it naturally defines a hybridization of the physical modeling problem. More specifically, this type of situation suggests that when one is interested in predicting the conformation or dynamics of the compound molecule that it should be physically acceptable to invoke a structural partitioning of the problem. In particular, one can apply a hybrid QM/MM modeling approach where the "passive" part (i.e., primarily receiving the influence) is characterized by MM techniques that are incapable of predicting electronic effects, and the "active" part (i.e., primarily dictating the response) is treated by a first principles QM-based approach. Such hybrid modeling approach allows one to study both the electronic structure and atomic motions of TMS-DNA conjugates.

A very simple example of a hybrid partitioning for TMS that is relevant to the problem under discussion is shown in Figure 5.3. Here, consider the task of modeling two conformations of the TMS molecule where the only change is produced by a 180 degree flip (see Figure 5.3a) of the top carbon ring relative to the bottom carbon ring. More specifically, in this example a double-bond originally pointing to the left (see left-side of Figure 5.3a) flips, or rotates about the next lower bond, such that it ends up pointing to the right (see right-side of Figure 5.3a). In this type of situation, it may be useful to execute a partitioning (see Figure 5.3b) where the portion of the molecule above the rotating bond (i.e., now shown on the right-side of Figure 5.3b) is to be treated primarily by a QM approach, and the portion of the molecule below the rotating bond (i.e., now shown on the left-side of Figure 5.3b) is to be treated primarily by a MM technique. Note the best choices for the QM and MM components will depend on which part of the molecule is primarily responsible for influencing the conformational change, and how to make this choice (or guess) will become more apparent when the entire TMS-DNA problem is discussed below. Also, since the hybridized simulation method will require an independent assessment for how the QM component (i.e., represented by spheres in Figure 5.3b) contributes to the total molecular potential energy (see discussion below), it will be necessary to calculate both the MM and QM solutions for this molecular partition where the broken bond on the left end is saturated by a hydrogen atom (see Figure 5.3c).

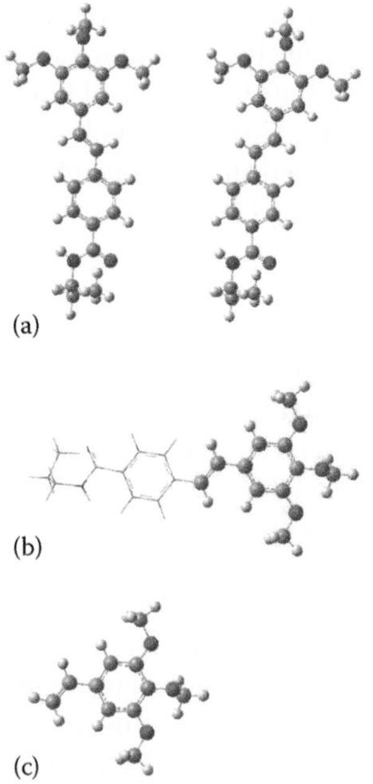

FIGURE 5.3 **(See color insert)** An example partitioning of the TMS molecule. (a) Two conformations of TMS that differ only by a 180 degree flip (i.e., compare left conformation to the right conformation) of the upper carbon ring relative to lower carbon ring in the chain. (b) A partitioning about the rotation bond in (a) where the upper portion (now shown on right) is treated primarily by a QM approach, and the lower portion (now shown on left) is treated primarily by a MM technique. (c) QM molecular partition (i.e., upper part from (a) and right part from (b)) where the interface bond has been saturated by hydrogen to enable both MM and QM simulations on this portion of the TMS molecule.

In the hybrid QM/MM computer simulations that are utilized for these studies, the active site (i.e., the TMS molecule combined with some number of bases) which represents a small portion of the entire interacting molecule region is considered on a QM level of theory using Hartree Fock (HF) or Density Functional Theory (DFT) methods. The excited states are obtained using time-dependent HF or DFT. The remainder of the system (i.e., a fragment of ss- or ds-DNA) is described using MM force fields or semi-empirical quantum mechanics. In studying these TMS-DNA conjugate structures, a partitioning method (i.e., called ONIOM) is utilized that followed the procedure described in [18]. A Gaussian-based approach is then used to calculate the optimized geometries and associated energies. This procedure involves calculating the energy of the combined structure at a lower level of theory and modeling of the active site using both higher and lower levels of theory. Here, the difference of the two active site calculations is combined with the results from the entire structure to determine a consistent overall energy balance (i.e., MM result for the active site is subtracted from the MM result for the combined structure and then added to the QM result for the active site). Hence, the recognized active TMS component is characterized by the more accurate QM theory and the entire structure (which includes the passive part) is first treated by the more computationally efficient MM theory and then double-counting of the active part is removed by subtracting the results from the MM treatment of the TMS component. Note that a partitioning across covalent bonds is required so link-atoms must be used to saturate the bonds that were broken to define the smaller active site molecule (e.g., analogous to what was done in Figure 5.3c). Hydrogen was used as a link atom in all simulations. In the studies presented here, it is reasonable to assume that only valence electrons participate in the forming of all chemical bonds. The Stevens/Basch/Krauss effective core potential (ECP) split-valence basis [19, 20] was used in both the HF and DFT simulations. In addition, to account for electron-electron interactions a long-range and van-der-Waals corrected hybrid DFT functional was applied [21,22].

Hybrid DFT/MM studies on the symmetric (TMS-TGCGCA)2,where each identical DNA strand is bonded to TMS, may be used to demonstrate that at least two stable conformations should exist (see Figure 5.4). The initial simulations utilized a partitioning where the active site includes only TMS and the first base (i.e., the active site molecule is TMS-T) because TMS has a structure similar to the molecule discussed in Figure 5.3 and therefore can be expected to exhibit a similar ring flip that should be somewhat independent of the base-pair. The common features of the two conformations include the proximity between the two -OCH3 groups (methoxy groups) of the stilbene derivative and the -CH2-group (methylene group) of the deoxynucleoside of the complementary DNA chain (see Figure 5.2a and b). As was expected, the two predicted conformations are distinguished primarily by 180 degree TMS ring-structure flips (see Figure 5.4c and d). Here, the conformation with the left-handed ring flip (i.e., Figure 5.4c) has a slightly lower energy. Previously, this TGCGCA duplex capped with TMS was fabricated and studied experimentally using NMR [13]. The authors concluded that the two sets of signals they observed in ^{1}H NMR spectrum of the structure are resulting from the two different orientations of the stilbene relative to the neighboring nucleobases. Structural details deduced by the authors of Ref. [13] included a 180° flip of the stilbene ring system between the major and minor conformations. Therefore, the hybrid DFT/MM modeling results presented in [11] agree well with NMR-based deductions. The switching between the two conformations was explored by freezing the torsion that is responsible for the ring flip, and then by optimizing the structure over the rest of internal coordinates. A transient structure that corresponds to a 90 degree (approximately half-way) turn in the torsion (within QM region) is plotted in Figure 5.5a. Notice that the proximity between the tail methoxy groups of TMS and the complementary DNA strand is retained, and the T base of the first DNA strand is pushed back to allow for the transition. The optimized energy versus torsion (within the QM region) for TDDFT/MM predicted structures of Figure 5.4 is plotted in Figure 5.5b. The minima corresponding to conformations I and II are highlighted with black arrows. Only the ground energy level and first three excited states are plotted for the QM part of the system. The predicted barrier for flipping is low, so the transitions can by activated by thermal effects.

In the Figure 5.6, the predicted vibrational absorption spectra of the electronic ground state of the symmetric $(TMS\text{-}TGCGCA)_2$ structure are plotted for the two conformations. The entire spectrum as given in Figure 5.6a demonstrates many

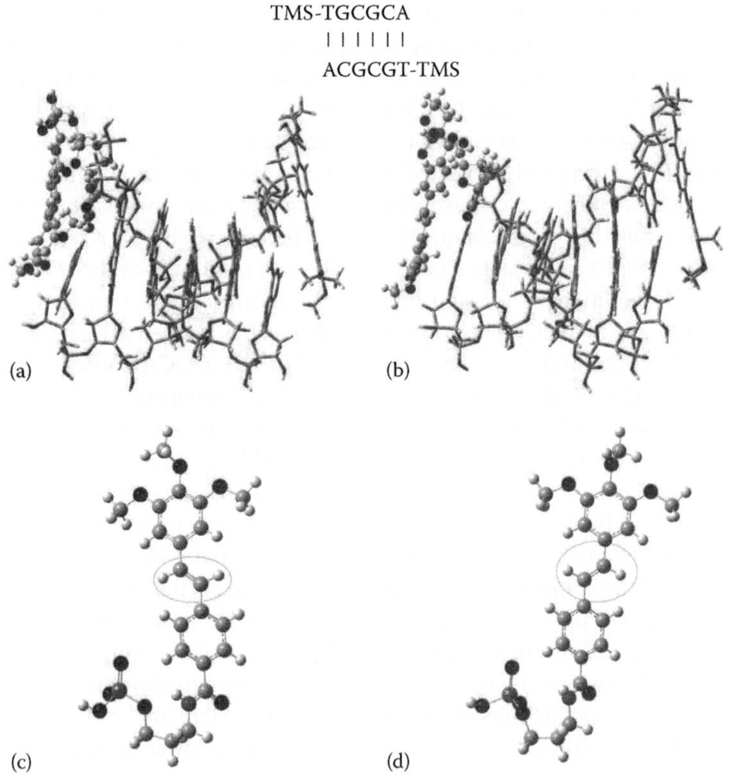

FIGURE 5.4 (See color insert) (Top) Two DFT/MM predicted conformations of the double-stranded TMS-DNA structure $(\text{TMS-TGCGCA})_2$. (a) conformation I, (b) conformation II. For each conformation, color-coded spheres represent the portion of a structure simulated with the DFT model and color-coded tubes represent the portion of a structure described only by the MM model. (Botton) Isolated views of the TMS components from the two conformations of $(\text{TMS-TGCGCA})_2$ given in (a) and (b). Here the two TMS substructures differ by a 180 degree flip of the end carbon ring (i.e., the upper ring in the illustration above) relative to the next carbon ring in the chain (i.e., the lower ring in the illustration above) where: (c) illustrates the ring flip to the left associated with the molecule from (a); and (d) illustrates the ring flip to the right associated with the molecule from (b).

similarities for both conformations. However, when one carefully considers subregions of the THz regime, distinctly unique spectral lines can be observed near ~67, 82, 90 and 100 cm^{-1} (see Figure 5.6b and c) and also near 280 cm^{-1} as discussed in Ref. [11]. These absorption features can be attributed to vibrations in torsions and bond angles. Finally, spectral signatures that are unique to one of the two conformations can be identified near 10, 12, 27, 33, 39, 54, 58 and 65, 68 cm^{-1} (see Figure 5.6b) with all of these residing in the low THz (collective motion) regime. In addition, the results for the six lowest electronic transitions (i.e., with state labeling 0, 1, … 6) that are associated with the two conformations (I, II) along with the corresponding oscillator strengths were given in Reference [11]. The predicted 0-6 transition is 0.13 eV

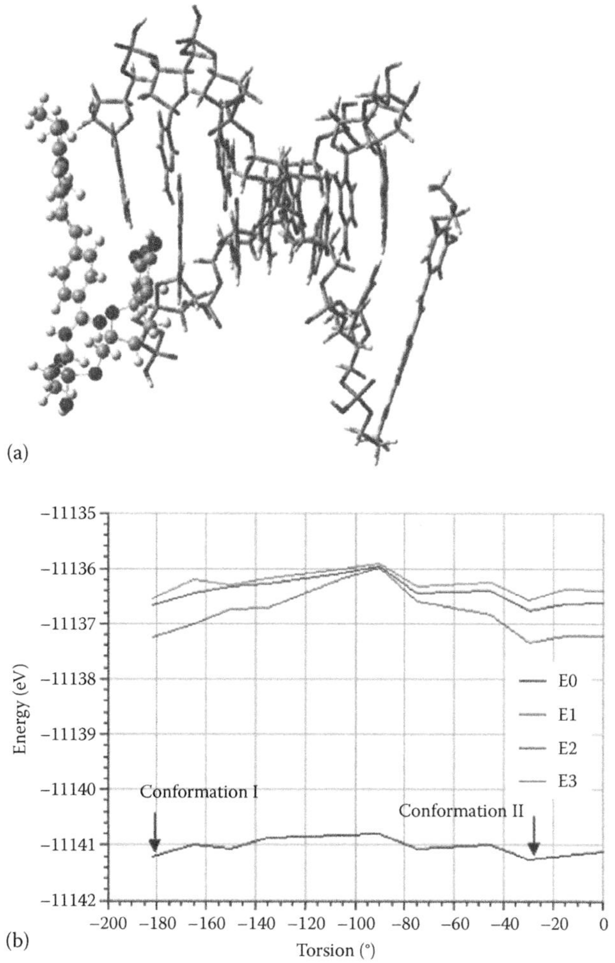

FIGURE 5.5 **(See color insert)** (a) Transient DFT/MM predicted structure that corresponds to a 90 degree (approximately half-way) turn in the torsion angle (within QM region) that is responsible for switching between conformations I and II of Fig. 2. The proximity between tail oxygen groups of TMS and the second (non-bonded to TMS) DNA strand is retained. Also, the T base of the first DNA strand is pushed back to allow for the transition. (b) Optimized energy versus torsion (within QM region) for TDDFT/MM predicted structures of Figs. 2. Minima corresponding to conformations I and II are highlighted with black arrows. Ground level and first three excited levels are plotted for the QM part of the system.

higher and substantially stronger for conformation II as compared to conformation I. The predicted 0-1 transition is also 19% stronger for II type conformation. Hence, both the spectral results and transition characteristics show potential for discriminating between the two conformations. Both conformations depicted in Figure 5.4 correspond to trans TMS. However, a cistrans conformation of (TMS-TGCGCA)2 was also predicted by a DFT/MM analysis. For this case one TMS is in cis conformation

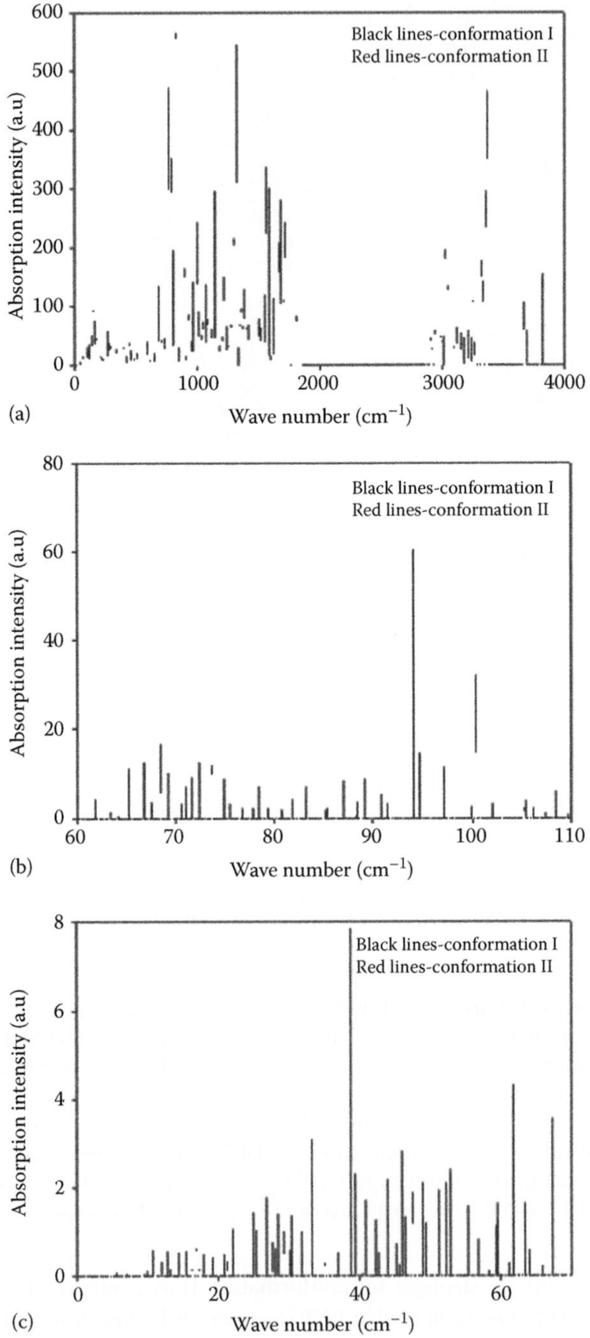

FIGURE 5.6 (a) Entire absorption spectra of the two conformations shown in Fig. 4. (b) Spectra for frequency range 2.0-3.5 THz, (c) Spectra for frequency range below 2.1 THz.

(i.e., top TMS molecule perpendicular to DNA strand) and another one is in trans (i.e., bottom TMS molecule is aligned to DNA strand). There also exists an all-cis conformation for this molecule with a slightly lower energy.

Additional studies were also performed to investigate single TMS capping of ds-DNA, i.e. TMS-$(TGCGCA)_2$. Since no experimental data were available to infer an approximate stable conformation as in the last case considered, initial simulations were performed using the less physically accurate but much more computationally efficient hybrid HF/MM model to obtain the two conformations given in Figure 5.7. Here, the active site for these simulations consists of the TMS molecule bound to a single T base, or TMS-T. The results in Figure 5.7a and b show that the major difference in the two conformations is related to the orientation of the TMS molecule with the ds-DNA, i.e., the result 7(a) has TMS aligned with the end base-pair T-A and the result in 7(b) has TMS aligned with the ds-DNA backbone. The accuracy of both results were confirmed by the DFT/MM model using the same active-site partition and an expanded active-site partition that included the first three bases adjacent to the binding site of the TMS (i.e., TMS-TGC). These results demonstrate conformations that maintain the same basic qualitative orientation between the TMS and ds-DNA molecules. Simulations that sampled a number of the available coordinates belonging to the TMS molecule revealed that changes to the torsion coordinate associated with the chemical bond between the TMS molecule and the DNA might be useful in defining a energy-space pathway between the two refined minimum energy conformations that possess the general geometries as illustrated in Figure 5.7a and b. The energy diagram in Figure 5.7c depicts this energy-space pathway for the case of the ground-state and a few excited-state conformations that were obtained from hybrid DFT/MM based simulations using an active-site partition of TMS-T. In particular, the energy profile for the 0th, 1st, 2nd and 9th electronic states are given as a function of the torsion angle that connects TMS to the DNA. As is clearly shown, all the energy profiles along the single torsion coordinate variation between the two stable conformations contain a significant barrier. This means that changes in other coordinate spaces will need to be accessed in order to realize light-induced switching between the positions of the two stable conformations (i.e., around −75 and −185 degrees) which have been highlighted with arrows in Figure 5.7c. Note that determining energy-space pathways for even moderately complicated molecules can be a significant challenge, and active work by our group on developing computationally methods for addressing this problem will be discussed in Section IV. While a light-based switching methodology is lacking at this point, TMS-$(TGCGCA)_2$ is an attractive BMS candidate because the spectral absorption results for these two stable conformations for the IR-to-THz regime (Figure 5.7d) and for the THz regime (Figure 5.7e) both show substantial differences in their individual spectral fingerprints [11]. Note that line broadening at THz frequencies is typically on the order of 0.5 cm^{-1} [21], therefore most of these THz absorption lines can be detected. Hence, there appears to be significant opportunity for utilizing the spectral absorption signatures associated with the multiple conformational states of stilbene-DNA conjugates to infer the base-pair sequences of DNA fragments.

These collective results demonstrate that first principle and DFT hybrid QM/MM methodologies can bridge a gap in understanding electronic and atomic structure

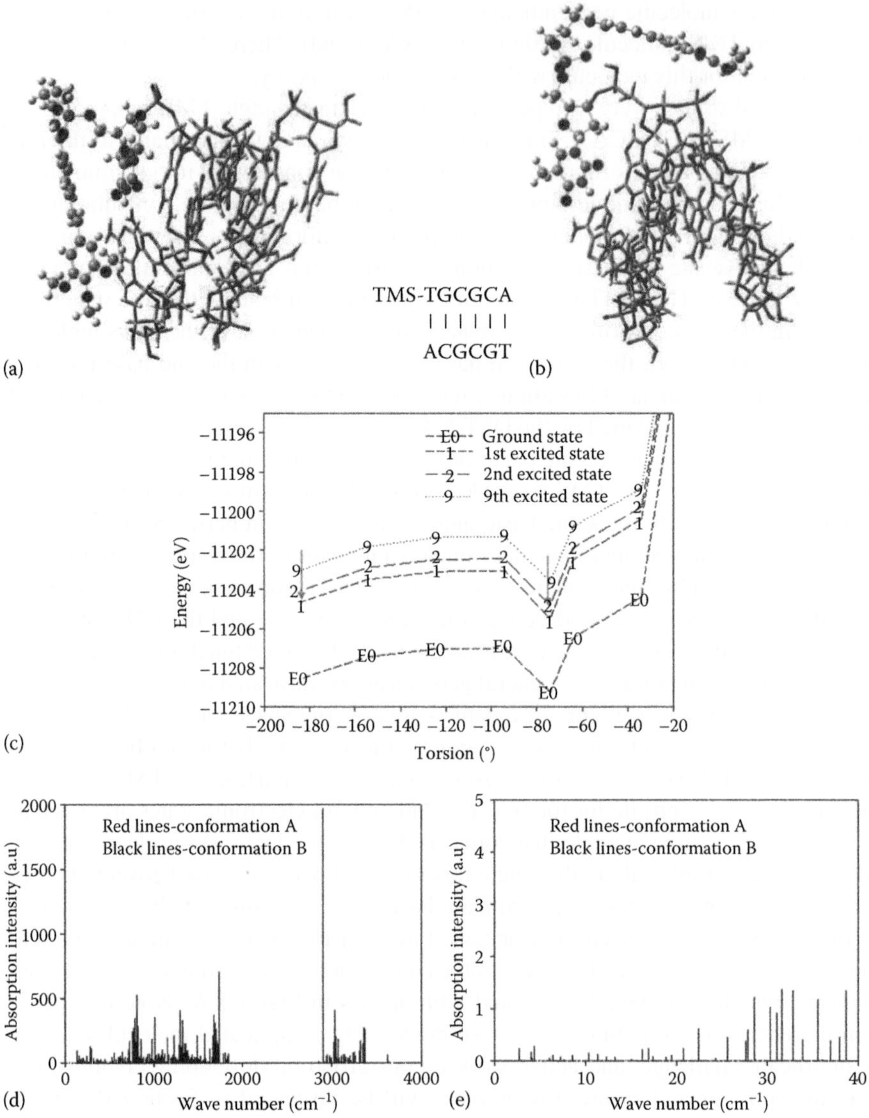

FIGURE 5.7 **(See color insert)** (Top) Two predicted conformations for double-stranded DNA fragment TGCGCA capped on one end with TMS obtained from hybrid HF/MM. (a) TMS chain is aligned along the end base pair T-A, and (b) TMS chain is aligned along the ds-DNA backbone. (Middle) Energy-scan profiles (from hybrid DFT/MM model) for the 0th, 1st, 2nd and 9th states of the TMS-TGCGCA molecule versus torsion angle. (c) The two stable conformations are highlighted with arrows where (a) is on left and (b) is on right. (Bottom) The DFT/MM predicted absorption spectra for each conformation is also shown for (d) the THz to IR regime, and (e) THz regime.

and light-induced interactions in complex bio-organic systems. In particular, these simulation techniques were used to successfully derive two distinct low-energy conformations of (TMS-TGCGCA)$_2$ that differ in the orientation of the stilbene ring system relative to the terminal base pair, and which are in agreement with experimental NMR data. In addition, the approach was able to predict the all-cis and cis-trans conformations, along with two distinct conformations for the asymmetric structure with only one TMS bonded to the TGCGCA duplex, i.e., TMS-(TGCGCA)2. Both base pair composition and conformation differences are predicted to impact vibrational absorption spectra especially in the terahertz-regime and to manifest themselves in conformation-dependent and sequence-dependent shifts in electronic transitions. Hence, this work suggests that stilbene derivatives may be useful for synthesizing DNA conjugates that facilitate the spectral characterization of the associated base pair sequences. More specifically, this research suggests that stilbene-DNA conjugates offer the type of switchable spectral characteristics that will be useful for THz-based and possibly IR-based detection and identification purposes. These facts are particularly important when one considers the development of advanced biologically-based architectures that use light-induced interactions and the resulting THz-frequency spectral signatures to extract information about the molecular structure and dynamics as discussed in [5]. The next section will discuss how these same modeling techniques can be used to facilitate the independent experimental analysis of component molecules (e.g., biotin) that will be required for the successful construction of novel biological architectures of the type discussed in Section III. Section IV will then conclude with a discussion of novel algorithm development for deriving light-induced energy-space pathways.

5.5 CHARACTERIZATION OF ACTIVE MOLECULES IN POLYETHYLENE (PE) MATRICES

As discussed earlier, the self-assembly of novel biological architectures for use in long-wavelength spectral sensing application will require the incorporation of active molecules into nanoscaffolds (e.g., the DNA cassettes shown in Figure 5.2) which will in turn be used to create the larger molecular matrices and supramolecular assemblies. These installation procedures for incorporating the functional polymeric components (e.g., stilbene-DNA BMSs) will require the use of other linker molecules (e.g., biotin/streptavidin) to facilitate the molecular construction. Hence, it will be very important to understand the dynamics of these individual molecular components and their subsequent influence on the larger nanostructures [3]. Biotin is an example of an important active molecule that is expected to play a key role in the molecular assembly of DNA-based nanostructures. The characterization of biotin is especially important since it is known to be optically active that therefore could strongly influence the functionality of the molecular switches to be used in DNA-based nanostructures, for example, in DNA origami. Therefore, it is very important to understand the fundamental dynamics of isolated biotin, as well as molecular compounds that contain biotin. Fortunately, materials such as polyethylene (PE) form a non-bonded (hydrogen-bonded) material matrix that is transparent to long-wavelength radiation and as such they are an important experimental tool for understanding the molecular dynamics of quasi-isolated molecules that are embedded into these matrices if one is able to accurately predict the interactions between the embedded molecules and the PE superstructure. Therefore, it will be very useful to explore the impact of the polyethylene (PE) matrix embedment on the THz spectra of these types of active biological molecules.

In very recent work, polyethylene was chosen as a matrix for the investigation of active biological or organic molecules. This material is weakly absorbing at THz frequencies and is widely used in spectroscopy [23]. It consists of (C_2H_4) polymer chains that are arranged in a crystal lattice [21]. The quality of polyethylene depends on processing. This study focuses on high-density polyethylene (HDPE) that has little branching. In the case of branching, main chains have minor chains bonded to them in places where a normal CH2 group is replaced by a CH thus creating one extra valence. Therefore, the HDPE structure is close to an ideal crystal lattice of the polyethylene (PE). Several atomic structures of polyethylene slabs suitable for incorporating in hybrid abinitio/MM structural models were simulated. In all structural models, orthorhombic PE lattice symmetry was used to generate a starting atomic arrangement. Polyethylene II was modeled which means that all polymer chains were terminated with methyl end groups (-CH3 or -Me). Also, no lattice defects were taken into account with the exception of defects that were related to an active molecule embedding. Previously, the vibrations and light absorption in PE lattice were obtained theoretically using specially developed MM force fields [23, 25]. In this work, all-purpose force fields such as Universal force field (UFF) and Amber's force field were employed in the MM portions of simulations. The THz-to-IR absorption spectra of polyethylene that were obtained with DFT/MM approach for the structural models of PE are in a good agreement with observed spectral lines that were reported in [23].

Biotin is a flexible biologically active molecule with a number of distinct low-THz spectral features [26]. It is a vitamin coenzyme and it functions as a bio-reactions catalyst, and the bio-activity of biotin is utilized for bio-chemical sensing [27]. Hartree-Fock (HF) ab-initio and DFT simulations of biotin were performed both in harmonic approximation and with corrections for anharmonicity. Specifically, 3rd and 4th order anharmonicity corrections in atomic displacements to harmonic vibrational states were calculated numerically for biotin within the first principles theoretical framework by the perturbation theory [28]. In Figure 5.8a, results for integrated line intensities that were obtained with HF/6-311G(2d,d,p) are plotted for low-THz together with observed low-temperature (4.2K) absorption peak positions from Reference [26]. Predicted lines in the harmonic approximation (solid red lines in Figure 5.8a) are in a good agreement with observed ones, in particular, a strongest line is close to 43 cm^{-1} (1.3 THz), 2nd strongest is at ~94 cm^{-1} (2.8 THz) followed by lines near 33 cm^{-1} (1 THz) and near 20 cm^{-1} (0.53 THz). But not all observed lines can be explained, since there are more of them (see black dots in Figure 5.8a), than the theory predicts. The inclusion of anharmonicity corrections (blue dashed lines in Figure 5.8a) does not introduce significant changes in line positions; as with or without anharmonicity, more lines are observed than predicted.

To explore this issue further, first-principle molecular dynamical (MD) simulations of biotin were performed using the Atom Centered Density Matrix Propagation

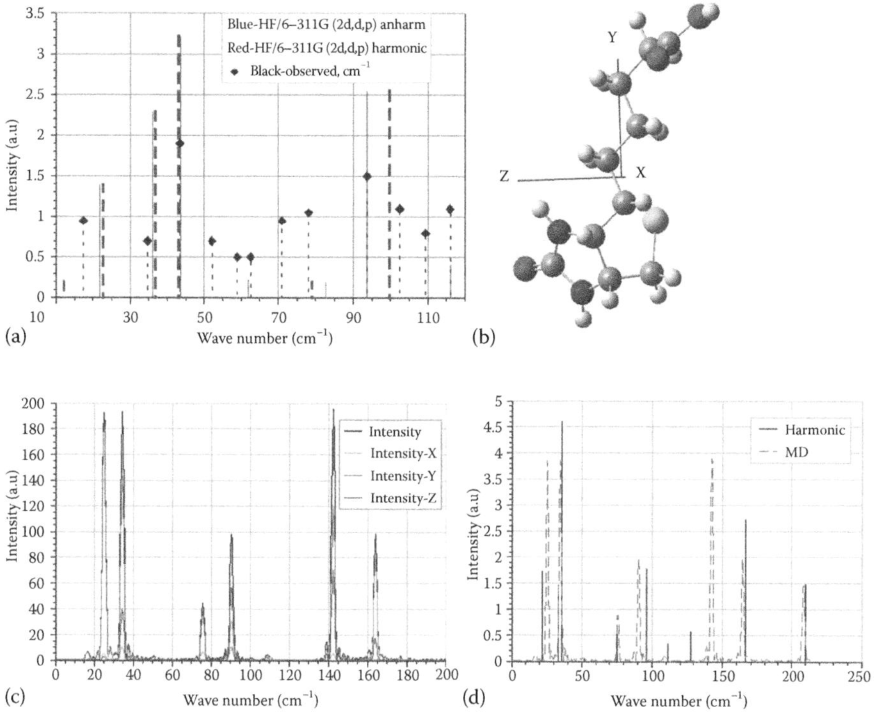

FIGURE 5.8 **(See color insert)** (a) THz absorption spectrum of biotin: harmonic (red lines); anharmonic (blue dashed lines); and, Experimental observations at 4.2K reported in [23] (black dots). Theory used HF/6-311G(2d,d,p) [25]. (b) A snapshot of biotin from the MD simulations. (c) Low-THz absorption spectra of biotin obtained from the DFT-based MD simulation. The entire spectrum (black line) and contributions from Cartesian components of dipole moment (green, red and blue lines respectively). (d) Low-THz absorption spectra of biotin obtained from the DFT-based MD simulation (red dashed line) and from the DFT-based energy minimization in harmonic approximation with the same density functional and basis (black lines).

molecular dynamics model (ADMP) that provides good energy conservation [29]. The Stevens/Basch/Krauss effective core potential (ECP) split-valence basis [30] was used in these MD simulations. In addition, to account for electron-electron interactions, a long-range and van-der-Waals corrected hybrid DFT functional was applied [22]. In the ADMP approach, the basis functions may be considered as "traveling" along with the classical nuclear coordinates. As a result of the ADMP simulations, the atomic trajectories, velocities and a low-temperature (~4K) dipole moment time history of biotin were obtained for about 15 picoseconds with a 1 femto-second time step. At each step, the dipole moment was computed quantum-mechanically using the wave functions that depend from atomic positions, so that both the electrical and mechanical anharmonicities were taken into account. The energy was well conserved during the course of simulations. The absorption spectrum was computed within the framework of linear response theory using the relationship between the equilibrium dipole moment fluctuations and the light absorption cross-section [31].

In this approach, the total absorption intensity is proportional to a power spectrum that is represented by a sum of squared Fourier transforms of Cartesian components of dipole moment history. In Figure 5.8b, the simulated biotin molecule is shown with Cartesian axes. In Figure 5.8c, the total intensity (black lines) is plotted along with its Cartesian contributions (green, red, blue) for the lower THz portion of the spectrum (0–6 THz or 0–200 cm^{-1}). Notice that the most significant contributions come from the Y-component of the dipole moment (red) for major lines around 30–35 cm^{-1} and 140 cm^{-1}, which corresponds to a direction along the carbon chain (see Figure 5.8b). In Figure 5.8d, the low THz spectrum of total intensity (red dash) from Figure 5.8c is plotted against the absorption lines (black) that were calculated for the optimized biotin structure in harmonic approximation using the same DFT potential and basis as in the ADMP approach. This comparison clearly demonstrates that the harmonic spectrum yields an overall reasonable THz fingerprint of biotin. Most major lines are well reproduced with a notable exception of the ADMP result at 143 cm^{-1}. Here, harmonic approximation yields a line that is shifted with respect to the ADMP result to lower frequencies at 128 cm^{-1}. Therefore, the ADMP simulation confirms the harmonic result and also predicts that there are only 5 absorption lines below 150 cm^{-1}.

Notice that the measured biotin was embedded in HDPE [26]. In order to study theoretically the effect of embedding, the QM/MM approach was used. Specifically, *first principle* QM methods and DFT were combined with MM empirical descriptions in order to perform large-scale atomistic spectral modeling of biotin in PE in a broad frequency range from THz to UV. The most important part of the system (i.e. biotin in a cavity) is described using first principles QM or DFT. In this case, PE strands were either in MM-only region or in a QM region without the introduction of any link atoms between MM and QM parts of the system [28]. Using this approach, the energy was calculated and geometry optimizations were performed for developed structural models to minimize the energy. Then atomic vibrations and absorption spectra were obtained in harmonic approximation. Excited electronic states were studied using time-dependent (TD) DFT. The two-step optimization process consisted of a MM step that optimized biotin's placement inside a PE matrix and a hybrid optimization step. In the second step, partitioning of a structure in a QM and MM-only regions (layers) was performed and equilibrium coordinates for the structure were obtained. Atomic vibrations were then analyzed in the harmonic approximation for the equilibrium structure and line intensities were calculated. In Figure 5.9, an example of a studied structure consisting of a biotin molecule (spheres) embedded in a polyethylene matrix (spheres and tubes) is shown. Here, tubes represent a MM-only region. In Figure 5.9a the QM region is extended to include surrounding PE strands. In this particular case, the MM region was immobilized so that only *ab-initio* atoms were allowed to move during a geometry optimization. In Figure 5.9b, a model with biotin-only QM region is shown. In Figure 5.9c, an optimization result for PE with cavity without a biotin molecule is presented.

The resulting low-THz absorption spectrum for the entire structure is shown in Figure 5.9d. Here, blue lines represent the theory, b3lyp/UFF, for model of Figure 5.9a, red lines the model of Figure 5.9b and black dots represent centers of observed lines at 4.2K as reported in [26]. The Stevens/Basch/Krauss effective core

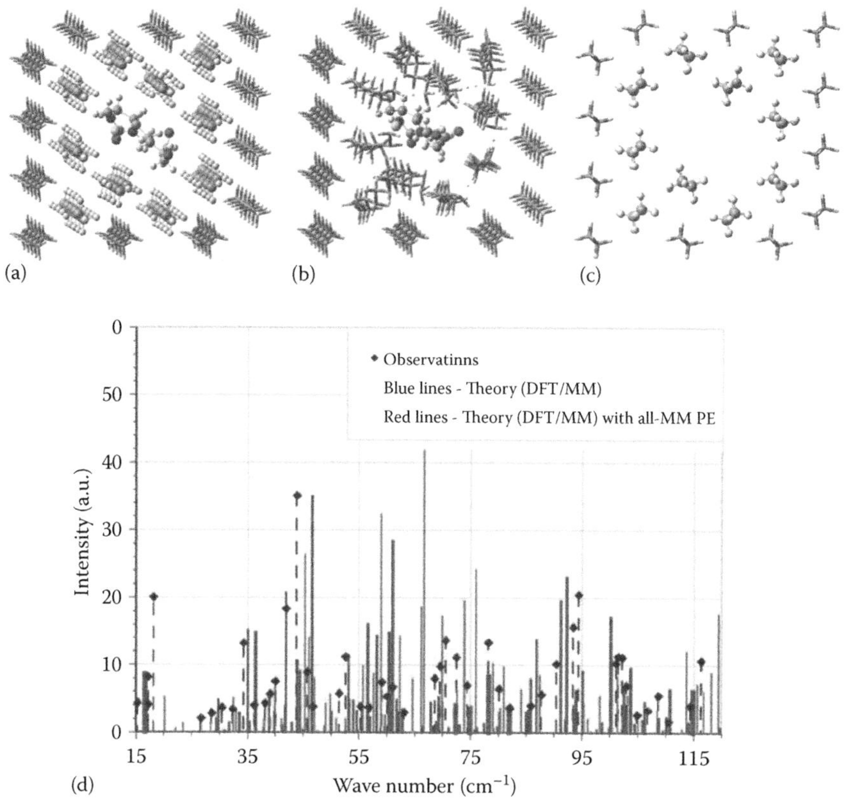

FIGURE 5.9 (See color insert) Simulation results for Biotin in a cavity inside a polyethylene matrix Test cases include: (a) Extended QM region; (b) Biotin-only QM region; and, (c) PE with cavity. Spheres represent atoms in QM region and tubes represent atoms in MM region. (d) Low-THz absorption spectrum of the biotin-PE structures. Theory for model of Fig. 18(a) (blue lines); theory for model of Fig. 18(b) (red lines); and experimental observations at 4.2K reported in [23] (black dots).

potential (ECP) split-valence basis was used in this simulation [21]. Additional low-THz lines involve atomic vibrations in biotin and PE both and reflect THz coupling between biotin and its PE cavity. While line widths were not calculated, the theoretical spectrum of biotin in HDPE obtained from the extended QM model agree well with those observed experimentally at 4.2K within 15–120 cm^{-1} range.

In Figure 5.10a, predicted QM/MM THz absorption spectra below 7 THz are plotted for the case of biotin in PE that is associated with the model of Figure 5.9a (see blue lines) and for the case of PE with a cavity (i.e., no biotin) that is associated with the model of Figure 5.9c (see black lines). Here, one can observe a major contribution of biotin to the overall THz absorption. In Figure 5.10b, the predicted broad spectrum THz-to-IR resonance absorption of biotin in PE (blue lines) and an isolated biotin molecule (red dashed lines) are plotted. Notice that major lines in the spectra are almost the same with exception for C-H stretch

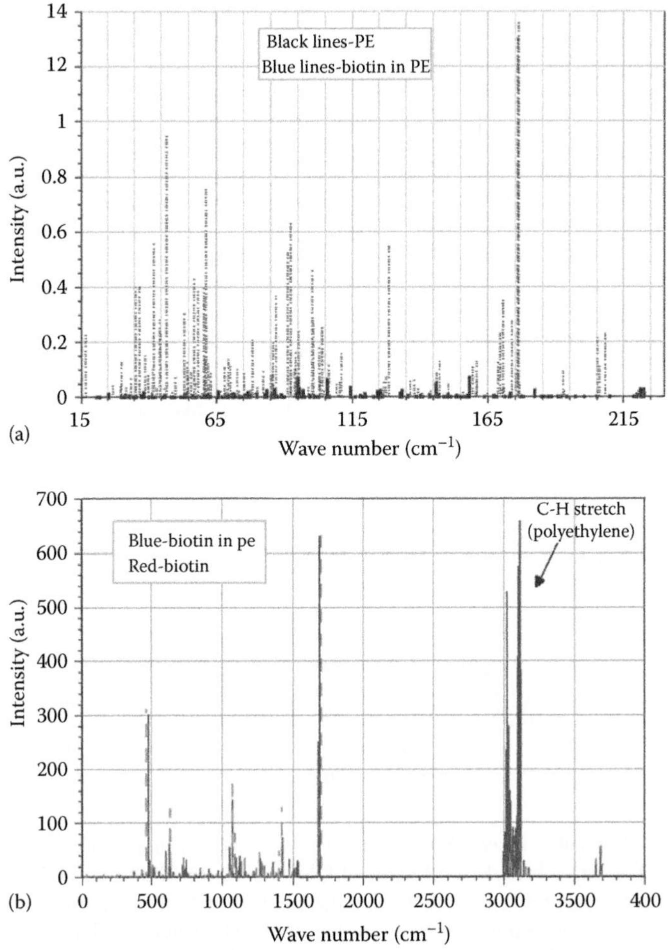

FIGURE 5.10 **(See color insert)** (a) Predicted THz absorption spectra below 7 THz. Biotin in PE as shown in Fig. 9(a) (blue lines) and PE with a cavity as shown in Fig. 9(c) (black lines). (b) Predicted resonance absorption spectra of biotin in PE (blue lines) and single molecule biotin (red dashed lines). The spectral features are very similar with the exception for the polyethylene C-H stretch modes occurring near 3000 cm^{-1}.

modes of polyethylene (see black arrow) that are visible around 3000 cm^{-1}. In Table 5.1 presents the predictions for the lowest-five non-forbidden (i = 1, 2, 3, 4, & 7) electronic excited-states (and transition intensities) for biotin and biotin in PE that correspond to the models of Figure 5.9a and b, to biotin taken from the conformation of model of Figure 5.9a, and to an isolated biotin molecule. Predicted changes in transition energies suggest conformational differences between isolated biotin and biotin in PE. In particular, the presence of PE is predicted to cause a line shift to higher energy by 37 meV with simultaneous decrease in transition intensity

TABLE 5.1
Predicted Electronic Excited-States (& Transition Intensities) for Biotin ant Biotin in PE

E(i)	Model of Fig. 9(a) eV (cm^{-1})	Model of Fig. 9(b) eV (cm^{-1})	Biotin-Only from Fig. 9(a) eV (cm^{-1})	Isolated Biotin eV (cm^{-1})
E(1)	4.9646 (0)	5.1343 (0)	5.0558 (0)	5.1120 (0)
E(2)	5.1145 (0.0004)	5.2460 (0.0001)	5.1282 (0)	5.1377 (0)
E(3)	5.3657 (0.0008)	5.3918 (0.0005)	5.3575 (0.0005)	5.3867 (0.0005)
E(4)	5.6819 (0)	5.9151 (0.0005)	5.7490 (0.0001)	5.7200 (0.0003)
E(7)	6.1983 (0.0165)	6.2480 (0.0268)	6.2158 (0.0240)	6.1616 (0.0234)

[1] **NMR** is an effect whereby magnetic nuclei (with non-zero spin) in a magnetic field absorb and re-emit electromagnetic (EM) energy. This energy is at a specific resonance frequency which depends on the strength of the magnetic field and atom surroundings.

by 42%. Other differences in the transition energies suggest a significant effect of PE cavity on the spectrum of biotin.

The molecule azobenzene, as well as its derivatives, is also of interest because its molecular properties (e.g., conformation, spectral absorption, etc.) are known to be dependent on excitation by light. More specifically, it can be switched by light between trans and cis forms [32]. Hence, hybrid QM/MM simulations were performed to study azobenzene embedded in PE matrices. In Figure 5.11a, an optimization result for trans-azobenzene in a PE matrix is shown. In Figure 5.11b, a detail of the simulated structure with azobenzene is plotted. No substantial strand rearrangement to fit trans-azobenzene in PE is required. The deformation region is localized within 10–12 A around azobenzene (in the center of Figure 5.11a). However, azobenzene is constrained inside PE. On the other hand, substantial strand rearrangement is required to fit cis-azobenzene in PE according to the preliminary QM/MM results. This could affect the relaxation mechanism in excited trans-azobenzene molecules and therefore have an impact on their trans-cis photo-switching and sensing properties. These types of studies on azobenzene, and their conjugates with DNA, will be continued in the future in order to define novel BMSs that will be useful for functionlizing DNA-derivative architectures.

Ab-initio, DFT and hybrid QM/MM methodologies have been shown here to be effective tool for assessing the impact of polymer matrices on the THz spectra of biological molecules such as biotin. While polyethylene is almost transparent at THz, additional low-THz lines are predicted in the biotin/PE system, which reflects a dynamic interaction between biotin and a surrounding PE cavity. The obtained results agree well with the available experimental data. Also, in accordance with the preliminary QM/MM results, trans-azobenzene can be trapped in a polyethylene matrix which could alter its light switching and sensing characteristics. Trans-cis photoisomerization of azobenzene in PE matrices is a subject of an ongoing research.

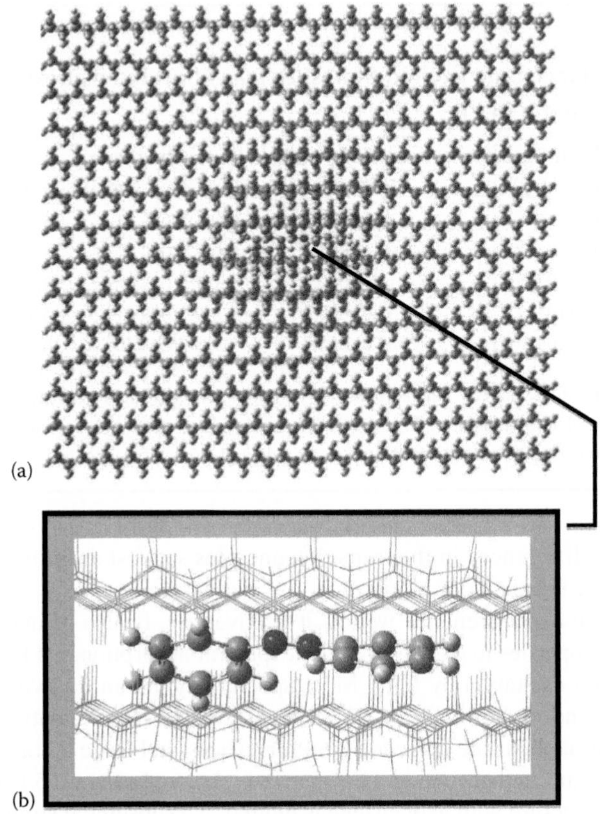

FIGURE 5.11 **(See color insert)** Azobenzene in polyethylene (PE) matrix. (a) Illustration of entire model structure, and (b) Detail showing the hybrid QM/MM model for the azobenzene (spheres) and PE (chains). Note that for these simulations azobenzene molecule was treated QM and all PE was MM.

5.6 LIGHT-INDUCED TRANSITIONS IN MULTIDIMENSIONAL MOLECULAR COORDINATE SPACES

Once DNA-based BMSs are identified that exhibit multiple stable conformation states with differing THz and/or far IR spectral signatures, it will be necessary to determine energy-space pathways that can be used to make transitions between these conformational states by application of light-based excitations. While a number of relatively small molecules are know the exhibit light-induced conformational transformations with simply defined energy-space pathways (e.g., the vision process in mammals is initiated by light-induced isomerizations of 11-cis retinal to all-trans retinal [33]), when one considers even moderately complex molecules the desired transformation trajectory can become dependent on multiple molecular coordinates and/or it may involve multiple excited-state transitions. In such cases, the computational burden for determining these energy-space trajectories and the required light excitation(s) can become excessive, or even practically intractable.

The basic issues associated with these types of energy-space transition problems are nicely illustrated by 2-butene, which is known to undergo light-induced transformations between trans-to cis-conformations [6]. Figure 5.12a illustrates the trans-2-butene conformation and the cis-2-butene conformation where the transition between the two has been defined as a rotation about the single molecular coordinate labeled as D5 (i.e., torsion rotation about the carbon-to-carbon double-bond). It is useful to note that the *trans*-conformation refers to the fact that the two CH_3 functional groups reside on opposite sides of the double-bonded carbon atoms, whereas in the cis-conformation the two CH_3 functional groups reside on the same side of the double-bonded carbon atoms. While 2-butene is amenable to multiple coordinate transformations when excited by light (e.g., D5 and D8 as shown in Figure 5.12a) it is possible to describe the trans-to cis-transformation most simply in terms of the D5 coordinate as shown in Figure 5.12b. As is illustrated symbolically by the green-line, a light-induced transition from the ground-state (S_0) to the first excited-state (S_1) modifies the charge distribution of the trans-2-butene molecule such that the lowest-energy of the first excited-state lies directly above the potential energy barrier in the ground-state. Hence, the molecule will first undergo a non-radiative conformation change (D5 changes from 180 to 90 degrees) that is followed by a radiative decay (i.e., from S_1 to S_0) where 50% of the excited molecules are placed into a position where they can naturally evolve (i.e., D5 changes from 90 to 0 degrees) into the cis-2-butene conformation. While this single-coordinate based physical picture provides an accurate light-based methodology for inducing the trans-to cis-conformation change, it may ignore dynamics that might be useful for identifying alternative coordinate-space pathways. Indeed, a more robust analysis [34] can be used to more accurately characterize this same transition in terms of two (i.e., D5 and D8) dynamical coordinates as illustrated in Figure 5.12c. As shown, when the steepest descent evolution is determined in terms of multiple coordinates, the molecule is also shown to exhibit variations in the D8 coordinate (which is also mirrored in the corresponding coordinate of opposing CH_3) where a complete cycle is executed (i.e., angular increase in D8 during the S_1 state evolution is matched exactly by angular decrease in D8 during the S_0 state evolution) such that the initial and final values of D8 end up being equal. While the D8 coordinate can be argued as irrelevant in this particular transformation of 2-butene, it illustrates the basic point that multiple-coordinate energy-space analyses will be of critical importance for defining the required BMSs.

At this time, active research is being conducted to develop new simulation tools that will be able to utilize the richness of the higher-dimensional coordinate-space transformations in combination with multiple frequency-and/or time-domain light-pumping to access complicated switching-trajectory pathways between known (and unknown) stable states of the target BMSs. As this is a very challenging quantum chemistry problem, it is expected that this will require new simulation tools that combine hybrid QM/MM modeling with efficient optimization algorithms and proper interfaces for human-guided analysis of light-induced transitions occurring within complex molecules. Both the computational burden and the potential opportunity for discovery can be illustrated by results from recent exploratory studies on the

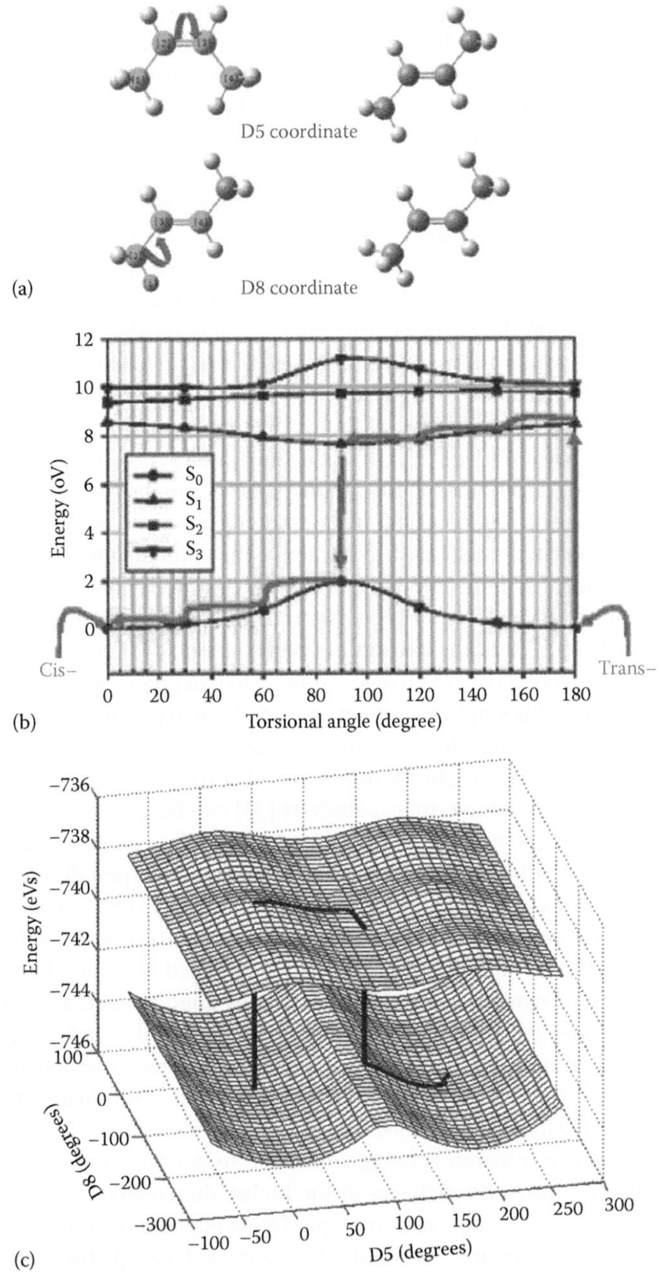

FIGURE 5.12 (See color insert) (a) 2-butene molecule with coordinate rotations for D5 (upper) and D8 (lower) labeled. (b) One-dimensional analysis (i.e. D5 coordinate only) that successfully predicts the trans-to cis-transition of 2-butene. (c) Actual two-dimensional transition path associated with conformational transformation defined by (b) that shows transversals (back and forth) in D8 space.

relatively simple molecule stilbene, which has already be discussed as an important switch-inducing component for use in defining DNA-based BMSs. Note that these particular studies sought to begin at a known ground-state conformation of stilbene (i.e., cis-stilbene) and then to use permutation sets of applied light-excitations and assumed radiative-decays (i.e., an example might be a light-induced excitation from S_0 to S_2, followed by the application of downward-decays when local minima are encountered at S_2 and S_1) in an attempt to find an alternative metastable-state within the ground-level.

Consider first the simulation of energy-space trajectories of the stilbene molecule where two coordinates were considered at-a-time, and steepest-descent was used to perform the optimization over smoothly-defined energy surfaces that were populated by fully-parallel calls to the program Gaussian (see Reference [35] for an example of this mathematical approach). Here, it is useful to note that the numerical-optimization trajectory calculations required negligible simulation time as compared to defining the energy surface. While these two-coordinate at-a-time numerical searches failed to identify any alternative light-induced molecular conformations for stilbene, they are useful for illustrating the significant computational burden (~1 hour for each $\{S_0, S_2, S_1, S_0\}$ permutation) even when each point of the potential energy mapping was calculated on a separate processor. These significant computational requirements and the realization that multiple coordinated interactions are important for light-induced stilbene transformations subsequently motivated the development and application of a more efficient optimization algorithm for testing.

Consider now the simulation of the energy-space trajectories for the stilbene molecule using this new algorithm where small incremental patches were dynamically defined to significantly reduce the computational burden of defining the required energy surfaces, and that enabled optimization over much larger coordinate spaces. A highly innovative aspect of this algorithm was the strategic use of Smolyak interpolation that can be made exact for second order systems using only 61 grid points per hypercube (see Reference [36] for an example of this mathematical approach). This simulation methodology was able to execute numerical searches that considered the evolution of five molecular coordinates at-a-time, and when those coordinates were those defined as shown in Figure 5.13a and used with the excitation/decay series as defined in Figure 5.13b, a light-induced pathway to a newly discovered alternative conformation was determined in only 27 hours. This is a very efficient simulation as compared to the prior algorithm, which would have used 20 weeks of simulation time to perform the required grid-space mapping. A study of the time evolution of the five molecular coordinates (see Figure 5.13c) reveals that the motions of the coordinate pairs (D_2, D'_2) and (D_3, D'_3) track each other exactly (which is why this labeling was chosen) and that the two CH_3 functional groups simultaneously undergo (i.e., in stereo) an identical circular-flapping type motion that occurs relative to the D_1 coordinate, during a period of time where D_1 first rotates by almost 180 degrees and then returns to its initial positions.

The evolution of the stilbene molecule into this very subtle difference in conformation (see Figure 5.14a) is clearly enabled by the unique multi-coordinate motions (see Figure 5.14b). More specifically, the two CH3 functional groups have assumed final positions where they are spread more laterally apart from

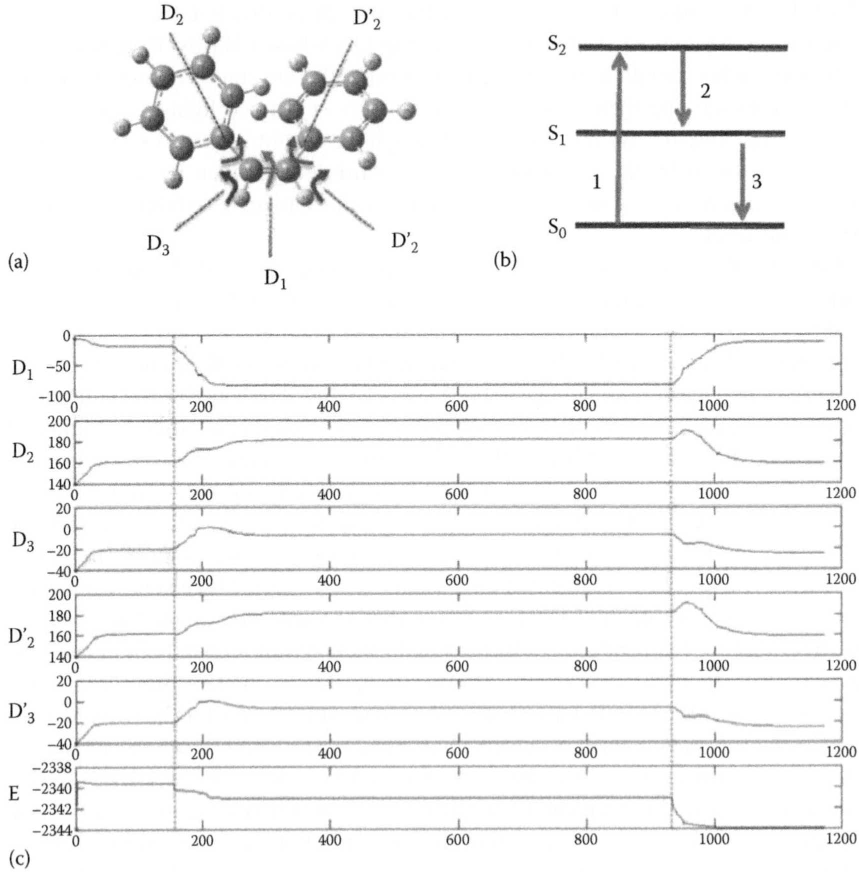

FIGURE 5.13 **(See color insert)** Five coordinate energy-space trajectory simulation information. (a) Illustration of the five coordinate angles D_1, D_2, D'_2, D_3 and D'_3. (b) Illustration of the excitation/decay series $\{S_0, S_1, S_2\}$. (c) Time evolution of the five coordinate angles (in angular degrees) and molecular energy (in eVs). Note that the red-dashed lines divide the three $\{S_2, S_1, S_0\}$ regions.

each other by virtue of the spiraling motions of D_1, D_2 (D'_2) and D_3 (D'_3) that occur most strongly when the molecule is decaying through the S1 and S0 energy spaces. It should be noted that the use of the incremental patch method prevents an immediate assessment of the energy barrier heights between this new metastable-state and the natural ground-state, so additional studies are now underway to determine the thermal stability of the metastable-state. In any event, these results demonstrate the power and potential of this new optimization algorithm for use in searching for new energy-space pathways. Hence, work is now being performed to develop a useful graphical interface that will facilitate human-guided simulations in the future, after which, the simulation tool will be used to design and optimize stilbene-derivative/DNA complexes for use as BMSs in sensing architectures.

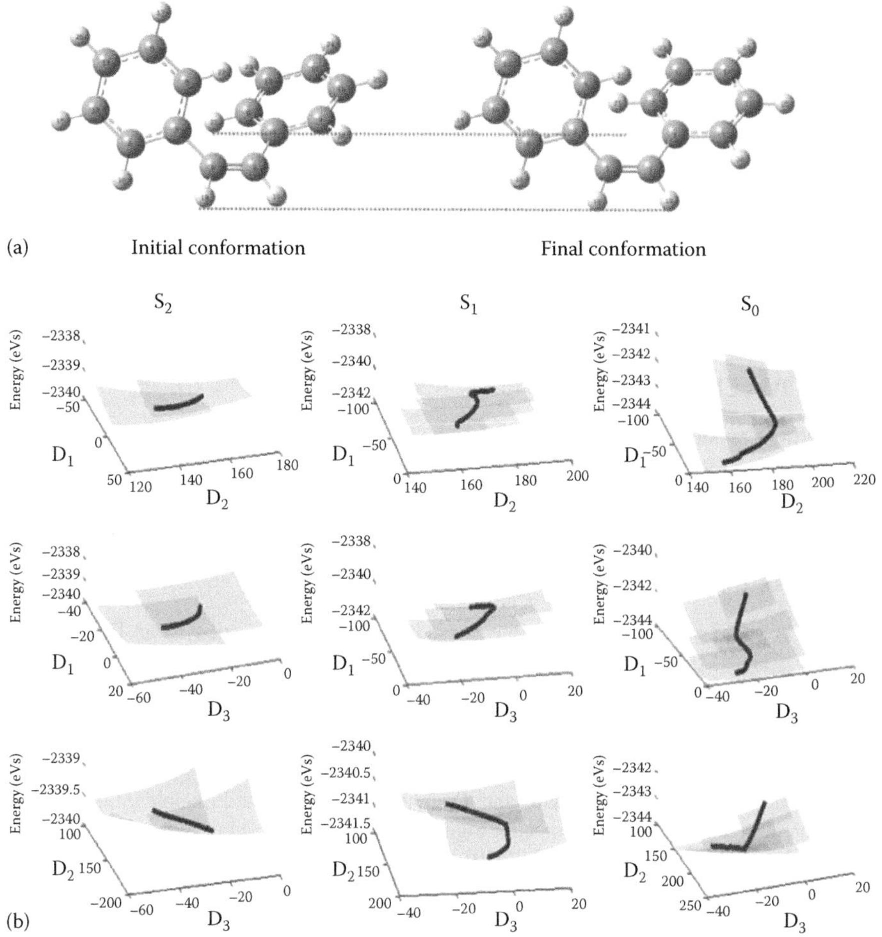

FIGURE 5.14 (a) Initial and final conformations of 2-butene from five coordinate energy-space trajectory simulation. Both views are oriented relative to the left CH_3 group and the red dashed lines are included for perspective. (b) Two-dimensional (2-D) trajectories (black lines) and energy-space patches (in blue) for pathways through each of the three energy-state spaces $\{S_2, S_1, S_0\}$. Note that a strong downward spiral within any 2-D plot (& separation of the trajectory line from the patch) indicates a strong influence of one or more other coordinate dimensions.

5.7 CONCLUSIONS

An overview has been provided on the physics and modeling of DNA-based molecular components that have potential for use in realizing DNA-derivative architectures with a capability for executing long-wavelength based bio-sensing. More specifically, research work was presented that is defining DNA-based biological molecule switches (BMSs) that will be useful for incorporating into larger DNA-based nanoscaffolds for the purposes of defining a novel class of smart materials for terahertz (THz) and/or very far-infrared (far-IR) based biological sensing. Recent progress was discussed on the theoretical design and optimization of a novel class of DNA conjugate molecules that are amenable light-induced switching of their conformation states. These type of smart sensing materials are of interest because may provide for the extraction of nanoscale information (e.g., composition, dynamics, conformation) through electronic/photonic transformations to the macroscale. Therefore, this research has relevance to long-wavelength bio-sensing applications, and for use in the characterization and study of analogous biological systems such as antibody and receptor mimics, which could define new recognition detection methodologies and synthetic medical treatments.

ACKNOWLEDGEMENT

The authors wish to recognize Prof. Michael Norton, of Marshall University, and Prof. Carl T. Kelley and Dr. David Mokrauer, of North Carolina State University for helpful discussions and information.

REFERENCES

1. D. Woolard, G. Recine, A. Bykhovski and W. Zhang, "Molecular-Level Engineering of THz/IR-Sensitive Materials for Future Biological Sensing Applications," *SPIE Proceedings* Vol. 7763 (2010).
2. Dwight L Woolard and James O. Jensen, "Functionalized DNA Materials for Sensing & Medical Applications," *SPIE Proceedings*, DSS11 SPIE Defense, Security, and Sensing (2011).
3. Alexei Bykhovski, Peiji Zhao, and Dwight Woolard, "First Principle Study of the Terahertz and Far-Infrared Spectral Signatures in DNA Bonded to Silicon Nano Dots," *IEEE Sensors,* **10** (3), 585–598 (2010).
4. P. Zhao and B. Woolard, "Influence of base pair interaction on vibrational spectrum of a poly-dG molecule bonded to Si substrates," *IEEE Sensors Journal,* **8** (6), 998–1003, 2008.
5. D. L. Woolard, E. R. Brown, M. Pepper M, "Terahertz Frequency Sensing and Imaging: A Time of Reckoning Future Applications?" *Proceedings of the IEEE*, **93** (10): 1722–1743 (2005); Y. Luo, B. Gelmont, and D. Woolard, "Bio-Molecular Devices for Terahertz Frequency Sensing," in Theoretical and Computational Chemistry Vol. 17, Molecular and Nano Electronics: Analysis, Design and Simulation, Ed. by J.M. Seminario, Elsevier, 55–81 (2007)
6. M. Rahman and M. L. Norton, "Two-Dimensional Materials as Substrates for the Development of Origami-Based Bionanosensors," *IEEE Trans. Nanotechnology,* **9**, 539–542 (2010).
7. W. Shen, H. Zhong D. Neff and M. Norton, "NTA Directed Protein Nanopatterning on DNA Origami Nanoconstructs," *JACS,* **131**, pp. 6660–6661 (2009).
8. W. Liu, H. Zhong, R. Wang and N. Seeman, "Crystalline Two-Dimensional DNA origami Arrays," *Angewandte Chemie,* **123** (1), 278–281 (2011).
9. Anton Simeonov, Masayuki Matsushita, Eric A. Juban, Elizabeth H. Z. Thompson, Timothy Z. Hoffman, Albert E. Beuscher IV, Matthew J. Taylor, Peter Wirsching, Wolfgang Rettig, James K. McCusker, Raymond C. Stevens, David P. Millar, Peter G. Schultz, Richard A. Lerner, Kim D. Janda, "Blue-Fluorescent Antibodies," *Science* **290**, 307–313 (2000)

10. Frederick D. Lewis, Jianqin Liu, Xiaoyang Liu, Xiaobing Zuo, Ryan T. Hayes, and Michael R. Wasielewski, "Dynamics and Energetics of Hole Trapping in DNA by 7-Deazaguanine," *Angew. Chem. Int. Ed.* 2002, **41** (6), 1026–1028.
11. Alexei Bykhovski and Dwight Woolard, "Hybrid Ab-Initio/Empirical Modeling of the Conformations and Light-Induced Transitions in Stilbene-Derivatives Bonded to DNA," *IEEE Transactions on Nanotechnology,* **9** (5) (2010).
12. Michael J. S. Dewar, Eve G. Zoebisch, Eamonn F. Healy, James J. P. Stewart, "Development and use of quantum mechanical molecular models. 76. AM1: a new general purpose quantum mechanical molecular model," *J. Am. Chem. Soc.,* 1985, **107** (13), 3902–3909 (1985).
13. Jennifer Tuma, Ralph Paulini, Jan A. Rojas Stutz, and Clemens Richert, How Much pi-Stacking Do DNA Termini Seek? Solution Structure of a Self-Complementary DNA Hexamer with Trimethoxystilbenes Capping the Terminal Base Pairs *Biochemistry* 2004, *43,* 15680–15687.
14. D. A. Case, D. A. Pearlman, J. W. Caldwell, T.E. Cheatham, J. Wang, W.S. Ross, C.L. Simmerling, T. A. Darden, K. M. Merz, R.V. Stanton, A.L. Cheng, J.J. Vincent, M. Crowley, V. Tsui, H. Gohlke, R.J. Radmer, Y. Duan, J. Pitera, I. Massova, G.L. Seibel, U.C. Singh, P.K. Weiner, and P.A. Kollman (2004), AMBER 8, University of California, San Francisco, http://amber.scripps.edu/, p.263.
15. W. D. Cornell, P. Cieplak, C. I. Bayly, I. R. Gould, K. M. Merz Jr., D. M. Ferguson, D. C. Spellmeyer, T. Fox, J. W. Caldwell, and P. A. Kollman, *J. Am. Chem. Soc.* 117, 5179 (1995).
16. C.I. Bayly, P. Cieplak, W.D. Cornell & P.A. Kollman, "A Well-Behaved Electrostatic Potential Based Method Using Charge Restraints For Determining Atom-Centered Charges: The RESP Model," *J. Phys. Chem.* **97**, 10269 (1993).
17. Alexei Bykhovski, Tatiana Globus, Tatyana Khromova, Boris Gelmont, and Dwight Woolard. "An Analysis of the THz Frequency Signatures in the Cellular Components of Biological Agents," *Int. J. High Speed Electr. Systems*, **17** (2), 225–237 (2007).
18. Alexei Bykhovski, Tatiana Globus, Tatyana Khromova, Boris Gelmont, and Dwight Woolard. Resonant Terahertz Spectroscopy of Bacterial Thioredoxin in Water: Simulation and Experiment, in Selected Topics in Electronics and Systems, World Scientific, Vol. 48, 367–375 (2008).
19. M. Svensson, S. Humbel, and K. Morokuma, "Energetics using the single point IMOMO (integrated molecular orbital+molecular orbital) calculations: Choices of computational levels and model system," *J. Chem. Phys.* **105**, 3654–3661 (1996).
20. Michael J. Frisch and Aeleen Frisch, Gaussian 09 User's Reference and IOps Reference, http://www.gaussian.com.
21. W. Stevens, H. Basch, and J. Krauss, "Compact effective potentials and efficient shared-exponent basis sets for the first-and second-row atoms," *J. Chem. Phys.* **81**, 6026–6033 (1984).
22. Jeng-Da Chai and Martin Head-Gordon, "Long-range corrected hybrid density functionals with damped atom-atom dispersion corrections," *Phys. Chem.,* 2008, **10**, 6615–6620 (2008).
23. C. Painter, Michael M. Coleman, Jack L. Koenig, The theory of vibrational spectroscopy and its application to polymeric materials, John Wiley & Sons, NY, p.323 (1982).
24. G. Avitabile, R. Napolitano, B. Pirozzi, K. D. Rouse, M. W. Thomas, B. T. M. Willis, Low Temperature Crystal Structure Of Polyethylene: Results From A Neutron Diffraction Study and from Potential Energy Calculations, *Polymer Letters Edition* **13**, 351–355 (1975).
25. M. Tasumi and T. Shimanouchi, Crystal Vibrations and Interatomic Forces in Polymethylene Crystals, *J. Chem. Phys.,* **43** (4), 1245–1258 (1965).
26. T.M. Korter, D.F. Plusquellic, Continuous-wave terahertz spectroscopy of biotin: vibrational anharmonicity in the far-infrared, *Chemical Physics Letters* **385**, 45–51 (2004).

27. David C. Schriemer and Liang Li, "Combining Avidin-Biotin Chemistry with Matrix-Assisted Laser Desorption/Ionization Mass Spectrometry," *Anal. Chem.*, **68**, 3382–3387 (1996).
28. Alexei Bykhovski and Dwight Woolard, "Spectra of Biological and Organic Molecules In Polymer Matrices: A Hybrid Ab-Initio/Empirical Study," 2010 International Symposium on Spectral Sensing Research (2010 ISSSR), Springfield, MO, 20–24 June (2010).
29. S. S. Iyengar, H. B. Schlegel, J. M. Millam, G. A. Voth, G. E. Scuseria, and M. J. Frisch, "Ab initio molecular dynamics: Propagating the density matrix with Gaussian orbitals. II. Generalizations based on mass-weighting, idempotency, energy conservation and choice of initial conditions," *J. Chem. Phys.*, **115**, 10291–302 (2001).
30. W. Stevens, H. Basch, and J. Krauss, "Compact effective potentials and efficient shared-exponent basis sets for the first-and second-row atoms," *J. Chem. Phys.* **81**, 6026–6033 (1984).
31. Peter H. Berens and Kent R. Wilson, "Molecular dynamics and spectra. I. Diatomic rotation and vibration," *J. Chem. Phys.* **74** (9), 4872–4882 (1981).
32. Kevin G. Yager and Christopher J. Barrett, "Azobenzene Polymers for Photonic Applications," *Smart Light-Responsive Materials.* Edited by Yue Zhao and Tomiki Ikeda, John Wiley & Sons, Inc. (2009).
33. D. L. Woolard, Y. Luo, B. L. Gelmont, T. Globus and J. O. Jensen, "Bio-molecular inspired electronic architectures for enhanced sensing of THz-frequency bio-signatures", *Int. J. High Speed Electron. Syst.,* **16** (2), 609–637 (2006).
34. Dave Mokrauer, C. T, Kelley and Alexei Bykhovski, "Efficient Parallel Computation of Molecular Potential Energy Surfaces for the Study of Light-Induced Transition Dynamics in Multiple Coordinates," IEEE Transactions on Nanotechnology, **10** (5) 70–74 (2010).
35. David Mokrauer, C. T. Kelley, Alexei Bykhovski, *Parallel Computation of Surrogate Models for Potential Energy Surfaces,* DCABES, pp.1–4, Ninth International Symposium on Distributed Computing and Applications to Business, Engineering and Science (2010).
36. David Mokrauer, C. T. Kelley, Alexei Bykhovski, *Simulations of Light-Induced Molecular Transformations in Multiple Dimensions with Incremental Sparse Surrogates,* submitted to Journal of Algorithms & Computational Technology (2011).

Part 2

Biosensors and Integrated Devices

6 *In Situ* Nanotechnology-Derived Sensors for Ensuring Implant Success

Sirinrath Sirivisoot and Thomas J. Webster

CONTENTS

6.1 INTRODUCTION

Osteoblasts and osteoclasts are located in bone, a natural nanostructured mineralized organic matrix. While osteoblasts make bone, osteoclasts decompose bone by releasing acid that degrades calcium (Ca) phosphate–based apatite minerals in an aqueous environment. The synthesis, deposition, and mineralization of this organic matrix, in which osteoblasts proliferate and mineralize (that is, deposit Ca), require the ordered expression of a number of osteoblast genes. Bone has the ability to self-repair or remodel routinely. However, osteoporosis (unbalanced bone remodeling) and other joint diseases (such as osteoarthritis, rheumatoid arthritis, and traumatic arthritis) can lead to bone fractures. These bone-associated disabilities lead to difficulties in performing common activities and may require an orthopedic implant. However, the average functional lifetime of, for example, a hip implant (usually composed of titanium [Ti]) is only 10–15 years. A lack of fixation into the surrounding bone eventually loosens the implant and is the most common cause of hip replacement failure.

Approximately 200,000 total hip replacements, 100,000 partial hip replacements, and 36,000 hip replacement revisions were performed in the United States in 2003 [1]. Unfortunately, only limited techniques, such as insensitive imaging (through x-rays or bone scans), exist to determine if sufficient bone growth is occurring next to a Ti implant after surgery. Thus, techniques used in orthopedic diagnostics need to more accurately identify musculoskeletal injuries and conditions after implantation. Although there have been improvements in implants to increase bone formation, the clinical diagnosis of new bone growth surrounding implants remains problematic, sometimes significantly increasing patients' hospital stays and decreasing the ability of doctors to quickly prescribe a change in action if new bone growth is insufficient.

Currently, a physical examination (e.g., palpation or laboratory testing) is performed before imaging techniques to inform a clinician about a patient's health. Although advanced imaging techniques, such as bone scans, computer tomography scans, and radiographs (x-rays), are important in a medical diagnosis, each technique has its own limitations and difficulties. A bone scan is used to identify areas of abnormal active bone formation, such as arthritis, infection, and bone cancer. However, a bone scan requires an injection of a radioactive substance (e.g., technetium) and a prolonged delay for absorbance before performing the scan. Computer tomography combines x-rays with computer technology to produce a two-dimensional cross-sectional image of the body on a computer screen. Although this technique produces more detail than an x-ray, in some cases (e.g., severe trauma to the chest, abdomen, pelvis, or spinal cord) a dye (e.g., barium sulfate) is injected in order to improve the clarity of the image. This often causes pain for the patient. Another technique, called electromyography, is used to analyze/diagnose nerve functions inside the body. Thin electrodes are placed in soft tissue to help analyze and record electrical activity in muscles. However, this electrode technique causes pain and discomfort for the patient. When the needles are removed, soreness and bruising can occur.

Electrochemical biosensors on the implant itself show promise for medical diagnostics *in situ* to possibly determine new bone growth surrounding the implant. Yet, this technology has not been fully explored. The incorporation of such electrochemical biosensors into current bone implants may be possible through nanotechnology; different types of nanoscale biomaterials have varying abilities to enhance *in vitro* and *in vivo* bone formation. Carbon nanotubes (CNTs) are macromolecules of carbon, which are classified as either single-walled CNTs (SWCNTs), with diameters of 0.4–2 nm, or multiwalled CNTs (MWCNTs), with diameters of 2–100 nm [2,3]. Due to their unique electrical, mechanical, chemical, and biological properties [4–11], CNTs have shown promise for bone implantation. For example, nanocomposites of polylactic acid and MWCNTs increased osteoblast proliferation by 46% and Ca production by greater than 300% when an alternating current was applied *in vitro* [12,13]. In addition, combining MWCNTs with precusor powders improved the mechanical properties of as-aligned hydroxyapatite (HA) composite coatings [9]. MWCNTs, reinforced with HA coatings, promoted human osteoblast proliferation *in vitro*, as observed in the appearance of cells near MWCNT regions [8]. Osteoblasts extended in all directions within the CNT scaffold on polycarbonate membranes [14]. Osteoblasts significantly enhanced their adhesion on vertically aligned MWCNT arrays, according to Giannona et al., by recognizing nanoscaled features [15]. In addition, Zanello et al. showed that CNTs are suitable for the proliferation of osteosarcoma ROS 17/2.8 cells on SWCNT and MWCNT scaffolds [7]. Human osteoblast-like (Saos2) cells grown on an MWCNT scaffold showed a higher cell density and transforming growth factor-β1 concentration compared with those grown on polystyrene and polycarbonate scaffolds [16].

Our previous studies have shown greater osteoblast differentiation on MWCNT-Ti than on Ti alone [17]. Moreover, MWCNTs are a promising material for electrochemical biosensors because they also possess relatively well-characterized behavior in terms of electron transport [18–21]. Coating Ti electrodes with CNTs also increases the active surface area and enhances direct electron transfer [22,23]. These studies encourage the use of MWCNTs to modify currently used Ti for orthopedic implants. CNTs can be produced by laser furnace arc and chemical vapor deposition (CVD) methods. Unlike the laser furnace arc, CVD is a scalable and controllable method of obtaining high-purity CNTs [24]. A strong mechanical connection between CNTs and Ti is needed for nanosensor applications [25], and the reduced resistance between Ti and CNTs [26] leads to increased current passage. For example, Sato et al. showed the potential for using CNTs in electrochemical biosensors by using CVD to grow MWCNTs, with Ti biometallic particles and cobalt (Co) as a catalyst. They revealed that Co has the ability to combine with Ti [27] and since the CNTs were grown using Co catalyzed in this study, the formation of a strong electrical contact between metallic Ti and MWCNTs was possible.

In order to form a more robust interconnection, CNTs were anchored in the pores of anodized nanotubular Ti in this study. Then, using CVD techniques, MWCNTs were grown out of the anodized Ti nanotubes (with diameters of 50–60 nm and depths of 200 nm) as a template.

In vivo, many cell processes rely on the redox reactions of various biomolecules and ions. The mechanisms of electron transfer reaction and the role of proteins in aiding the electron transfer of redox processes can be examined by electrochemical analysis. In electronic theory, when two different materials come in contact with each other, electron transfer will occur in an attempt to balance Fermi levels, causing the formation of a double layer of electrical charge at the interface [28]. Because the formation of an electric contact between the redox proteins and the electrode surface is the fundamental challenge of electrochemical biosensor devices, in this study we investigated the redox reactions of iron (II/III) and the osteoblast extracellular components at the surface of Ti, anodized Ti, and MWCNT-Ti.

6.2 MATERIALS AND METHODS

6.2.1 Preparation of Nanotubular Anodized Titanium

Ti was used to develop a nanotubular thin layer of Ti dioxide (TiO_2) by an anodization technique adapted from a previous study [17]. Briefly, 99.2% commercially pure Ti sheets (Alfa Aesar) were cut into squares (1 cm^2) and cleaned with acetone and 70% ethanol under sonication for 10 min each. After rinsing with deionized water, Ti was etched for 10 s in a solution of 1.5% by weight of nitric acid and 0.5% by weight of hydrofluoric acid to remove the existing oxidized layer on Ti. Immediately after etching, Ti was used as an anode electrode and a high-purity platinum (Pt) mesh (Alfa Aesar) was used as a cathode. In a Teflon beaker, both electrodes were immersed in an electrolyte solution of 1.5% by weight hydrofluoric acid in deionized water. The distance of Ti from the Pt mesh was 1 cm. A direct current power supply (3645A; Circuit Specialists) fed 20 V between the anode and the cathode for 10 min to create uniform nanotubes of TiO_2 on commercially pure Ti in this study.

6.2.2 Cobalt-Catalyzed Chemical Vapor Deposition

Afterward, MWCNTs were grown from the nanoporous TiO_2 using the CVD method. A plasma-enhanced CVD (PECVD) system (Applied Science and Technology Inc.) was used to grow MWCNTs from nanotubular anodized Ti. The anodized Ti samples were soaked in a solution of 5% by weight cobaltous nitrate (Allied Chemical) in methanol for 5 min prior to the CVD process. Then, Co-catalyzed anodized Ti was rinsed with distilled water and dried with compressed air. The samples were then placed in a CVD chamber and air was pumped out to a base pressure below 10 mTorr. The samples were heated to 700°C in a flow of 100 sccm hydrogen gas (H_2) for 20 min. The gas composition was changed to 40 sccm H_2 and 160 sccm acetylene gas (C_2H_2) for 30 min to grow MWCNTs. The MWCNT-Ti was cooled in a 100 sccm argon flow before studying its cellular responses by examination under a scanning electron microscope (SEM; Leo 1530VP FE-4800) and investigation of its electrochemical behaviors. The dense and entangled MWCNTs formed a three-dimensional structure on the Ti surface with exceptional electrical conductivity and surface area.

6.2.3 Cell Culture and Morphology

In vitro osteoblast cytocompatibility assays were determined on commercially pure Ti, anodized Ti, MWCNT-Ti, and carbon nanopaper (Buckypaper; NanoLab). Samples were sterilized by ultraviolet (UV) light exposure for 4 h on each side. The substrates were immediately rinsed three times with phosphate-buffered saline (PBS; 8 g NaCl, 0.2 g KCl, 1.5 g Na_2PO_4, and 0.2 g KH_2PO_4 in 1000 mL deionized water adjusted to a pH of 7.4; Sigma-Aldrich) and placed in 12-well plates. Osteoblasts (hFOB 1.19; CRL-11372; ATCC; population number = 5–12) were used to study cell adhesion and differentiation. For adhesion assays, 3500 osteoblasts/cm^2 were seeded in Dulbecco's modified eagle medium (DMEM; Gibco) supplemented with 10% fetal bovine serum (FBS; Hyclone) and 1% penicillin/streptomycin (P/S; Hyclone) onto the substrates under standard incubator conditions (a humidified 5% CO_2 and 95% air environment at 37°C) for 4 h, and then washed three times with PBS to remove nonadherent cells. The osteoblasts were fixed with acetate buffer containing 10% formalin (Fisher) for 10 min and the nuclei were stained with, 4′,6-diamidino-2-phenylindole (DAPI; Sigma), diluted at 300 nM. Cells were counted in five random fields per sample under a fluorescence microscope (Zeiss Axiovert 200 M). The cell culture was repeated three different times. After seeding for 4 h on commercially pure Ti, anodized Ti, and MWCNT-Ti, the morphology of the osteoblasts was evaluated by SEM (Hitachi 2700). Before SEM analysis, all samples were fixed with an acetate buffer containing 10% formalin (Fisher) for 10 min, dehydrated with a series of ethanol (30%, 50%, 70%, 90%, and 100%; 20 min each), critical point dried (CPD, LADD Research), and sputter-coated with gold-palladium to a thickness of 20 nm.

6.2.4 Cellular Assays (Alkaline Phosphatase and Calcium Deposition)

To study osteoblast differentiation between non-Ca- and Ca-depositing cells, osteoblasts were seeded at a density of 40,000 cells/cm^2 in DMEM supplemented with 10% FBS, 1% P/S, 50 nM β-glycerophosphate (Sigma), and 50 μg/mL ascorbic acid (Sigma) under standard cell culture conditions (a humidified 5% CO_2 and 95% air environment at 37°C) for 7, 14, and 21 days. The cell media were changed every other day. At the end of each time point, an alkaline/acid phosphatase assay kit (Upstate) was used to determine the concentration of alkaline phosphatase in cell lysates. The cell lysates were prepared by first rinsing all samples three times with Tris-buffered saline (TBS; 42 mM Tris-HCl, 8 mM Tris base, and 0.15 M NaCl; a pH of 7.4; Sigma-Aldrich) and then subjecting the cells to three freeze–thaw cycles using distilled water. A Ca quantification kit (Sigma) was used to determine the amount of Ca deposited by the osteoblasts seeded on each sample. An acidic supernatant solution for a Ca deposition assay was prepared by incubating all the samples with 0.6 N HCl (Sigma) for 24 h. The light absorbance was measured by a spectrophotometer (SpectraMAX 340PC[384]; Molecular Devices) at 650 nm for alkaline phosphatase activity and 570 nm for Ca deposition. Long-term cytocompatible testing was conducted in triplicate and was repeated three different times. The numerical data were analyzed using standard Student's *t*-test.

6.2.5 Osteoblast Extracellular Component Solutions

Osteoblasts were cultured in DMEM supplemented with 10% FBS, 1% P/S, 50 nM β-glycerophosphate, and 50 μg/mL ascorbic acid for 7, 14, and 21 days on commercially pure Ti with an initial cell density of 40,000 cells/cm^2 under standard cell culture conditions. After each time period, and three freeze–thaw cycles with distilled water to lyse the cells, Ti was soaked in 0.6 N HCl (Sigma) for 24 h to dissolve the extracellular components of the osteoblasts. A Ca quantification kit (Sigma) was used to determine the amount of Ca deposited by the osteoblasts. The light absorbance of Ca in the supernatants was measured by a spectrophotometer at 570 nm. The chemical composition on the surfaces of the Ti substrates after osteoblast culturing for 21 days was evaluated by energy dispersive spectroscopy (SEM-EDS; Leo 1530VP FE-4800).

6.2.6 Electrochemical Measurements

The cyclic voltammetry experiments in this study were performed on an epsilon EC electrochemical workstation and Digisim software (Bioanalytical). Commercially pure Ti, anodized Ti, and MWCNT-Ti were used as working electrodes with geometric areas of 1 mm^2 and 1 cm^2. A silver/silver chloride (Ag/AgCl; MW-2052; Bioanalytical) and a Pt (MW-1032; Bioanalytical) wire were used as the reference and counter electrodes, respectively. The electrodes were cleaned with deionized water before measuring. The electrodes were connected to the electrochemical workstation and immersed in an electrolyte solution. The electrolytes were a 10 mM $K_3Fe(CN)_6$ (potassium ferricyanide) solution with 1 M KNO_3, and a solution of 0.6 N HCl dissolved the extracellular components formed on the conventional Ti after 7, 14, and 21 days. Cyclic voltammograms (CVs) were generated by applying a linear sweep potential at several scan rates. The electrochemical responses at the three types of working electrodes were compared with the CVs in both electrolyte solutions. All measurements were carried out at room temperature and in oxygen-containing electrolyte solutions.

6.3 RESULTS AND DISCUSSION

6.3.1 Electrode Topography

The conventional Ti surface, shown in Figure 6.1a, exhibits a smooth Ti oxide. After anodization, nanopores of Ti oxide were uniformly formed on the Ti surface with diameters of 50–60 nm and depths of 200 nm, as shown in Figure 6.1b. MWCNTs covering the anodized Ti are shown in Figure 6.1c (side view) and Figure 6.1d (top view).

6.3.2 Osteoblast Responses

The results of this study showed that after 4 h a similar number of osteoblasts adhered on unanodized Ti, anodized nanotubular Ti without MWCNTs, and anodized nanotubular Ti with MWCNTs (Figure 6.2); all amounts were greater than those that adhered to the carbon nanopaper. Although MWCNTs on anodized Ti showed fewer cells adhering, osteoblasts were observed with their typical cytoplasmatic extension (Figure 6.3c and d), resulting in stronger adhesion (higher density of cell points of contact) [17]. Osteoblasts on the anodized Ti (without MWCNTs) were less spread than on the MWCNTs grown out of Ti (Figure 6.3a and b). Additionally, Zanello et al. observed a thin neurite-like cytoplasmic prolongation of osteoblasts that reached the nanotube bundles after cell culturing for 5 days. The lower osteoblast growth in Figure 6.2 may be due, in part, to the loss of MWCNTs during the process of fixing and staining the osteoblasts. More importantly, the results of the present study showed for the first time that the biological qualification assays, specifically alkaline phosphatase activity and Ca deposition by osteoblasts, increased on MWCNTs grown from anodized nanotubular Ti compared with anodized nanotubular Ti without MWCNTs, unanodized Ti, and carbon nanopaper after 21 days (Figure 6.4).

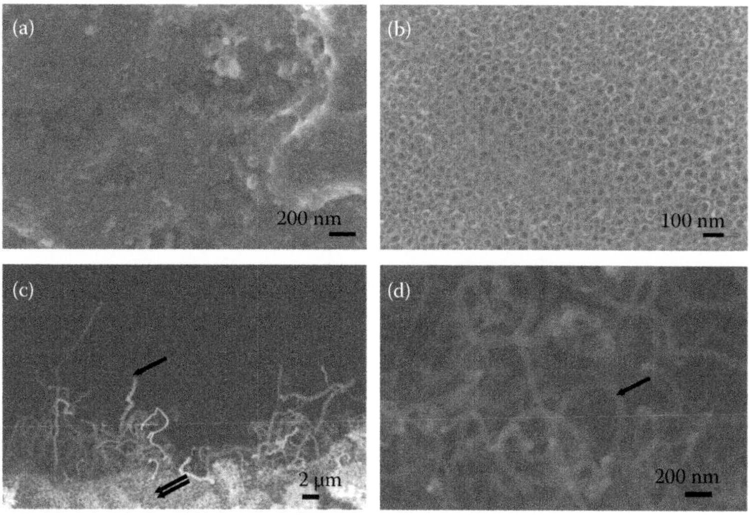

FIGURE 6.1 SEM micrographs of the electrode surfaces: (a) conventional Ti; (b) anodized Ti; and (c) side view and (d) top view of MWCNT-Ti. Single arrows show MWCNTs, while double arrows show the anodized Ti template.

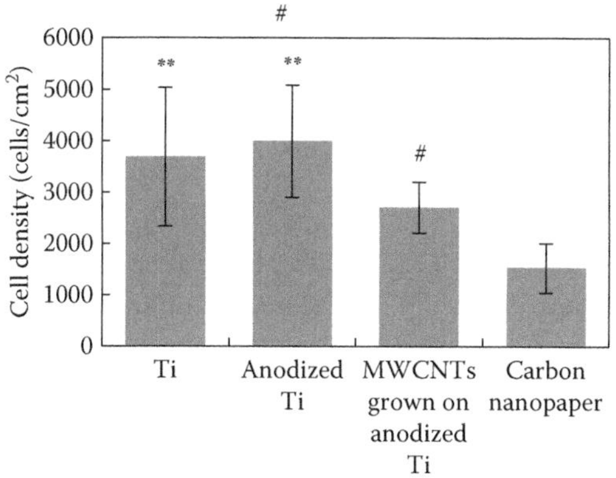

FIGURE 6.2 Osteoblast adhesion for 4 h on Ti, anodized Ti, MWCNTs grown on anodized Ti, and carbon nanopaper. Values are mean ± SEM; $n = 3$; $^{**}p < 0.05$ compared with carbon nanopaper; $\#p < 0.05$ compared with MWCNTs grown on anodized Ti.

6.3.3 Osteoblast Extracellular Component Analysis

After osteoblasts were cultured for 21 days, the results from EDS performed on commercial Ti substrates revealed the presence of various minerals indicating newly formed bone. Figure 6.5 shows the peaks of many inorganic substances, such as magnesium (Mg), phosphorus (P), sulfur (S), potassium (K), and Ca. The Ca/P weight ratio of the minerals deposited by osteoblasts on Ti (1.34) was less than that on MWCNT-Ti (1.52) in this study. However, the Ca/P ratio of

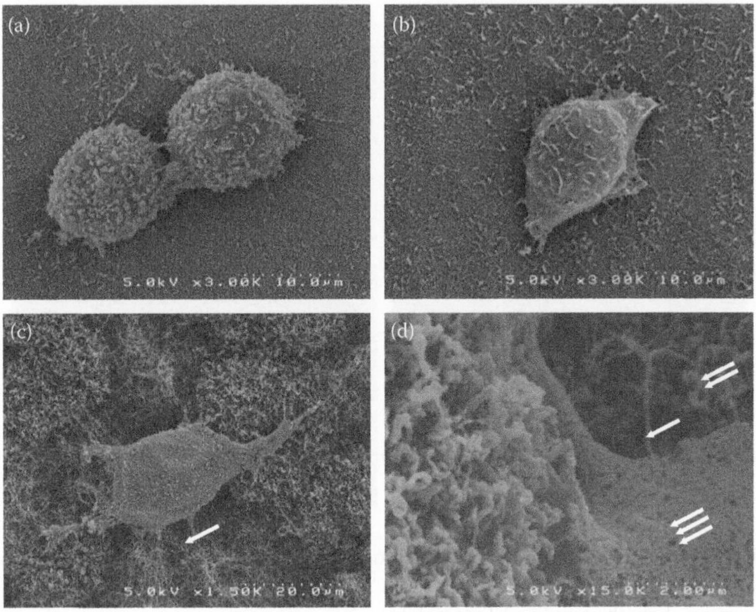

FIGURE 6.3 Osteoblast morphology after 4 h of adhesion on: (a, b) anodized Ti (scale bars = 10 μm). Round osteoblasts are on the anodized Ti substrate without MWCNTs; (c, d) MWCNTs grown on anodized Ti (scale bars = 20 μm left and 2 μm right). Single arrows show the cytoplasmatic extensions of osteoblasts. Double arrows show MWCNTs, while triple arrows show the osteoblast membrane.

HA, the main Ca/P crystallite in bone, is typically about 1.67 [29,30]. This study demonstrated that the minerals deposited by osteoblasts on MWCNT-Ti were more similar to natural bone than the minerals deposited on Ti. X-ray diffraction (XRD) analysis also showed that more HA was deposited on MWCNT-Ti than on both the conventional and anodized Ti after osteoblasts were cultured for 21 days, as shown in Figure 6.6. In addition, the amount of Ca deposited by osteoblasts, as determined by a Ca quantification assay kit, was 1.481 $\mu g/cm^2$ for 7 days, 1.597 $\mu g/cm^2$ for 14 days, and 2.483 $\mu g/cm^2$ for 21 days on conventional Ti, as shown in Figure 6.7b. These results imply a greater deposition of Ca by osteoblasts on MWCNT-Ti.

Bone resorption and remodeling involve the secretion of hydrochloric acid (HCl) by osteoclasts [31]. Osteoclasts dissolve bone mineral by isolating a region of the matrix and then secreting HCl and proteinases on the bone surface, resulting in the bone acting as a reservoir of Ca^{2+}, PO_4^{3-}, and OH^- [32]. This may be the reason why HCl is routinely used in laboratories to dissolve Ca minerals deposited on osteoblast-seeded scaffolds *in vitro* as a supernatant for further use in a Ca deposition assay. Thus, after dissolving with 0.6 N HCl, it is likely that Ca^{2+}, PO_4^{3-}, and OH^- are contained in the solution of osteoblast extracellular components.

The formation of bone matrix minerals first depends on achieving a critical concentration of Ca and P. Then, phospholipids, anionic proteins, as well as Ca and P

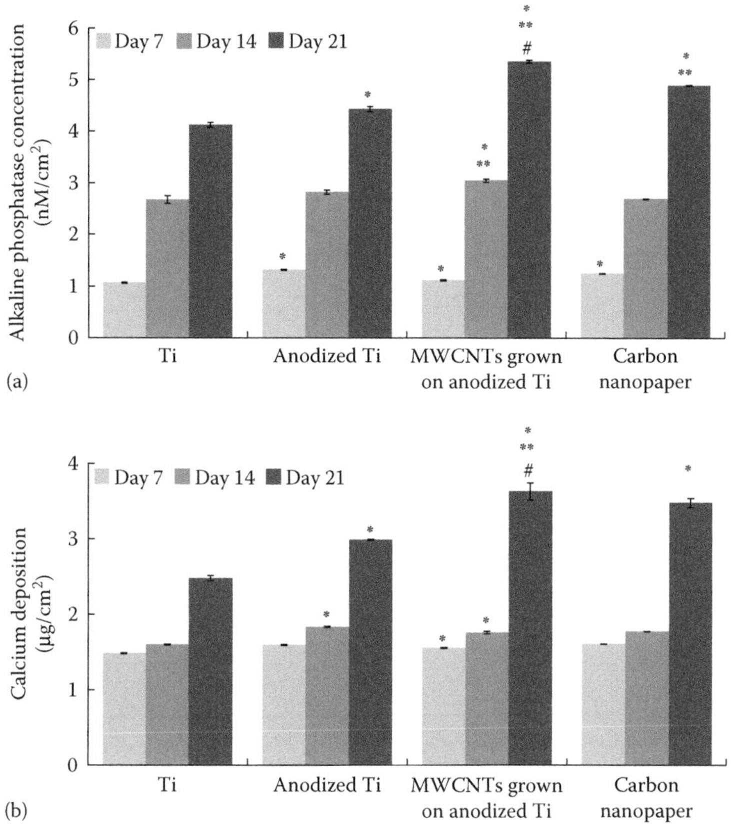

FIGURE 6.4 Osteoblast long-term functions: (a) alkaline phosphatase activity; values are mean ± SEM; $n = 3$; $^*p < 0.05$ compared with Ti; $^{**}p < 0.1$ compared with anodized Ti; and $\#p < 0.1$ compared with carbon nanopaper; (b) calcium deposition on the areas of interest.

aggregate in nucleation pores that are in the 35 nm "hole zone" between collagen molecules [33]. The addition of Ca, P, and hydroxyl ions contributes to the growth of crystalline HA. However, the crystals are not pure since they also contain carbonate, sodium, potassium, citrate, and traces of other elements, such as strontium and lead. The imperfection of the HA contributes to its minerals' solubility, which plays a role in the ability of osteoclasts to resorb the mineral phases. Although osteoblasts synthesize Type I collagen, which is the predominant organic component of bone, Type III/V/VI collagens also exist in bone. Moreover, noncollagenous proteins of bone (such as growth factors, osteocalcin, osteopontin, and osteonectin) and proteins in serum and other tissues (such as fibronectin, vitronectin, and laminin) are absorbed into the mineral component during bone growth. These components are directly involved in the genesis and maintenance of bone. Thus, not only minerals but also other noncollagenous proteins are broken down after dissolution with HCl. As such, unspecific proteins in the supernatant when cyclic voltammetry is performed with

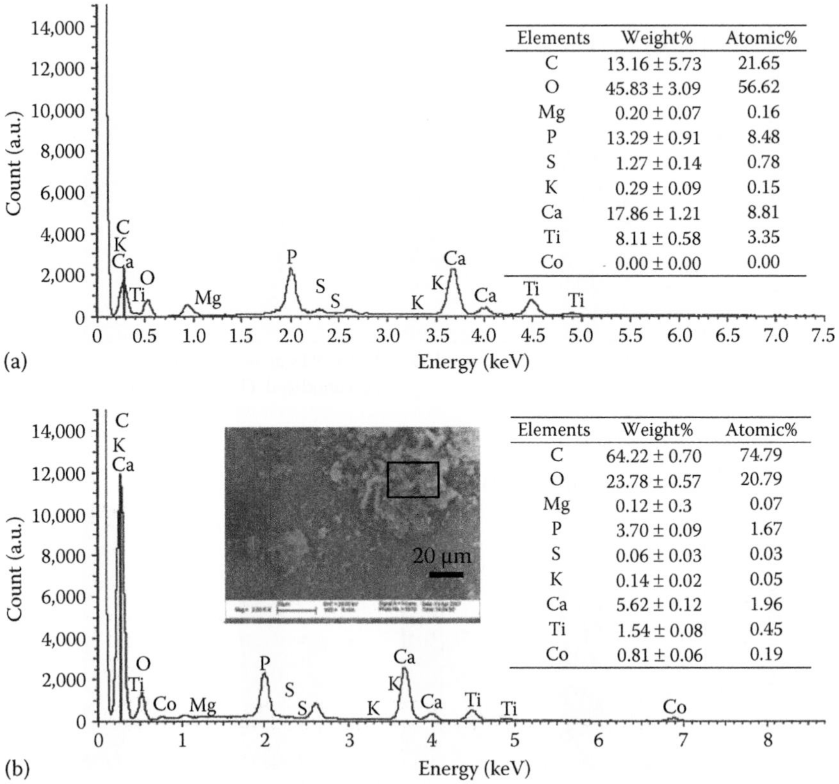

(a)

Elements	Weight%	Atomic%
C	13.16 ± 5.73	21.65
O	45.83 ± 3.09	56.62
Mg	0.20 ± 0.07	0.16
P	13.29 ± 0.91	8.48
S	1.27 ± 0.14	0.78
K	0.29 ± 0.09	0.15
Ca	17.86 ± 1.21	8.81
Ti	8.11 ± 0.58	3.35
Co	0.00 ± 0.00	0.00

(b)

Elements	Weight%	Atomic%
C	64.22 ± 0.70	74.79
O	23.78 ± 0.57	20.79
Mg	0.12 ± 0.3	0.07
P	3.70 ± 0.09	1.67
S	0.06 ± 0.03	0.03
K	0.14 ± 0.02	0.05
Ca	5.62 ± 0.12	1.96
Ti	1.54 ± 0.08	0.45
Co	0.81 ± 0.06	0.19

FIGURE 6.5 EDS analysis of osteoblasts cultured for 21 days on (a) Ti and (b) MWCNT-Ti. SEM micrograph of inset (b) shows the analyzed area. For the MWCNT-Ti, more Ca and P deposited by osteoblasts were observed. The tables in (a) and (b) show the composition of the mineral deposits after osteoblasts were cultured for 21 days. The Ca/P weight ratio on bare Ti was 1.32 and on MWCNT-Ti it was 1.52.

MWCNT-Ti may contribute to the observed redox process in CV. On an electrode surface, the native conformation of a protein may be retained (reversible or diffusion controlled) or distorted (irreversible or adsorption controlled), depending on the extent of the interactions (such as the electrostatic or covalent bonds) between them.

The type of electrolyte is important to the redox reaction because the capacitance gained and the scan rate window are dependent on it [34]. Since neither nitrogen nor argon were degassed when the cyclic voltammetry was performed, the redox reactions in the CV were related to the H_2O decomposition, which includes H^+ reduction and OH^- oxidation. The pH of the electrolyte also affects the H_2O decomposition. In the case of the acidic 1 N HCl solution (without other electroactive molecules), the H^+ ions induced the formation of H_2 gas at −0.2 V as a result of the increased reduction current [34]. In the alkaline 1 M KOH solution, OH^- ions resulted in the formation of O_2 gas at 0.7 V by the oxidation process. This fact should be noted because *in vivo*, these decompositions may occur at the interface of the Ti implant due to pH changes and the presence of H_2O and O_2 around the implant.

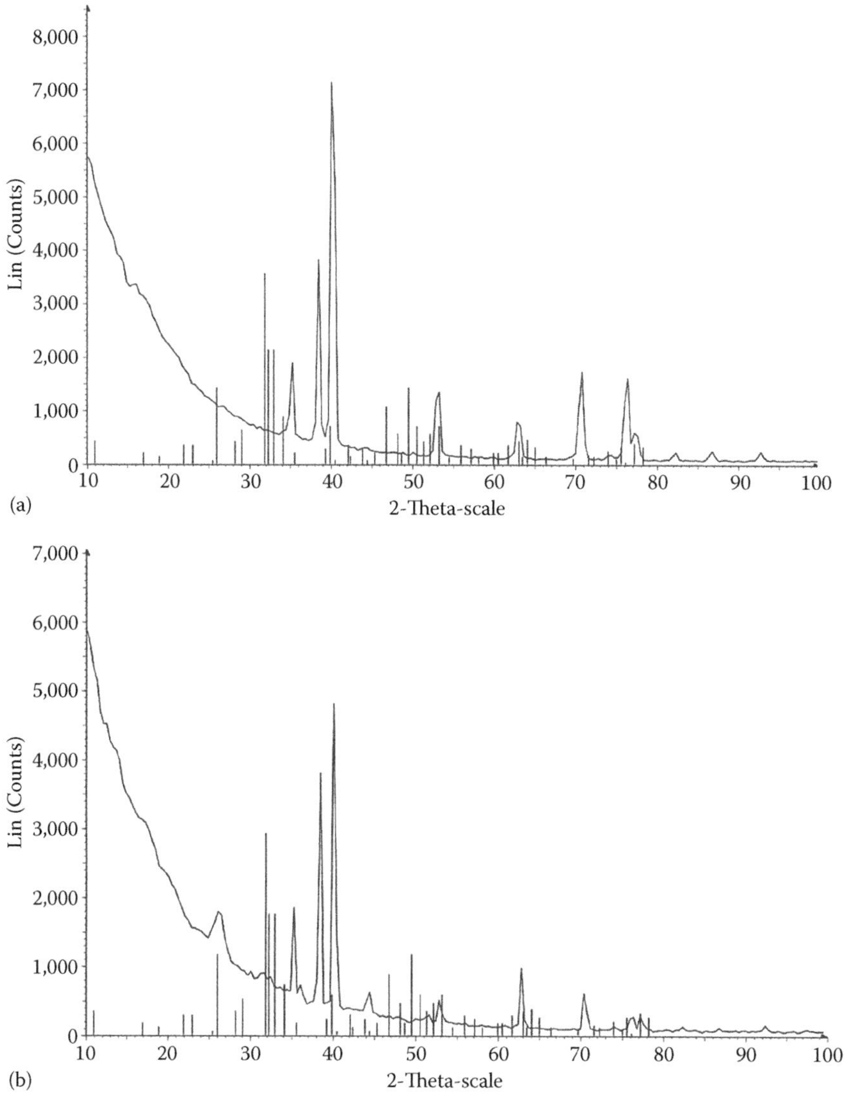

FIGURE 6.6 XRD analysis of hydroxyapatite-like (HA; $Ca_5(PO_4)_3OH$) deposited minerals after osteoblasts were cultured for 21 days on (a) Ti and (b) MWCNT-Ti. The micrographs show that the peak pattern of HA more closely matches that of the mineral deposited by osteoblasts when cultured on MWCNT-Ti than Ti. The y axis is the number of counts at each angular position that gives the line intensity of the x-ray.

6.3.4 Electrochemical Behavior of $Fe^{2+/3+}$ Redox Couple at Ti, Anodized Ti, and MWCNT-Ti

The $Fe(CN)_6^{4-/3-}$ redox system, which exhibits heterogeneous one-electron transfer ($n = 1$), is one of the most extensively studied redox couples in electrochemistry [35].

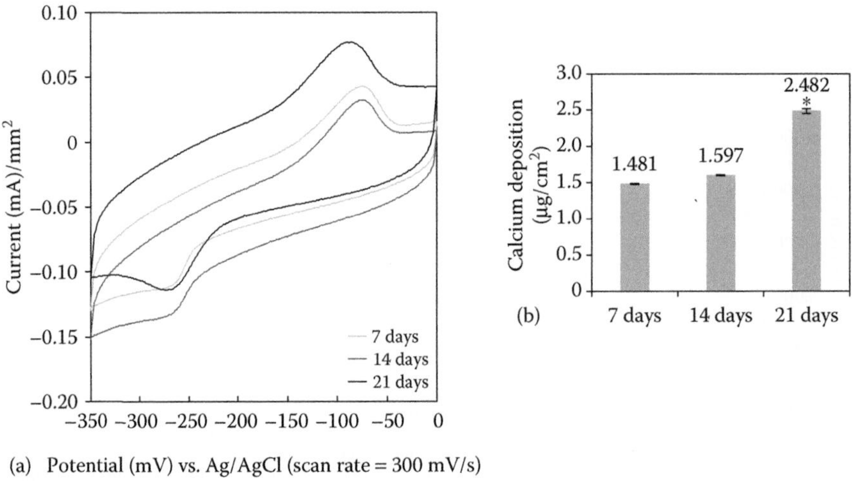

FIGURE 6.7 (a) Cyclic voltammograms of MWCNT-Ti electrodes in an electrolyte solution of the extracellular matrix secreted by osteoblasts cultured on conventional Ti after 21 days. (b) Results of a calcium deposition assay determining the calcium concentrations in an electrolyte solution of the extracellular matrix secreted by osteoblasts on conventional Ti after 7, 14, and 21 days of culture. Data = mean ± SEM; $n = 3$; $*p < 0.01$ (compared with 7 and 14 days).

We performed cyclic voltammetry experiments of the $Fe^{2+/3+}$ redox couple by placing MWCNT-Ti in an electrolyte solution of 10 mM $K_3Fe(CN)_6$ and 1 M KNO_3. In $K_3Fe(CN)_6$, the reduction process is Fe^{3+} ($Fe(CN)_6^{3-} + e^- \rightarrow Fe(CN)_6^{4-}$) followed by the oxidation of Fe^{2+} ($Fe(CN)_6^{4-} \rightarrow Fe(CN)_6^{3-} + e^-$) under a sweeping voltage. The iron (II/III) redox couple did not exhibit any observable peaks for bare Ti or anodized Ti electrodes, as shown in Figure 6.8a and b. This implies that the electrochemical reaction was slow on both these electrodes. However, the highly direct electron transfers at the MWCNT-Ti electrode were observed as redox peaks, shown in Figure 6.7c. At a scan rate of 100 mV/s in Figure 6.8c, a well-defined redox peak appears with the anodic (E_{pa}) and cathodic (E_{pc}) potentials at 175 and 345 mV, respectively. Moreover, in Figure 6.8e, the relationship is linear between the anodic and cathodic peak currents versus the square root of the scan rate, while the ratio of I_{pa}/I_{pc} is about 1, corresponding to the Randles–Sevcik equation (Equation 6.1). Because the root scan rate has this linear relation with the peak currents, the mass transport in this process must be by diffusion. Zhang et al. found that the heterogeneous charge-transfer rate constant (k) of $Fe(CN)_6^{4-/3-}$ complex with H_2O as a solvent is 0.05 cm/s [36]. Since the k value is in the range of 10^{-4}–10^{-1} cm/s and $\Delta E_p > 59/n$ mV (in this case $n = 1$ and $\Delta E_p \sim 170$ mV), this process is quasi-reversible.

To analyze the electrochemical behavior at the surface of the MWCNT-Ti electrode, we used the Randles–Sevcik equation for quasi-reversible processes (Equation 6.1). Hence, the peak current (I_p) is given by

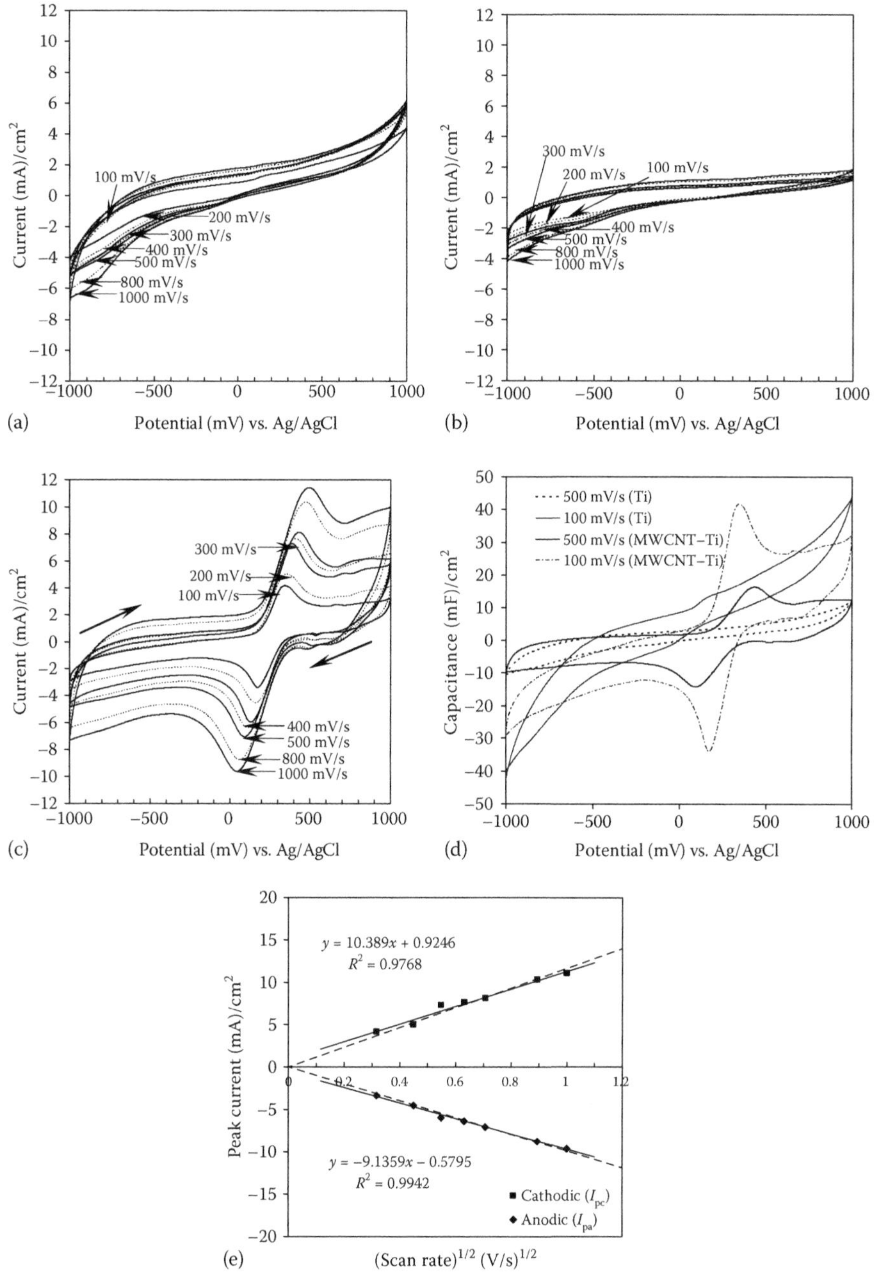

FIGURE 6.8 Cyclic voltammograms with an electrolyte solution of 10 mM $K_3Fe(CN)_6$ in 1 M KNO_3 for (a) conventional Ti; (b) anodized Ti; and (c) MWCNT-Ti. (d) The capacitance of all the electrodes in comparison. (e) A plot of the square root of the scan rates with anodic peak currents (I_{pa}) and cathodic peak currents (I_{pc}).

$$I_p = 2.99 \times 10^5 A D^{1/2} n (n_a \gamma)^{1/2} C \tag{6.1}$$

where:

- n is the number of electrons participating in the redox process
- n_a is the number of electrons participating in the charge-transfer step
- A is the area of the working electrode (cm^2)
- D is the diffusion coefficient of the molecules in the electrolyte solution (cm^2/s)
- C is the concentration of the probe molecule in the bulk solution (molar)
- γ is the scan rate of the sweep potential (V/s)

When the $Fe(CN)_6^{4-/3-}$ redox system exhibits heterogeneous one-electron transfer ($n = n_a = 1$) and the concentration C is equal to 10 mM, the diffusion coefficient D is equal to $6.7 \pm 0.02 \times 10^{-6}$ cm^2/s [37,38]. Hence, the quasi-reversible redox of iron (II/III) is truly enhanced by MWCNTs.

Other studies have shown promising results when using CNT-modified electrodes in biosensing. Since MWCNTs have good electrochemical characteristics as electron mediators and adsorption matrices [39], they may further enhance applications in biosensor systems. For example, Harrison et al. showed that CNTs offer a promising method to enhance detection sensitivity because they have high signal-to-noise ratios [11]. The structure-dependent metallic character of CNTs should allow them to promote electron-transfer reactions for redox processes, which can provide the foundation for unique biochemical sensing systems at low overpotentials [19]. The electrolyte–electrode interface barriers are reduced by CNTs because they facilitate double-layer effects [40]. Typically, when the supporting electrolyte is at least a hundredfold greater than the active electrolyte [41], the charge in the electrolyte solution causes the Debye layer to be more compact. Therefore, this compact layer can rapidly exchange electrons between electroactive proteins and the surface of an MWCNT-Ti electrode. This is likely the reason why we observed the sharpened cathodic and anodic peaks in CV, as shown in Figures 6.8a and c, 6.9c, and 6.10b.

Although the surface area of the working electrode is constant in the presence of MWCNTs, the overall surface area and the electroactivity of the surface increased. MWCNTs act as a nanobarrier between the TiO_2 layer, which is its growth template, and the electrolyte solution. Due to their nanosize effects, the surface atoms or molecules of MWCNTs play an important role in determining their bulk properties [42]. A large surface, corresponding to a greater electrocatalytic activity, confers the catalytic role on MWCNTs in the chemical reaction. Well-defined and persistent redox peaks are shown in Figure 6.8c, confirming the increases in electron transfers and higher electrochemical activities at the MWCNT-Ti surface. Cyclic voltammetry was also performed with a glassy carbon electrode (GCE; Bioanalytical) and a Pt electrode (PTE; Bioanalytical), which showed $\Delta E_p > 59/n$ mV (data not shown). Importantly, for the same scan rate, a CV of GCE and PTE showed redox reactions with similar but more widely separated anodic and cathodic peaks, confirming the performance of the MWCNT-Ti electrode.

However, Figure 6.8a and b illustrate that the Ti and anodized Ti do not show any redox peaks, which is likely due to the inert inhibiting property of TiO_2 on electron transfer through them. Figure 6.8a and b show that a bare Ti has less capacitance

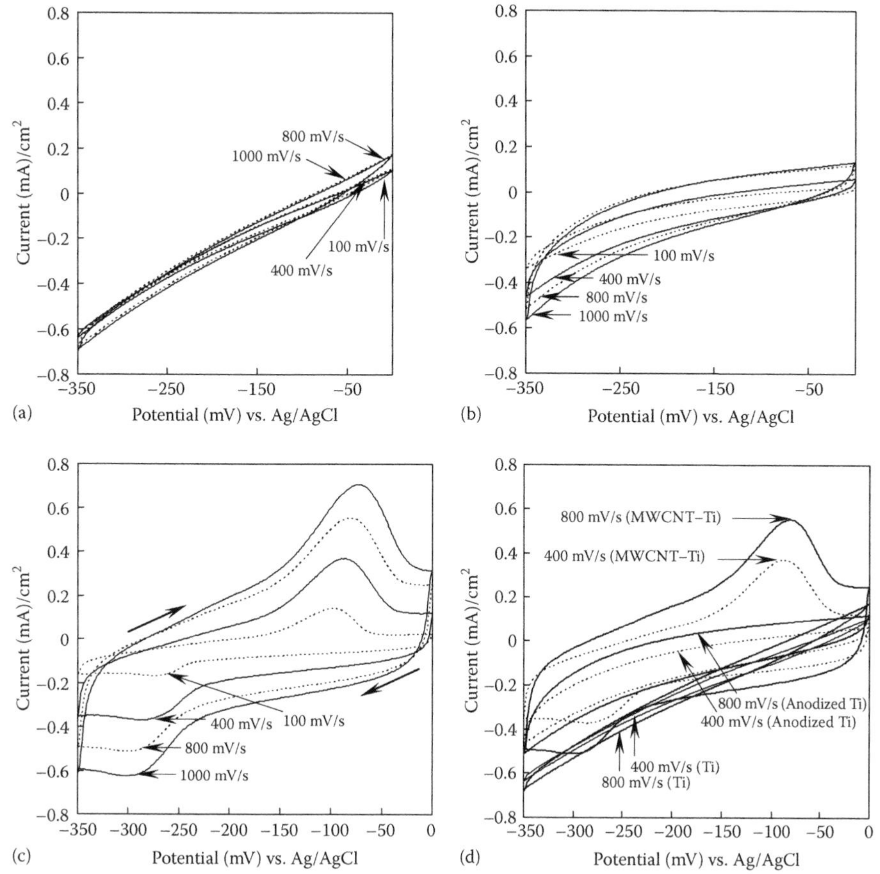

FIGURE 6.9 Cyclic voltammograms with an electrolyte solution of the extracellular matrix secreted by osteoblasts after 21 days of culture for (a) conventional Ti; (b) anodized Ti; and (c) MWCNT-Ti with a working area of 1 cm^2. (d) CV of all three electrodes in comparison. Only MWCNT-Ti possesses the quasi-reversible redox potential, while conventional Ti and anodized Ti do not.

than an MWCNT-Ti. The capacitance relationship is derived from $i = C(dv/dt)$, where dv/dt is a scan rate [35]. Hence, the capacitance between the working and reference electrodes during cyclic voltammetry is calculated by dividing the current in a CV with respect to the specific scan rate, as shown in Figures 6.8d and 6.10d. In summary, the electrochemical response of the MWCNT-Ti electrode promotes a higher charge transfer than the Ti and anodized Ti electrodes.

6.3.5 Electrochemical Behavior at Ti, Anodized Ti, and MWCNT-Ti in the Solution of Osteoblast Extracellular Components

The result from the previous section shows the utility of MWCNT-Ti as an electrode for detecting $Fe^{2+/3+}$ redox couples. For biological applications, it is also

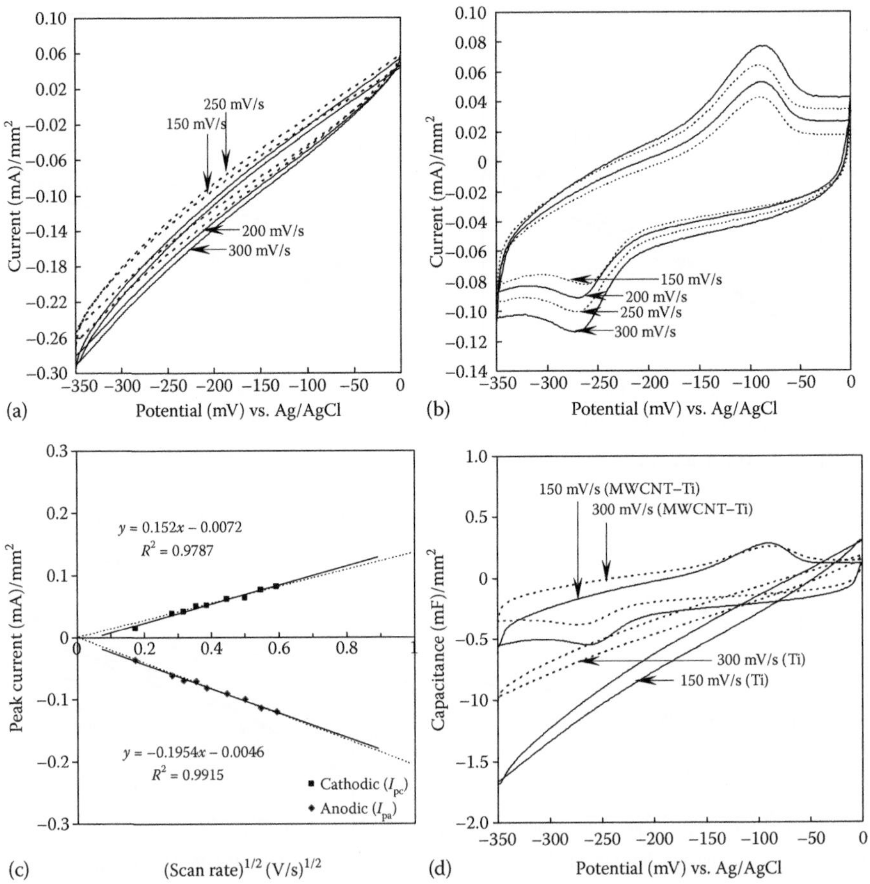

FIGURE 6.10 Cyclic voltammograms with an electrolyte solution of the extracellular matrix secreted by osteoblasts after 21 days of culture for (a) conventional Ti and (b) MWCNT-Ti with a working area of 1 mm^2. (c) Plot of the experimental cathodic and anodic peak currents, obtained from (b), versus the square root of the scan rates; and (d) the compared capacitance of MWCNT-Ti and Ti.

necessary to show that MWCNT-Ti can detect redox reactions associated with osteoblast differentiation. We used extracellular components from osteoblasts in an HCl solution to mimic the biological environment around an orthopedic implant. Figure 6.9a and b show that bare Ti and anodized Ti electrodes cannot detect any redox processes, while the results in Figure 6.9c confirm that MWCNTs enhance direct electron transfer through Ti by adding the high-conductivity surface of MWCNTs.

After decreasing the surface area of an MWCNT-Ti electrode from 1 cm^2 to 1 mm^2, the redox potentials were also decreased, as shown in Figures 6.9c and 6.10b. The faradic current of the oxidation process dropped approximately 10 times with respect to the decrease in the surface area A, corresponding to Equation 6.1.

Corresponding to the Randles–Sevcik equation, the peak current was linearly proportional to the area. When plotting the anodic (I_{pa}) and cathodic peak (I_{pc}) currents of an MWCNT-Ti, a linear relationship to the square root of the scan rates was observed, as shown in Figure 6.10c.

A CV showed well-defined redox peaks in the osteoblast supernatants at all concentrations. The peaks detected in the solution at the Ca concentrations of 1.481 μg/cm^2 (after 7 days) and 1.597 μg/cm^2 (after 14 days) have less faradic current and I-V graph area than the concentration of 2.48 μg/cm^2 (after 21 days), as shown in Figure 6.7a and b. Typically, the more HA deposited and synthesized by osteoblasts, the higher the Ca and protein concentrations is. Therefore, it is likely that the supernatant of 21 days had higher protein concentrations than the solutions of 7 and 14 days, leading to the stronger redox reactions and higher capacitance between the electrode surfaces.

In addition, an interpretation of the CV results must consider other factors. In particular, the measurement of the redox reactions when oxygen was dissolved in the electrolyte solution showed that the peak currents were shifted toward the negative. Furthermore, a change in the pH of the electrolyte solution and the presence of water may also shift the current and affect the potential of the redox peaks. Thus, cyclic voltammetry with the potential range between −1 and 1 V of MWCNT-Ti was performed in a pure solution of 0.6 N HCl without other proteins or ions. A H^+ reduction peak at a potential between −0.5 and −0.8 V was found (data not shown). Moreover, the H^+ reduction peak still appeared when window ranges from −5 to 5 V and −0.35 to 0 V were used, confirming that H_2O decomposition occurred in our CV experiments.

The MWCNT-Ti electrodes showed the potential for better performance than commercially available electrodes, such as GCE and PTE. When using the sweep voltage from −1 to 1 V, a set of mixed redox peaks with two oxidations and two reductions appeared (data not shown). These two redox peaks in the CV indicated that the process likely involved more than one electron transfer. For each couple, the amplitude and position of the oxidation peak in the CV were unequal to those of the reduction peak. The second oxidation peak (omitted from Figures 6.6b or 6.9c) was observed at 236 mV for the MWCNT-Ti electrode, 405 mV for the GCE, and −33 mV for the PTE, all at a scan rate of 100 mV/s. However, for the GCE and PTE electrodes, a second reduction did not appear, while the MWCNT-Ti showed two complete and well-defined redox reactions. These two peaks reflect that more than one electron transfer is involved, since one occurs at a positive potential and one at a negative potential in the CV. Further, the current amplitude for the MWCNT-Ti peaks was noticeably larger than that for GCE and PTE. Thus, it can be concluded that the MWCNT-Ti electrode has a higher electron transfer as well.

At first, we hypothesized that these redox reactions happened due to the redox of inorganics such as Ca or P. In the case of Ca, the reduction of Ca^{2+} to Ca crystal (nucleation) is typically expected with the reduction current for a highly negative potential. A positive feedback IR compensation (IRC) needs to be applied to the cyclic voltammetry in order to drive the oxidation current to dissolve the Ca crystal into Ca^{2+}. For example, an overpotential of more than −1 V, an IRC of 0.771 Ω, a melting temperature of 900°C, and a scan rate of 50 mV/S are required [43]. As IRC was not used in this study, the diffusion-controlled behavior, which electrochemically depletes Ca cations near the electrode surface, was impossible to achieve. In addition, the reduction and oxidation of titanium oxides, which are the reaction counterparts of Ca redox, appeared at a positive potential between 0 and 0.25 V, respectively, which did not appear in our CV. In the case of P, further cyclic voltammetry experiments were performed in PBS. A final concentration of 1 M PBS, which was prepared in-house, had 137 mM NaCl, 10 mM phosphate, 2.7 mM KCl, and a pH of 7.4. Metal orthophosphate ions exhibit high stability to reduce from P(V) (such as PO_4^{3-}, PO_4^-) to P(III), or further to P(I). However, no redox peaks appeared in the potential range of −1 to 1 V (data not shown). It is therefore likely that neither the redox reactions of Ca ions nor phosphate occurred due to the absence of the peaks. Instead, we hypothesized that the appearance of the redox peaks when using an MWCNT-Ti electrode was likely because of the reduction and oxidation of proteins from the osteoblast supernatant.

The observed quasi-reversible redox reaction in this study reflected the fact that MWCNT-Ti promotes electrochemical reactions. However, an irreversible process can occur at the interface of an electrode due to the denaturation of proteins, or ion reduction. Typically, when the carbon bonds of proteins are changed, irreversible organic reduction or oxidation happens [44]. However, unchanged ion configurations (Ca^{2+}, PO_4^{3-}, OH^-) can also lead to irreversible processes, solely through reduction or oxidation. Importantly, without MWCNTs on the Ti surface, nonspecific protein adsorption may readily occur, leading to an electrochemically insulating surface layer. Nevertheless, the redox at MWCNT-Ti did not decrease its oxidation and reduction currents after applying the sweep voltage for some time, so protein adsorption did not seem to occur significantly at the MWCNT-Ti electrode. Indeed, in the osteoblast supernatant, the MWCNT-Ti electrode provides a relatively specific surface to which the proteins can bind reversibly by diffusion while retaining their function. The protein redox is also dependent on the pH of the electrolyte solution (acidic or basic) [45]. This shift appears due to the influence of 0.6 N HCl with H_2O as a solvent influences the proteins, leading the peak shift into the negative potential, as shown in Figures 6.7a, 6.9b and 6.10c.

We hypothesized that proteins were involved in these redox reactions at the surface of MWCNT-Ti in the solution of osteoblast extracelluar components. If these proteins maintain their function and are detected by the MWCNT-Ti electrode, the excellent electrochemical behavior observed here may show MWCNT-Ti as a superior candidate for biosensor applications. However, investigating the types and sizes of acidic-resistant proteins is still in progress.

6.3.6 Applications and Limitations

As shown here, the performance of the MWCNT-Ti electrode, but not the bare Ti, provided excellent redox reactions. However, one has to consider that there are some specific conditions used in this study. As mentioned, the electrochemical experiments were performed without degassing with bubble nitrogen gas.

This omission may have resulted in an apparent background signal, which was mixed and appeared in our current–voltage curves. Moreover, the redox reaction at the MWCNT-Ti electrode in this study may depend on various ion sources, including reversible protein adsorption, dissolved oxygen, and the acidic nature of HCl. However, the redox reactions of iron (II/III) and the osteoblast extracellular components occurred only when using MWCNT-Ti, and not when using Ti or anodized Ti. Importantly, we showed that MWCNTs performed as electrocatalysts for the oxidation and reduction of iron (II/III) and of extracellular components, which included various proteins and inorganic substances that were synthesized and deposited by osteoblasts.

The electrolyte solution of osteoblast extracellular components was composed of inorganic and organic substances. We did not classify the types of molecules in the solution, but rather, all components were dissolved together in the acidic solution, and were used in all cyclic voltammetry experiments in this study. In future work, we aim to classify the protein types in our osteoblast supernatant.

However, it is likely that *in vivo* the capacitance and the electroactive surface of Ti will be enhanced after bone tissue is formed without the presence of MWCNTs. After a hip implant, blood and body fluids surround the Ti implant, creating a specific capacitance for a Ti electrode. Within 1 month, as osteoblasts deposit Ca around the implant, the capacitance of the implant and bone tissue increases. Cell membranes also have electric potentials and perform redox processes [46,47]; we found that after 14 days *in vitro* these redox processes at the surface of bare Ti had increased (data not shown).

In vitro, MWCNTs have been used to modify electrodes to detect other biomolecules and enhance their redox reactions, such as nicotinamide adenine dinucleotide (NADH) [48], hydrogen peroxide [21,48], glucose [22], putrescine [49], and DNA [50]. Specifically, the growth of MWCNTs from nanoporous metal electrodes can enhance many diverse sensor applications. For example, MWCNTs functionalized with thiol groups have been used for sensing aliphatic hydrocarbons (such as methanol and ethanol), forming unique electrical identifiers [18]. Moreover, MWCNTs grown on silicon substrates that enhanced ionic conductivity were integrated with unmodified plant cellulose as a film, and used as a cathode electrode in a lithium ion battery, or as a supercapacitor with bioelectrolytes (such as biological fluid and blood) [51]. Interestingly, both the supercapacitor and battery, derived from an MWCNTs nanocomposite film, can be integrated together to build a hybrid or dual-storage battery-in-supercapacitor device. The cytocompatible MWCNTs device may be useful to integrate with an orthopedic-implanted biosensor as passive energy storage.

Not only can an implanted electronics circuit be powered by an implant battery (fabricated or self-integrated) within the implant material, but it can also be induced by an external power supply. For orthopedic application, the use of telemetric devices began in 1966 and telemetric orthopedic implants became imperative in order to transmit a signal to an external device [52–57]. For example, Graichen et al. developed a 9-channel telemetry transmitter used for *in vivo* load measurements in three patients with shoulder Ti endoprostheses [58]. The low radio frequency (RF) of the externally inductive power source generates power transmission through the implanted metal. At the end of the Ti implant, the pacemaker feedthrough rounds a single-loop antenna to transmit the pulse interval signal, which is modulated at a higher RF by the microcontroller, to an external device. Then, the RF receiver of the external device synchronizes with the modulated pulse interval to recover the data stream and report to a clinician. Thus, either an implanted battery or a telemetry system can supply the electric power to a Ca-detectable chip inside orthopedic implants for clinical use.

The real-time sensing concept in orthopedic implants for bone growth can also apply in cases of infection and scar tissue formation. An energy source energizes a telemetric-implanted circuit and allows it to transmit data to an external receiver to determine bone growth. Indeed, the electrochemical biosensor may reduce the complexity of imaging techniques and patient difficulties.

6.4 CONCLUSIONS

The capabilities of Ti as an electrochemical electrode increased remarkably by growing MWCNTs out of anodized Ti nanopores (MWCNT-Ti). MWCNTs improved the sensitivity of the bare Ti electrode displaying redox peaks in CVs and interestingly high capacitance. These results provide evidence that MWCNT-Ti can serve as a novel electrochemical electrode with great electrocatalytic properties due to its increased surface area and conductivity. Moreover, MWCNTs were shown to be more highly electroactive in chemical transformations than the metallic (Ti) surface, which typically undergoes electrochemical oxidation or dissolution of metal oxides, and is chemically susceptible to corrosion. MWCNTs promoted a redox reaction by enhancing the direct electron transfer through their electrically conductive surface surrounded by ionic solutions, which contained the electroactive species, herein ferri/ferricyanide and the extracellular components from osteoblasts. Moreover, MWCNTs are cytocompatible, promote osteoblast differentiation after 21 days, and can be integrated into a supercapacitor or battery to enhance the functionalities of biosensing systems *in vivo*. Therefore, MWCNT-Ti constitutes an exceptional candidate as an electrochemical electrode to determine *in situ* new bone growth surrounding an orthopedic implant. Further, the ability of MWCNT-Ti to sense osteoblast extracellular components by detecting their redox reaction profiles in specific concentrations may improve the diagnosis of orthopedic implant success or failure, and thus improve clinical efficacy.

ACKNOWLEDGMENTS

The authors would like to thank the Coulter Foundation for funding, Professor Brian W. Sheldon for the chemical vapor deposition (CVD) facility, Anthony W. McCormick and Geoffrey Williams for their help with the scanning electron microscope, and Maryam Jouzi for her help with an electrochemical workstation.

REFERENCES

1. Zhan C, Kaczmarek R, Loyo-Berrios N, Sangl J, Bright RA. Incidence and short-term outcomes of primary and revision hip replacement in the United States. *Journal of Bone and Joint Surgery – American* 89(3):526–533, 2007.

2. Iijima S. Helical microtubules of graphitic carbon. *Nature* 354(6348):56–58, 1991.
3. Lin Y, Taylor S, Li H, Fernando KAS, Qu L, Wang W, et al. Advances toward bioapplications of carbon nanotubes. *Journal of Materials Chemistry* 14(4):527–541, 2004.
4. Webster TJ, Waid MC, McKenzie JL, Price RL, Ejiofor JU. Nano-biotechnology: Carbon nanofibres as improved neural and orthopaedic implants. *Nanotechnology* 15(1):48–54, 2004.
5. Smart SK, Cassady AI, Lu GQ, Martin DJ. The biocompatibility of carbon nanotubes. *Toxicology of Carbon Nanomaterials* 44(6):1034–1047, 2006.
6. Zanello LP. Electrical properties of osteoblasts cultured on carbon nanotubes. *Micro and Nano Letters* 1(1):19–22, 2006.
7. Zanello LP, Zhao B, Hu H, Haddon RC. Bone cell proliferation on carbon nanotubes. *Nano Letters* 6(3):562–567, 2006.
8. Balani K, Anderson R, Laha T, Andara M, Tercero J, Crumpler E, et al. Plasma-sprayed carbon nanotube reinforced hydroxyapatite coatings and their interaction with human osteoblasts in vitro. *Biomaterials* 28(4):618–624, 2007.
9. Chen Y, Zhang TH, Gan CH, Yu G. Wear studies of hydroxyapatite composite coating reinforced by carbon nanotubes. *Carbon* 45(5):998–1004, 2007.
10. Wei W, Sethuraman A, Jin C, Monteiro-Riviere NA, Narayan RJ. Biological properties of carbon nanotubes. *Journal of Nanoscience and Nanotechnology* 7:1284–1297, 2007.
11. Harrison BS, Atala A. Carbon nanotube applications for tissue engineering. *Cellular and Molecular Biology Techniques for Biomaterials Evaluation* 28(2):344–353, 2007.
12. Supronowicz PR, Ajayan PM, Ullmann KR, Arulanandam BP, Metzger DW, Bizios R. Novel current-conducting composite substrates for exposing osteoblasts to alternating current stimulation. *Journal of Biomedical Materials Research* 59(3):499–506, 2002.
13. Ciombor DM, Aaron RK. The role of electrical stimulation in bone repair. *Orthobiologics* 10(4):579–593, 2005.
14. Aoki N, Yokoyama A, Nodasaka Y, Akasaka T, Uo M, Sato Y, et al. Cell culture on a carbon nanotube scaffold. *Journal of Biomedical Nanotechnology* 1:402–405, 2005.
15. Giannona S, Firkowska I, Rojas-Chapana J, Giersig M. Vertically aligned carbon nanotubes as cytocompatible material for enhanced adhesion and proliferation of osteoblast-like cells. *Journal of Nanoscience and Nanotechnology* 7:1679–1683, 2007.
16. Tsuchiya N, Sato Y, Aoki N, Yokoyama A, Watari F, Motomiya K, et al. Evaluation of multi-walled carbon nanotube scaffolds for osteoblast growth. In: Tohji K, Tsuchiya N, Jeyadevan B, (eds), *Water Dynamics: 4th International Workshop on Water Dynamics.* AIP Conference Proceedings, pp. 166–169. 7–8 November 2006, Sendai, Japan, 2007.
17. Sirivisoot S, Yao C, Xiao X, Sheldon BW, Webster TJ. Greater osteoblast functions on multiwalled carbon nanotubes grown from anodized nanotubular titanium for orthopedic applications. *Nanotechnology* 18(36):365102, 2007.
18. Padigi SK, Reddy RKK, Prasad S. Carbon nanotube based aliphatic hydrocarbon sensor. *Biosensors and Bioelectronics* 22(6):829–837, 2007.
19. Roy S, Vedala H, Choi W. Vertically aligned carbon nanotube probes for monitoring blood cholesterol. *Nanotechnology* 17(4):S14–S18, 2006.
20. Tang H, Chen JH, Huang ZP, Wang DZ, Ren ZF, Nie LH, et al. High dispersion and electrocatalytic properties of platinum on well-aligned carbon nanotube arrays. *Carbon* 42(1):191–197, 2004.
21. Kurusu F, Koide S, Karube I, Gotoh M. Electrocatalytic activity of bamboo-structured carbon nanotubes paste electrode toward hydrogen peroxide. *Analytical Letters* 39:903–911, 2006.
22. Liu Y, Wang M, Zhao F, Xu Z, Dong S. The direct electron transfer of glucose oxidase and glucose biosensor based on carbon nanotubes/chitosan matrix. *Biosensors and Bioelectronics* 21(6):984–988, 2005.

23. Liu S, Lin B, Yang X, Zhang Q. Carbon-nanotube-enhanced direct electron-transfer reactivity of hemoglobin immobilized on polyurethane elastomer film. *Journal of Physical Chemistry B* 111(5):1182–1188, 2007.
24. Robertson J. Realistic applications of CNTs. *Materials Today* 7(10):46–52, 2004.
25. Talapatra S, Kar S, Pal SK, Vajtai R, Ci L, Victor P, et al. Direct growth of aligned carbon nanotubes on bulk metals. *Nature Nanotechnology* 1:112–116, 2006.
26. Ngo Q, Petranovic D, Krishnan S, Cassell AM, Ye Q, Li J, et al. Electron transport through metal-multiwall carbon nanotube interfaces. *IEEE Transactions on Nanotechnology* 3:311–317, 2004.
27. Sato S, Kawabata A, Kondo D, Nihei M, Awano Y. Carbon nanotube growth from titanium-cobalt bimetallic particles as a catalyst. *Chemical Physics Letters* 402(1–3):149–154, 2005.
28. Shulz F, Bond AM. *Electroanalytical Methods: Guide to Experiments and Applications.* Berlin: Springer, 2002.
29. Calafiori A, Di Marco G, Martino G, Marotta M. Preparation and characterization of calcium phosphate biomaterials. *Journal of Materials Science: Materials in Medicine* 18:2331–2338, 2007.
30. Wang H, Lee J-K, Moursi A, Lannutti JJ. Ca/P ratio effects on the degradation of hydroxy-apatite in vitro. *Journal of Biomedical Materials Research Part A* 67A(2):599–608, 2003.
31. Zaidi M, Moonga BS, Huang CL-H. Calcium sensing and cell signaling processes in the local regulation of osteoclastic bone resorption. *Biological Reviews* 79(1):79–100, 2004.
32. Blair HC, Zaidi M, Schlesinger PH. Mechanisms balancing skeletal matrix synthesis and degradation. *Biochemical Journal* 364(2):329–341, 2002.
33. Bronner F, Farach-Carson MC. *Bone Formation.* New York: Springer, 2003.
34. Hong MS, Lee SH, Kim SW. Use of KCl aqueous electrolyte for 2 V manganese oxide/activated carbon hybrid capacitor. *Electrochemical and Solid-State Letters* 5(10):A227–A230, 2002.
35. Tamir G, Moti B-D, Itshak K, Raya S, Ze'ev RA, Eshel B-J, et al. Electro-chemical and biological properties of carbon nanotube based multi-electrode arrays. *Nanotechnology* 18(3):035201, 2007.
36. Zhang Y, Baer CD, Camaioni-Neto C, O'Brien P, Sweigart DA. Steady-state voltammetry with microelectrodes: Determination of heterogeneous charge transfer rate constants for metalloporphyrin complexes. *Inorganic Chemistry* 30(8):1682–1685, 1991.
37. Hrapovic S, Liu Y, Male KB, Luong JHT. Electrochemical biosensing platforms using platinum nanoparticles and carbon nanotubes. *Analytical Chemistry* 76(4):1083–1088, 2004.
38. Hrapovic S, Luong JHT. Picoamperometric detection of glucose at ultrasmall platinum-based biosensors: Preparation and characterization. *Analytical Chemistry* 75(14):3308–3315, 2003.
39. Sotiropoulou S, Gavalas V, Vamvakaki V, Chaniotakis NA. Novel carbon materials in biosensor systems. *Biosensors and Bioelectronics* 18(2–3):211–215, 2003.
40. Fang WC, Sun CL, Huang JH, Chen LC, Chyan O, Chen KH, et al. Enhanced electrochemical properties of arrayed CNx nanotubes directly grown on Ti-buffered silicon substrates. *Electrochemical and Solid State Letters* 9(3):A175–A178, 2006.
41. Christian GD. *Analytical Chemistry.* New York: Wiley, 1980.
42. Cao G. *Nanostructures and Nanomaterials: Synthesis, Properties and Applications.* London: Imperial College Press, 2004.
43. Chen GZ, Fray DJ. Voltammetric studies of the oxygen-titanium binary system in molten calcium chloride. *Journal of the Electrochemical Society* 149(11):E455–E467, 2002.
44. Schuring J. *Redox Fundamentals, Processes, and Application.* New York: Springer, 2000.

45. Battistuzzi G, Loschi L, Borsari M, Sola M. Effects of nonspecific ion-protein interactions on the redox chemistry of cytochrome c. *Journal of Biological Inorganic Chemistry* 4(5):601–607, 1999.
46. Jeansonne BG, Feagin FF, McMinn RW, Shoemaker RL, Rehm WS. Cell-to-cell communication of osteoblasts. *Journal of Dental Research* 58(4):1415–1423, 1979.
47. Jeansonne BG, Feagin FF, Shoemaker RL, Rehm WS. Transmembrane potentials of osteoblasts. *Journal of Dental Research* 57(2):361–364, 1978.
48. Lin Y, Yantasee W, Wang J. Carbon nanotubes (CNTs) for the development of electrochemical biosensors. *Frontiers in Bioscience: A Journal and Virtual Library* 10:492–505, 2005.
49. Luong JHT, Hrapovic S, Wang D. Multiwall carbon nanotube (MWCNT) based electrochemical biosensors for mediatorless detection of putrescine. *Electroanalysis* 17(1):47–53, 2005.
50. He P, Xu Y, Fang Y. Applications of carbon nanotubes in electrochemical DNA biosensors. *Microchimica Acta* 152(3):175–186, 2006.
51. Pushparaj V, Shaijumon M, Kumar A, Murugesan S, Ci L, Vajtai R, et al. Flexible energy storage devices based on nanocomposite paper. *Proceedings of the National Academy of Sciences* 104(34):13574–13577, 2007.
52. Kaufman K, Kovacevic N, Irby S, Colwell C. Instrumented implant for measuring tibiofemoral forces. *Journal of Biomechanics* 29(5):667–671, 1996.
53. Bergmann G, Deuretzbacher G, Heller M, Graichen F, Rohlmann A, Strauss J, et al. Hip contact forces and gait patterns from routine activities. *Journal of Biomechanics* 34(7):859–871, 2001.
54. Bergmann G, Graichen F, Rohlmann A, Verdonschot N, van Lenthe GH. Frictional heating of total hip implants. Part 1: Measurements in patients. *Journal of Biomechanics* 34(4):421–428, 2001.
55. D'Lima DD, Townsend CP, Arms SW, Morris BA, Colwell CW. An implantable telemetry device to measure intra-articular tibial forces. *Journal of Biomechanics* 38(2):299–304, 2005.
56. Graichen F, Bergmann G. Four-channel telemetry system for in vivo measurement of hip joint forces. *Journal of Biomedical Engineering* 13(5):370–374, 1991.
57. Graichen F, Bergmann G, Rohlmann A. Hip endoprosthesis for in vivo measurement of joint force and temperature. *Journal of Biomechanics* 32(10):1113–1117, 1999.
58. Graichen F, Arnold R, Rohlmann A, Bergmann G. Implantable 9-channel telemetry system for in vivo load measurements with orthopedic implants. *IEEE Transactions on Bio-Medical Engineering* 54(2):253–261, 2007.

7 Nucleic Acid and Nucleoprotein Nanodevices

Steven S. Smith

CONTENTS

7.1 INTRODUCTION

Since its beginnings and popularization in computer science, nanotechnology has moved from the realm of fantasy [1] as an understanding of the chemistry of the nanoscale took hold [2], to the realm of gee-whiz proofs of concepts that have generated very few of the promised revolutionary approaches to biology and biotechnology. That is beginning to change as these early research efforts mature and provide new tools and fresh insights into biology. Here, I focus on a group of self-assembling nucleic acid and nucleoprotein nanodevices. In order to bring this branch of nanotechnology into focus, it is important to define both the nature of the scale implied and the meaning of the term *device*.

The term *nanoscale* implies that the device itself does not exceed 100 nm in any dimension. Since the devices under consideration are composed of molecules made up of multiple atoms with bond lengths in the 100 pm range, in general, they exceed

1 nm in at least one dimension. On this scale, nucleic acid and nucleoprotein nanodevices are essentially macromolecules or macromolecular assemblies. In the strictest sense, most biologically functional molecules fit the definition of a device since the concept encompasses that of both the tool and the machine. Tools are defined as physical items that can be "used to carry out a particular function" [3] and machines are defined as "assemblages of parts that impart force, motion or energy to one another in a predetermined fashion" [4]. Nanodevices composed of synthetic, redesigned, or repurposed nucleic acids and proteins have been successful in applications involving binding, catalysis, and molecular dynamics.

7.2 BINDING

7.2.1 Aptamers

Aptamers are short nucleic acids that have been selected using the principles of directed molecular evolution [5,6] for their ability to bind specific ligands [7,8]. In this process, a random-sequence library of short DNA sequences is prepared by chemical synthesis and generally amplified by DNA or RNA synthesis. Once amplified, the library is exposed to the desired ligand, and those members of the library that bind to the ligand are retained and copied to produce a simpler library that is subjected to additional rounds of binding, selection, and amplification until only a few sequences remain. In general, one finds that during the selection process the affinity of the selected pool for the ligand increases at each round until a tight-binding molecule is obtained. Unconjugated nucleic acid aptamers have been used in clinical trials to target thrombin as an anticoagulant [9] or nucleolin as an anticancer agent [10]. The thrombin aptamer is an unmodified 26-mer designated NU172. The nucleolin aptamer is also an unmodified 26-mer designated AS1411. It is composed largely of TGG triplet repeats. These two aptamers self-assemble from synthetic DNA made up of unmodified nucleotides. Aptamers for additional targets such as platelet-derived growth factor have been developed; however, these generally require modified nucleic acid linkages and chemical coupling to polyethylene glycol (see, e.g., [11]) in order to be effective.

7.2.2 Aptamer-Targeted Small Interfering RNA (siRNA)

While the availability of transfection has permitted siRNA to provide numerous advances in somatic cell genetics, its effectiveness as a therapeutic has been limited. In general, the key problems with siRNA delivery have been the stability of RNA in the bloodstream and the requirement for cellular internalization. Over the past few years, aptamers have been used to improve the delivery of siRNA to living cells and organisms. The anti-prostate-specific membrane antigen (PSMA) aptamers developed by Coffey's group [12] are among the most widely employed in this area. PSMA or glutamate carboxypeptidase II is overexpressed in prostatic tissue [13] compared with other tissue types. As a cell-surface glycoprotein it has the interesting property of being internalized [14], making it capable of selectively delivering bound ligands to prostate tissues.

The original aptamers selected for PSMA binding [12] were designated A10 and A9. Both have been used in full-length and truncated forms to deliver various siRNAs to the LNCaP prostate cancer cell line since this cell line retains the ability to express PSMA. For example, the PSMA aptamer covalently linked to siRNA against eukaryotic elongation factor 2 (EEF2) was cytotoxic to LNCaP cells, and this cytotoxicity could be enhanced by placing two copies of the PSMA aptamer in the synthetic RNA [15]. The resulting bivalent siRNA-aptamer system self-assembled into an active RNA nanodevice that could be readily taken up by LNCaP cells. Moreover, the bivalent aptamer was internalized more rapidly than the univalent control.

Refinements in the development of this system have yielded a number of additional approaches to siRNA delivery. Aptamer A10 has been modified to carry siRNA in the form of a short hairpin (shRNA). The passenger strand of siRNA against DNA protein kinase (DNAPK) was linked to the truncated A10 aptamer (A10-3) [12] through the 3′ terminus of the aptamer [16]. Since DNAPK downregulation induces radiosensitivity, the authors were able use nude mouse xenografts to demonstrate enhanced sensitivity to ionizing radiation in human tumor cell lines LNCaP and PC3-PIP (cell lines expressing PSMA). *Ex vivo* human tissues were also radiosensitized by this aptamer conjugate. In these experiments, the self-assembling aptamer was produced by T7 transcription from a DNA template. Thus, it lacked RNA stabilizing modifications and was directly injected into tumor sites in mouse xenografts. Systemic injection was not tested so the stability of the aptamer conjugate and its ability to concentrate at a tumor site were not tested [16].

Systemic delivery via tail vein injection was achieved [17] in a mouse model in which CT26 cell lines that expressed PSMA or control CT26 cells that did not express PSMA, xenografted onto the same mouse, were exposed to a PSMA-targeted siRNA against *Smg1*. Knockdown of *Smg1* inhibits nonsense-mediated mRNA decay (NMD) by blocking the formation of the mRNA surveillance complex. This results in the accumulation of nonsense RNA and the generation of an autologous immune response to *Smg1* knockdown cells. The use of the PSMA-directed nanodevice is especially noteworthy in that systemic delivery was successful in an immunocompetent mouse. Clinical trials of PSMA aptamer–targeted siRNAs have not yet been completed.

7.2.3 Multivalent Targeting

7.2.3.1 Tetravalent phi29 pRNA Targeting Device

The phi29 packaging motor contains a hexameric RNA held together by a kissing interaction between each of six monomers. Various RNA assemblies are possible based on this motif [18]. A stable four-way junction can be constructed by modifying the sequence of the monomer [19]. Each of the four arms of the monomer can be functionalized, and the central core RNA:RNA duplex region is exceptionally stable. The effect of multivalency on siRNA potency [20] could be elegantly demonstrated by transfection with mono-, di-, tri-, and tetrasubstitution with an siRNA against luciferase [19]. A biological effect was achieved with a folate/Alexa-647 substituted tetramer. In this case, only two of the arms were substituted, with the other arms available for further modification. When nude mice bearing folate receptor–expressing xenografts were given systemic injections of the device via the tail vein, they rapidly populated the xenograft [19], suggesting that therapeutic applications of this system may be possible.

7.2.3.2 Trivalent Nucleoprotein Targeting Device

Self-assembling nanodevices composed of DNA and protein have been prepared using DNA methyltransferase fusion proteins targeted to specific sites on oligodeoxynucleotides [21–24]. This technology relies on the capacity of DNA methyltransferases to form covalent linkages with modified DNA. This property coupled with their selectivity for defined nucleic sequences allows them to link fusion proteins to preselected sites on the DNA scaffold [21–23].

A trivalent device that links thioredoxin-*Eco*RII DNA methyltransferase fusion protein to each of the three arms of a DNA Y-junction was found to be selectively taken up by certain prostate and breast cancer cell lines. Moreover, the fluorescently labeled Y-junction DNA was internalized by DU145 cells and appeared in the nucleus within 2 min of incubation [25].

Thioredoxin targeting was observed to be cell-type selective [25], in that only DU145, LNCaP, MCF-7, and HK293 cells bound significant amounts of the trivalent device, while COS-7, PC3, and primary prostate epithelial cells (PrEC) did not bind the device. The fluorescently labeled Y-junction targeted by three copies of the thioredoxin fusion protein yielded much higher (approximately eightfold) bound fluorescence than the fluorescently labeled thioredoxin protein itself. This effect was traced to multivalency since the fluorescently labeled thioredoxin was not significantly reduced in bioactivity, and the target proteins appeared to be dimers [26].

When the trivalent targeting device was used to target human tissues, it was found to bind selectively to the reactive stroma adjacent to malignant prostate cancer [26]. The suggestion that thioredoxin partners such as thioredoxin reductase 2, whose primary function is to reduce hydrogen peroxide to water, might be induced in reactive stroma in order to protect stromal cells from the mutagenic effects of hydrogen peroxide produced in the neutrophil- and macrophage-associated inflammatory responses [26] was borne out by the ability of the nanodevice to distinguish benign prostatic hyperplasia from prostate cancer [27].

7.3 CATALYSIS

One of the most important findings in enzymology is that most enzymes stabilize the transition state of the reaction by having a higher affinity for it than they have for either the reactants or the products. Given the wealth of evidence that both DNA and RNA can fold into aptamers with high affinities for various compounds, it should come as no surprise that these nucleic acids can catalyze chemical reactions by developing high-efficiency binding to transition states. The discovery and subsequent design of ribozymes and DNAzymes are now the subject of a rather large body of literature. Here, I describe examples of each that function as effective nucleic acid devices.

7.3.1 Ribozymes

In general, therapeutic applications of ribozymes lie in the realm of genetic therapy since they require both transfection of and expression within the targeted cell population (see, e.g., [28]). In effect, therapeutic ribozymes have not yet been shown to function as freestanding nucleic acid devices. They are, on the other hand, contributing significantly to our understanding of how the RNA world might have looked as prebiotic evolution moved nucleic acids toward a Darwinian threshold. The RNA ligases form a freestanding class of such nucleic acids that has yielded significant new information not only on molecular evolution but also on catalytic mechanism. Directed evolution from random libraries was able to produce ribozymes that ligate two RNA molecules through a 3′-5′ linkage with a rate enhancement of 7×10^6 over the uncatalyzed rate [29]. Several ribozyme ligases have been extensively characterized [30–34]. The structure of one of them—the L1 ligase—has been determined, and with it a very plausible reaction mechanism [35]. In the proposed mechanism [35], adenine 39 in a Mg^{2+}-stabilized cage around the ligation point serves as a general base that abstracts the proton from the 3′ hydroxyl of the attacking ribose. The nucleophile thus generated is well positioned by the docked ribozyme to attack the α-phosphate of the terminal triphosphate of the ligatable oligonucleotide to produce the 3′-5′ linkage. Refinements of these ligases have introduced credence to Francis Crick's [36] proposal that the first enzymes were RNA replicases. Although a true replicase is yet to be discovered, ligase systems that can undergo a form of autocatalysis [34,37] have been isolated, although the feedstock requires a pool of molecules that are complex.

7.3.2 DNAzymes

Like ribozymes, DNAzymes are catalytic nucleic acids. However, the DNA backbone offers additional stability to these macromolecules. DNAzymes are generally selected from DNA libraries by directed evolution. Since they are more stable than ribozymes in biofluids, they have been tested for their therapeutic effects by direct perfusion in animal studies. Direct perfusion after an experimentally induced heart injury in rats with a DNAzyme against *Egr-1* RNA decreased the size of the resulting infarction produced after the release of the coronary artery ligature [38]. More recently, similar studies in pigs confirmed that the DNAzyme against *Erg-1* RNA reduced the infarct size and improved recovery. Further, direct intravitreal injection of a DNAzyme directed against *c-Jun* prevented experimentally induced neovascularization of the eye in mice [39].

In addition to these therapeutic applications, DNAzymes have been utilized in a variety of nanoscale sensing systems. These molecules possess a G-quadruplex structure that binds hemin, catalyzes the breakdown of hydrogen peroxide and can be coupled to the oxidation of either luminol or 2,2′-azino-bis(3-ethylbenzthiazoline-6-sulfonic acid) to produce light or a colored product, respectively [40,41]. Among the numerous applications of this DNAzyme are the monitoring of a polymerase chain reaction (PCR) [42], and metal ion detection. In the latter application, K^+ could be detected at micromolar concentration [43] and Hg^+ [44] and Ag^+ [45] at nanomolar concentrations. Additional applications involve direct quantification of the folding of the DNA structure or its prevention and are discussed in Section 7.4.2.

7.3.3 Ordered Functionality

There is often considerable overlap between the concepts proposed for molecular evolution in the nucleic acid world and many concepts developed in nucleic acid nanodevice technology. The suggestion that nucleic acid complementarity could provide a means of both organizing and ordering metabolic functionalities [46] in primitive metabolic systems is an excellent example of this overlap. In their proposal, Gibson and Lamond [46] suggested that short nucleic acid adaptors with base complementarity to specific sites along a longer nucleic acid chain would align linked substrate and catalytic nucleic acids in a specific order. In this way, the central nucleic acid chain would not only confine, but also organize the metabolic system into what was termed a *metabolosome*. Niemeyer and coworkers [47] first implemented this concept in nucleic acid nanodevice construction. Adapter oligodeoxynucleotides covalently attached to streptavidin were used to capture immunoglobulin G or alkaline phosphatase. The protein nucleic acid conjugates were then hybridized to a DNA template that permitted the formation of a supramolecular assembly as demonstrated by gel retardation [47]. Subsequently, NAD(P)H:FMN oxidoreductase was linked 20 nt away from bacterial luciferase in an attempt to detect substrate channeling. In this system, the oxidoreductase uses nicotinamide adenine dinucleotide (NADH) to produce flavin mononucleotide (FMN), one of the substrates for bacterial luciferase, and fluorescence is measured. The effectiveness of channeling by this system in free solution was not reported. Instead, the two-enzyme couple was tethered to a streptavidin microtiter plate.

NADH-coupled enzyme systems are routinely used in biochemistry. In these reactions, the coupling enzyme is generally at a relatively high concentration in order to prevent kinetic lag between product production and detection. Substrate channeling concepts [48] elucidated with multisubunit enzymes suggest highly evolved mechanisms for passing substrates without diffusion in the bulk phase. Based on enzyme coupling concepts, one would expect to detect an enhancement of the coupling efficiency at lower concentrations of the coupling enzyme (luciferase) with reactions performed in solution. Thus, proximity should have an effect. In this case, the enzymes were positioned so as to be approximately on the same side of a short DNA molecule, with a center-to-center distance of about 6.8 nm. The system appears to have yielded a twofold enhancement in the reaction rate when the two-enzyme nanodevice was constrained on a streptavidin-coated microtiter plate compared with stippling the streptavidin-coated surface with like concentrations of the separate DNA-linked component enzymes that form the device.

Degradation of enzyme activity by biotinylation may account for some of the problem, or it may be that diffusion and convection override proximity in such a system. Degradation during biotinylation seems unlikely given the results of a follow-on study with glucose oxidase and horseradish peroxidase, both of which retained normal activity even after the nucleic acid coupling step [49]. Using this second enzyme couple arrayed on microtiter plates as previously described, the DNA-linked device seeded on a streptavidin plate was compared with its components stippled on a streptavidin plate. Since the intermediary substrate produced by glucose oxidase is hydrogen peroxide, Muller and Niemeyer were able to test the two systems with catalase to determine whether or not the presumed substrate channeling could protect hydrogen peroxide more effectively when the DNA-linked device was in operation. The coupled device gave an approximately 1.7-fold increase in production of the downstream fluorescent product; however, the activity of the DNA-linked device was strongly affected by catalase, which reduced the production of the downstream product by about fivefold. While this effect does not rule out channeling, it demonstrates that the diffusion corridor for the hydrogen peroxide intermediate exposes it to the solvent. Moreover, the surface attachment suggests that the intermediate may also be subject to convection.

When the glucose oxidase/horseradish peroxidase couple was arrayed on a linear rolling circle amplification product containing multiple repetitive elements, each of which could house a copy of the DNA-linked device, a strong enhancement of the reaction catalyzed by the arrayed devices was observed over equal concentrations of the free enzymes in solution [50]. A similar result was obtained when planar DNA origami molecules, comprising hexagonal subunits, were designed to link the glucose oxidase/horseradish peroxidase couple. In this case, the distance between the two enzymes was varied from 13 to 33 nm with very little effect (an approximately 20% increase at 13 nm spacing) on the solution reaction rate.

Based on these experiments, one can conclude that placing evolutionarily disparate enzymes in close proximity produces weaker channeling effects than have been seen in highly evolved polyfunctional enzyme systems [48]. Even so, primitive metabolosomes appear possible given the wealth of DNAzymes and RNAzymes now available, but their construction as originally proposed by Gibson and Lamond [46] and Levisohn and Spiegelman [6] remains to be accomplished.

7.4 DYNAMICS

7.4.1 I-Switch

I-switches are based on the capacity of the i-motif in DNA to form C-rich quadruplex structures. This C-rich motif occurs in nature as the complement to the G-rich regions that can form the G-quadruplex. However, in contrast to the G-quadruplex, which is formed of stacked planar tetrads formed by Hoogsteen-paired guanines, i-DNA, as the C-rich quadruplex is called, forms between protonated cytosine pairs that intercalate so that pairs of strands are joined by base pairing in an interlocking structure. Since protonation at N3 of cytosine facilitates the formation of the structure, its formation is favored by low pH.

This system has been successfully applied to the determination of intracellular pH [51,52]. The device is composed of a gapped duplex carrying C-rich sequences at either end. Each of the C-rich sequences is modified with a chromophore. Fluorescence is quenched when the switch closes at low pH when protonation of cytosine favors the formation of i-DNA between the two C-rich single strands, resulting in the juxtaposition of the two chromophores. The i-switch carrying Alexa-488 and Alexa-647 was found to have the best dynamic range in the pH 6–7 range. The device was taken up by cells and was able to report pH changes associated with endosomal maturation in both cultured cells [51] and coelomocytes of living *Caenorhabditis elegans* [52].

7.4.2 G-Switch

G-switches utilize the capacity of short guanine-rich sequences to adopt quadruplex structures under certain conditions. The ability to stabilize the folded or open structure has yielded several useful detection methods. In general, these have lower sensitivity than the DNAzyme methods, since they rely on the ability of the folded structure to bind a chromophore with enhanced fluorescence and have seen limited application in biofluids or living cells (see, e.g., [53]).

REFERENCES

1. Drexler, K. E. 1986. *Engines of Creation: The Coming Era of Nanotechnology*. New York: Random House.
2. Smalley, R. and E. Drexler. 2003. Nanotechnology. *Chemical and Engineering News* 81 (48):37–42.
3. Stevenson, A. and C. A. Lindberg (eds). 2010. *New Oxford American Dictionary*. 3rd edn., New York: Oxford University Press. Available at: www.oxforddictionaries.com.
4. Grove, P. (ed.). 1986. *Webster's Third New International Dictionary*. Springfield, MA: Merriam Webster.
5. Mills, D. R., R. L. Peterson, and S. Spiegelman. 1967. An extracellular Darwinian experiment with a self-duplicating nucleic acid molecule. *Proc Natl Acad Sci USA* 58 (1):217–224.
6. Levisohn, R. and S. Spiegelman. 1969. Further extracellular Darwinian experiments with replicating RNA molecules: Diverse variants isolated under different selective conditions. *Proc Natl Acad Sci USA* 63 (3):805–811.
7. Tuerk, C. and L. Gold. 1990. Systematic evolution of ligands by exponential enrichment: RNA ligands to bacteriophage T4 DNA polymerase. *Science* 249 (4968):505–510.
8. Ellington, A. D. and J. W. Szostak. 1990. In vitro selection of RNA molecules that bind specific ligands. *Nature* 346 (6287):818–822.
9. Ni, X., M. Castanares, A. Mukherjee, and S. E. Lupold. 2011. Nucleic acid aptamers: Clinical applications and promising new horizons. *Curr Med Chem* 18 (27):4206–4214.
10. Bates, P. J., D. A. Laber, D. M. Miller, S. D. Thomas, and J. O. Trent. 2009. Discovery and development of the G-rich oligonucleotide AS1411 as a novel treatment for cancer. *Exp Mol Pathol* 86 (3):151–164.
11. Gragoudas, E. S., A. P. Adamis, E. T. Cunningham Jr., M. Feinsod, and D. R. Guyer. 2004. Pegaptanib for neovascular age-related macular degeneration. *N Engl J Med* 351 (27):2805–2816.
12. Lupold, S. E., B. J. Hicke, Y. Lin, and D. S. Coffey. 2002. Identification and characterization of nuclease-stabilized RNA molecules that bind human prostate cancer cells via the prostate-specific membrane antigen. *Cancer Res* 62 (14):4029–4033.
13. Israeli, R. S., C. T. Powell, J. G. Corr, W. R. Fair, and W. D. Heston. 1994. Expression of the prostate-specific membrane antigen. *Cancer Res* 54 (7):1807–1811.

14. Goodman Jr. O. B., S. P. Barwe, B. Ritter, P. S. McPherson, A. J. Vasko, J. H. Keen, D. M. Nanus, N. H. Bander, and A. K. Rajasekaran. 2007. Interaction of prostate specific membrane antigen with clathrin and the adaptor protein complex-2. *Int J Oncol* 31 (5):1199–1203.
15. Wullner, U., I. Neef, A. Eller, M. Kleines, M. K. Tur, and S. Barth. 2008. Cell-specific induction of apoptosis by rationally designed bivalent aptamer-siRNA transcripts silencing eukaryotic elongation factor 2. *Curr Cancer Drug Targets* 8 (7):554–565.
16. Ni, X., Y. Zhang, J. Ribas, W. H. Chowdhury, M. Castanares, Z. Zhang, M. Laiho, T. L. DeWeese, and S. E. Lupold. 2011. Prostate-targeted radiosensitization via aptamer-shRNA chimeras in human tumor xenografts. *J Clin Invest* 121 (6):2383–2390.
17. Pastor, F., D. Kolonias, P. H. Giangrande, and E. Gilboa. 2010. Induction of tumour immunity by targeted inhibition of nonsense-mediated mRNA decay. *Nature* 465 (7295):227–230.
18. Shu, Y., M. Cinier, D. Shu, and P. Guo. 2011. Assembly of multifunctional phi29 pRNA nanoparticles for specific delivery of siRNA and other therapeutics to targeted cells. *Methods* 54 (2):204–214.
19. Haque, F., D. Shu, Y. Shu, L. S. Shlyakhtenko, P. G. Rychahou, B. M. Evers, and P. Guo. 2012. Ultrastable synergistic tetravalent RNA nanoparticles for targeting to cancers. *Nano Today* 7 (4):245–257.
20. Nakashima, Y., H. Abe, N. Abe, K. Aikawa, and Y. Ito. 2011. Branched RNA nanostructures for RNA interference. *Chem Commun (Camb)* 47 (29):8367–8369.
21. Smith, S. S., L. Niu, D. J. Baker, J. A. Wendel, S. E. Kane, and D. S. Joy. 1997. Nucleoprotein-based nanoscale assembly. *Proc Natl Acad Sci USA* 94 (6):2162–2167.
22. Smith, S. S. 2001. A self-assembling nanoscale camshaft: Implications for nanoscale materials and devices constructed from proteins and nucleic acids. *Nano Letters* 1 (2):51–56.
23. Smith, S. S. 2002. Designs for the self-assembly of open and closed macromolecular structures and a molecular switch using DNA methyltransferase to order proteins on nucleic acid scaffolds. *Nanotechnology* 13:413–419.
24. Clark, J., T. Shevchuk, P. M. Swiderski, R. Dabur, L. E. Crocitto, Y. I. Buryanov, and S. S. Smith. 2003. Mobility-shift analysis with microfluidics chips. *Biotechniques* 35 (3):548–554.
25. Singer, E. M. and S. S. Smith. 2006. Nucleoprotein assemblies for cellular biomarker detection. *Nano Lett* 6 (6):1184–1189.
26. Singer, E. M., L. E. Crocitto, Y. Choi, S. Loera, L. M. Weiss, S. A. Imam, T. G. Wilson, and S. S. Smith. 2011. Biomarker identification with ligand-targeted nucleoprotein assemblies. *Nanomedicine (Lond)* 6 (4):659–668.
27. Singer, E., J. Linehan, G. Babilonia, S. A. Imam, D. Smith, S. Loera, T. Wilson, and S. Smith. 2013. Stromal response to prostate cancer: Nanotechnology-based detection of thioredoxin-interacting protein partners distinguishes prostate cancer associated stroma from that of benign prostatic hyperplasia. *PLoS One* 8 (6):e60562.
28. He, P., D. Zhu, J. J. Hu, J. Peng, L. S. Chen, and G. X. Lu. 2010. pcDNA3.1(-)-mediated ribozyme targeting of HER-2 suppresses breast cancer tumor growth. *Mol Biol Rep* 37 (3):1597–1604.
29. Bartel, D. P. and J. W. Szostak. 1993. Isolation of new ribozymes from a large pool of random sequences [see comment]. *Science* 261 (5127):1411–1418.
30. Robertson, M. P. and A. D. Ellington. 1999. In vitro selection of an allosteric ribozyme that transduces analytes to amplicons. *Nat Biotechnol* 17 (1):62–66.
31. Rogers, J. and G. F. Joyce. 1999. A ribozyme that lacks cytidine. *Nature* 402 (6759):323–325.
32. Jaeger, L., M. C. Wright, and G. F. Joyce. 1999. A complex ligase ribozyme evolved in vitro from a group I ribozyme domain. *Proc Natl Acad Sci USA* 96 (26):14712–14717.

33. Ikawa, Y., K. Tsuda, S. Matsumura, and T. Inoue. 2004. De novo synthesis and development of an RNA enzyme. *Proc Natl Acad Sci USA* 101 (38):13750–13755.
34. Ferretti, A. C. and G. F. Joyce. 2013. Kinetic properties of an RNA enzyme that undergoes self-sustained exponential amplification. *Biochemistry* 52 (7):1227–1235.
35. Robertson, M. P. and W. G. Scott. 2007. The structural basis of ribozyme-catalyzed RNA assembly. *Science* 315 (5818):1549–1553.
36. Crick, F. H. 1968. The origin of the genetic code. *J Mol Biol* 38 (3):367–379.
37. Lincoln, T. A. and G. F. Joyce. 2009. Self-sustained replication of an RNA enzyme. *Science* 323 (5918):1229–1232.
38. Bhindi, R., L. M. Khachigian, and H. C. Lowe. 2006. DNAzymes targeting the transcription factor Egr-1 reduce myocardial infarct size following ischemia-reperfusion in rats. *J Thromb Haemost* 4 (7):1479–1483.
39. Fahmy, R. G., A. Waldman, G. Zhang, A. Mitchell, N. Tedla, H. Cai, C. R. Geczy, C. N. Chesterman, M. Perry, and L. M. Khachigian. 2006. Suppression of vascular permeability and inflammation by targeting of the transcription factor c-Jun. *Nat Biotechnol* 24 (7):856–863.
40. Li, Y. and D. Sen. 1997. Toward an efficient DNAzyme. *Biochemistry* 36 (18):5589–5599.
41. Travascio, P., Y. Li, and D. Sen. 1998. DNA-enhanced peroxidase activity of a DNA-aptamer-hemin complex. *Chem Biol* 5 (9):505–517.
42. Cheglakov, Z., Y. Weizmann, M. K. Beissenhirtz, and I. Willner. 2006. Ultrasensitive detection of DNA by the PCR-induced generation of DNAzymes: The DNAzyme primer approach. *Chem Commun* (30):3205–3207.
43. Nakayama, S. and H. O. Sintim. 2009. Colorimetric split G-quadruplex probes for nucleic acid sensing: Improving reconstituted DNAzyme's catalytic efficiency via probe remodeling. *J Am Chem Soc* 131 (29):10320–10333.
44. Kong, D. M., N. Wang, X. X. Guo, and H. X. Shen. 2010. "Turn-on" detection of Hg^{2+} ion using a peroxidase-like split G-quadruplex-hemin DNAzyme. *Analyst* 135 (3):545–549.
45. Kong, D. M., L. L. Cai, and H. X. Shen. 2010. Quantitative detection of Ag(+) and cysteine using G-quadruplex-hemin DNAzymes. *Analyst* 135 (6):1253–1258.
46. Gibson, T. J. and A. I. Lamond. 1990. Metabolic complexity in the RNA world and implications for the origin of protein synthesis. *J Mol Evol* 30 (1):7–15.
47. Niemeyer, C. M., T. Sano, C. L. Smith, and C. R. Cantor. 1994. Oligonucleotide-directed self-assembly of proteins: Semisynthetic DNA–streptavidin hybrid molecules as connectors for the generation of macroscopic arrays and the construction of supramolecular bioconjugates. *Nucleic Acids Res* 22 (25):5530–5539.
48. Huang, X., H. M. Holden, and F. M. Raushel. 2001. Channeling of substrates and intermediates in enzyme-catalyzed reactions. *Annu Rev Biochem* 70:149–180.
49. Muller, J. and C. M. Niemeyer. 2008. DNA-directed assembly of artificial multienzyme complexes. *Biochem Biophys Res Commun* 377 (1):62–67.
50. Wilner, O. I., S. Shimron, Y. Weizmann, Z. G. Wang, and I. Willner. 2009. Self-assembly of enzymes on DNA scaffolds: En route to biocatalytic cascades and the synthesis of metallic nanowires. *Nano Lett* 9 (5):2040–2043.
51. Modi, S., M. G. Swetha, D. Goswami, G. D. Gupta, S. Mayor, and Y. Krishnan. 2009. A DNA nanomachine that maps spatial and temporal pH changes inside living cells. *Nat Nanotechnol* 4 (5):325–330.
52. Surana, S., J. M. Bhat, S. P. Koushika, and Y. Krishnan. 2011. An autonomous DNA nanomachine maps spatiotemporal pH changes in a multicellular living organism. *Nat Commun* 2:340.
53. Zhang, Z., E. Sharon, R. Freeman, X. Liu, and I. Willner. 2012. Fluorescence detection of DNA, adenosine-5′-triphosphate (ATP), and telomerase activity by zinc(II)-protoporphyrin IX/G-quadruplex labels. *Anal Chem* 84 (11):4789–4797.

8 2-D Nanofluidic Bioarray for Nucleic Acid Analysis

Abootaleb Sedighi, Lin Wang, and Paul C.H. Li

CONTENTS

8.1 INTRODUCTION TO NANOFLUIDIC DNA BIOARRAY HYBRIDIZATION

Nucleic acid hybridization techniques feature the use of a probe nucleic acid molecule and a target nucleic acid molecule. Here, probe molecules are usually short single-stranded nucleic acids (DNA or RNA) or oligonucleotides with known sequences; whereas target molecules are prepared from polymerase chain reaction (PCR) amplification of genomic extracts. Probe-target hybridization leads to the formation of a double-stranded molecule, called a duplex. The method of DNA bioarray hybridization evolved from Southern blotting technology based on solid-phase hybridization in the early 1990s [1]. This method relies on the immobilization of the probe molecules onto a solid surface to recognize their complementary DNA target sequences by hybridization. Millions of features have been integrated onto a standard glass or silicon slide by microprinting or *in situ* synthesis of oligonucleotides [2,3]. The relative abundance of nucleic acid sequences in the target solution can be measured from chip hybridization results optically, electrochemically, or radiochemically, with proper detection labels [4]. DNA bioarrays have dramatically accelerated many types of investigations including gene expression profiling, comparative genomic hybridization, protein–DNA interaction studies (chromatin immunoprecipitation), single-nucleotide polymorphism (SNP) detection, as well as nucleic acid diagnostic applications. The progress of DNA bioarray technology during the last couple of years has been summarized in many books and reviews [4–8].

With the rapid growth of microelectromechanical systems (MEMS), microfluidic/nanofluidic technology has been developed rapidly with many applications over the past decade. By combining the fields of nanofluidics and DNA bioarrays, the advantages of both fields can be exploited simultaneously [9–11]. Nanofluidics deals with the transfer and control of small amounts of fluids in nanoscale flow configurations. One obvious advantage of using a nanofluidic system is the dramatic reduction in the sample volume. Instead of handling a volume of milliliters, a nanofluidic system could deal with a volume of nanoliters down to as low as 1 pL [11]. Many interesting biological samples such as proteins and DNA extraction are rare or unavailable in large amounts; the nanofluidic method therefore provides a way to study these materials efficiently. The small volumes of the nanofluidic system make it possible to develop compact and portable lab-on-a-chip systems that can be manufactured cheaply by mass production.

The second advantage of using nanofluidic technology is that the liquid movement in microchannels with a large surface-to-volume ratio facilitates target diffusion to the probes anchored to the substrate surface. In conventional DNA bioarray hybridization, ~30 μL samples are applied to the biochip and the solutions are distributed across the probe arrays by covering them with a glass coverslip [12]. This hybridization is static and so after the reaction-limited initial time period, probe depletion renders this process diffusion limited [13,14]. Since diffusion coefficients (D) for nucleic acids in aqueous solutions are on the order of 10^{-7} cm^2 s^{-1} [15], the typical distance ($L = \sqrt{Dt}$) moved by the target nucleic acid in 24 h solely due to diffusion is ~1 mm. Considering that the horizontal length scale of a bioarray is on the order of a few centimeters, hemispherical depletion volumes form around each probe [16]. Pappaert et al. found that for an analysis time of 24 h, the maximum binding efficiency was less than 0.2%; to increase that to 2%, a 6-day analysis would be required [17]! Therefore, the conventional static method is a very inefficient means of achieving hybridization.

Nanofluidics also offers the advantage of multisample capabilities on one chip. Conventional bioarray experiments usually allow one sample to be applied on one glass chip [12]. However, in the application of genetic mutation analysis, clinical diagnostics, or microorganism identifications, a direct comparison between different samples is made using one chip but not on many different chips, because the quality of chips with probe arrays varies from batch to batch [5]. Nanofluidics allows for the delivery of controlled volumes of samples and reagents to the DNA bioarray. By integrating multiple channels onto one chip, a high-throughput multisample analysis has been achieved [18–21]. Moreover, DNA hybridization assays depend on different parameters such as temperature and the stringency of the hybridization and washing buffer solutions. The parameter space could be too large to be addressed efficiently using conventional approaches [22]. On the contrary, since the nanofluidic method has the capability of multiple sample injection as well as accurate control of the liquid flow and temperature, this method has been developed for the automated selection of optimal assay parameters [23,24].

In conventional bioarray hybridization protocols, sample solutions are loaded directly onto the glass chip surface and incubated for hours. Therefore, an intuitive way is to use a large microfluidic chamber covering an area consisting of spotted probes, where sample DNA solutions are delivered to, and hybridized with, the probes. The chamber is usually detachable from the glass chip and is made of an elastomer, such as polydimethylsiloxane (PDMS) [25]. Different groups have successfully combined the microfluidic chamber approach with high-density bioarrays [16,26–29]. The hybridization time was successfully reduced from overnight to less than 2 h.

Although the nanofluidic system offers a significant improvement over the conventional bulk solution method, two inherent problems arise with the use of large chamber hybridization on high-density bioarrays. First, hybridization is most efficient when each target nucleic acid molecule can move throughout the solution and encounter every probe molecule. If the size of the chip consisting of the microfluidic chamber is comparable to the size of a standard microscope slide, the liquid flows have to be carefully designed to achieve an equally distributed liquid movement within the chamber as well as to avoid any trapped air bubbles during filling [16,17,26,27,31–38]. Second, since a large number of probe features are packed into a small area, great efforts are necessary for data processing, normalization, and interpretation [30].

In a gene expression analysis, many thousands of genes are simultaneously monitored to create a global picture of their cellular functions and high-density bioarrays are thus needed. On the contrary, in many gene diagnostic applications, once a relatively small number of genes are identified using high-density DNA bioarrays, low-density bioarrays can be designed to screen these genes across many patients or to detect SNPs [39]. This low-density approach has been proved reliable, cost-effective, and fast in data analysis and interpretation [40–46].

8.2 NANOFLUIDIC DNA HYBRIDIZATION WITH LOW-DENSITY PROBE ARRAYS

8.2.1 Low-Density Bioarray Method with Pin-Spotted Probes

The nanofluidic method is suitable for conducting a low-density DNA microarray analysis. With micromachining techniques, the hybridization microchannels are aligned with rows of probe spots. Both straight [47–55] and serpentine microchannels have been designed [23,56–60]. Wang et al. used a bioarray chip for a mutational analysis (Figure 8.1a and b) [48]. The authors found that the nanofluidic flow enhanced the kinetics of the hybridization reaction between the target DNA and the arrayed probe spots. As shown in Figure 8.1c and d, the time required for the signals to reach the intensities close to saturation drops from ~3 h in the static condition to ~24 s when performing hybridization on-chip. Figure 8.2 shows an electrochemical sensor developed by Lenigk et al. for genotyping PCR products [58]. In their device, probe spots were contained in a serpentine channel where an oscillating flow, created by an integrated air pump, accelerated the DNA hybridization of the PCR products in such a way that the first reading of the hybridization signal corresponded to a signal-to-noise (S/N) ratio of 20, which was already sufficient for determining the genotypes of the samples (Figure 8.2b). In terms of probe spotting, either commercial arrayers [53,56–58] or micropipettors [47,48,52] have been used. The use of microchannels instead of large hybridization chambers alleviates the complicated design needed for homogeneous hybridization across the entire chamber area.

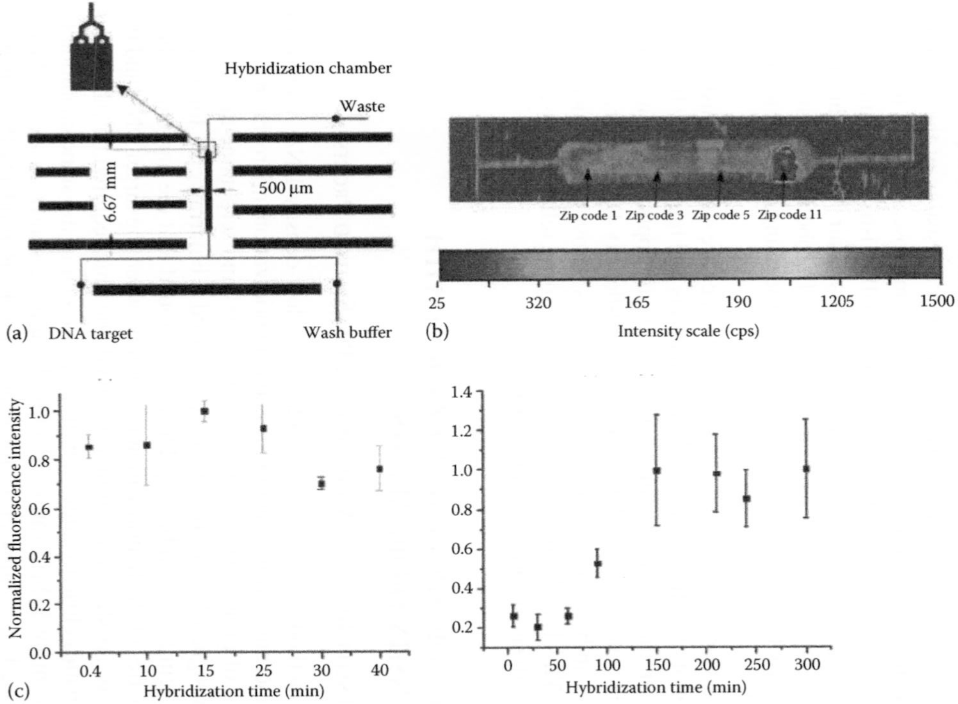

FIGURE 8.1 (a) Schematic of the hybridization chamber surrounding the probe spots. The chamber is fed by two nanofluidic channels from the bottom. One delivers the DNA target solutions and the other carries the wash buffer. (b) The image shows the result in the hybridization chamber after the mutation analysis. Dye-labeled target molecules hybridize to the probe spot containing their complementary probes called zip codes. (c) Kinetics of the hybridization using nanofluidic channels (left) and in a control experiment in which the probe array was soaked in a target solution instead of using nanofluidic channels to deliver the target solution (right). (From Wang, Y. et al., *Analytical Chemistry*, 75, 1130–1140, 2003. With permission.)

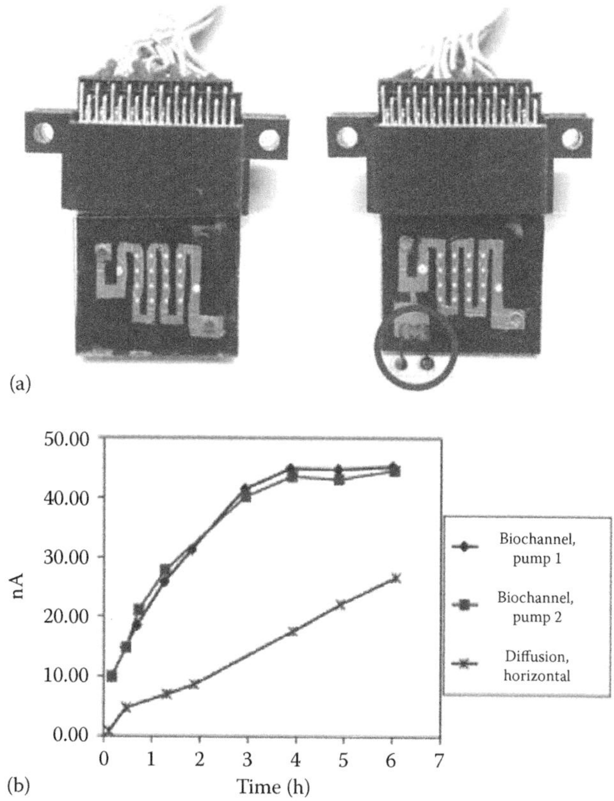

FIGURE 8.2 (a) An electrochemical sensor with the spotted probes covered by a serpentine channel with (right) and without (left) the integrated air pump to provide agitation of the hybridization mixture. (b) Comparison of hybridization kinetics in a nanofluidic channel with convection (with pump) and diffusion controlled (without pump). (From Lenigk, R. et al., *Analytical Biochemistry*, 311, 40–49, 2002. With permission.)

8.2.2 Low-Density Bioarray Method with 2-D Intersection Approach

In low-density DNA bioarray analysis, the conventional pin-spotting method has been used to create dot-like probe arrays. The performance of the subsequent hybridization assay is thus heavily influenced by the quality of the immobilized probe spots. In practice, since the spotting solutions are exposed to air and the spotting pins are very close to each other, the pin-spotting method may suffer from splashing, evaporation, and cross contamination [61]. In addition, probe spots of high homogeneity are beneficial, since this simplifies image analysis and considerably enhances the accuracy of signal detection. When a spotting buffer such as saline sodium citrate (SSC) is used, spot uniformity is often poor due to the hydrophobic properties of a chemically modified glass surface. Supplementary chemicals such as dimethyl sulfoxide (DMSO) can improve spot uniformity, but this could introduce new problems such as swelling the spot size [62]. Moreover, during the blocking and cleaning procedures after spotting, the remaining unreacted probe molecules could diffuse away and smear the chip to form a comet-like spot [12]. Furthermore, for subsequent use of the microchannel hybridization with the spotted bioarray, additional devices such as steel clamps must be used to ensure that the entire hybridization microchannel is sealed and well aligned to the probe rows [58].

A simple and effective microspotting or printing method is to use the nanofluidic network to create probe line arrays on a surface. Patterning biomolecules on a variety of solid substrates (gold, glass, or polymer) has been successfully achieved with PDMS-based microchannel plates [62–69]. Because the immobilization reaction between the biomolecules and the surface is confined in microchannels, solution evaporation and splashing are thus prevented. In our practice, only nanoliter volumes of reagent are needed to fill through each microchannel and the solution can be kept under room temperature for hours without drying out [70]. Figure 8.3a shows an example of a PDMS microchannel plate assembled with a glass chip, in which an individual channel is shown in Figure 8.3b. Chemically modified DNA probes are introduced into the microchannels and are immobilized on the glass chip. The bonding between the PDMS channel plate and the glass chip is reversible. After peeling off the PDMS plate, a homogeneous distribution of the probe molecules along the

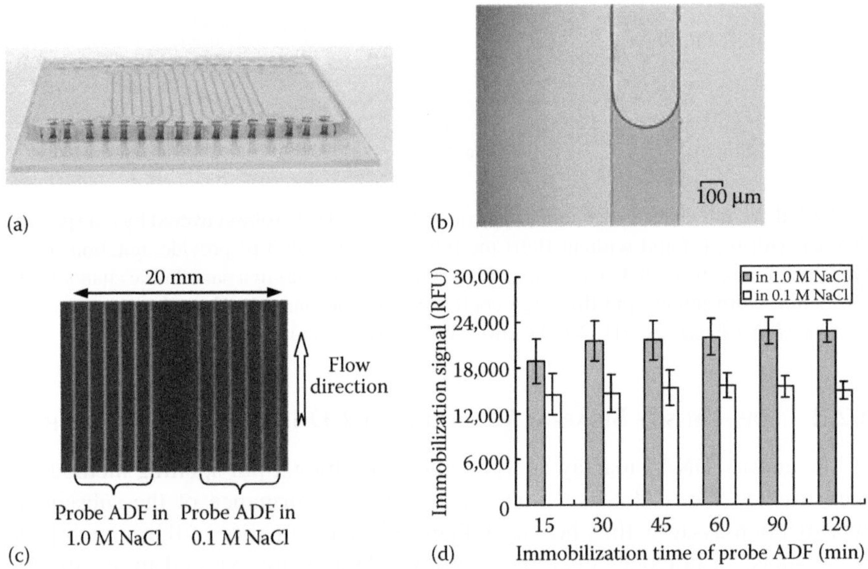

FIGURE 8.3 (a) An image of the assembly of a 2″×2″ PDMS channel plate on a 3″×2″ glass chip. The 16 channels were filled with blue-dyed solutions. (b) A microscopic view of a straight channel partially filled with the dye solution. (c) A fluorescent image of immobilized DNA probe lines (vertical stripes) printed on the glass chip and in buffer solutions consisting of two different salt concentrations. The DNA probe molecules were labeled with the fluorescein dye. (d) The corresponding fluorescent signals from the immobilized DNA probe lines in (c). (From Wang, L. and Li, P.C.H., *Journal of Agricultural and Food Chemistry*, 55, 10509–10516, 2007. With permission.)

area enclosed by the microchannels is achieved with high resolution, with the fluorescent image shown in Figure 8.3c. Since the flow operation in each microchannel could be manipulated individually, the probe immobilization step and subsequent processes such as washing and blocking can be independently conducted, and the individual channels can thus be used for condition optimization tests (Figure 8.3d). Moreover, cross contamination from the diffusion of unreacted agents is avoided during the washing steps. The nanofluidic approach is simple and flexible for printing low-density probe arrays.

Low-density probe line arrays created using the nanofluidic method have been used successfully for nucleic acid analysis [67–78]. In these applications, a 2-D intersection approach was used, which resembled the use for immunoassays reported previously [79,80]. As shown in Figure 8.4, probe lines are printed onto a chemically modified glass chip with the first microchannel plate as depicted before. After peeling off Plate 1, Channel plate 2, which is used for hybridization, is then assembled with the same glass chip. Here, nanofluidic channels are arranged horizontally and they are orthogonal to the preprinted probe lines. Sample DNA molecules that flow through the microchannels will intersect with the probe line arrays. If the complementary probe molecules are encountered, the target molecules are retained and rectangular hybridization patches are thus formed (Figure 8.5). These patches have been detected by surface plasmon resonance imaging [67] or, more often, by using a fluorometric flatbed scanner [70–78,81]. As compared with a pin-spotted bioarray, the final hybridization results obtained from the nanofluidic bioarray is more organized for subsequent image analysis.

The 2-D bioarray format is well suited for parallel sample hybridizations. Unlike a pin-spotted, low-density DNA bioarray, the use of long and narrow probe line arrays alleviates the need for alignment between hybridization channels and probe dots [23]. Parallel hybridizations can be used not only for the investigation

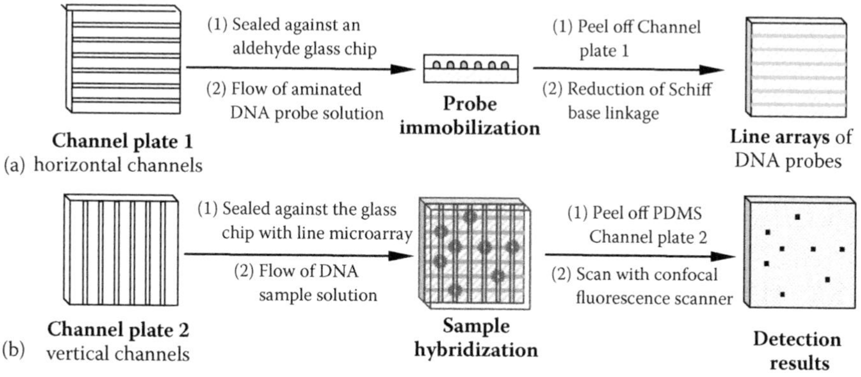

FIGURE 8.4 The 2-D bioarray method using straight microchannels. (a) The creation of a DNA probe line array on an aldehyde-modified glass chip via straight microchannels. (b) The hybridization of DNA samples in straight channels orthogonal to the straight probe lines previously printed on the glass chip. (From Wang, L. and Li, P.C.H., *Journal of Agricultural and Food Chemistry*, 55, 10509–10516, 2007. With permission.)

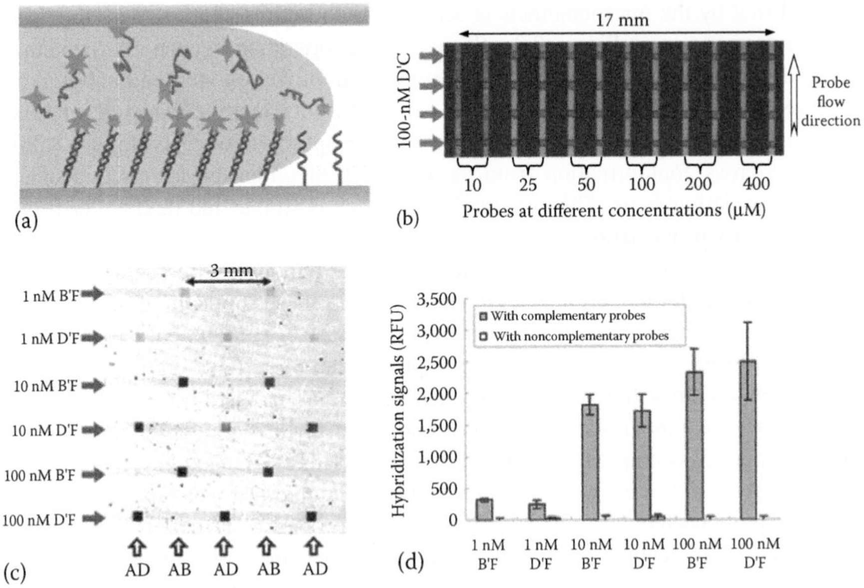

FIGURE 8.5 (a) Fluorescently labeled target DNAs contained in the flow stream hybridize with their complementary probe strands on the surface. (b) Dual-channel fluorescent images of DNA hybridization results using the 2-D bioarray method. The overlaid images from the same glass chip show both printed probe lines (vertical lines) and square hybridization patches at intersections. (c) Fluorescent images of DNA hybridization results from two sets of complementary probe-target strands. (d) A histogram showing that the targets specifically bind to their complementary probes. (From Wang, L. and Li, P.C.H., *Journal of Agricultural and Food Chemistry*, 55, 10509–10516, 2007. With permission.)

or comparison of optimal assay conditions, but also for high-throughput multisample analysis. Benn et al. modeled the hybridization rates of 60-mer oligonucleotides in eight concentrations from 10 pM to 10 nM [72]. Situma et al. used a 16 × 16 array to detect two different low-abundant DNA point mutations in oncogenes [73]. Wang et al. also used a 16 × 16 bioarray chip to test the effect of salt concentrations as well as hybridization times. With optimized conditions, they successfully discriminated PCR product samples extracted from two fungal pathogen genomes [70,71].

8.2.3 High-Throughput Parallel Hybridizations with Centrifugal Force-Driven Nanoflows

The achievement of parallel sample hybridizations in multiple microchannels raised the question of finding an effective way of simultaneous liquid delivery. The conventional liquid pumping method used in microchannels is a pressure-driven flow provided by syringe pumping or vacuum suction. Since each microchannel has to be connected to pump tubing and synchronization has to be considered to ensure parallel flows, this arrangement could be complicated when many channels are used.

Another drawback of this pumping method is that high pressure is required for liquid delivery in long and narrow microchannels, and this in turn requires a very tight seal between the channel plate and the substrate. For example, a steel clamp is used to tighten the bioarray chip assembly [59]. Electroosmotic flow (EOF) is another microflow method that has been used for parallel liquid pumping within multiple channels [82,83]. However, the flow control of the EOF method depends not only on the applied voltage across the microchannel, but also on the physicochemical properties of the microchannel surface as well as the ionic strength of the buffered solutions [84]. The high ionic concentration typical of DNA hybridization buffer [12,85,86] may result in excessive Joule heating and electrolysis [47,87–89]. These effects will result in dynamic changes in the liquid temperature and pH value [90], causing instability of hybridization and SNP discrimination performance. Therefore, only a few reports in terms of applying an EOF to DNA bioarray analysis have been published [47].

An alternative liquid pumping method is to utilize the body force of the liquid column itself, and such a force can be created under a centrifugal force field. Compared with other methods, centrifugal pumping is easy to implement and is not sensitive to the physicochemical properties of the liquid. Therefore, fluids can be moved in a parallel manner in multiple channels of a wide range of sizes.

The centrifugal platform has been developed in many applications including environmental assays, cell lysis, separation and extraction, immunoassay in biomedical diagnosis, and nucleic acid analysis [16,51–53,91–108]. Flow dynamics studies as well as nanofluidic operation of the liquid in rotating radial microchannels were also investigated [109–120]. More studies regarding the application of centrifugal pumping to nanofluidics can be found in several reviews [21,121–125]. Moreover, the compact disk (CD) and related industries, which include disk materials, disk fabrication, signal reading and error correction, as well as the rotor-driven/control system, have been well developed over the past decades. Combined with MEMS technology, the centrifugal platform could fit into this CD system to develop a portable and point-of-care analysis system [125–127].

Although centrifugal pumping has been successfully exploited in many nanofluidic assays, those assays usually provide only 1-D analysis [16, 51–55]. Bynum and Gordon used extra rotors on a centrifugal platform to induce reciprocal flows inside hybridization chambers [16]. After a 17-h reaction, the signals measured were 10 times higher than those of the conventional static method [16]. Peytavi et al. developed a chamber hybridization system, where the assembly of a glass slide and a PDMS microfluidic chamber is placed on a CD support that can hold up to five slides [51]. The authors demonstrated the discrimination of four clinically relevant *Staphylococcus* species by a 15-min automated hybridization process performed at room temperature [51]. Li et al. [53] presented a CD-like device capable of generating the reciprocating flow of DNA samples within microchannels and demonstrated an application in rapid DNA hybridization assay [53]. Here, the centrifugal force was used to drive the sample solution to flow through the hybridization channel into a temporary collection reservoir while the capillary force pulled the solution back into the hybridization channel during the stopping period (Figure 8.6). The sample hybridization time was reduced to 90 s and the sample volume was as low

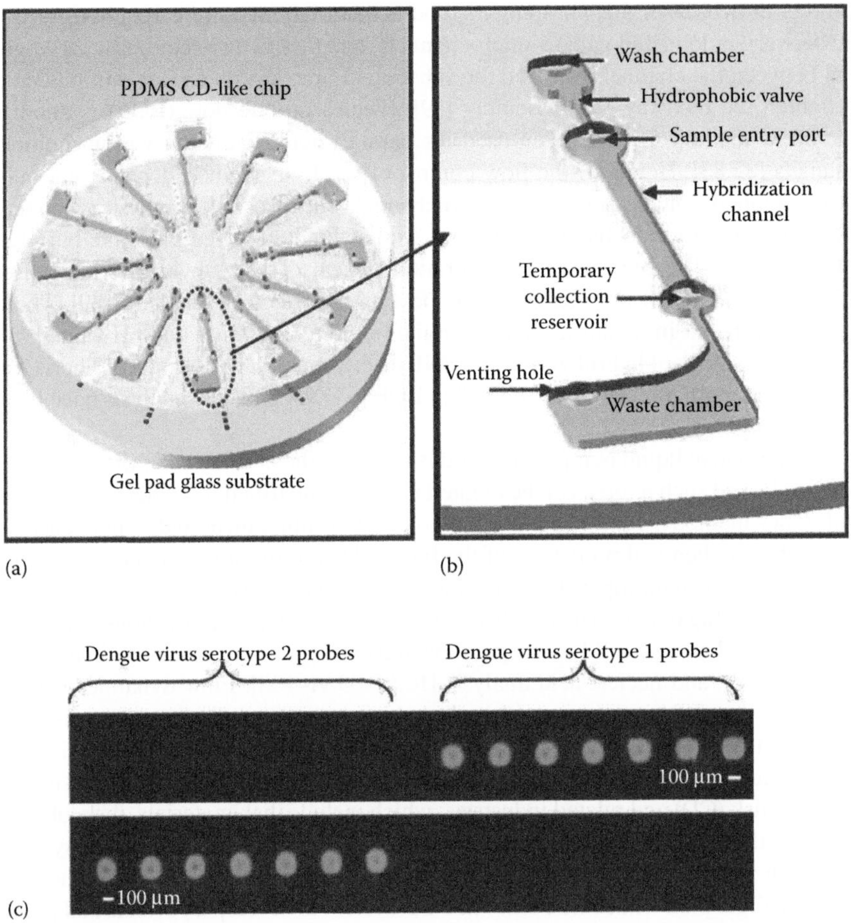

FIGURE 8.6 (a) Schematic representation of a CD device for DNA hybridization. It consists of a PDMS CD slab containing 12 DNA hybridization assay units sealed with a glass substrate with immobilized DNA probe arrays. (b) Schematic diagram of a single DNA hybridization assay unit. (c) Hybridization specificity tests with the CD device. Top: Dengue virus serotype 1 targets bind only to serotype 1 probes. Bottom: Dengue virus serotype 2 targets bind only to serotype 2 probes. The CD device was rotated at 22 Hz for 3 s and then stopped for 3 s during the reciprocating process. The duration of the reciprocating process was 90 s. (From Li, C.Y. et al., *Analytica Chimica Acta*, 640, 93–99, 2009.)

as 350 nL [53]. Brøgger and coworkers have recently developed a CD device for the detection of chromosome translocation [54]. As shown in Figure 8.7, they employed a series of capillary burst microvalves to control the stepwise fluid flow movement from the center toward the periphery of the CD.

The limited applications of centrifugal pumping to 2-D nucleic acid bioarray analysis could be attributed to the radial-only structure design. Currently, almost all of the fluidic patterns in centrifugal platforms are fabricated in the radial orientation. However, there is not enough space to accommodate the fluid structures in the

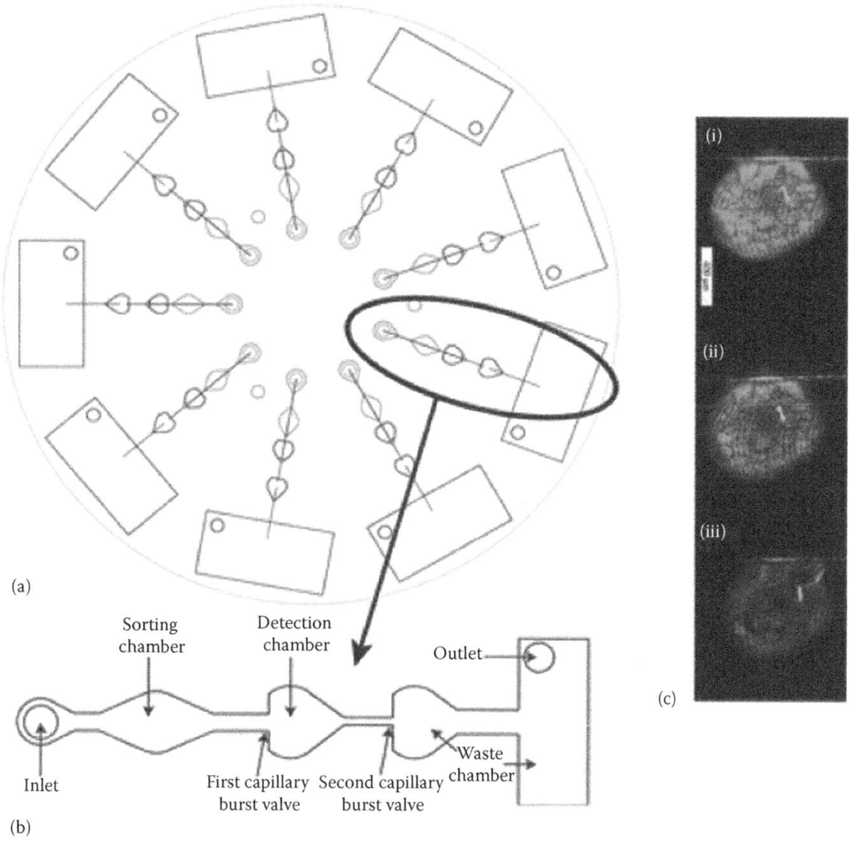

FIGURE 8.7 (a) Schematics of the CD device. (b) The magnified diagram shows the sorting chamber, the detection chamber, and a waste reservoir separated by capillary burst valves with different widths. There are 24 channels, representing the 24 different chromosomes in the human karyotype, and each channel can detect one specific translocation. (c) The signals that resulted from the spots at different stages: (i) immobilization of the FITC-labeled DNA probe on PMMA; (ii) subsequent hybridization of the Cy3-labeled target DNA; and (iii) denaturation of the double-stranded DNA. (From Brøgger, A.L. et al., *Lab on a Chip*, 12, 4628–4634, 2012. With permission.)

radial format. For example, if a centrifugal platform is built on a 120 mm regular CD with a 15 mm center spindle hole, the length of a nanofluidic structure cannot be more than 53 mm. In addition, for such a short microchannel, the capillary effect may dominate the liquid flow process and the flow velocity cannot be easily controlled. Furthermore, centrifugal force increases with the increase of the distance of the liquid column from the center of the CD. As characterized in the work of Duffy et al. and Madou et al., the average velocity of the liquid is proportional to the radial extent of the fluid in a microchannel [93,101]. Such an increasing flow velocity in the microchannel produces challenges and difficulties in the design of a fluidic structure [112,115,120].

In a 2-D low-density bioarray analysis, the probe molecules are spotted in line arrays and the orientation of the sample flows has to be orthogonal to the probe lines. This intersection approach cannot be applied to a radial-only nanofluidic structure because centrifugal pumping can only be exploited in one direction. Therefore, spotted probe arrays have to be used for microchannel hybridization, which limits sample throughput, and wastes a lot of disk space.

Recently, a 2-D DNA bioarray method has been reported, in which the centrifugal force advantages were exploited twice, first in the radial microchannels, and then in the spiral channels [76]. As shown in Figure 8.8, the bioarray assembly consists of two CD-like PDMS chips as microchannel plates and one 92 mm glass wafer as the substrate [75–77]. In the first step, Channel Plate 1 is assembled with the glass wafer for printing the radial probe line arrays; during the second step, spiral Channel Plate 2 is sealed against the wafer with printed probe lines after removing Channel Plate 1.

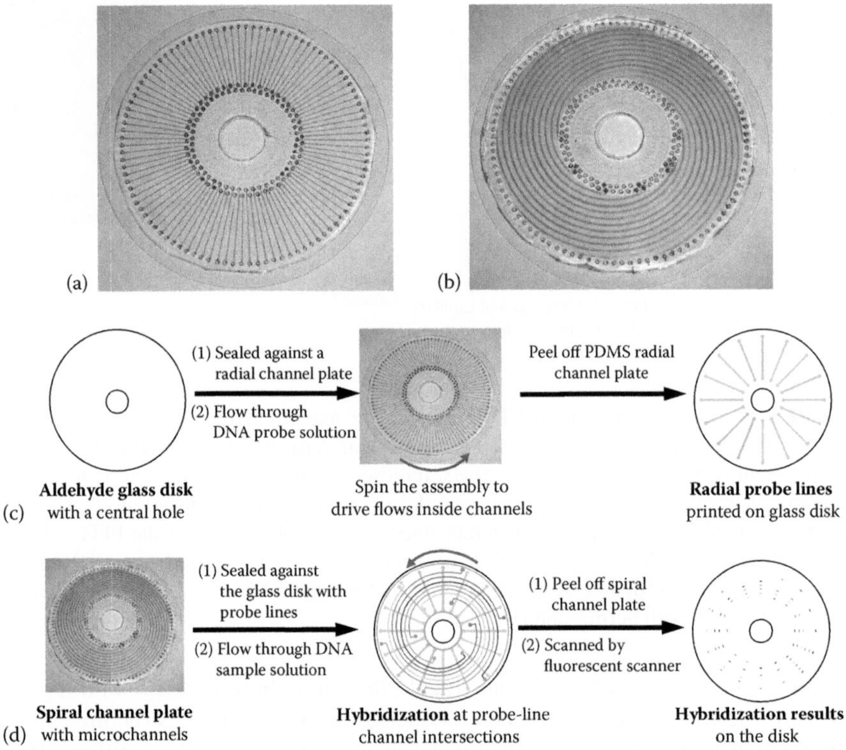

FIGURE 8.8 (a) Images of the assembly of a radial PDMS channel plate with a glass disk. Each channel was filled with dyed solutions. (b) Images of the assembly of a spiral PDMS channel plate with the glass disk. Ten groups of channels, three in each group, were filled with dyed solutions. (c) Probe line printing with the radial channel plate. (d) Hybridization procedure with the spiral channel plate. Hybridization occurs at the intersections of the spiral channels and the radial probe lines, shown as dark patches in the rightmost disk. (From Wang, L. and Li, P.C.H., *Analytical Biochemistry*, 400, 282–288, 2010. With permission.)

In both steps, liquid flows in the channels are driven by centrifugal pumping obtained by spinning the assembly. Because the spiral channels are nearly orthogonal to the radial direction, flows of samples (containing complementary targets) thus intersect with the probe lines printed on the disk and result in hybridization patches. The spiral channel design has been reported in detail elsewhere [127]. Figure 8.9 shows the fluorescent images from a 3-min hybridization of oligonucleotide samples of different concentrations [76]. The hybridization patches are rectangular and easy to analyze. As shown in Figure 8.9b insets, DNA molecules are not retained by the noncomplementary probe molecules at the intersection areas. For a disk of 92 mm in diameter, up to 100 samples had been analyzed in parallel and each sample volume was less than 1 μL [75–77].

The characterization of centrifugal flow hybridization in spiral microchannels was also studied [76]. Figure 8.10a shows a fluorescent image comparing the hybridization results under two conditions: continuous flow and stop flow. Hybridization intensities from the 3-min continuous flow are very close to those from the 2-h stop flow [77]. Moreover, as shown in Figure 8.10a, nonspecific binding is negligible in the continuous-flow method because the sample flow from centrifugal pumping continually removes

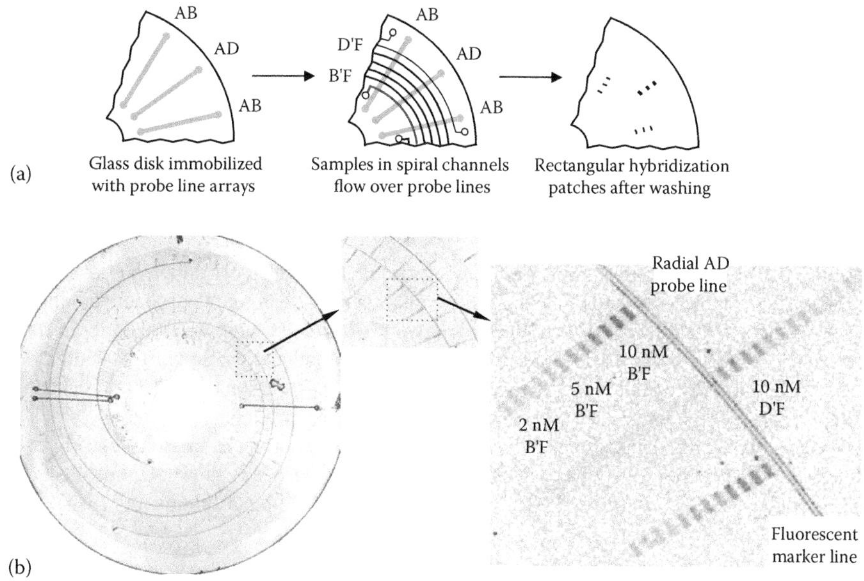

FIGURE 8.9 DNA hybridization on a CD-like chip. (a) The schematic diagram shows hybridization patches formed at the intersections between six sample channels (in curve) and three probe lines (in radial). (b) A fluorescent image of a glass disk with hybridization patches. The radial or spiral traces resulted from the flow of marker solutions for easy positioning. The two right insets show the magnified images of rectangular patches formed near the disk center, which resulted from the hybridization of different concentrations of oligonucleotide samples with their complementary probe lines. Oligonucleotide hybridizations were achieved at room temperature in 3-min spinning at 700 rpm, and then dried out at 3600 rpm. (From Wang, L. et al., *Analytica Chimica Acta*, 610, 97–104, 2008. With permission.)

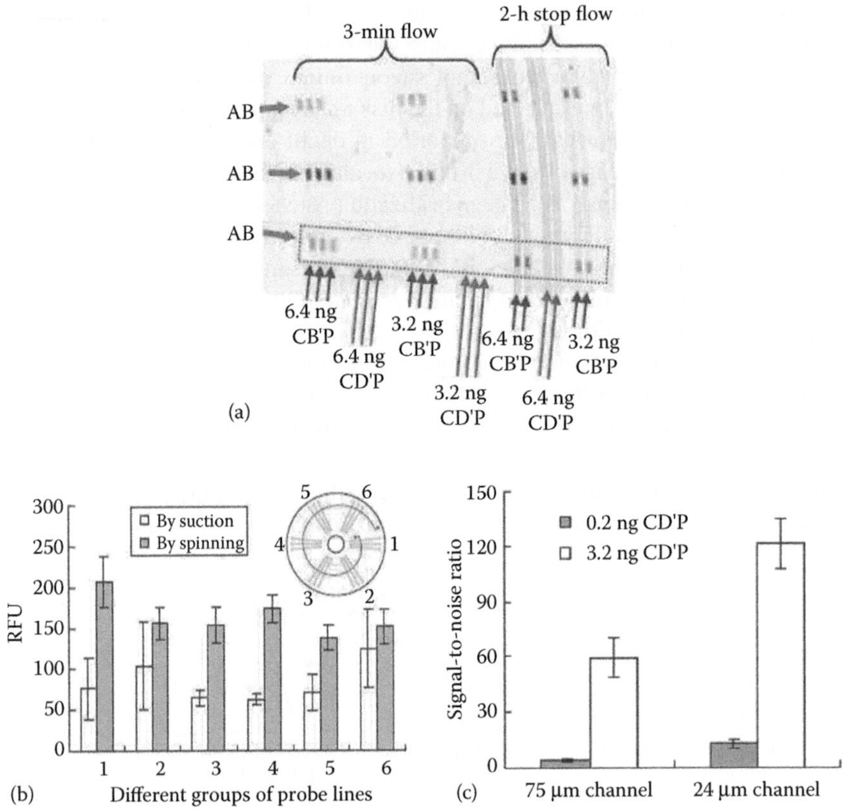

FIGURE 8.10 (a) Fluorescent images showing the PCR product hybridizations conducted with two different flow methods. The dark rectangular patches represent the specific binding of the complementary targets. CD′P and CB′P are two PCR products, and only the latter is complementary to the probe AB molecule. (b) Hybridization signal comparison of two continuous flow methods (driven by either centrifugal force or vacuum suction) on the same microchannel plate on six groups of oligo probes. The bar numbers in the histogram match the numbers shown in the inset, indicating the positions of hybridization along the spiral channels. (c) The effect of a shallower channel depth on higher hybridization signal intensities is shown. (From Wang, L. and Li, P.C.H., *Analytical Biochemistry*, 400, 282–288, 2010. With permission.)

unhybridized DNA molecules to prevent them from accumulating and binding nonspecifically onto the glass surface. On the contrary, in the stop-flow method, when the DNA samples are incubated in microchannels for a longer time (2 h), the nonspecific binding of DNA that is observed along the spiral channels is shown as strips, and posthybridization channel washing is thus required (Figure 8.10a). As depicted in Figure 8.10b, the signals obtained from centrifugal pumping (by spinning) show a higher level of consistency, as compared with those obtained by vacuum suction. It was also found that the shallower microchannel gave a better sensitivity [77,129]. The S/N ratios from the 24 μm microchannel chip are twice as high as those from the 75 μm microchannel chip. The reduction of the microchannel height enhanced the mass transport of the target DNA to the capture probes and thus generated higher hybridization signals. Moreover, the DNA molecules were stretched in nanoflows, which offered further advantages for hybridization efficiency [49,50,128].

Thereafter, PCR product samples from the genomic extraction of three fungal pathogens were successfully detected with the CD-like bioarray assembly under optimized conditions [77]. The influences of the chip spinning speed and the temperature on the sensitivity as well as the specificity of the signals were investigated (Figure 8.11). The hybridization time was determined by the chip spinning speed during the hybridization test. A lower spinning speed (or a higher residence time in the microchannel) would result in stronger hybridization signals (Figure 8.11a). In addition, a higher hybridization temperature resulted in an improved sensitivity, that is, 0.2 ng PCR products already gave an S/N ratio of ~20 at 42°C, as shown in Figure 8.11b. In addition, the method has been used for the discrimination of two

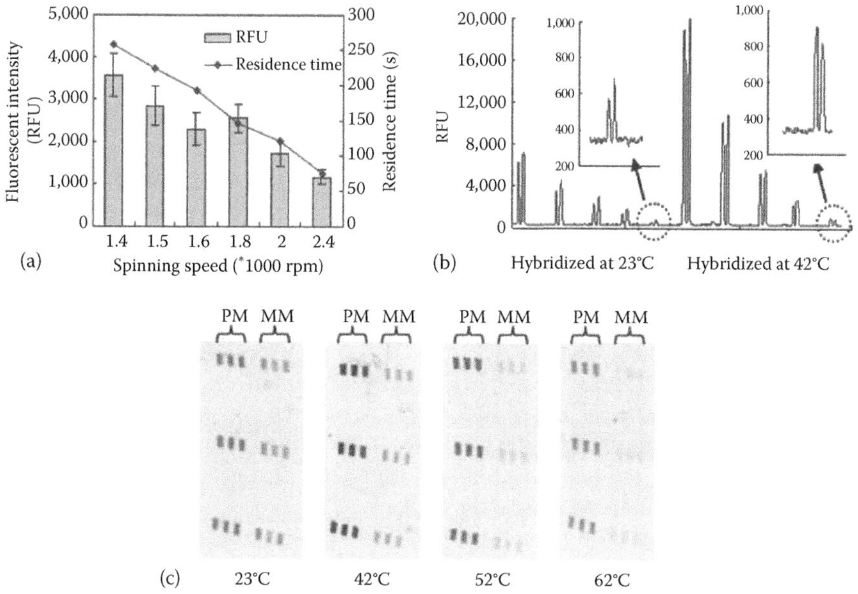

FIGURE 8.11 (a) A histogram showing the hybridization signals from the same amount of DNA samples hybridized under different disk spinning speeds. The line graph shows the corresponding sample residence time in the spiral microchannels. (b) Fluorescent intensities along a probe line. Five PCR product samples of different amounts were detected and each sample was conducted in duplicate. From left to right, five samples (6.4, 3.2, 1.6, 0.8, and 0.2 ng) were first hybridized at room temperature for 3 min. After drying out of these channels, another group of five samples was hybridized at 42°C for 3 min. In all cases, the volume of the sample solutions was 1.0 μL. The insets show the magnified graphs of the hybridization signals of the 0.2 ng PCR product samples. (c) Differentiation of PCR products with a single base-pair difference at various hybridization temperatures. Each image was obtained from the hybridization of sample solutions in three spiral channels intersecting with three probe lines at the specified temperature. Two PCR products, perfectly matched (PM) and mismatched (MM), were tested in the experiment. (From Wang, L. and Li, P.C.H., *Analytical Biochemistry*, 400, 282–288, 2010. With permission.)

related fungal subspecies, *Botrytis cinerea* and *B. squamosa*. The two PCR amplicons differ in only one base pair in the middle of the 264 bp long sequences. With temperature control, the two DNA samples were discriminated by using a 3-min flow hybridization as shown in Figure 8.11b.

In addition to the radial–spiral approach where the 360° spiral channels intersected with the radial probe lines, Chen et al. also developed the double-spiral format in which the 180° clockwise spiral channels intersected with the 180° anticlockwise spiral probe lines [78]. In this manner, a higher spot density was achieved for the high-throughput bioarray assay on a 92 mm circular glass disk, which was the same size as the one used previously in the radial–spiral format. The four-replicate arrangement, where each inlet channel branches into four spiral channels, helps the bioarray experiment to be self-corrected (Figure 8.12a inset). For instance, when one spiral channel was blocked and the sample solution could not flow through it to produce the hybridization spots (a false-negative error), this error could be corrected by observing the hybridization spots in the other three spiral channels. Moreover, because the group of 16 spots

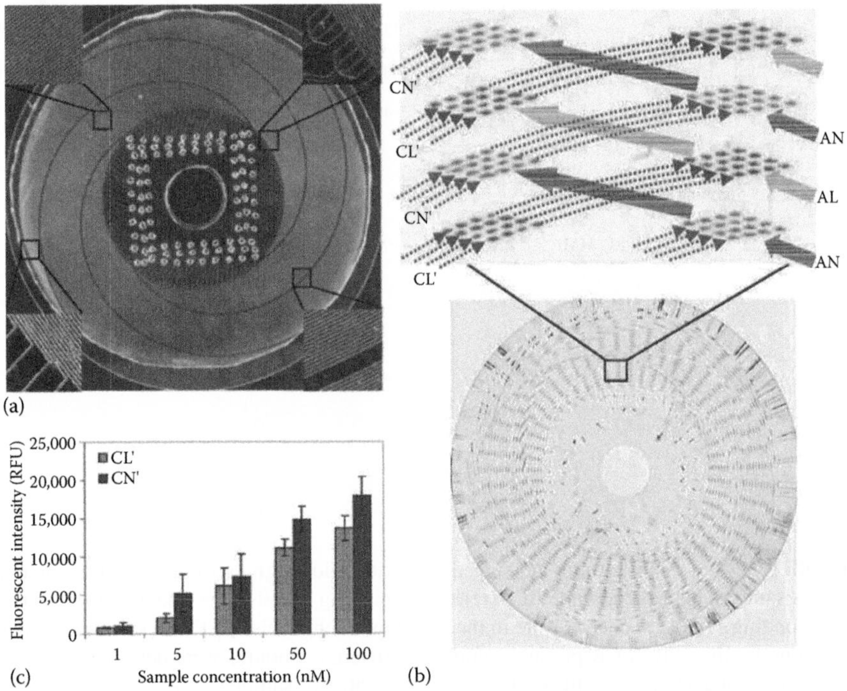

FIGURE 8.12 The double-spiral nanofluidic bioarray method. (a) An image of the anticlockwise spiral channel plate. The four anticlockwise lines represent the gap zone in which there are no channels. The four insets zoom in the regions near the inlet zone, gap zone, channel zone, and outlet zone. (b) The pattern of 192 × 192 hybridizations achieved on one glass disk. The inset shows a close-up of the hybridization spots. (c) Spot intensities of different concentrations of CL and CN hybridized with complementary probe lines (AL and AN) printed at the probe concentration of 50 μM. (From Chen, H. et al., *Lab on a Chip*, 8, 826–829, 2008. With permission.)

comes from the same batches of probe and sample solutions, fake signals from small surface defects, which are common in the conventional bioarray method, can also be corrected [12]. The four-replicate design increases the confidence of target detection. Using this double-spiral format, up to 384 assays (on 96 different samples) can be performed at the same time onto 384 probe lines (96 probes in 4 replicates) [78].

8.3 SUMMARY AND CONCLUSIONS

Bioarray-based hybridization has been an important technique in genomic research. Because of their flexibility and low production cost, bioarrays with low-density probes have been used in SNP detection as well as nucleic acid diagnostic applications. Recently, nanofluidic technology was combined with the DNA bioarray method through covering the spotted probe area with chambers or microchannels for sample hybridization.

Probe molecules could also be spotted by the nanofluidic method and line arrays thus formed on the solid support. A 2-D bioarray method has been developed in which the hybridization microchannels are orthogonal to these line arrays, and complementary DNA molecules will be retained at the intersection areas. This 2-D method alleviates the need for the exact alignment of the hybridization microchannels and probe spots printed from the pin-spotted method. The spot shape in this 2-D method is more regular for convenient image analysis. A 16×16 bioarray device has been used to successfully discriminate multiple PCR product samples from two fungal pathogen genomes.

In the nanofluidic bioarray method, different techniques have been developed to generate nanoflows. Centrifugal force was employed to drive liquid movement in microchannels without additional solution interface contacts. The centrifugal force–driven flow also facilitates parallel reactions. However, due to the radial-only structure design, the applications of centrifugal pumping to 2-D nucleic acid bioarray analysis are limited. To address this problem, CD-like PDMS chips with radial or spiral microchannels were used. This design enabled the use of centrifugal pumping for the 2-D DNA bioarray method, and the centrifugal advantages were exploited twice, first in the radial microchannels, and then in the spiral channels. This technique has been successfully applied to the fast detection and discrimination of up to 100 DNA samples. The method was also developed in a double-spiral way to obtain a higher probe/sample density with self-correction benefits. As summarized in this chapter, the nanofluidic bioarray method shows the advantages of less sample usage and fast reaction kinetics, as well as multiple sample capabilities for high-throughput analysis.

REFERENCES

1. U. Maskos and E. M. Southern, Oligonucleotide hybridizations on glass supports: A novel linker for oligonucleotide synthesis and hybridization properties of oligonucleotides synthesized Insitu, *Nucleic Acids Research*, 20(7), 1679–1684, 1992.
2. T. R. Hughes, M. Mao, A. R. Jones et al., Expression profiling using microarrays fabricated by an ink-jet oligonucleotide synthesizer, *Nature Biotechnology*, 19(4), 342–347, 2001.
3. R. J. Lipshutz, S. P. A. Fodor, T. R. Gingeras et al., High density synthetic oligonucleotide arrays, *Nature Genetics*, 21, 20–24, 1999.

4. A. Sassolas, B. D. Leca-Bouvier, and L. J. Blum, DNA biosensors and microarrays, *Chemical Reviews*, 108(1), 109–139, 2008.
5. S. Russell, L. A. Meadows, and R. R. Russell, *Microarray Technology in Practice*, Amsterdam: Academic Press/Elsevier, 2009.
6. M. Schena, *Microarray Analysis*, New York: Wiley-Liss, 2003.
7. M. J. Buck and J. D. Lieb, ChIP-chip: Considerations for the design, analysis, and application of genome-wide chromatin immunoprecipitation experiments, *Genomics*, 83(3), 349–360, 2004.
8. J. D. Hoheisel, Microarray technology: Beyond transcript profiling and genotype analysis, *Nature Reviews Genetics*, 7(3), 200–210, 2006.
9. A. Khademhosseini, Chips to hits: Microarray and microfluidic technologies for high-throughput analysis and drug discovery, *Expert Review of Molecular Diagnostics*, 5(6), 843–846, 2005.
10. C. Situma, M. Hashimoto, and S. A. Soper, Merging microfluidics with microarray-based bioassays, *Biomolecular Engineering*, 23(5), 213–231, 2006.
11. R. H. Liu, K. Dill, H. S. Fuji et al., Integrated microfluidic biochips for DNA microarray analysis, *Expert Review of Molecular Diagnostics*, 6(2), 253–261, 2006.
12. D. Bowtell and J. Sambrook, *DNA Microarrays: A Molecular Cloning Manual*, Cold Spring Harbor, NY: Cold Spring Harbor Laboratory Press, 2003.
13. K. Pappaert, P. Van Hummelen, J. Vanderhoeven et al., Diffusion-reaction modelling of DNA hybridization kinetics on biochips, *Chemical Engineering Science*, 58(21), 4921–4930, 2003.
14. H. Y. Dai, M. Meyer, S. Stepaniants et al., Use of hybridization kinetics for differentiating specific from non-specific binding to oligonucleotide microarrays, *Nucleic Acids Research*, 30(16), e86, 2002.
15. G. L. Lukacs, P. Haggie, O. Seksek et al., Size-dependent DNA mobility in cytoplasm and nucleus, *Journal of Biological Chemistry*, 275(3), 1625–1629, 2000.
16. M. A. Bynum and G. B. Gordon, Hybridization enhancement using microfluidic planetary centrifugal mixing, *Analytical Chemistry*, 76(23), 7039–7044, 2004.
17. K. Pappaert, J. Vanderhoeven, P. Van Hummelen et al., Enhancement of DNA micro-array analysis using a shear-driven micro-channel flow system, *Journal of Chromatography A*, 1014(1–2), 1–9, 2003.
18. J. Hong, J. B. Edel, and A. J. deMello, Micro- and nanofluidic systems for high-throughput biological screening, *Drug Discovery Today*, 14(3–4), 134–146, 2009.
19. G. Keramas, G. Perozziello, O. Geschke et al., Development of a multiplex microarray microsystem, *Lab on a Chip*, 4(2), 152–158, 2004.
20. Y. Sun, I. Perch-Nielsen, M. Dufva et al., Direct immobilization of DNA probes on non-modified plastics by UV irradiation and integration in microfluidic devices for rapid bioassay, *Analytical and Bioanalytical Chemistry*, 402(2), 741–748, 2012.
21. A. M. Foudeh, T. Fatanat Didar, T. Veres et al., Microfluidic designs and techniques using lab-on-a-chip devices for pathogen detection for point-of-care diagnostics, *Lab on a Chip*, 12(18), 3249–3266, 2012.
22. A. N. Rao, C. K. Rodesch, and D. W. Grainger, Real-time fluorescent image analysis of DNA spot hybridization kinetics to assess microarray spot heterogeneity, *Analytical Chemistry*, 84(21), 9379–9387, 2012.
23. M. Dufva, J. Petersen, and L. Poulsen, Increasing the specificity and function of DNA microarrays by processing arrays at different stringencies, *Analytical and Bioanalytical Chemistry*, 395(3), 669–677, 2009.
24. N. C. H. Le, V. Gubala, E. Clancy, T. Barry, T. J. Smith, and D. E. Williams, Ultrathin and smooth poly(methyl methacrylate) (PMMA) films for label-free biomolecule detection with total internal reflection ellipsometry (TIRE), *Biosensors and Bioelectronics*, 36(1), 250–256, 2012.

25. J. C. McDonald, D. C. Duffy, J. R. Anderson et al., Fabrication of microfluidic systems in poly(dimethylsiloxane), *Electrophoresis*, 21(1), 27–40, 2000.
26. J. Liu, B. A. Williams, R. M. Gwirtz et al., Enhanced signals and fast nucleic acid hybridization by microfluidic chaotic mixing, *Angewandte Chemie-International Edition*, 45(22), 3618–3623, 2006.
27. R. H. Liu, R. Lenigk, R. L. Druyor-Sanchez et al., Hybridization enhancement using cavitation microstreaming, *Analytical Chemistry*, 75(8), 1911–1917, 2003.
28. R. C. Anderson, X. Su, G. J. Bogdan et al., A miniature integrated device for automated multistep genetic assays, *Nucleic Acids Research*, 28(12), e60, 2000.
29. M. Noerholm, H. Bruus, M. H. Jakobsen et al., Polymer microfluidic chip for online monitoring of microarray hybridizations, *Lab on a Chip*, 4(1), 28–37, 2004.
30. P. Stafford, *Methods in Microarray Normalization*, Boca Raton, FL: CRC Press, 2008.
31. N. B. Adey, M. Lei, M. T. Howard et al., Gains in sensitivity with a device that mixes microarray hybridization solution in a 25-μm-thick chamber, *Analytical Chemistry*, 74(24), 6413–6417, 2002.
32. J. Vanderhoeven, K. Pappaert, B. Dutta et al., Comparison of a pump-around a diffusion-driven, and a shear-driven system for the hybridization of mouse lung and testis total RNA on microarrays, *Electrophoresis*, 26(19), 3773–3779, 2005.
33. J. Vanderhoeven, K. Pappaert, B. Dutta et al., DNA microarray enhancement using a continuously and discontinuously rotating microchamber, *Analytical Chemistry*, 77(14), 4474–4480, 2005.
34. P. K. Yuen, G. S. Li, Y. J. Bao et al., Microfluidic devices for fluidic circulation and mixing improve hybridization signal intensity on DNA arrays, *Lab on a Chip*, 3(1), 46–50, 2003.
35. M. K. McQuain, K. Seale, J. Peek et al., Chaotic mixer improves microarray hybridization, *Analytical Biochemistry*, 325(2), 215–226, 2004.
36. J. M. Hertzsch, R. Sturman, and S. Wiggins, DNA microarrays: Design principles for maximizing ergodic, chaotic mixing, *Small*, 3(2), 202–218, 2007.
37. H. H. Lee, J. Smoot, Z. McMurray et al., Recirculating flow accelerates DNA microarray hybridization in a microfluidic device, *Lab on a Chip*, 6(9), 1163–1170, 2006.
38. C. G. Cooney, D. Sipes, N. Thakore et al., A plastic, disposable microfluidic flow cell for coupled on-chip PCR and microarray detection of infectious agents, *Biomedical Microdevices*, 14(1), 45–53, 2012.
39. A. Bouchie, Shift anticipated in DNA microarray market, *Nature Biotechnology*, 20(1), 8, 2002.
40. J. P. Gillet, T. J. Molina, J. Jamart et al., Evaluation of a low density DNA microarray for small B-cell non-Hodgkin lymphoma differential diagnosis, *Leukemia and Lymphoma*, 50(3), 410–418, 2009.
41. G. Meneses-Lorente, F. de Longueville, S. Dos Santos-Mendes et al., An evaluation of a low-density DNA microarray using cytochrome P450 inducers, *Chemical Research in Toxicology*, 16(9), 1070–1077, 2003.
42. F. de Longueville, D. Surry, G. Meneses-Lorente et al., Gene expression profiling of drug metabolism and toxicology markers using a low-density DNA microarray, *Biochemical Pharmacology*, 64(1), 137–149, 2002.
43. J. P. Gillet, T. Efferth, D. Steinbach et al., Microarray-based detection of multidrug resistance in human tumor cells by expression profiling of ATP-binding cassette transporter genes, *Cancer Research*, 64(24), 8987–8993, 2004.
44. D. Tejedor, S. Castillo, P. Mozas et al., Reliable low-density DNA array based on allele-specific probes for detection of 118 mutations causing familial hypercholesterolemia, *Clinical Chemistry*, 51(7), 1137–1144, 2005.
45. P. Alvarez, P. Saenz, D. Arteta et al., Transcriptional profiling of hematologic malignancies with a low-density DNA microarray, *Clinical Chemistry*, 53(2), 259–267, 2007.

46. C. A. Koch, P. C. H. Li, and R. S. Utkhede, Evaluation of thin films of agarose on glass for hybridization of DNA to identify plant pathogens with microarray technology, *Analytical Biochemistry*, 342(1), 93–102, 2005.
47. D. Erickson, X. Z. Liu, U. Krull et al., Electrokinetically controlled DNA hybridization microfluidic chip enabling rapid target analysis, *Analytical Chemistry*, 76(24), 7269–7277, 2004.
48. Y. Wang, B. Vaidya, H. D. Farquar et al., Microarrays assembled in microfluidic chips fabricated from poly(methyl methacrylate) for the detection of low-abundant DNA mutations, *Analytical Chemistry*, 75(5), 1130–1140, 2003.
49. Y. C. Chung, Y. C. Lin, M. Z. Shiu et al., Microfluidic chip for fast nucleic acid hybridization, *Lab on a Chip*, 3(4), 228–233, 2003.
50. J. H. S. Kim, A. Marafie, X. Y. Jia et al., Characterization of DNA hybridization kinetics in a microfluidic flow channel, *Sensors and Actuators B-Chemical*, 113(1), 281–289, 2006.
51. R. Peytavi, F. R. Raymond, D. Gagne et al., Microfluidic device for rapid (<15 min) automated microarray hybridization, *Clinical Chemistry*, 51(10), 1836–1844, 2005.
52. G. Y. Jia, K. S. Ma, J. Kim et al., Dynamic automated DNA hybridization on a CD (compact disc) fluidic platform, *Sensors and Actuators B-Chemical*, 114(1), 173–181, 2006.
53. C. Y. Li, X. L. Dong, J. H. Qin et al., Rapid nanoliter DNA hybridization based on reciprocating flow on a compact disk microfluidic device, *Analytica Chimica Acta*, 640(1–2), 93–99, 2009.
54. A. L. Brøgger, D. Kwasny, F. G. Bosco et al., Centrifugally driven microfluidic disc for detection of chromosomal translocations, *Lab on a Chip*, 12(22), 4628–4634, 2012.
55. H.-I. Peng, C. M. Strohsahl, and B. L. Miller, Microfluidic nanoplasmonic-enabled device for multiplex DNA detection, *Lab on a Chip*, 12(6), 1089–1093, 2012.
56. Y. J. Liu, C. B. Rauch, R. L. Stevens et al., DNA amplification and hybridization assays in integrated plastic monolithic devices, *Analytical Chemistry*, 74(13), 3063–3070, 2002.
57. Y. C. Chung, Y. C. Lin, C. D. Chueh et al., Microfluidic chip of fast DNA hybridization using denaturing and motion of nucleic acids, *Electrophoresis*, 29(9), 1859–1865, 2008.
58. R. Lenigk, R. H. Liu, M. Athavale et al., Plastic biochannel hybridization devices: A new concept for microfluidic DNA arrays, *Analytical Biochemistry*, 311(1), 40–49, 2002.
59. C. W. Wei, J. Y. Cheng, C. T. Huang et al., Using a microfluidic device for 1 μl DNA microarray hybridization in 500 s, *Nucleic Acids Research*, 33(8), e78, 2005.
60. C.-H. Tai, J.-W. Shin, T.-Y. Chang et al., An integrated microfluidic system capable of sample pretreatment and hybridization for microarrays, *Microfluidics and Nanofluidics*, 10(5), 999–1009, 2012.
61. M. Campas and I. Katakis, DNA biochip arraying, detection and amplification strategies, *TrAC-Trends in Analytical Chemistry*, 23(1), 49–62, 2004.
62. V. Le Berre, E. Trevisiol, A. Dagkessamanskaia et al., Dendrimeric coating of glass slides for sensitive DNA microarrays analysis, *Nucleic Acids Research*, 31(16), e88, 2003.
63. E. Delamarche, A. Bernard, H. Schmid et al., Microfluidic networks for chemical patterning of substrate: Design and application to bioassays, *Journal of the American Chemical Society*, 120(3), 500–508, 1998.
64. M. Geissler, E. Roy, G. A. Diaz-Quijada et al., Microfluidic patterning of miniaturized DNA arrays on plastic substrates, *ACS Applied Materials and Interfaces*, 1(7), 1387–1395, 2009.
65. E. Delamarche, D. Juncker, and H. Schmid, Microfluidics for processing surfaces and miniaturizing biological assays, *Advanced Materials*, 17(24), 2911–2933, 2005.
66. E. Delamarche, A. Bernard, H. Schmid et al., Patterned delivery of immunoglobulins to surfaces using microfluidic networks, *Science*, 276(5313), 779–781, 1997.

67. H. J. Lee, T. T. Goodrich, and R. M. Corn, SPR imaging measurements of 1-D and 2-D DNA microarrays created from microfluidic channels on gold thin films, *Analytical Chemistry*, 73(22), 5525–5531, 2001.
68. H. Arata, H. Komatsu, A. Han et al., Rapid microRNA detection using power-free microfluidic chip: Coaxial stacking effect enhances the sandwich hybridization, *The Analyst*, 137(14), 3234–3237, 2012.
69. J. Wen, X. Shi, Y. He et al., Novel plastic biochips for colorimetric detection of biomolecules, *Analytical and Bioanalytical Chemistry*, 404(6–7), 1935–1944, 2012.
70. L. Wang and P. C. H. Li, Flexible microarray construction and fast DNA hybridization conducted on a microfluidic chip for greenhouse plant fungal pathogen detection, *Journal of Agricultural and Food Chemistry*, 55(26), 10509–10516, 2007.
71. L. Wang and P. C. H. Li, Gold nanoparticle-assisted single base-pair mismatch discrimination on a microfluidic microarray device, *Biomicrofluidics*, 4(3), 32209, 2010.
72. J. A. Benn, J. Hu, B. J. Hogan et al., Comparative modeling and analysis of microfluidic and conventional DNA microarrays, *Analytical Biochemistry*, 348(2), 284–293, 2006.
73. C. Situma, Y. Wang, M. Hupert et al., Fabrication of DNA microarrays onto poly(methyl methacrylate) with ultraviolet patterning and microfluidics for the detection of low-abundant point mutations, *Analytical Biochemistry*, 340(1), 123–135, 2005.
74. Y. C. Li, Z. Wang, L. M. L. Ou et al., DNA detection on plastic: Surface activation protocol to convert polycarbonate substrates to biochip platforms, *Analytical Chemistry*, 79(2), 426–433, 2007.
75. X. Y. Peng, P. C. H. Li, H. Z. Yu et al., Spiral microchannels on a CD for DNA hybridizations, *Sensors and Actuators B-Chemical*, 128(1), 64–69, 2007.
76. L. Wang, P. C. H. Li, H. Z. Yu et al., Fungal pathogenic nucleic acid detection achieved with a microfluidic microarray device, *Analytica Chimica Acta*, 610(1), 97–104, 2008.
77. L. Wang and P. C. H. Li, Optimization of a microfluidic microarray device for the fast discrimination of fungal pathogenic DNA, *Analytical Biochemistry*, 400(2), 282–288, 2010.
78. H. Chen, L. Wang, and P. C. H. Li, Nucleic acid microarrays created in the double-spiral format on a circular microfluidic disk, *Lab on a Chip*, 8(5), 826–829, 2008.
79. C. A. Rowe, L. M. Tender, M. J. Feldstein et al., Array biosensor for simultaneous identification of bacterial, viral, and protein analytes, *Analytical Chemistry*, 71(17), 3846–3852, 1999.
80. A. Bernard, B. Michel, and E. Delamarche, Micromosaic immunoassays, *Analytical Chemistry*, 73, 8–12, 2001.
81. A. Sedighi and P. C. H. Li, Kras gene codon 12 mutation detection enabled by gold nanoparticles conducted in a nanobioarray chip, *Analytical Biochemistry*, 448, 58–64, 2014.
82. G. Krishnamoorthy, E. T. Carlen, H. L. deBoer et al., Electrokinetic lab-on-a-biochip for multi-ligand/multi-analyte biosensing, *Analytical Chemistry*, 82(10), 4145–4150, 2010.
83. S. C. Jacobson, T. E. McKnight, and J. M. Ramsey, Microfluidic devices for electrokinetically driven parallel and serial mixing, *Analytical Chemistry*, 71(20), 4455–4459, 1999.
84. P. C. H. Li, *Microfluidic Lab-on-a-Chip for Chemical and Biological Analysis and Discovery*, Boca Raton, FL: Taylor & Francis/CRC Press, 2006.
85. P. Gong and R. Levicky, DNA surface hybridization regimes, *Proceedings of the National Academy of Sciences of the United States of America*, 105(14), 5301–5306, 2008.
86. R. Levicky and A. Horgan, Physicochemical perspectives on DNA microarray and biosensor technologies, *Trends in Biotechnology*, 23(3), 143–149, 2005.
87. G. Y. Tang, D. G. Yan, C. Yang et al., Assessment of Joule heating and its effects on electroosmotic flow and electrophoretic transport of solutes in microfluidic channels, *Electrophoresis*, 27(3), 628–639, 2006.

88. X. C. Xuan, B. Xu, D. Sinton et al., Electroosmotic flow with Joule heating effects, *Lab on a Chip*, 4(3), 230–236, 2004.
89. I. Rodriguez and N. Chandrasekhar, Experimental study and numerical estimation of current changes in electroosmotically pumped microfluidic devices, *Electrophoresis*, 26(6), 1114–1121, 2005.
90. B. J. Kirby and E. F. Hasselbrink, Zeta potential of microfluidic substrates: 1. Theory, experimental techniques, and effects on separations, *Electrophoresis*, 25(2), 187–202, 2004.
91. A. S. Watts, A. A. Urbas, E. Moschou et al., Centrifugal microfluidics with integrated sensing microdome optodes for multiion detection, *Analytical Chemistry*, 79(21), 8046–8054, 2007.
92. R. D. Johnson, I. H. A. Badr, G. Barrett et al., Development of a fully integrated analysis system for ions based on ion-selective optodes and centrifugal microfluidics, *Analytical Chemistry*, 73(16), 3940–3946, 2001.
93. I. H. A. Badr, R. D. Johnson, M. J. Madou et al., Fluorescent ion-selective optode membranes incorporated onto a centrifugal microfluidics platform, *Analytical Chemistry*, 74(21), 5569–5575, 2002.
94. J. Steigert, M. Grumann, T. Brenner et al., Fully integrated whole blood testing by real-time absorption measurement on a centrifugal platform, *Lab on a Chip*, 6(8), 1040–1044, 2006.
95. M. Gustafsson, D. Hirschberg, C. Palmberg et al., Integrated sample preparation and MALDI mass spectrometry on a microfluidic compact disk, *Analytical Chemistry*, 76(2), 345–350, 2004.
96. J. Steigert, T. Brenner, M. Grumann et al., Integrated siphon-based metering and sedimentation of whole blood on a hydrophilic lab-on-a-disk, *Biomedical Microdevices*, 9(5), 675–679, 2007.
97. J. L. Zhang, Q. Q. Guo, M. Liu et al., A lab-on-CD prototype for high-speed blood separation, *Journal of Micromechanics and Microengineering*, 18(12), 125025, 2008.
98. D. C. Duffy, H. L. Gillis, J. Lin et al., Microfabricated centrifugal microfluidic systems: Characterization and multiple enzymatic assays, *Analytical Chemistry*, 71(20), 4669–4678, 1999.
99. P. Andersson, G. Jesson, G. Kylberg et al., Parallel nanoliter microfluidic analysis system, *Analytical Chemistry*, 79(11), 4022–4030, 2007.
100. S. W. Lee, J. Y. Kang, I. H. Lee et al., Single-cell assay on CD-like lab chip using centrifugal massive single-cell trap, *Sensors and Actuators A: Physical*, 143(1), 64–69, 2008.
101. S. Haeberle, T. Brenner, R. Zengerle et al., Centrifugal extraction of plasma from whole blood on a rotating disk, *Lab on a Chip*, 6(6), 776–781, 2006.
102. S. Lai, S. N. Wang, J. Luo et al., Design of a compact disk-like microfluidic platform for enzyme-linked immunosorbent assay, *Analytical Chemistry*, 76(7), 1832–1837, 2004.
103. A. Penrose, P. Myers, K. Bartle et al., Development and assessment of a miniaturised centrifugal chromatograph for reversed-phase separations in micro-channels, *Analyst*, 129(8), 704–709, 2004.
104. A. P. Wong, M. Gupta, S. S. Shevkoplyas et al., Egg beater as centrifuge: Isolating human blood plasma from whole blood in resource-poor settings, *Lab on a Chip*, 8(12), 2032–2037, 2008.
105. B. S. Lee, J. N. Lee, J. M. Park et al., A fully automated immunoassay from whole blood on a disc, *Lab on a Chip*, 9(11), 1548–1555, 2009.
106. L. G. Puckett, E. Dikici, S. Lai, M. Madou, L. G. Bachas, and S. Daunert, Investigation into the applicability of the centrifugal microfluidics development of protein-platform for the ligand binding assays incorporating enhanced green fluorescent protein as a fluorescent reporter, *Analytical Chemistry*, 76(24), 7263–7268, 2004.
107. M. Grumann, J. Steigert, L. Riegger et al., Sensitivity enhancement for colorimetric glucose assays on whole blood by on-chip beam-guidance, *Biomedical Microdevices*, 8(3), 209–214, 2006.

108. Y. K. Cho, J. G. Lee, J. M. Park et al., One-step pathogen specific DNA extraction from whole blood on a centrifugal microfluidic device, *Lab on a Chip*, 7(5), 565–573, 2007.
109. J. M. Chen, P. C. Huang, and M. G. Lin, Analysis and experiment of capillary valves for microfluidics on a rotating disk, *Microfluidics and Nanofluidics*, 4(5), 427–437, 2008.
110. S. Haeberle, T. Brenner, H. P. Schlosser et al., Centrifugal micromixer, *Chemical Engineering and Technology*, 28(5), 613–616, 2005.
111. T. Glatzel, C. Litterst, C. Cupelli et al., Computational fluid dynamics (CFD) software tools for microfluidic applications—A case study, *Computers and Fluids*, 37(3), 218–235, 2008.
112. T. Brenner, T. Glatzel, R. Zengerle et al., Frequency-dependent transversal flow control in centrifugal microfluidics, *Lab on a Chip*, 5(2), 146–150, 2005.
113. M. Liu, J. Zhang, Y. Liu et al., Modeling of flow burst, flow timing in lab-on-a-CD systems and its application in digital chemical analysis, *Chemical Engineering and Technology*, 31(9), 1328–1335, 2008.
114. D. S. Kim and T. H. Kwon, Modeling, analysis and design of centrifugal force driven transient filling flow into rectangular microchannel, *Microsystem Technologies*, 12(9), 822–838, 2006.
115. D. S. Kim and T. H. Kwon, Modeling, analysis and design of centrifugal force-driven transient filling flow into a circular microchannel, *Microfluidics and Nanofluidics*, 2(2), 125–140, 2006.
116. J. Siegrist, M. Amasia, N. Singh et al., Numerical modeling and experimental validation of uniform microchamber filling in centrifugal microfluidics, *Lab on a Chip*, 10(7), 876–886, 2010.
117. J. Kim, H. Kido, R. H. Rangel et al., Passive flow switching valves on a centrifugal microfluidic platform, *Sensors and Actuators B-Chemical*, 128(2), 613–621, 2008.
118. J. Ducree, S. Haeberle, T. Brenner et al., Patterning of flow and mixing in rotating radial microchannels, *Microfluidics and Nanofluidics*, 2(2), 97–105, 2006.
119. L. Wang and P.C.H. Li, Microfluidic DNA microarray analysis: A review, *Analytica Chimica Acta*, 687, 12–27, 2011.
120. M. S. Al-Qahtani and M. N. Basha, Prediction of flow and heat transfer in narrow rotating rectangular channels using Reynolds stress model, *Heat and Mass Transfer*, 44(5), 505–516, 2008.
121. J. V. Zoval and M. J. Madou, Centrifuge-based fluidic platforms, *Proceedings of the IEEE*, 92(1), 140–153, 2004.
122. D. D. Nolte, Invited review article: Review of centrifugal microfluidic and bio-optical disks, *Review of Scientific Instruments*, 80(10), 1011101, 2009.
123. D. Mark, S. Haeberle, G. Roth et al., Microfluidic lab-on-a-chip platforms: Requirements, characteristics and applications, *Chemical Society Reviews*, 39(3), 1153–1182, 2010.
124. R. A. Potyrailo, W. G. Morris, A. M. Leach et al., Analog signal acquisition from computer optical disk drives for quantitative chemical sensing, *Analytical Chemistry*, 78(16), 5893–5899, 2006.
125. S. A. Lange, G. Roth, S. Wittemann et al., Measuring biomolecular binding events with a compact disc player device, *Angewandte Chemie-International Edition*, 45(2), 270–273, 2006.
126. M. Madou, J. Zoval, G. Y. Jia et al., Lab on a CD, *Annual Review of Biomedical Engineering*, 8, 601–628, 2006.
127. X. Y. Peng and P. C. H. Li, Centrifugal pumping in the equiforce spiral microchannel, *Canadian Journal of Pure and Applied Sciences*, 2(3), 551–556, 2008.
128. J. W. Larson, G. R. Yantz, Q. Zhong et al., Single DNA molecule stretching in sudden mixed shear and elongational microflows, *Lab on a Chip*, 6(9), 1187–1199, 2006.
129. L. Wang, M.-C. Kropinski, and P. C. H. Li, Analysis and modeling of flow in rotating spiral microchannels: Towards math-aided design of microfluidic systems using centrifugal pumping, *Lab on the Chip*, 11, 2097–2108, 2011.

9 Optical Oxygen Sensors for Micro- and Nanofluidic Devices

Volker Nock, Richard J. Blaikie, and Maan M. Alkaisi

CONTENTS

9.1 INTRODUCTION

Ever since the emergence of photosynthesis and subsequent appearance of eukaryotic organisms, oxygen and life have been connected through a complicated interdependence. The use of oxygen as substrate for energy production, albeit very efficient, is not without risk [1]. A few electrons combining prematurely with an O_2 molecule can lead to the formation of reactive oxygen species (ROS), such as H_2O_2. The presence of these ROS in turn can lead to the oxidation of lipids, nucleic acids, and proteins and hence result in cellular dysfunction or even cell death, something exacerbated even further by the fact that acute increases or decreases in cellular oxygen (hyperoxia/hypoxia) can induce further the generation of excess ROS. One of the most important lessons to be drawn from this intricate relationship is that efficient cellular function only occurs within a narrow range of oxygen concentrations.

This indicates that efficient control of the cellular microenvironment is of utmost importance in in vitro cell culture. One successful approach to increased environmental control is to reduce the dimensions and thus sample size in a system to reduce intermixing and increase the influence of diffusion-limited processes predominant at the micron scale. The characteristics of laminar flow, as observed in microfluidic devices, allows one to, for example, generate parallel multistream flows with stable interstream interfaces in a single channel. Material transport across these interfaces is by diffusion only and can be controlled using the flow speed of the individual streams. In cell biology, this phenomenon can be applied to produce controlled chemical microenvironments down to subcellular dimensions [2,3]. To this day, the use of multiple parallel flow streams has been explored mostly for the delivery of biochemical reagents to cells [2–4]. Recently though, their potential for the generation of cellular microenvironments with controlled oxygen concentrations has begun to attract increasing interest.

In cell-based microfluidic applications in particular, the oxygen concentration of a sample stream itself represents a parameter with significant effect on cellular development and function. For example, the concentration of dissolved oxygen (DO) a cell is exposed to has been found to be intimately linked to cell survival, metabolism, and function [5]. Devices capable of exposing defined areas of a cell culture, individual cells, and regions on the surface of an individual cell to controlled DO levels therefore have the potential to yield novel insights into cell biology and tissue formation [6–9]. Furthermore, measuring and controlling the DO concentrations of sample streams will also increase the relevance of existing small-molecule delivery applications, which to this day have been performed with media equilibrated under atmospheric oxygen conditions only [2–4,10,11]. Common to all these examples is the need for a simple and robust means of measuring DO in the respective microfluidic devices. In the following, we aspire to demonstrate that thin-film optical oxygen sensors provide such a means and discuss their fabrication, calibration, and use for spatially resolved measurement of DO_2 in microfluidic devices related to the control of cellular microenvironments.

9.2 DISSOLVED OXYGEN IN BIOLOGICAL APPLICATIONS

In a natural environment, such as in mammalian organs, cellular oxygen concentration is maintained to normoxic (12% to <0.5% O_2) conditions [12]. Regulation to within this relatively narrow range of normoxia is necessary in vivo to prevent oxidative damage to the cell from excess oxygen (hyperoxia) and metabolic demise from insufficient oxygen (hypoxia) [13]. Absolute normoxic values for a specific cell are furthermore dependent on the cell localization within a particular organ, such as the liver for example. In the liver, localization is exhibited in form of oxygen-modulated zonation and gradients along the length of a sinusoid result in regionally dominant metabolic and detoxification functions [14]. This natural sensitivity to local oxygen concentration can be replicated in vitro to selectively increase specific liver cell function like urea synthesis [15].

Mounting evidence indicates that in vitro cell-culture experiments performed in air (≈21% O_2) may introduce excessive stress on cells due to exposure to unnaturally high oxygen concentrations [12]. Fibroblasts exposed to different oxygen concentrations were found to adjust to high oxygen levels by reversible growth inhibition and differentiation. For neural and other stem cells in comparison, hypoxic conditions were observed to promote growth and influence differentiation [16]. Measurement of oxygen concentration is therefore of special importance when in vitro results are to be compared to cell behavior observed in vivo. In summary, it can be stated that oxygen uptake of cells is a powerful marker for metabolic status, health, and response to exogenous and endogenous stimuli [17].

Beyond simple cellular constructs, microfluidic devices are also increasingly used to handle higher organisms such as nematodes and mammalian embryos. The health and behavior of these organisms has preimplantation been found to be strongly linked to the oxygen microenvironment. For preimplantation mammalian embryos, a direct link exists between oxygen consumption and their development status [17], while *Caenorhabditis elegans* nematodes show strong behavioral preference for 5%–12% O_2 [18]. All these examples indicate that biomedical and agricultural applications would profit extensively from a reliable, nonintrusive tool to investigate in vitro oxygen concentrations.

9.2.1 Principles of Oxygen Sensing

Traditional laboratory procedures for the measurement of oxygen in solution require the extraction of a sample volume for external analysis. This sampling approach is limited by the difficulty and time needed for extraction and analysis. With fluid volumes in the range of microliters and below, as found frequently in current microfluidic Lab-on-a-Chip (LOC) devices, analyte sampling constitutes a major disturbance of the system to be measured. This is a particular problem for sensing of less stable solutes such as oxygen, where retrieval is likely to significantly alter the sample characteristics [19].

9.2.2 Amperometric vs Optical Oxygen Sensing

Two main technologies, amperometric electrochemical and optical sensing, currently constitute the bulk of integrated sensors for the measurement of DO_2 in biological applications. Both are tolerant to liquid exposure and exhibit the high sensitivity needed to detect the small changes of oxygen encountered in the cellular environment. The first principle, amperometric electrochemical sensing, has been applied in a variety of biomedical LOC devices [20–22]. Despite continuous interest, several significant limitations have been identified concerning the use of amperometric sensors. When exposed to organic matter sensor, lifetime has been found to decrease through membrane fouling. Oxidation of the electrode surface area after prolonged usage and electrolyte depletion can be responsible for inconsistency between measurement data. A second, more important concern relates to the fundamental method of operation of the amperometric electrode [23]. The Clarke sensor operates by reducing oxygen and thus is prone to show signal dependence on flow rate in a low flow environment like a bioreactor device. Additional problems are posed by miniaturization itself, mainly due to the need to integrate a reference electrode [19].

9.2.3 Fluorescence-Based Optical Oxygen Sensing

In contrast, fluorescent dye-based optical sensing does not exhibit analyte depletion, and has therefore emerged as a promising alternative in biomedical applications. The principle of measurement with these types of sensors relies on the quenching of either the intensity or lifetime of light emitted by a dye in the presence of molecular oxygen. Fluorescence occurs upon excitation of the dye molecule (fluorophores*) to an exited state S^* and subsequent relaxation back to the ground state S_0 via the emission of light [24]:

$$S_0 + h \times \upsilon_{ex} \rightarrow S^* \tag{9.1}$$

$$S^* \rightarrow S_0 + h \times \upsilon_{em} \tag{9.2}$$

$$S^* + O_2 \rightarrow S_0 \,|_K \tag{9.3}$$

where

h is Planck's constant
$\upsilon_{ex} > \upsilon_{em}$ are the frequencies of the excitation and emission light, respectively
K is the molecular rate constant for oxygen quenching

One pathway for relaxation from the exited state of the fluorophore is via interaction with a secondary molecule. In case of oxygen, fluorescence quenching has a pronounced effect on the quantum yield as a result of the triplet ground state of the O_2 molecule. The fluorescence intensity, I, and lifetime, τ, can be obtained for dissolved or gaseous oxygen from the Stern–Volmer equation for intensity [25]:

$$\frac{I}{I_0} = \frac{\tau}{\tau_0} = 1 + K_{sv}^{s} \times [O_2] = 1 + K_{sv}^{G} \times pO_2 \tag{9.4}$$

where

K_{sv}^{s} and K_{sv}^{G} are the Stern–Volmer constants for solution and gas, respectively
I_0 and τ_0 are the reference values in absence of oxygen
$[O_2]$ is the oxygen concentration in solution
pO_2 is the gaseous partial pressure of oxygen

Figure 9.1 illustrates the experimental setup and measurement principle of oxygen sensing based on fluorescence intensity quenching. As illustrated in Figure 9.1a, the intensity signal of a sensor pattern when no oxygen is present (0% O_2) decreases

* Although also often referred to as a phosphor, the terms fluorophore and fluorescence are used in context with PtOEPK here to emphasize fluorescence microscopy as the tool for sensor readout.

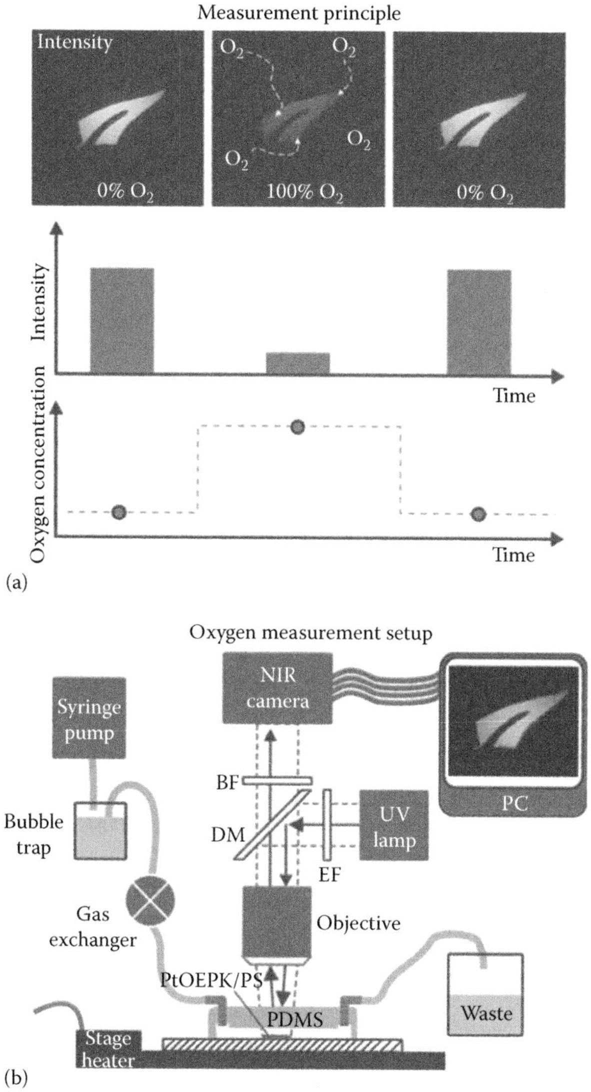

FIGURE 9.1 Principle of optical oxygen sensing: (a) the fluorescent signal intensity of a sensor film pattern (0% O_2) is reversibly quenched with exposure to oxygen molecules (100% O_2). Using a prerecorded calibration curve, the change in intensity can be related to oxygen concentration. (b) Experimental setup for spatially resolved measurement of O_2 in microfluidic devices. A standard fluorescence microscope with a custom filter set and a digital camera sensitive in the NIR is used for data acquisition. (Adapted from Nock, V. and Blaikie, R.J., *IEEE Sensors J,* 10, 1813, 2010.)

upon exposure to gaseous or DO molecules (100% O_2). This decrease in intensity is recorded using a standard fluorescence microscope (see Figure 9.1b) and translated into oxygen concentration by comparison to a prerecorded reference value. Since the fluorescence quenching is reversible, the intensity reverses to its initial value once the oxygen is removed again (0% O_2).

As indicated in Equation 9.4, the change in fluorescence lifetime can also be used to measure oxygen. However, if spatial information is to be extracted this significantly complicates the experimental setup [26]. The following sections will thus focus on intensity-based oxygen measurement.

9.2.4 Optical Sensor Materials

While oxygen-dependent quenching is exhibited by the majority of fluorescent dyes, a small subgroup has been found to be especially suited due to high sensitivity and long fluorescence lifetimes. These dyes are commonly categorized into organic fluorescent or organometallic compound probes [25] and used in solution [17,26,27] or immobilized on a support matrix [19,28–31] for the detection of DO_2. An excellent overview of the different forms of sensor application can be found in [32]. For use with low-cost LOCs, dye immobilization has the advantages of increasing the ease of handling and reducing the amount of fluorescent dye required. Table 9.1 shows a

TABLE 9.1
List of Organometallic Luminescence Probes for Optical Oxygen Sensing

Probe	λ_{max} (abs) [nm]	λ_{max} (em) [nm]	Suitable Solvents for Immobilizing Polymer Films	I_0/I_{100} Depending on Polymer Matrix
$Ru(dpp)_3^{2+}$	337, 457	610	Dichloromethane	4.4 in silicone film 1.1 in PS film 3.5 in PVC film
$Os(dpp)_3^{2+}$	454, 500, 580, 650	729	Dichloromethane	4.5 in poly(DMS) film
$Ir(ppy)_3$	376	512	THF, dichloromethane	1.2 in PS film
PtOEP	381, 535	646	THF, dichloromethane, toluene	4.5 in PS film
PtTFPP	395, 541	648	THF, dichloromethane, toluene	3.0 in PS film
PtOEPK	398, 592	758	Toluene, chloroform	2.0 in PVC film 20 in PS film
PdOEP	393, 512, 546	663	THF, dichloromethane, toluene	11.5 in PS film
PdOEPK	410, 603	790	Toluene, chloroform	8.0 in PVC film 28 in PS film

Source: Adapted from Amao, Y., *Microchim. Acta*, 143, 1, 2003.
Note: Both palladium OEPK and platinum OEPK exhibit very high intensity ratios I_0/I_{100} immobilized in PS films.

list of organometallic complexes used as oxygen sensors together with absorption/emission wavelengths, suitable solvents for immobilizing polymer films, and relative intensity change depending on the polymer matrix.

Out of these, platinum(II) octaethylporphyrin ketone (PtOEPK) suspended in a polystyrene (PS) matrix has been identified as well suited for use in LOC devices due to its desirable optical properties, compatibility with standard optical components, and being readily available commercially. The PtOEPK dye molecule shows strong phosphorescence with a high quantum yield and long lifetime at room temperature. Sensors immobilized in PS have been shown to diminish only 12% under continuous illumination for 18 h, detected by absorbance measurement [25]. It further exhibits both a long-wave shift and an extended long-term photostability compared to other fluorescent dyes [33]. This allows for the use of standard optical filters and makes sample handling less critical. Choosing PS as the polymer matrix complements, the advantages of PtOEPK are providing good oxygen permeability, biocompatibility, and low autofluorescence.

9.3 INTEGRATION OF OPTICAL SENSORS INTO MICROFLUIDIC DEVICES

For any sensor system to find widespread use in microfluidic devices, a straightforward integration process compatible with common low-cost fabrication techniques is essential. In this section, the PtOEPK/PS optical sensor system is used as an example to discuss the limitations of existing fabrication methods and two alternative processes for the patterning of polymer-encapsulated oxygen sensors are introduced. The process flow and patterning results for each are described in detail, as well as methods for the integration of the patterns into devices. While developed for PtOEPK/PS, the fundamental fabrication principles can be applied to other oxygen sensors and sensing applications.

9.3.1 Deposition of Sensor Films

Beyond the advantages mentioned earlier, PtOEPK/PS shares one major limitation with most other material systems, namely the challenges to fabricate micron-scale patterns of the material and integrate them with commonly used microfluidic devices. Organic solvents such as acetone readily dissolve PS, making it impossible to subsequently apply photoresist on PtOEPK/PS films for standard lithographic patterning [34]. As an alternative to direct lithography, Vollmer et al. [29] proposed photoresist liftoff in combination with pipetting of individual PtOEPK/PS patches to overcome this problem. They demonstrated an integrated oxygenator device with two large manually deposited sensor patches and successively calibrated these for DO_2 measurement. Recently, Sinkala and Eddington [35] proposed imprinting into cast PtOEPK/PS films with polydimethyl-siloxane (PDMS) stencils as a method to fabricate oxygen-sensitive microwells. Due to limited film homogeneity or not fully removing the sensor material, the sensor patterns fabricated with both these methods have rather limited applicability. To improve the

PtOEPK/PS integration into microfluidic devices, we have thus developed two novel fabrication methods combining deposition by spin coating with either patterning by soft lithography [36] or a sacrificial metal mask [37], which will be introduced in the following.

9.3.2 Soft-Lithography Process

Figure 9.2 shows a schematic of the sensor fabrication process using spin coating and soft lithography [36]. The sensor film is deposited with high film thickness uniformity and repeatability on a standard spin coater. When used as oxygen sensor, these film properties lead to increased homogeneity and stability of the signal intensity. Thus improved films, in turn, allow one to implement fluorescent intensity-based, laterally resolved measurement of oxygen concentration. While spin-coated films can be applied to various substrates and directly used as sensors, their integration into microfluidic and PDMS-based devices in particular requires a means to produce design-specific sensor patterns. Due to differences in surface chemistry, a substrate fully covered in a thin film of PS makes it significantly more difficult to achieve a permanent bond to PDMS devices using the established process of surface- activated bonding. This problem can be avoided by limiting the sensor film to areas within a particular fluidic

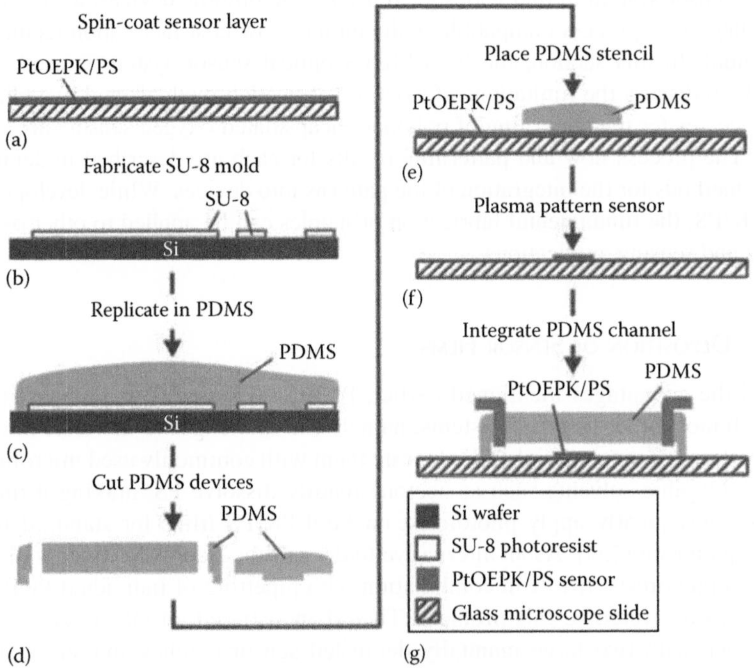

FIGURE 9.2 (a–g) Schematic of the device fabrication and sensor patterning process using soft lithography. (Adapted from Nock, V. and Blaikie, R.J., *IEEE Sensors J,* 10, 1813, 2010. With permission.)

feature, such as a channel or reactor chamber. The PDMS device can thus be bonded to the exposed glass substrate surrounding the patterned sensor patches.

Oxygen-permeable films of the PtOEPK/PS sensor material are prepared by pipetting PtOEPK/PS dissolved in toluene onto clean glass microscope slides, as indicated in Figure 9.2a. The final film thickness is controlled through the spin speed and depends on the concentration of the solution and the solvent used. Coated PtOEPK/PS forms a solid film by solvent evaporation when left overnight at room temperature. Being a common prototyping material in microfluidics, PDMS is used to form the microfluidic channels and as a masking material to pattern the sensor films. The latter takes advantage of the flexibility of cured PDMS and poor adhesion of it on PS. This allows one to use pre-patterned PDMS stamps to temporarily protect certain areas on the sensor film. Both positive and negative elastomer stamps can be fabricated. The negative stamps (stencil masks) contain vertical vias for the reactive-ion etch (RIE) patterning, which can be problematic for isolated features. Positive stamps are typically made up of a 100–200 μm thick pattern detail part and a thicker backing that laterally overhangs the pattern by several millimeters. The extent of this overhang is determined by cutting away excess PDMS around the pattern areas and simplifies handling during stamp placement.

As depicted in Figure 9.2b, both microchannels and stamps are produced by fabricating a master with the inverse of the desired design in SU-8 negative tone photoresist on a silicon wafer. To facilitate later removal of the cured replica, a fluoropolymer layer is deposited onto the resist mould. Meanwhile, liquid PDMS prepolymer is prepared and degassed in vacuum to remove any trapped air bubbles. The prepolymer is poured onto the resist master and cured on a hotplate. Once cured, the PDMS replicas are peeled off, the backing cut to size using a scalpel and, in case of microchannels, access holes are cut using a hole punch. To produce negative stamps, the process is modified by using the process of exclusion molding [38]. For patterning, the stamps are brought into conformal contact with the PtOEPK/PS layer (Figure 9.2e) and placed in a RIE tool. Dry-etching conditions are based on data published for PS films [39] and adapted for use with the RIE setup by increasing the etch pressure and power. After the etching step, the stamps are peeled off and can be reused for further patterning.

In a final step, illustrated in Figure 9.2g, the patterned sensor films are integrated with a microfluidic network. Depending on the intended application, integration can be performed by either irreversible O_2 plasma bonding or pressure clamping. For the former, both the surface of the PDMS channel network and the glass substrate with the sensor pattern are activated in a RIE using oxygen plasma, manually aligned and brought into conformal contact [38]. Whilst short, this surface activation step will lead to some isotropic etching of the sensor layer by the O_2 plasma. Under certain circumstances, such as for characterization of different substrates or sensor systems, it can be advantageous to be able to remove the channel structure from the sample after use. Reversible clamping can be achieved using a custom-machined plate of transparent polymer, such as poly-methylmethacrylate (PMMA). The seal is formed by sandwiching the glass substrate with oxygen sensor and microfluidic channels between two polymer plates or the metal plate of a microscope stage warmer. The stack is then secured using screws, which can be individually adjusted to provide the best possible seal. By applying this method, different sensor samples can be calibrated using a single channel structure.

A range of PtOEPK/PS test patterns fabricated using positive and negative stamps are shown in Figure 9.3. The minimum feature size achievable is limited to around 25 μm, mainly by the high aspect ratio SU-8 lithography required for the stamp mold master. Due to the thick PDMS, backing supporting the positive stamps enclosed features cannot be fabricated easily. However, arrays of lines and dots, as well as other complex shapes such as text, can be readily replicated. Negative stencils made of 100 μm thick PDMS films (see Figure 9.3e) provide an alternative to positive stamps, but require more delicate handling and also have problems with the replication of enclosed features. In spite of these limitations and with typical microfluidic channels 50 μm and wider, the use of soft-lithographic stamps provides a straightforward method for the integration of oxygen sensing into microfluidic devices. Once molded off the primary master, individual stamps can be reused to pattern multiple substrates. The latter in particular makes the process well suited for biological laboratories as the stamp mold master only needs to be fabricated once. For applications requiring higher-resolution sensor features, a second, photolithographic patterning method has been developed, which is introduced in the following section.

9.3.3 Sacrificial Metal Mask Process

Extending the sensor patterning to smaller dimensions is desirable in particular for cell biological applications. The introduction of LOC devices in the field of cell biology has revolutionized assay techniques and made it possible to perform experiments on individual cells as opposed to the Petri dish-sized population studies [2,40–43]. Assuming cells with typical attached diameters in the 10–50 μm range, a decrease in the pattern size to below 5 μm would allow one to integrate a single discreet sensor patch or whole arrays there of subjacent to each individual cell. To facilitate such single-cell oxygen measurements, we have developed a modified process based on the previous one by replacing the limiting PDMS stamps with sacrificial metal masks [37]. The process retains the homogeneous film characteristics obtained through spin coating the sensor material, while eliminating the need to fabricate separate polymer stamps. This simplifies the process significantly and makes the achievable minimum resolution mainly dependent on the lithography step used. A schematic of the modified sensor fabrication process using spin coating and sacrificial metal masks is shown in Figure 9.4.

The process begins with the application of an initial PtOEPK/PS sensor film to a substrate as outlined in the description of the soft-lithographical process. Once the solvent has evaporated, the sensor film is thermally annealed to prevent cracking of subsequent layers when the PtOEPK/PS film is heated over the glass temperature of PS. Following this, a 100 nm thick metal layer (i.e., tungsten) is sputter-deposited onto the sensor, thus protecting the encapsulating PS matrix from solvents used during lithographic processing. Due to this sacrificial

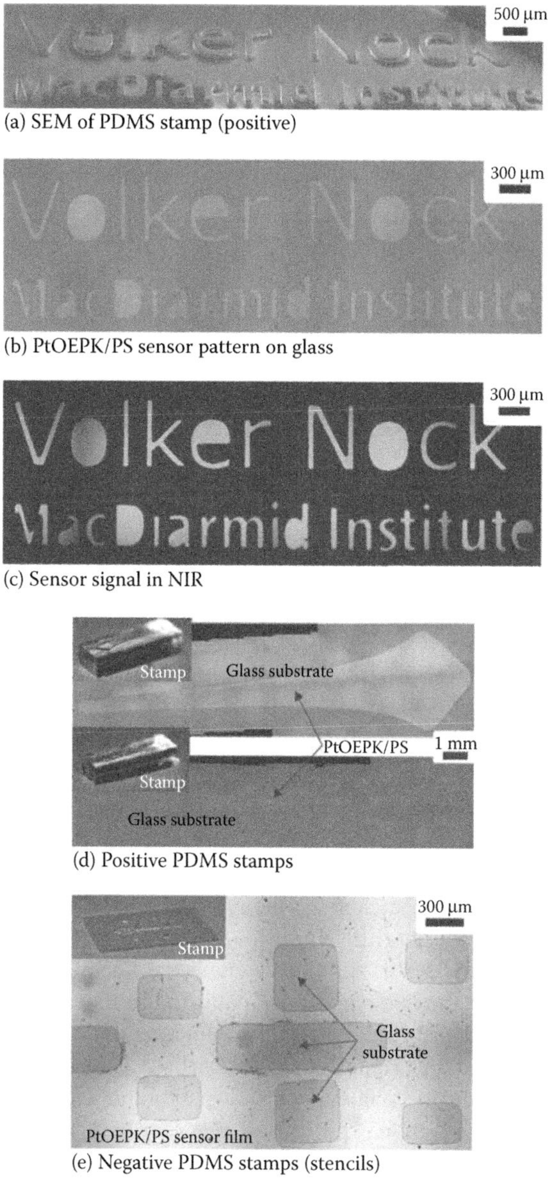

FIGURE 9.3 Results of the sensor film patterning using soft lithography. (a) SEM micrograph of a PDMS stamp (positive). (b) Optical micrograph of the corresponding PtOEPK/PS pattern on glass after transfer by reactive ion etching. Enclosed features such as in *a*, *e*, *o*, and *d* cannot be fabricated using positive stamps. (c) Fluorescence intensity micrograph of the sensor film response for exposure to air. (d) and (e) Both positive and negative PDMS stamps can be used for patterning. (Adapted from Nock, V., Control and measurement of oxygen in microfluidic bioreactors, PhD thesis, Department of Electrical and Computer Engineering, University of Canterbury, Christchurch, New Zealand, 2009. With permission.)

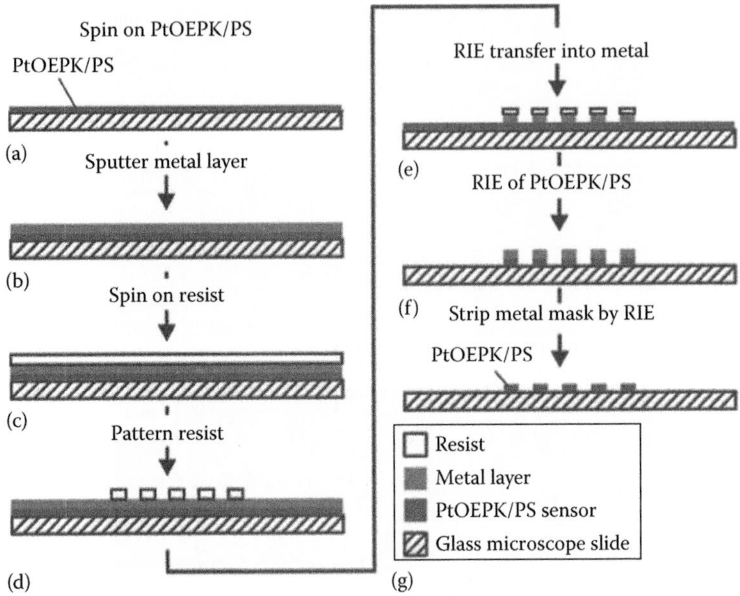

FIGURE 9.4 (a–g) Schematic of the oxygen sensor patterning process using optical or electron beam lithography (EBL). (Adapted from Nock, V. et al., *Microelectron. Eng.*, 87, 814, 2009. With permission.)

cover, a variety of conventional resists can now be applied and patterned using conventional photo- [37] or electron beam lithography (EBL) [44]. After exposure and development of the chosen resist, RIE is used to transfer the pattern into the subjacent tungsten layer. The etch gas is then switched to O_2 to transfer the pattern in the metal layer into the PtOEPK/PS film (see Figure 9.4f) and simultaneously strip the remainder of the resist. A final switch back to the metal etch gas removes the masking layer and thus concludes the patterning process. At this stage, the substrate and sensor patterns are ready to be integrated with prefabricated microfluidic devices using the techniques described earlier.

By using this method in combination with photolithographic chrome-on-glass masks and a standard projection exposure tool, we were able to fabricate large-scale arbitrary shapes and regular arrays of oxygen sensor patches with minimum feature sizes down to 3 μm [37]. Smaller patterns can be resolved by simply changing to a higher-resolution lithography technique such as interference lithography or EBL and the corresponding resist material. Figure 9.5 shows an example of a test pattern fabricated using EBL and the sacrificial metal mask process. Multiple arrays of oxygen sensor patches with edge lengths down to 500 nm were successfully replicated into a 600 nm thick PtOEPK/PS film. As opposed to the soft-lithographic process, enclosed features can be readily fabricated using this process and, in case of EBL, the amplification available for sensor readout becomes the limiting factor for the minimum practical pattern size.

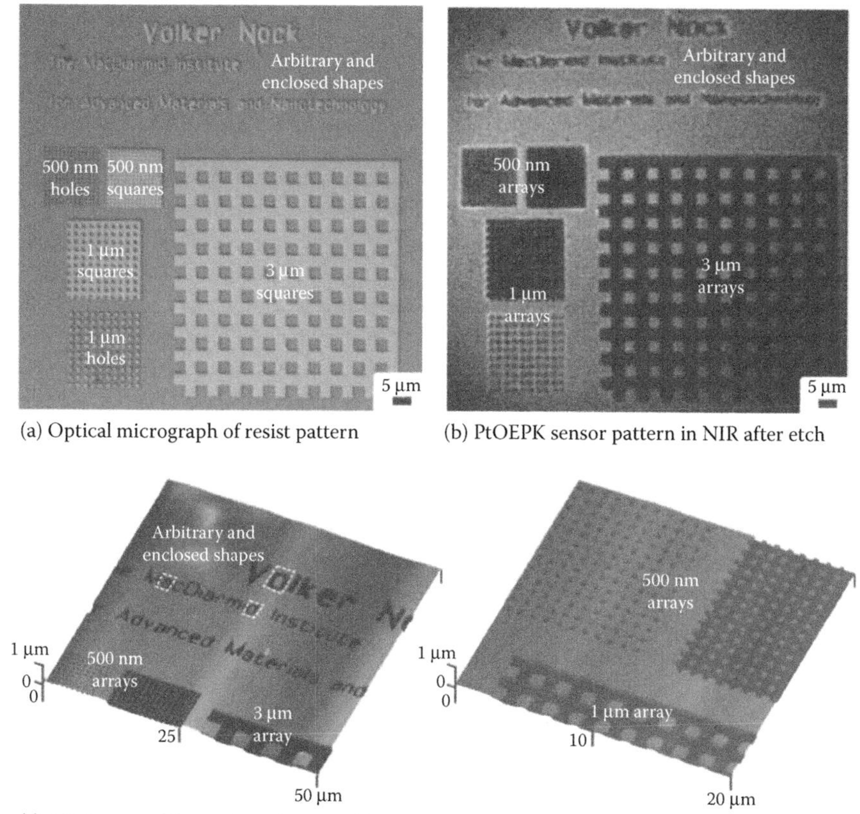

(a) Optical micrograph of resist pattern

(b) PtOEPK sensor pattern in NIR after etch

(c) AFM scans of the patterned sensor film

FIGURE 9.5 Results of the sensor film patterning using electron beam lithography. (a) Optical micrograph of the resist after e-beam exposure and development showing the replicated high-resolution patterns. (b) Fluorescence intensity micrograph of the patterned sensor film response for exposure to air. (c) AFM micrographs showing the replication of arbitrary enclosed and regular patterns in PtOEPK/PS. (Adapted from Nock, V. et al., Patterning of polymer-encapsulated optical oxygen sensors by electron beam lithography, in *Third International Conference on Nanoscience and Nanotechnology* (*ICONN 2010*), Sydney, New South Wales, Australia, 2011, pp. 237–240. With permission.)

9.4 LAB-ON-A-CHIP APPLICATIONS

With the full range of sensor patterning now available, integrated oxygen sensing can be realized in a variety of microfluidic devices and novel applications. In the following, three specific examples of the use of PtOEPK/PS sensor films in LOC-type devices are described to illustrate the versatility of the technology. Common to all potential application scenarios is the need to characterize and calibrate the response of the sensor for exposure to different oxygen concentrations. This is illustrated in the first part through an overview of the calibration process and a discussion of the intrinsic characteristics of the PtOEPK/PS system observed. Use of the calibrated sensors is then demonstrated for visualization and measurement of DO in multi-stream laminar flow and in contact with attached cancer cells.

9.4.1 Characterization and Calibration of Sensor Films

Prior to use, the integrated sensors have to be calibrated for measurement of gaseous or DO concentration. This is achieved by comparing the change in fluorescent sensor signal intensity for different solutions of oxygen concentration to a reference value provided by an external sensor [34]. After the calibration curve is recorded, oxygen concentration is determined by comparing the change of sensor intensity for an unknown concentration to the corresponding value determined during calibration. Due to the noncontact nature of the optical measurement, sensor readout can be performed on any fluorescence microscope fitted with the appropriate filter combination and camera (see Figure 9.2b). For fluid actuation, the device under test is interfaced with a syringe- or peristaltic pump and an external or integrated gas exchanger providing custom DO concentrations. A bubble trap placed before the gas exchanger and bioreactor chip removes potential bubbles in the fluidic circuit. The inlet oxygen concentration is determined using a flow-through Clarke-type oxygen sensor, which provides a reference value for calibration of the integrated oxygen sensor. LOC devices undergoing calibration are further mounted on a temperature-controlled microscope stage to provide stable conditions.

Typically, the sensor signal intensity of a freshly prepared PtOEPK/PS films is first measured for exposure to different gaseous oxygen concentrations directly after spin coating and patterning. For this, partial oxygen pressures of 0% (I_0) and 100% (I_{100}) are produced by blowing oxygen-free nitrogen and industrial grade oxygen, respectively, through a microfluidic PDMS channel and thus onto the integrated sensor patterns. During this characterization, several parameters were found to affect the signal ratio [36]. These parameters include, amongst others, temperature, molecular weight (M_w) of the PS used, and the PS concentration (% w/w) of the initial toluene solution. By far, the strongest influence on the intensity ratio I_0/I_{100} was found to be related to the thickness of the sensor films. From an engineering point of view this is very advantageous, as the final thickness can be easily adjusted via the spin speed during coating of the sensor layer.

While the thickness of the PtOEPK/PS films decreases linearly with increasing spin speed, the sensor signal ratio actually increases with decreasing film thickness. Figure 9.6a summarizes results for two PtOEPK/PS mixtures with 5% and 7% w/w PS at various thicknesses. For the 7% PS solution, I_0/I_{100} increases almost twofold from 3.6 at 1.3 μm film thickness to 6.8 at 0.6 μm. One possible explanation for this increase in intensity ratio is the corresponding increase in the surface area-to-volume ratio. With decreasing film thickness, the overall intensity contribution of dye close to the surface will increase and, simultaneously, the permeability-limited contribution from dye molecules in the PS bulk is reduced. In addition, a thinner film also requires less time for oxygen to migrate inside the film so that equilibrium with the environmental oxygen pressure is reached faster The second observation from Figure 9.6a is the dependence of the increase in intensity ratio on PS solution. A 0.7 μm thick film in 7% PS solution (M_w = 280 kD) shows a threefold

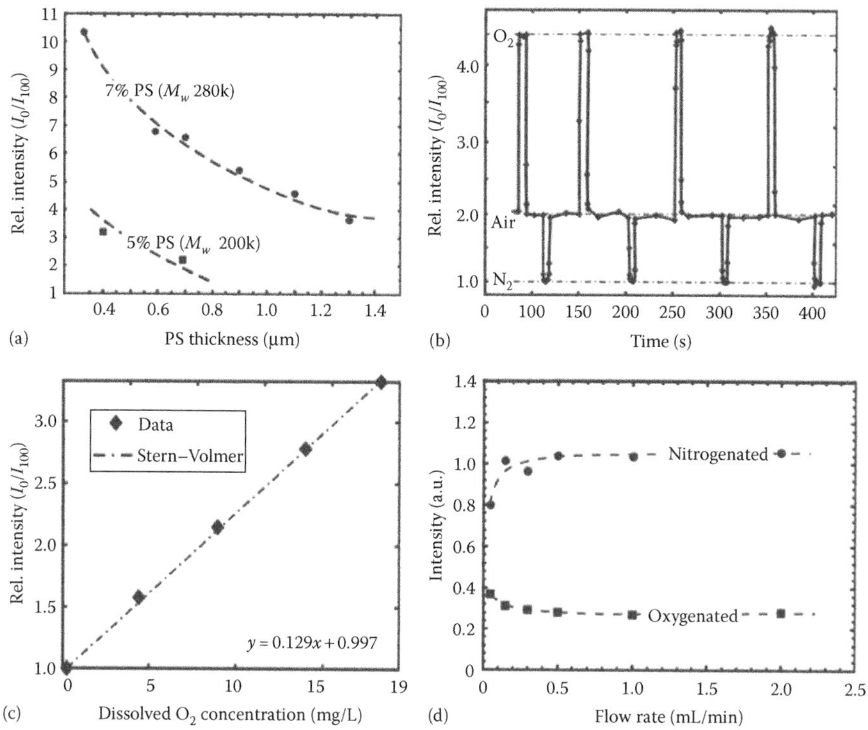

FIGURE 9.6 Sensor film characterization and calibration plots. (a) Relative intensity ratio of the PtOEPK/PS sensor films as a function of film thickness for two different PS concentrations and exposure to gaseous oxygen. (b) Dynamic sensor response of a 1.1 μm thick film for exposure to alternating gaseous oxygen concentrations. (c) Stern–Volmer curve for detection of dissolved oxygen in DI water with a 600 nm thick 7% w/w PtOEPK/PS sensor at a flow rate of $Q = 1$ mL/min ($T = 37.4$°C). (d) Plot of the sensor intensity data corresponding to flows of nitrogenated and oxygenated water across the tested dynamic operating range. At low flow rates, the operating range decreases slightly due to parasitic convective losses to the surrounding PDMS matrix. (Adapted from Nock, V. et al., *Lab on a Chip*, 8, 1300, 2008.)

increase in I_0/I_{100} compared to a 5% PS solution ($M_w = 200$ kD) of equal thickness. This significant difference is mainly attributed to the higher-molecular-weight PS used in the 7% solution. It is thought that during solvent evaporation the system of PS chains strives to attain a state of minimal energy by contraction. Molecules above the entanglement molecular weight of PS (Me, PS = 18 k) contract more slowly with increasing molecular weight and freeze in place before complete solvent evaporation This formation of disordered molecular chains can lead to an increase in oxygen permeability in PS films of higher M_w.

In general, these results indicate further possibilities for optimization of the sensor sensitivity through, for example, the use of thinner films and PS of different molecular weights. A fourth possible parameter not investigated here is the influence of the substrate material on sensor signal intensity. The maximum intensity ratio of 11 for a 350 nm thick 7% PtOEPK/PS film is the highest reported on a LOC-compatible glass substrate. Even higher values have been reported for films on a Mylar substrate [47]. This indicates that the choice of substrate material can provide an additional parameter in optimizing the signal ratio. However, for PDMS-based devices this has to be balanced with the processing compatibility of the substrate material, especially if surface-activated plasma bonding is to be used. Notwithstanding the potential for further optimization, it should be noted that all current sensor films described here exceed the minimum intensity ratio of 3 necessary to be considered suitable for oxygen sensing [25].

Aside from a high sensitivity, a second major advantage of luminescent probes is the excellent signal reversibility and a nearly instantaneous sensor response. This is a direct consequence of the electronic nature of the probe-analyte interaction and illustrated in Figure 9.6b for exposure to different gaseous oxygen concentrations. For measurement of DO, the sensor films follow the linear Stern-Volmer calibration model given in Equation 9.4. An example of a calibration at a constant flow rate of 1 mL/min and five different oxygen concentrations is shown in Figure 9.6c. The maximum DO concentration of 18.4 mg/L produced by the gas exchanger caused a factor of 3.4 change in total fluorescent intensity for this specific film. For flow rates below 0.5 mL/min, the dynamic operating range decreases slightly due to parasitic convective losses to the surrounding PDMS [29]. Depending on the initial concentration entering the device, oxygen is either added or removed from the liquid by mass transfer through the device walls, which is illustrated in Figure 9.6d by the decrease in intensity below 0.5 mL/min. This effect is due to the high permeability of oxygen in PDMS and demonstrates the high sensitivity of the sensor films. The obvious differences in signal decrease between hypoxic and hyperoxic water stem from the permeation coefficient of N_2 being a factor 2 smaller in PDMS than that of O_2 [48].

Finally, the overall stability and repeatability of the sensor response has also been found to be excellent. Dynamic measurements performed under 1 h illumination typically show no significant change in sensor intensity. Continuous illumination of PtOEPK/PS over 3 weeks with only minor signal degradation and the retaining of spectral and quenching characteristics for up to 2 years under storage have been reported [33]. These durations should be more than sufficient for most extended cell-culture studies where data acquisition usually is distributed over several shorter sessions per day.

9.4.2 Visualization of DO Concentrations

Essential to all experiments concerned with measuring and controlling oxygen concentrations is a reliable means of generating oxygenated flow. In microfluidic devices in particular biological cells need to be protected from excessive shear caused by gas bubbles [49], thus excluding the use of non- diffusive gas exchangers. PDMS-based versions on the other hand prevent the formation of inflow bubbles by limiting gas transfer to diffusion through a thin membrane [29]. This and their integral design make them ideally suited for integration with complex microfluidic LOCs. Photographs of such a PDMS gas exchanger and corresponding hydrodynamic focusing and multistream devices used in the following for DO visualization experiments are shown in Figure 9.7.

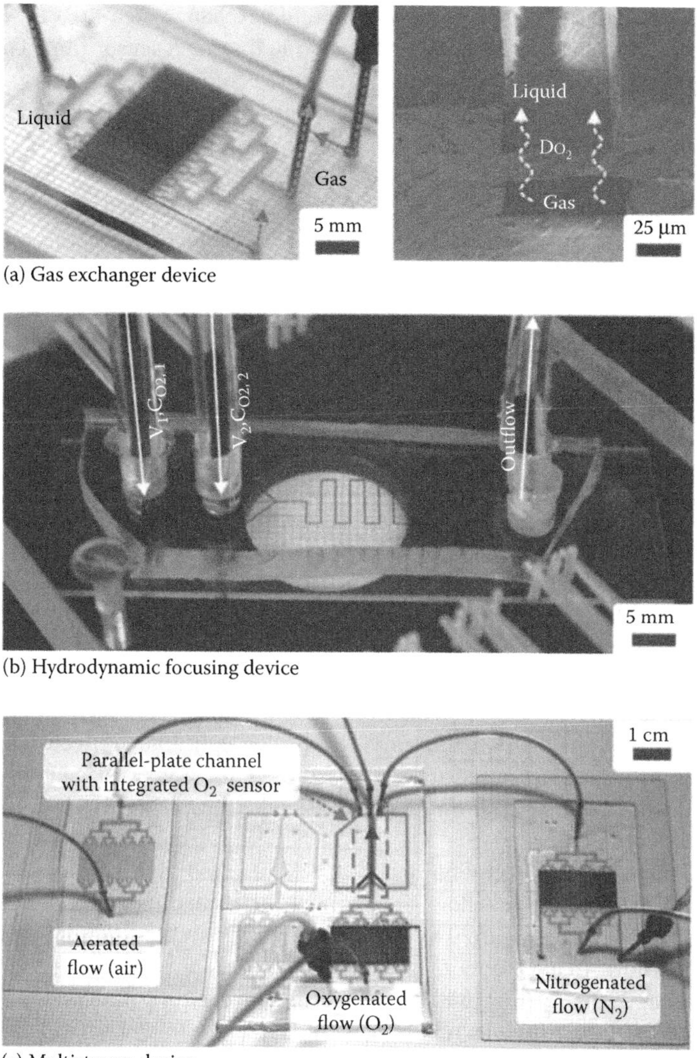

FIGURE 9.7 Photographs showing the microfluidic devices used to demonstrate oxygen measurement. (a) Close-ups of the integrated membrane-based PDMS gas exchanger used to generate flow with customdissolved oxygen concentrations. (b) Device used to generate hydrodynamically focused flow. Gas exchangers are not shown. (c) Device with three attached gas exchangers used to generate parallel flow streams. (Adapted from Nock, V. and Blaikie, R.J., *IEEE Sensors J.*, 10, 1813, 2010.)

The hydrodynamic focusing device uses a buffer flow inlet, which is divided into two channels to provide the side sheath flows for focusing and a central sample stream. Buffer and sample streams are then focused in a 200 μm wide rectangular microchannel with a total length of 100 mm and a common outlet. The multistream device on the other hand combines two streams from external versions of the PDMS gas exchanger with a third stream from an on-chip exchanger in a 1.6 mm wide rectangular parallel-plate microchannel with integrated oxygen sensor film. The output flows with varying oxygen concentration from the three gas exchangers are combined in this channel to yield the parallel laminar flow streams.

In the following, the use of these two devices with integrated sensors for oxygen control and visualization is demonstrated. The inlet flows provided by the PDMS gas exchangers can be varied from hypoxic (0 mg/L O_2) and aerated (8.6 mg/L O_2) to fully saturated hyperoxic water (34 mg/L O_2). Depending on the application, additional concentrations within this range can be produced by simply adjusting the gas mixture in the gas exchangers. The local oxygen concentration in the devices is visualized using fluorescence microscopy on the subjacent PtOEPK/PS layer.

9.4.3 Multi-Stream Flow

Due to the absence of convective mixing at low Reynolds numbers, diffusion is the predominant transport mechanism between two or more input streams. Several schemes for solute sorting and detection, such as the H-filter and T-sensor [50,51], operate using this principle. The T-sensor in particular has been used extensively to measure the diffusion coefficients of solutes between adjacent flow streams [52] and this is demonstrated for oxygen in this section. A second application of multistream flow can be found in form of chip-based multistream assays for cell-cell networks [4]. The capability to selectively expose certain regions of an interconnected 2-D cell culture to different environmental conditions and stimuli such as reduced or increased oxygen has the potential to yield novel insight into cell-cell interactions and signaling [53]. As opposed to well-based assays, colonies of cells in a multistream flow device remain in direct physical contact while experiencing different oxygen concentrations.

An example of the generation and spatially resolved visualization of multiple equal-width parallel laminar flow streams with individually controllable DO concentrations is shown in Figure 9.8. To demonstrate the oxygen sensor applicability, the resulting flow streams were imaged and analyzed using an integrated sensor film on the bottom of the channel [54]. Upon entering the central parallel-plate chamber, the three parallel flow streams remain separated over the total length of 18 mm due to the predominantly laminar flow regime. This results in three distinct oxygen concentration levels being generated across the width of the microchamber. Figure 9.8b shows the corresponding intensity images recorded via the integrated sensor film for the reactor inlet, midpoint, and outlet. Aerated water enters the chamber through the top inlet, hyperoxic through the central, and hypoxic through the bottom inlet. The difference in oxygen concentration between the individual streams is easily discernible and remains stable over the full length of the device. Using the recorded intensity images, the local oxygen concentration and profiles across the reactor width can now be quantified.

Cross-width plots of the DO concentration at the inlet and outlet obtained from the intensity images are shown in Figure 9.8c. The oxygen concentration levels of hyperoxic, aerated, and hypoxic water are determined by pre-calibration of the sensors as described earlier. As can be observed, the width of the individual streams or

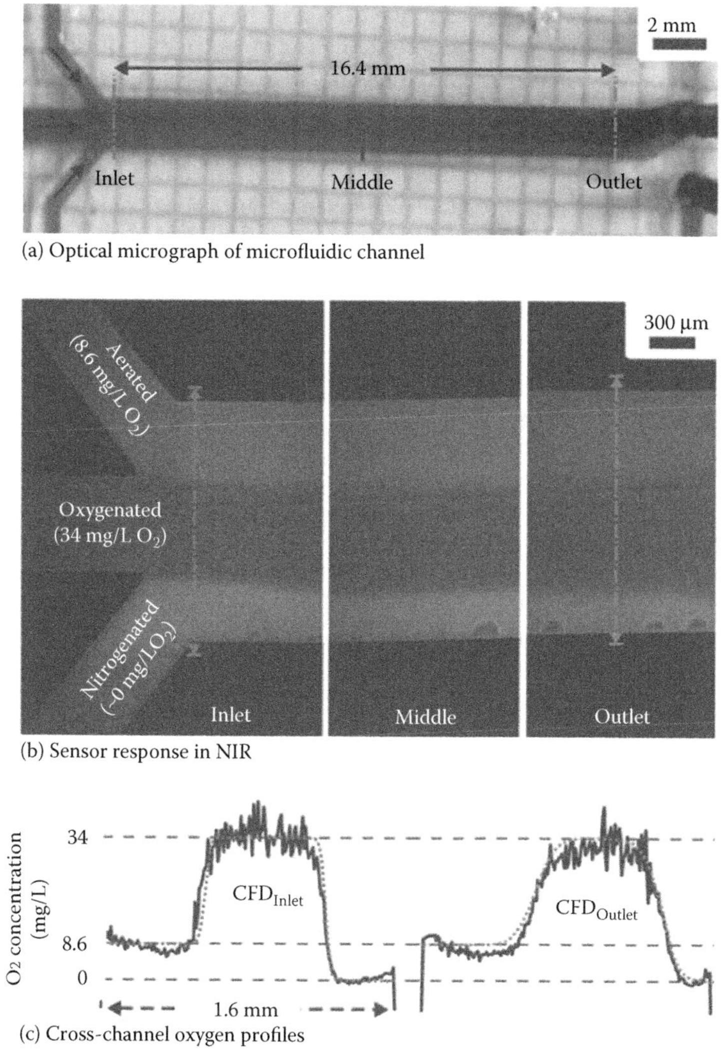

FIGURE 9.8 Demonstration of oxygen visualization and measurement in hydrodynamically focused flow. (a) Optical micrograph of the channel showing the central sample stream (dark gray) and buffer streams (light gray). Dashed rectangles indicate the imaging locations at the inlet and ≈ 6 cm downstream. (b) Oxygen-dependent optical intensity images of the sensor film at the locations indicated in (a). Regions of higher and lower intensity indicate lower and higher oxygen concentrations, respectively. In the two leftmost images the central sample stream corresponds to a DO concentration of 34 mg/L (hyperoxia) and the buffer streams to 0 mg/L (hypoxia), whereas in the image on the right the concentrations are inverted to yield a hypoxic sample stream. (c) Calibrated plots of the oxygen concentration across the channel at the locations indicated by the arrows in (b). The oxygen-rich sample stream remains focused to a width of ≈ 20 μm over the full length of 64 mm, while the generation of the oxygen-depleted stream indicates device versatility. (Adapted from Nock, V. and Blaikie, R.J., *IEEE Sensors J.*, 10, 1813, 2010.)

oxygen levels varies over the reactor length. These differences in level width from inlet to outlet of the individual streams are indicative of lateral diffusion of oxygen from the central oxygenated stream to regions of lower oxygen concentration (aerated and hypoxic streams). Since the device dimensions, initial oxygen concentrations of the three streams and the flow conditions are known; this phenomenon can be used to deduce the diffusion coefficient of oxygen in the fluid using Fick's law [6]. By solving the diffusion equation, the coefficient of diffusion is found as a function of the characteristic length:

$$DO_2 = \frac{x^2}{4 \times t} \tag{9.5}$$

where

DO_2 is the coefficient of diffusion of oxygen in the medium
t is the residence time in the channel
x is the diffusion length perpendicular to the flow direction

For a total flow rate of 0.3 mL/min in this example, the residence time $t = 1.05$ s. The diffusion length x can then be deduced from the difference of the slopes of the profiles at the inlet and outlet. By combining these values and solving Equation 9.5, the coefficient of diffusion for oxygen in water at a temperature of 37°C can be determined to $DO_2 = 2.57 \times 10^5$ cm^2/s. This value compares well with oxygen-diffusion coefficients in water of 2.52×10^5 cm^2/s at 35.1°C and 2.78×10^{-5} cm^2/s at 40.1°C published in literature [55], demonstrating the applicability of the integrated oxygen sensor films.

To further validate measurement result, computational fluid dynamic simulations of a microchannel with dimensions corresponding to the fabricated geometry can be performed. The simulation uses the experimental flow conditions and the measured coefficient of diffusion as model parameters [34]. Flow conditions are modeled using a Navier-Stokes application mode with three parallel inlet streams of equal flow rate. Cross-stream species transport is modeled using convection and diffusion. The measured DO_2 of 2.57×10^{-9} m^2/s is used as the isotropic diffusion coefficient of the liquid. Each inlet is assigned a constant species concentration and the boundary condition of the outlet is set to convective flux. To solve the model system, flow and species transport are coupled via the fluid velocity u parallel to the long axis of the reactor chamber.

For the multistream flow device in this example, simulated cross-chamber oxygen concentration profiles were evaluated at two points. The simulation results are plotted as dotted lines superimposed onto the measured profiles in Figure 9.8c. As can be seen, good agreement exists between the shape of the measured and the simulated concentration profiles at both the inlet and outlet and considering the noise on the measured data. In addition, the slopes of the profiles coincide indicating the validity of the diffusion model for analysis, as well as the value of the diffusion coefficient measured using the integrated optical sensor film. This demonstration illustrates the potential of the integrated sensors to be used to catalog oxygen-diffusion coefficients in a variety of biologically relevant liquids either as a stand-alone device or inline on a LOC prior to on-chip microfluidic cell-culture bioreactors.

9.4.4 Hydrodynamically Focused Flow

In two-dimensional symmetric hydrodynamic flow focusing, non-mixing streams can be further constrained laterally within the center of a microchannel by neighboring sheath flows from side channels [56]. The generation and spatially resolved detection of oxygen concentrations in such a setup are shown in Figure 9.9. With a single pump, typical buffer/sample flow rate combinations from 0.1 to 0.5 mL/min can be produced and oxygen-dependent fluorescence intensity images recorded at different positions along the meandering microchannel using the integrated oxygen sensor film [57]. Intensity image of the sensor film inside the microchannel at the inlet and 64 mm downstream are shown on the left and in the middle in Figure 9.9b. In this example, the oxygen concentration of the buffer streams was set to hypoxia and that of the sample stream to hyperoxia. The outline of the microchannel in the images indicates the capability of the sensor to resolve the different DO levels inside the channel boundaries with a lateral resolution of $<1\ \mu m$.

With the sample streams in this configuration, one would for example be able to study spatially resolved, the reversible growth inhibition and differentiation exhibited by individual fibroblasts when exposed to high oxygen levels [12]. By limiting the hyperoxic region to the narrow sample stream, the behavior of individual cells in a culture can be studied. For other types of cells, such as neural and some stem cells, hypoxic conditions promote growth and directly influence cell differentiation [16]. Switching the setup to these conditions is easily realized in this device by simply changing the gases used in the gas exchangers to yield a hypoxic sample stream and oxygen-saturated buffer streams. The intensity image corresponding to this particular configuration is shown on the right in Figure 9.9b. In both cases, the oxygen concentration of the buffer and sample streams can be finely tuned to the desired experimental conditions within the full range produced by the gas exchangers.

In addition to visualizing the DO content of the flow streams, the intensity response of the sensor film can also be used to measure the absolute local O_2 concentration. This is achieved by pre-calibration of the intensity change as described earlier. The calibration curve is then used to convert the change in intensity for an unknown concentration to the corresponding oxygen value in mg/L. Due to the homogeneous films obtained using spin coating and by using fluorescent intensity quenching for detection, the sensor can thus be used for spatially resolved oxygen sensing [36]. Pre-calibrated plots of the DO concentration profile across the channel width close to the focus point and 64 mm downstream are shown in Figure 9.9c. As can be observed from the measured profiles, hydro-dynamic focusing produces a dimensionally well-defined stream with stable oxygen concentration over long stream lengths. In addition, the excellent lateral resolution of the sensor film means that the submicron wide interstream boundary between the hypoxic and hyperoxic streams is well defined in the intensity images and the oxygen concentration profiles. Only minor broadening of the hyperoxic sample stream occurs, which is due to lateral oxygen diffusion over the extended channel length and can be controlled via the overall flow rate and thus the residence time of the liquid in the channel. The cross-channel plot for a hypoxic sample stream is shown on the right of Figure 9.9c and demonstrates the range of conditions the device can generate and detect.

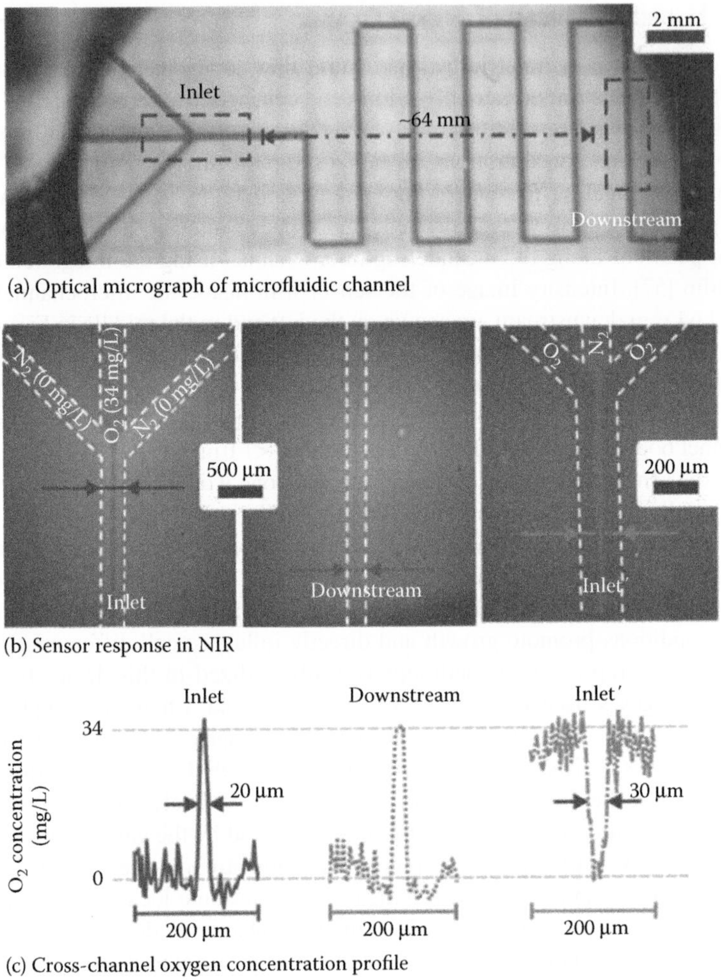

FIGURE 9.9 Visualization and measurement of oxygen in multistream laminar flow. (a) Optical micrograph of the parallel-plate rectangular microchannel with integrated optical oxygen sensor film indicating device dimensions, measurement locations, and flow layout. (b) Oxygen-dependent fluorescent intensity images recorded at the inlet, middle, and outlet. The intensity response for the top stream corresponds to aerated (8.6 mg/L O_2), the middle stream to hyperoxic (34 mg/L O_2), and the bottom stream to hypoxic (0 mg/L O_2) conditions. Interstream boundaries remain stable over the full channel length of 18 mm. (c) Oxygen concentration plots across the channel width at positions indicated by the dashed rectangles in (b). Concentration levels (dashed lines) were obtained through calibration of the sensor prior to use and coincide well with the individual streams. By applying diffusion theory, a coefficient of diffusion of oxygen in water of $DO_2 = 2.57 \times 10^{-5}$ cm²/s can be calculated from the decrease in slope from the inlet to outlet profile. This compares well with CFD simulation results obtained using the measured coefficient as parameter and overlaid as dotted lines in the plot. (Adapted from Nock, V. and Blaikie, R.J., *IEEE Sensors J.*, 10, 1813, 2010.)

Due to the intended use of the devices with hepatocyte liver cells (Ø ~ 20 μm) [58] or endometrial cancer cells (Ø ~ 35 μm) [59], the flow conditions shown are optimized to yield a sample stream width of around 20 μm. However, if needed, this can be further reduced significantly by adjusting the buffer/sample stream flow ratio. Stable sample streams of widths as small as 50 nm have been reported for mixing applications [56]. While sample streams of these dimensions enable high-resolution stimulation of certain areas on the surface of a single cell, the resulting fluid shear forces will have to be closely monitored so as not to influence the cell physiology and thereby reduce the relevance of the delivered stimuli.

9.4.5 Oxygen Sensing in Cell Culture

Both examples of sensor use introduced earlier localize the solid-state optical oxygen sensor films on the microchannel base either by simply covering the whole substrate [6] or at certain positions in the device by pre-patterning the sensor [36]. The latter in particular allows one to physically separate oxygen sensors and cell-culture areas inside a single LOC [9,28]. However, with biological applications trending toward higher resolution, this separation of sensor and cells and thus averaging nature of the oxygen measurement will no longer be sufficient to study effects on a single-cell level. One possible solution to overcome this limitation is to directly culture cells on the sensor film and this final section will summarize recent progress made toward this goal.

In general, the PtOEPK/PS material system is well suited for this since the PS matrix used to encapsulate the probe molecules is a common cell-culture substrate and shows good general biocompatibility [60]. Initial tests with cell adhesion enhancing extracellular matrix proteins indicated that the long-term oxygen sensor capability of a PtOEPK/PS film remains unaffected by a covering layer of type I collagen [61]. In total, the signal ratio decreases by only 8% over a measurement time of 14 days during which the sensor film was illuminated for an accumulated duration of 5 h or 20 min per day. The result indicates that an overlying collagen film does not significantly affect the sensor films. Furthermore, taking into account photobleaching and measurement accuracy, it can be said that the reduction of intensity is small enough to make the sensor suitable for long-term observation of oxygen concentrations.

The application of the sensor in this form is however only an intermediate step on the way to its use in conjunction with live cells. Ideally, cells are either cultured in an alternating pattern with interspersed sensor stripes or directly on top of the collagen-covered PtOEPK/PS layer, as shown schematically in Figure 9.10a. In both cases, chemicals secreted by the cells as part of metabolism and signaling have the potential to cause biofouling of the sensor [62]. In addition, diffusion of oxygen through the cell/collagen sandwich to the sensor layer underneath may be reduced due to oxygen uptake of the superimposed cells. To demonstrate the compatibility of the PtOEPK/PS oxygen sensor with direct cell culture, oxygen measurements were performed with cancer cells seeded onto a sensor layer [63].

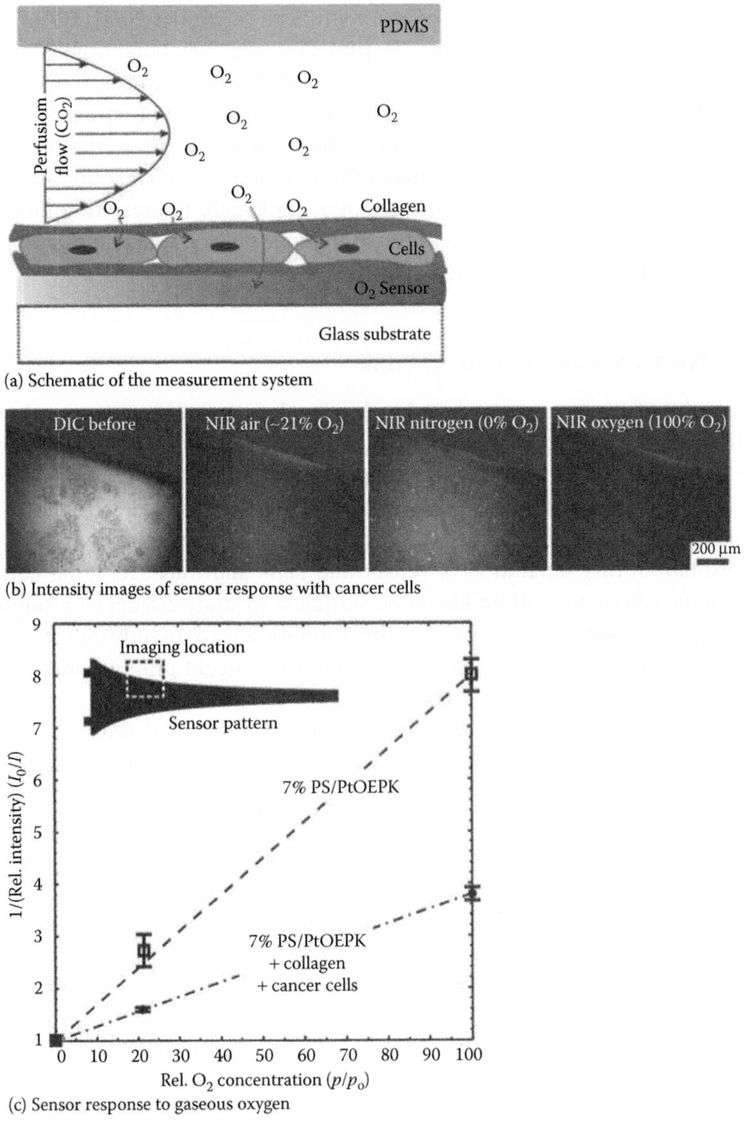

FIGURE 9.10 Demonstration of oxygen visualization and measurement in cell culture. (a) Schematic of the measurement system indicating the perfusion flow, sensor location, and oxygen transport. (b) Fluorescent intensity micrographs showing a region of the bioreactor indicated by the outline on the top right. From left to right: DIC microscopy of Ishikawa cancer cells growing on the sensor film prior to application of oxygen. Consecutive images of fluorescent sensor response of the same area in air (≈21% O_2), pure nitrogen (0% O_2), and pure oxygen (100% O_2). (c) Graph showing a plot of the relative sensor intensity as a function of oxygen concentration with and without cell culture (T = 37.4°C). (Adapted from Nock, V. et al., Oxygen control for bioreactors and in-vitro cell assays, in *Fourth International Conference on Advanced Materials and Nanotechnology,* Dunedin, New Zealand, pp. 67–70, 2009.)

Oxygen concentration-dependent fluorescence micrographs of a PtOEPK/PS sensor patch seeded with *Ishikawa* human endometrial cancer cells [64] after 3 days in culture are shown in Figure 9.10b. Prior to testing of the sensor function, all remaining culture media was rinsed off the cells and the sample was exposed alternatively to air, nitrogen, and oxygen. A graph of the relative intensity vs. the relative gaseous oxygen concentration for the sensor sample is shown in Figure 9.10c. The measured intensity ratio I_0/I_{100} exhibited by the sensor when covered with a film of type I collagen and cancer cells was measured to be 3.8. The second line in the graph indicates the sensor calibration result prior to deposition of collagen and culturing of the cells. In this case, the intensity ratio $I_0/I_{100} = 8.2$, approximately double the value recorded with superimposed cells and collagen.

Despite the significant decrease in intensity, these results provide a successful initial demonstration of the applicability of the PtOEPK/PS sensor films to in situ measurement of oxygen in active biological environments. The observed reduction in measured intensity ratio is likely due to reduced diffusion of oxygen through the cell/collagen sandwich. Further studies comparing areas of plain sensor with such covered in cells should be able to help shed light on this and simultaneously provide a tool to model the oxygen permeability of cell layers.

In addition to live cell assays in oxygen-sensitive microwells [35] and microchannel-covering sensor films [63], the newly available sensor patterning techniques discussed earlier further allow one to integrate complete arrays of micro- or nanoscale sensor patches subjacent to individual cells [37,44]. At sensor patch edge lengths of 1 μm and below, several independent patches can be localized underneath a single cell and thus provide local oxygen concentration information at a sub-cellular level. The number of patches per cell and hence the spatial resolution of the array can be increased by simply decreasing the size of the patches. Initial tests with *Ishikawa* endometrial cancer cells cultured in a 1 μm PtOEPK/PS sensor array demonstrate the feasibility of unobstructed sensor readout even through the overlying layer of biological material [44].

Further work is clearly required to fully classify sensor interaction with cells and other biological material. While changes in biologically relevant oxygen levels have been measured on a several hundred to a few tens of cells basis using PtOEPK/PS sensors [35], oxygen tension measurements at a single-cell level have yet to be demonstrated. Caution should also be applied to the interpretation of such localized measurements and the extrapolation of results obtained at a two-dimensional interface into the three-dimensional space earlier, as this will depend on a variety of factors such as the flow characteristics, boundary layer effects, and the cell type used.

9.5 OUTLOOK AND CONCLUSIONS

The deposition of solid-state polymer-encapsulated optical oxygen sensors has been demonstrated using simple casting and spin-coating techniques. Homogeneous spin-coated films allow for better thickness control and thus can be used for spatially resolved measurement of oxygen. Deposited films can be patterned using soft-lithographic stamps or photo- and EBL combined with reactive ion etching. Sensing characteristics of the films can be optimized by adjusting the composition and thickness of the encapsulating polymer matrix. Substrates with patterned sensor films can be integrated into common PDMS microfluidic devices by plasma bonding and the sensor signal can be recorded using a fluorescence microscope. Application of the integrated optical oxygen sensors to microfluidic LOC devices capable of generating microenvironments with controlled oxygen concentrations has been demonstrated. Spatially resolved in situ measurements of DO have been shown for parallel laminar and hydrodynamically focused streams and for oxygen concentrations ranging from 0 to 34 mg/L using low-cost reusable sensor films. Water-based sample streams with controlled oxygen concentration have been focused to widths of 20 μm and could be maintained over lengths exceeding 6 cm in these devices. Visualization of DO in multistream laminar flow with stream-independent control of oxygen concentration has also been demonstrated and devices based on this have been used to determine the coefficient of diffusion of oxygen in water by measuring diffusive stream broadening.

While optical oxygen sensor are increasingly being commercialized as water-soluble bioassay kits such as MitoXpress by Luxcel Biosciences [65], the current work in solid-state optical oxygen sensors has largely been of an academic nature. Apart from the use of sensor films as active layers on handheld laboratory optrodes, thin-film oxygen sensor have not yet found their way into commercially available standardized cell-culture wells or plates. However, the advantages of these thin-film sensors combined with their good biocompatibility should see increasing interest in the future. Compared to DO probes, polymer-encapsulated dye sensor films such as PtOEPK/PS are less expensive due to them being reusable and, through encapsulation in the PS matrix, less likely to influence the measurement by interacting with cells and microorganisms cultured in close proximity. Sensor films have been shown to remain fully functional under cell-culture conditions and in direct contact with live cells. Their solid-state nature combined with the noncontact optical readout further allows for oxygen measurements to be performed inside fully enclosed devices. This in particular will enable the study of the effects of oxygen concentration on cell development and function in general. In enclosed microfluidic LOC devices with reduced sample volumes, solutes like oxygen are less prone to interaction with the surrounding atmosphere and thus these devices provide higher sensitivity which should enable oxygen tension assays on a single-cell level. The sensors will further allow one to provide better environmental control for molecular delivery experiments, in particular by adding the ability to visualize and control the oxygen concentration of the perfusion media used to deliver other chemical stimuli.

REFERENCES

1. G. L. Semenza, Life with oxygen, *Science,* 318, 62–64, 2007.
2. S. Takayama, E. Ostuni, P. LeDuc, K. Naruse, D. E. Ingber, and G. M. Whitesides, Laminar flows: Subcellular positioning of small molecules, *Nature,* 411, 1016–1016, 2001.
3. S. Takayama, E. Ostuni, P. LeDuc, K. Naruse, D. E. Ingber, and G. M. Whitesides, Selective chemical treatment of cellular microdomains using multiple laminar streams, *Chemistry & Biology,* 10, 123–130, 2003.
4. H. Kaji, M. Nishizawa, and T. Matsue, Localized chemical stimulation to micropatterned cells using multiple laminar fluid flows, *Lab on a Chip,* 3, 208–211, 2003.

5. H. Zhang and G. Semenza, The expanding universe of hypoxia, *Journal of Molecular Medicine,* 86, 739–746, 2008.
6. V. Nock and R. J. Blaikie, Spatially resolved measurement of dissolved oxygen in multistream microfluidic devices, *IEEE Sensors Journal,* 10, 1813–1819, 2010.
7. P. Abbyad, P.-L. Tharaux, J.-L. Martin, C. N. Baroud, and A. Alexandrou, Sickling of red blood cells through rapid oxygen exchange in microfluidic drops, *Lab on a Chip,* 10, 2505–2512, 2010.
8. M. Polinkovsky, E. Gutierrez, A. Levchenko, and A. Groisman, Fine temporal control of the medium gas content and acidity and on-chip generation of series of oxygen concentrations for cell cultures, *Lab on a Chip,* 9, 1073–1084, 2009.
9. R. H. W. Lam, M.-C. Kim, and T. Thorsen, Culturing aerobic and anaerobic bacteria and mammalian cells with a microfluidic differential oxygenator, *Analytical Chemistry,* 81, 5918–5924, 2009.
10. B. Kuczenski, W. C. Ruder, W. C. Messner, and P. R. LeDuc, Probing cellular dynamics with a chemical signal generator, *PLoS ONE,* 4, e4847, 2009.
11. F. Wang, H. Wang, J. Wang, H.-Y. Wang, P. L. Rummel, S. V. Garimella, and C. Lu, Microfluidic delivery of small molecules into mammalian cells based on hydrodynamic focusing, *Biotechnology and Bioengineering,* 100, 150–158, 2008.
12. S. Roy, S. Khanna, A. A. Bickerstaff, S. V. Subramanian, M. Atalay, M. Bierl, S. Pendyala, D. Levy, N. Sharma, M. Venojarvi, A. Strauch, C. G. Orosz, and C. K. Sen, Oxygen sensing by primary cardiac fibroblasts: A key role of p21Waf1/Cip1/Sdi1, *Circulation Research,* 92, 264–271, 2003.
13. G. L. Semenza, HIF-1, O2, and the 3 PHDs: How animal cells signal hypoxia to the nucleus, *Cell,* 107, 1–3, 2001.
14. K. Jungermann and T. Kietzmann, Oxygen: Modulator of metabolic zonation and disease of the liver, *Hepatology,* 31, 255–260, 2000.
15. J. W. Allen and S. N. Bhatia, Formation of steady-state oxygen gradients in vitro—Application to liver zonation, *Biotechnology and Bioengineering,* 82, 253–262, 2003.
16. L.-L. Zhu, L.-Y. Wu, D. Yew, and M. Fan, Effects of hypoxia on the proliferation and differentiation of NSCs, *Molecular Neurobiology,* 31, 231–242, 2005.
17. F. C. O'Mahony, C. O'Donovan, J. Hynes, T. Moore, J. Davenport, and D. B. Papkovsky, Optical oxygen microrespirometry as a platform for environmental toxicology and animal model studies, *Environmental Science and Technology,* 39, 5010–5014, 2005.
18. J. M. Gray, D. S. Karow, H. Lu, A. J. Chang, J. S. Chang, R. E. Ellis, M. A. Marietta, and C. I. Bargmann, Oxygen sensation and social feeding mediated by a *C. elegans* guanylate cyclase homologue, *Nature,* 430, 317-322, 2004.
19. D. A. Chang-Yen and B. K. Gale, An integrated optical oxygen sensor fabricated using rapid-prototyping techniques, *Lab on a Chip,* 3, 297–301, 2003.
20. C.-C. Wu, T. Saito, T. Yasukawa, H. Shiku, H. Abe, H. Hoshi, and T. Matsue, Microfluidic chip integrated with amperometric detector array for in situ estimating oxygen consumption characteristics of single bovine embryos, *Sensors and Actuators, B: Chemical Sensors and Materials,* 125, 680–687, 2007.
21. J. Karasinski, L. White, Y. C. Zhang, E. Wang, S. Andreescu, O. A. Sadik, B. K. Lavine, and M. Vora, Detection and identification of bacteria using antibiotic susceptibility and a multi-array electrochemical sensor with pattern recognition, *Biosensors and Bioelectronics,* 22, 2643–2649, 2007.
22. E. Akyilmaz, A. Erdogan, R. Ozturk, and I. Yasa, Sensitive determination of L-lysine with a new amperometric microbial biosensor based on *Saccharomyces cerevisiae* yeast cells, *Biosensors andBioelectronics,* 22, 1055–1060, 2007.
23. Y.-J. Chuang, F.-G. Tseng, J.-H. Cheng, and W.-K. Lin, A novel fabrication method of embedded micro- channels by using SU-8 thick-film photoresists, *Sensors and Actuators A: Physical,* 103, 64–69, 2003.

24. S. Fischkoff and J. M. Vanderkooi, Oxygen diffusion in biological and artificial membranes determined by the fluorochrome pyrene, *Journal of General Physiology,* 65, 663–676, 1975.
25. Y. Amao, Probes and polymers for optical sensing of oxygen, *Microchimica Acta,* 143, 1–12, 2003.
26. D. Sud, G. Mehta, K. Mehta, J. Linderman, S. Takayama, and M.-A. Mycek, Optical imaging in micro- fluidic bioreactors enables oxygen monitoring for continuous cell culture, *Journal of Biomedical Optics,* 11, 050504–3, 2006.
27. J. Alderman, J. Hynes, S. M. Floyd, J. Kruger, R. O'Connor, and D. B. Papkovsky, A low-volume platform for cell-respirometric screening based on quenched-luminescence oxygen sensing, *Biosensors and Bioelectronics,* 19, 1529–1535, 2004.
28. P. Roy, H. Baskaran, A. W. Tilles, M. L. Yarmush, and M. Toner, Analysis of oxygen transport to hepatocytes in a flat-plate microchannel bioreactor, *Annals of Biomedical Engineering,* 29, 947–955, 2001.
29. P. Vollmer, R. F. Probstein, R. Gilbert, and T. Thorsen, Development of an integrated microfluidic platform for dynamic oxygen sensing and delivery in a flowing medium, *Lab on a Chip,* 5, 1059–1066, 2005.
30. X. Xiong, D. Xiao, and M. M. F. Choi, Dissolved oxygen sensor based on fluorescence quenching of oxygen-sensitive ruthenium complex immobilized on silica-Ni-P composite coating, *Sensors and Actuators B: Chemical,* 117, 172–176, 2006.
31. S. Lee, B. L. Ibey, G. L. Cote, and M. V. Pishko, Measurement of pH and dissolved oxygen within cell culture media using a hydrogel microarray sensor, *Sensors and Actuators B: Chemical,* 128, 388–398, 2008.
32. S. M. Grist, L. Chrostowski, and K. C. Cheung, Optical oxygen sensors for applications in microfluidic cell culture, *Sensors,* 10, 9286–9316, 2010.
33. D. B. Papkovsky, G. V. Ponomarev, W. Trettnak, and P. O'Leary, Phosphorescent complexes of porphyrin ketones: Optical properties and application to oxygen sensing, *Analytical Chemistry,* 67, 4112–4117, 1995.
34. V. Nock, Control and measurement of oxygen in microfluidic bioreactors, PhD thesis, Department of Electrical and Computer Engineering, University of Canterbury, Christchurch, New Zealand, 2009.
35. E. Sinkala and D. T. Eddington, Oxygen sensitive microwells, *Lab on a Chip,* 10, 3291–3295, 2010.
36. V. Nock, R. J. Blaikie, and T. David, Patterning, integration and characterisation of polymer optical oxygen sensors for microfluidic devices, *Lab on a Chip,* 8, 1300–1307, 2008.
37. V. Nock, M. Alkaisi, and R. J. Blaikie, Photolithographic patterning of polymer-encapsulated optical oxygen sensors, *Microelectronic Engineering,* 87, 814–816, 2009.
38. B. Jo, L. M. Van Lerberghe, K. M. Motsegood, and D. J. Beebe, Three-dimensional micro-channel fabrication in polydimethylsiloxane (PDMS) elastomer, *Journal of Microelectromechanical Systems,* 9, 76–81, 2000.
39. G. N. Taylor, T. M. Wolf, and J. M. Moran, Organosilicon monomers for plasma-developed x-ray resists, *Journal of Vacuum Science and Technology,* 19, 872–880, 1981.
40. S. Kobel, A. Valero, J. Latt, P. Renaud, and M. Lutolf, Optimization of microfluidic single cell trapping for long-term on-chip culture, *Lab on a Chip,* 10, 857–863, 2010.
41. A. Salehi-Reyhani, J. Kaplinsky, E. Burgin, M. Novakova, A. J. deMello, R. H. Templer, P. Parker, M. A. A. Neil, O. Ces, P. French, K. R. Willison, and D. Klug, A first step towards practical single cell proteomics: A microfluidic antibody capture chip with TIRF detection, *Lab on a Chip,* 11, 1256–1261, 2011.
42. M. C. Park, J. Y. Hur, H. S. Cho, S.-H. Park, and K. Y. Suh, High-throughput single-cell quantification using simple microwell-based cell docking and programmable time-course live-cell imaging, *Lab on a Chip,* 11, 79–86, 2011.

43. X. Li, Y. Chen, and P. C. H. Li, A simple and fast microfluidic approach of same-single-cell analysis (SASCA) for the study of multidrug resistance modulation in cancer cells, *Lab on a Chip,* 11, 13781384, 2011.
44. V. Nock, L. Murray, M. M. Alkaisi, and R. J. Blaikie, Patterning of polymer-encapsulated optical oxygen sensors by electron beam lithography, in *Third International Conference on Nanoscience and Nanotechnology (ICONN 2010),* Sydney, New South Wales, Australia, 2011, pp. 237–240.
45. Y. Amao, K. Asai, T. Miyashita, and I. Okura, Novel optical oxygen pressure sensing materials: Platinum porphyrin-styrene-trifluoroethylmethacrylate copolymer film, *Chemistry Letters,* 28, 1031–1032, 1999.
46. J. Zhao, S. Jiang, Q. Wang, X. Liu, X. Ji, and B. Jiang, Effects of molecular weight, solvent and substrate on the dewetting morphology of polystyrene films, *Applied Surface Science,* 236, 131–140, 2004.
47. P. Hartmann and W. Trettnak, Effects of polymer matrices on calibration functions of luminescent oxygen sensors based on porphyrin ketone complexes, *Analytical Chemistry,* 68, 2615–2620, 1996.
48. M. Ohyanagi, H. Nishide, K. Suenaga, and E. Tsuchida, Oxygen-permselectivity in new type polyor- ganosiloxanes with carboxyl group on the side chain, *Polymer Bulletin,* 23, 637–642, 1990.
49. J. Sung and M. Shuler, Prevention of air bubble formation in a microfluidic perfusion cell culture system using a microscale bubble trap, *Biomedical Microdevices,* 11, 731–738, 2009.
50. A. E. Kamholz, E. A. Schilling, and P. Yager, Optical measurement of transverse molecular diffusion in a microchannel, *Biophysical Journal,* 80, 1967–1972, 2001.
51. E. Kamholz and P. Yager, Theoretical analysis of molecular diffusion in pressure-driven laminar flow in microfluidic channels, *Biophysical Journal,* 80, 155–160, 2001.
52. M. S. Munson, K. R. Hawkins, M. S. Hasenbank, and P. Yager, Diffusion based analysis in a sheath flow microchannel: The sheath flow T-sensor, *Lab on a Chip,* 5, 856–862, 2005.
53. Michiels, T. Arnould, and J. Remacle, Endothelial cell responses to hypoxia: Initiation of a cascade of cellular interactions, *Biochimica et Biophysica Acta (BBA)— Molecular Cell Research,* 1497, 1–10, 2000.
54. V. Nock, R. J. Blaikie, and T. David, Generation and detection of laminar flow with laterally-varying oxygen concentration levels, in *12th International Conference on Miniaturized Systems for Chemistry and Life Sciences,* San Diego, CA, 2008, pp. 299–301.
55. P. Han and D. M. Bartels, Temperature dependence of oxygen diffusion in H_2O and D_2O, *Journal of Physical Chemistry,* 100, 5597–5602, 1996.
56. J. B. Knight, A. Vishwanath, J. P. Brody, and R. H. Austin, Hydrodynamic focusing on a silicon chip: Mixing nanoliters in microseconds, *Physical Review Letters,* 80, 3863, 1998.
57. V. Nock and R. J. Blaikie, Visualization and measurement of dissolved oxygen concentrations in hydro- dynamic flow focusing, in *Sensors, 2009 IEEE,* Christchurch, New Zealand, 2009, pp. 1248–1251.
58. V. Nock, R. J. Blaikie, and T. David, Microfluidics for bioartificial livers, *New Zealand Medical Journal,* 120, 2–3, 2007.
59. V. Nock, L. Murray, F. Samsuri, M. M. Alkaisi, and J. J. Evans, Microfluidics-assisted photo nanoimprint lithography for the formation of cellular bioimprints, *Journal of Vacuum Science & Technology B,* 28, C6K17-C6K22, 2010.
60. K. E. Geckeler, R. Wacker, and W. K. Aicher, Biocompatibility correlation of polymeric materials using human osteosarcoma cells, *Naturwissenschaften,* 87, 351–354, 2000.
61. V. Nock, R. J. Blaikie, and T. David, In-situ optical oxygen sensing for bio-artificial liver bioreactors, in *13th International Conference on Biomedical Engineering,* Singapore, 2009, pp. 778–781.

62. Starly and A. Choubey, Enabling sensor technologies for the quantitative evaluation of engineered tissue, *Annals of Biomedical Engineering,* 36, 30–40, 2008.
63. V. Nock, R. J. Blaikie, and T. David, Oxygen control for bioreactors and in-vitro cell assays, in *Fourth International Conference on Advanced Materials and Nanotechnology,* Dunedin, New Zealand, 2009, pp. 67–70.
64. M. Nishida, The Ishikawa cells from birth to the present, *Human Cell,* 15, 104–117, 2002.
65. C. Diepart, J. Verrax, P. B. Calderon, O. Feron, B. F. Jordan, and B. Gallez, Comparison of methods for measuring oxygen consumption in tumor cells in vitro, *Analytical Biochemistry,* 396, 250–256, 2010.

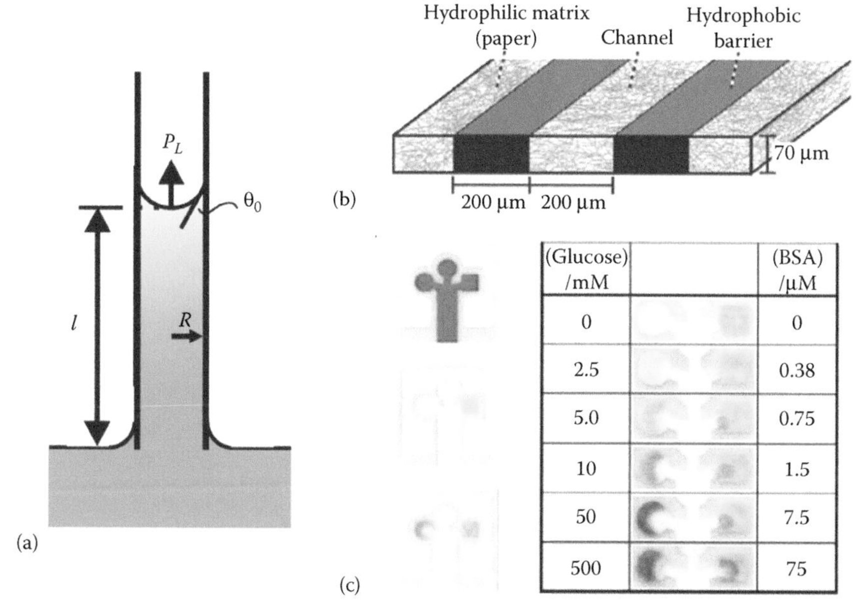

FIGURE 1.7 (a) Illustration of the spontaneous capillary rise of water in a hydrophilic capillary. (b, c) An example of "paper microfluidics," where liquid is guided via patterned wettability to reaction sites (circle and square terminus regions) by a capillary-driven flow in the porous paper: (b) Construction of the paper microfluidic device. The hydrophobic barriers consist of printed, then melted wax. (c) Detection of glucose and bovine serum albumin (BSA) at different concentrations (indicated by the table) using two branches of the same paper microfluidic device. (Adapted with permission from Martinez, A.W. et al., *Anal. Chem.*, 82, 3–10. Copyright 2009 American Chemical Society.)

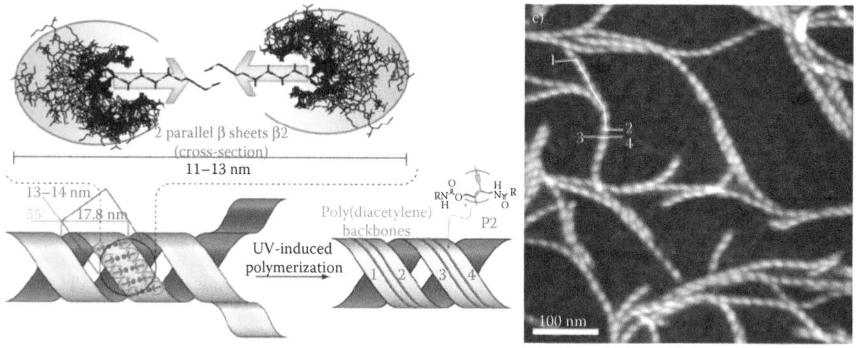

FIGURE 4.4 Self-assembly of oligopeptides linked with coil polymer segments and diacetylenes positions the alkynes in favorable geometries to facilitate the topochemical polymerization into polydiacetylene ultimately leading to a helical nanostructured material. Modified figure taken from reference 17 with permission. Copyright 2006, Wiley-VCH Verlag GmbH & Co. KGaA.

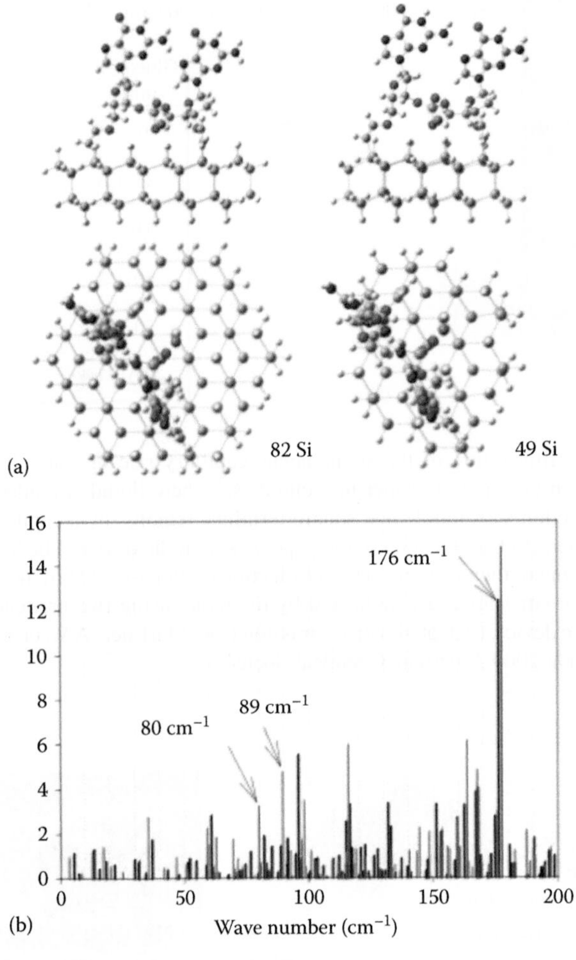

FIGURE 5.1 Simulation results for two compound structures dGG/Si82 and dGG/Si49. (a) Snapshots of the conformation for dGG/Si82 (left) and dGG/Si49 (right). (b) Spectral absorption for dGG/Si82 (black lines) and dGG/Si49 (red lines) below 200 cm^{-1} (6 THz).

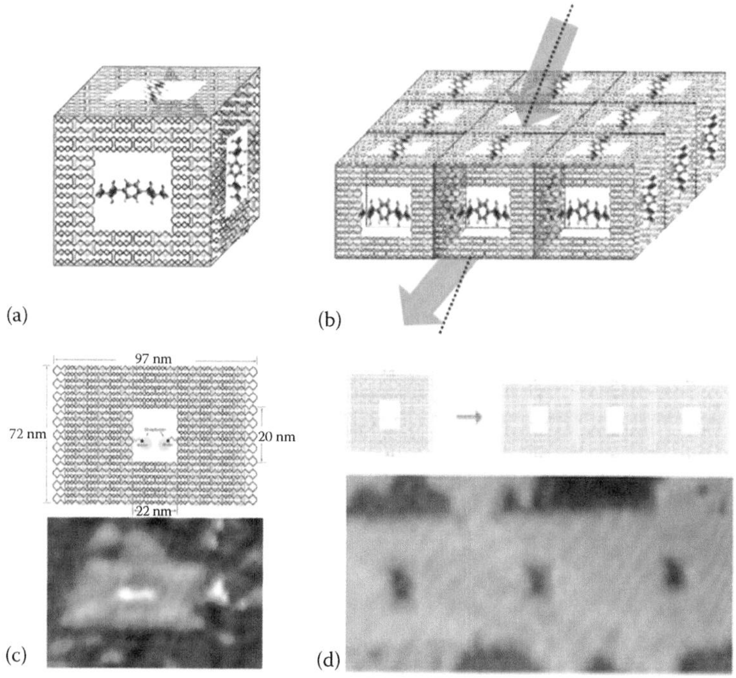

FIGURE 5.2 (a) DNA-based unit cell that spans 3-D coordinate space. (b) Smart material using the unit cell from (a) with measurable absorption/refraction. (c) An actual 97 × 72 nm DNA origami cassette design with biotinbiotin/streptavidin adapter molecules (upper) and AFM image of assembled structure (lower). (d) Linear array design (upper) using DNA cassette from (c) and AFM image of assembled array (lower).

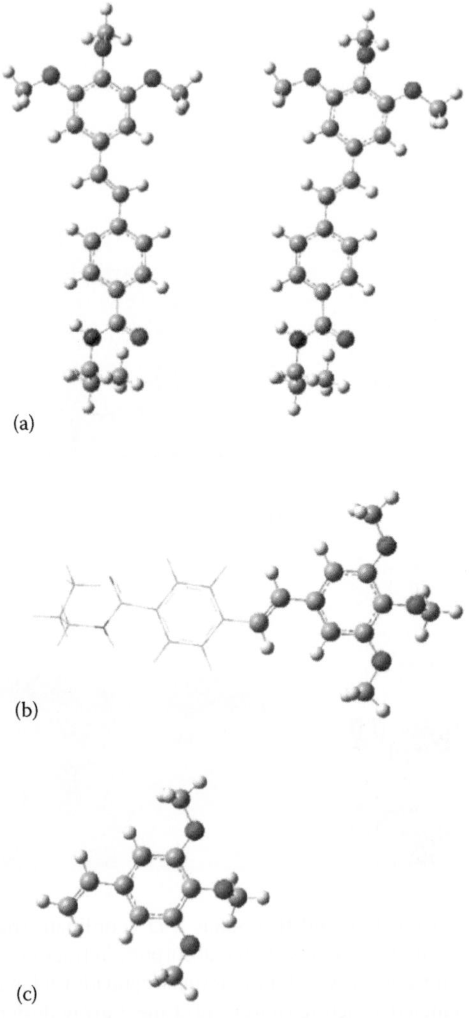

FIGURE 5.3 An example partitioning of the TMS molecule. (a) Two conformations of TMS that differ only by a 180 degree flip (i.e., compare left conformation to the right conformation) of the upper carbon ring relative to lower carbon ring in the chain. (b) A partitioning about the rotation bond in (a) where the upper portion (now shown on right) is treated primarily by a QM approach, and the lower portion (now shown on left) is treated primarily by a MM technique. (c) QM molecular partition (i.e., upper part from (a) and right part from (b)) where the interface bond has been saturated by hydrogen to enable both MM and QM simulations on this portion of the TMS molecule.

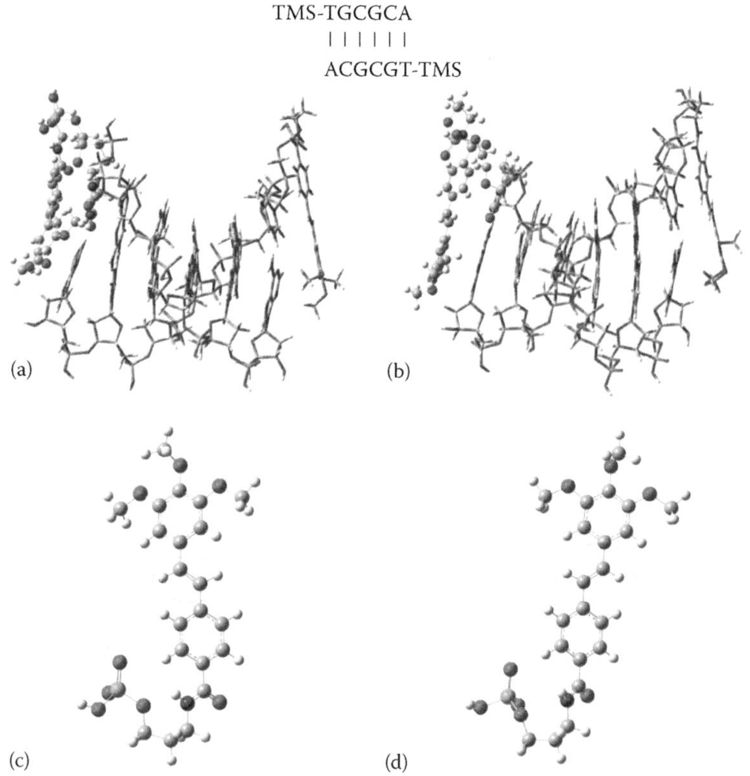

FIGURE 5.4 (Top) Two DFT/MM predicted conformations of the double-stranded TMS-DNA structure $(TMS\text{-}TGCGCA)_2$. (a) conformation I, (b) conformation II. For each conformation, color-coded spheres represent the portion of a structure simulated with the DFT model and color-coded tubes represent the portion of a structure described only by the MM model. (Botton) Isolated views of the TMS components from the two conformations of $(TMS\text{-}TGCGCA)_2$ given in (a) and (b). Here the two TMS substructures differ by a 180 degree flip of the end carbon ring (i.e., the upper ring in the illustration above) relative to the next carbon ring in the chain (i.e., the lower ring in the illustration above) where: (c) illustrates the ring flip to the left associated with the molecule from (a); and (d) illustrates the ring flip to the right associated with the molecule from (b).

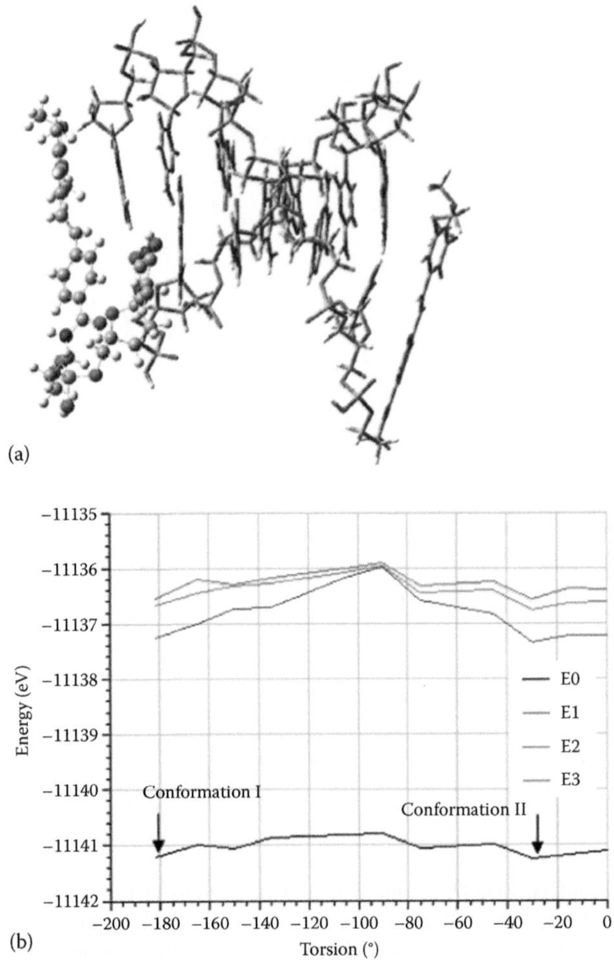

FIGURE 5.5 (a) Transient DFT/MM predicted structure that corresponds to a 90 degree (approximately half-way) turn in the torsion angle (within QM region) that is responsible for switching between conformations I and II of Fig. 2. The proximity between tail oxygen groups of TMS and the second (non-bonded to TMS) DNA strand is retained. Also, the T base of the first DNA strand is pushed back to allow for the transition. (b) Optimized energy versus torsion (within QM region) for TDDFT/MM predicted structures of Figs. 2. Minima corresponding to conformations I and II are highlighted with black arrows. Ground level and first three excited levels are plotted for the QM part of the system.

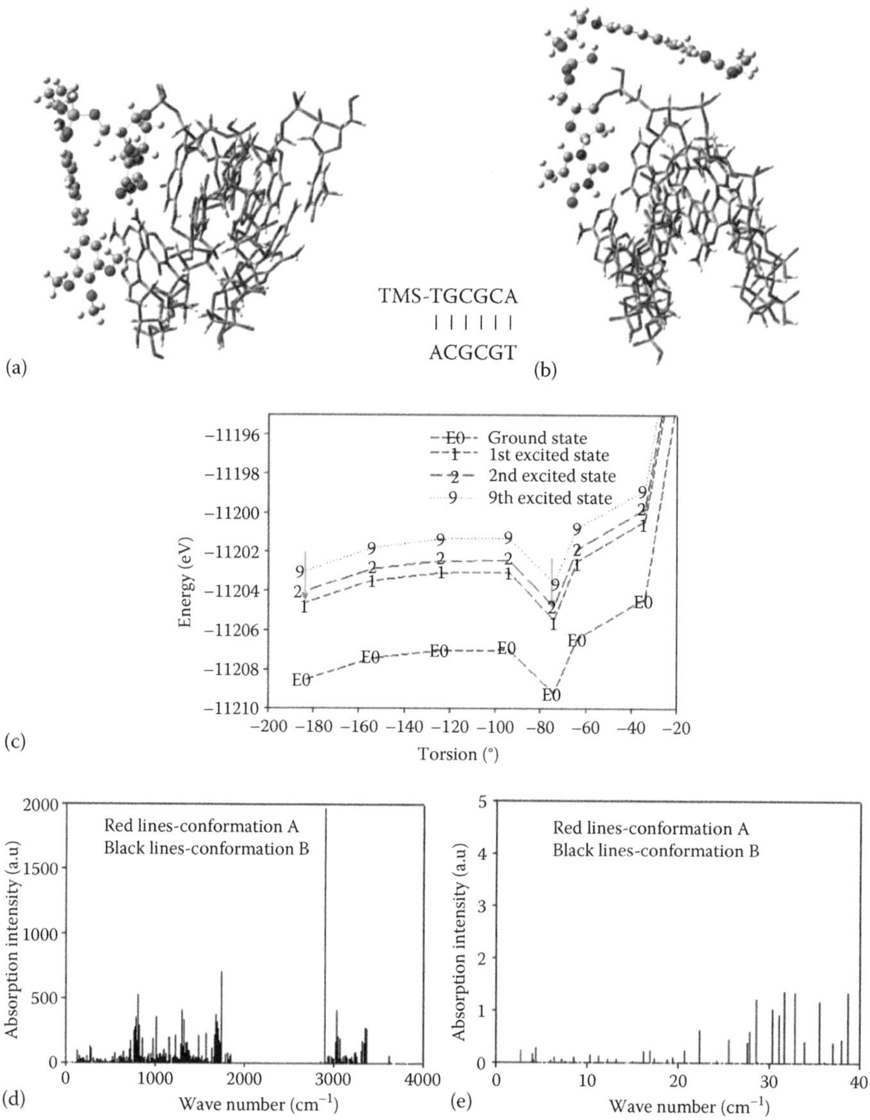

FIGURE 5.7 (Top) Two predicted conformations for double-stranded DNA fragment TGCGCA capped on one end with TMS obtained from hybrid HF/MM. (a) TMS chain is aligned along the end base pair T-A, and (b) TMS chain is aligned along the ds-DNA backbone. (Middle) Energy-scan profiles (from hybrid DFT/MM model) for the 0th, 1st, 2nd and 9th states of the TMS-TGCGCA molecule versus torsion angle. (c) The two stable conformations are highlighted with arrows where (a) is on left and (b) is on right. (Bottom) The DFT/MM predicted absorption spectra for each conformation is also shown for (d) the THz to IR regime, and (e) THz regime.

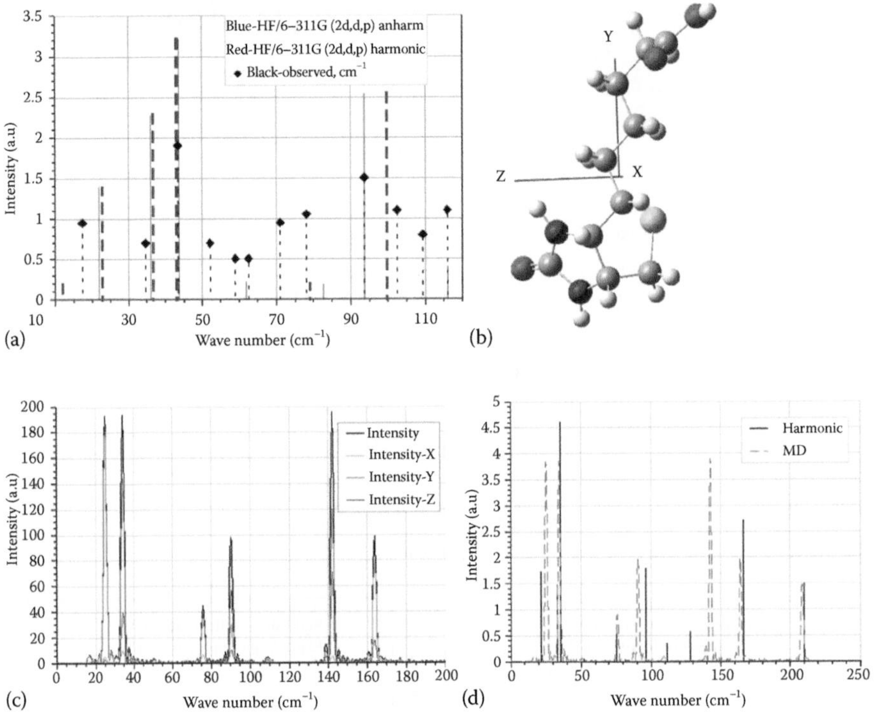

FIGURE 5.8 (a) THz absorption spectrum of biotin: harmonic (red lines); anharmonic (blue dashed lines); and, Experimental observations at 4.2K reported in [23] (black dots). Theory used HF/6-311G(2d,d,p) [25]. (b) A snapshot of biotin from the MD simulations. (c) Low-THz absorption spectra of biotin obtained from the DFT-based MD simulation. The entire spectrum (black line) and contributions from Cartesian components of dipole moment (green, red and blue lines respectively). (d) Low-THz absorption spectra of biotin obtained from the DFT-based MD simulation (red dashed line) and from the DFT-based energy minimization in harmonic approximation with the same density functional and basis (black lines).

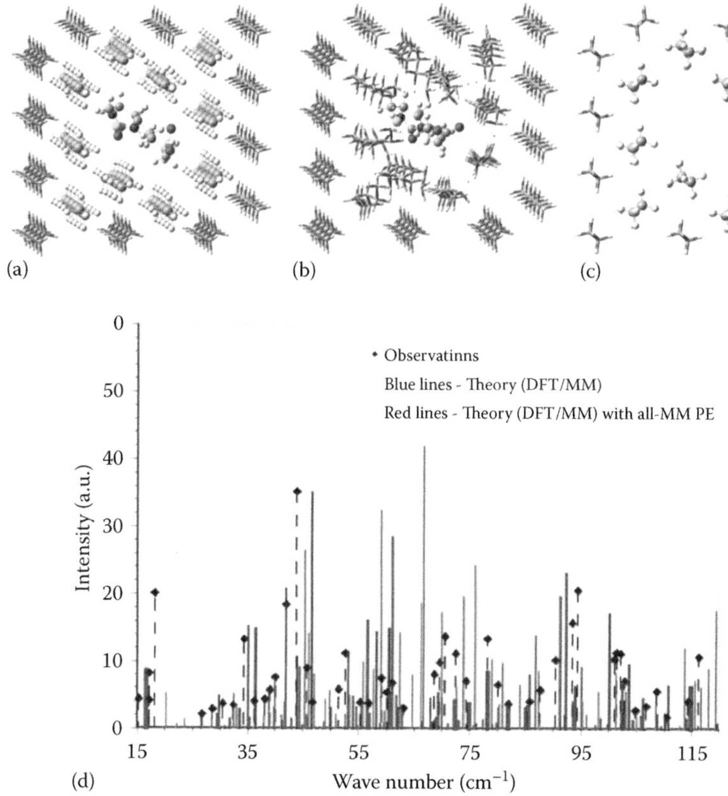

FIGURE 5.9 Simulation results for Biotin in a cavity inside a polyethylene matrix Test cases include: (a) Extended QM region; (b) Biotin-only QM region; and, (c) PE with cavity. Spheres represent atoms in QM region and tubes represent atoms in MM region. (d) Low-THz absorption spectrum of the biotin-PE structures. Theory for model of Fig. 18(a) (blue lines); theory for model of Fig. 18(b) (red lines); and experimental observations at 4.2K reported in [23] (black dots).

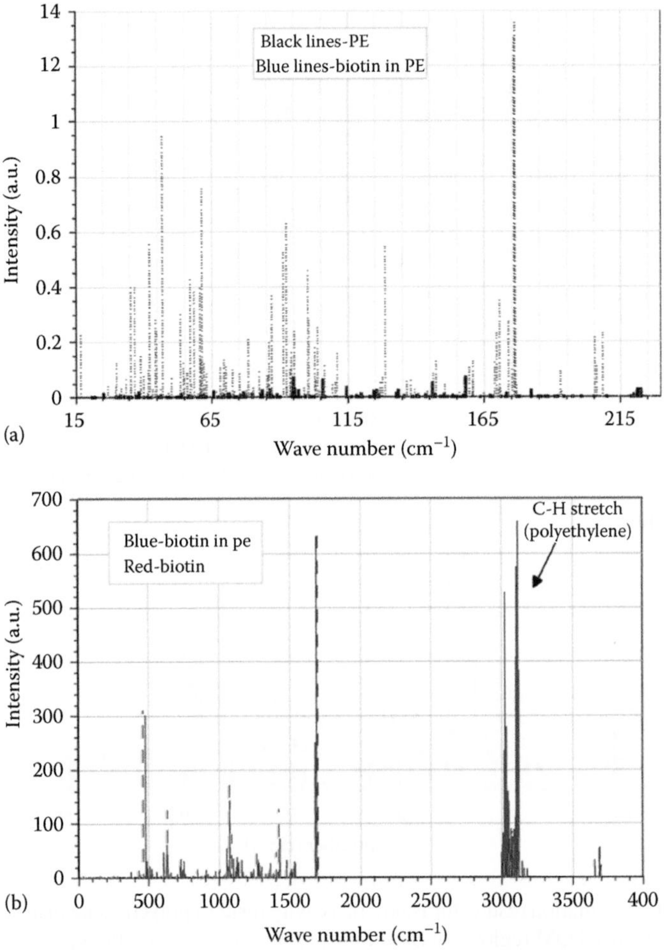

FIGURE 5.10 (a) Predicted THz absorption spectra below 7 THz. Biotin in PE as shown in Fig. 9(a) (blue lines) and PE with a cavity as shown in Fig. 9(c) (black lines). (b) Predicted resonance absorption spectra of biotin in PE (blue lines) and single molecule biotin (red dashed lines). The spectral features are very similar with the exception for the polyethylene C-H stretch modes occurring near 3000 cm^{-1}.

(a)

(b)

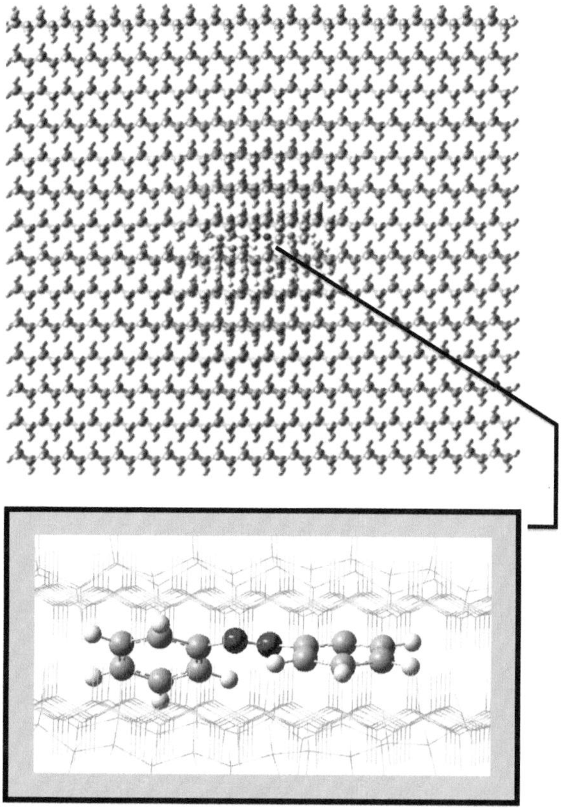

FIGURE 5.11 Azobenzene in polyethylene (PE) matrix. (a) Illustration of entire model structure, and (b) Detail showing the hybrid QM/MM model for the azobenzene (spheres) and PE (chains). Note that for these simulations azobenzene molecule was treated QM and all PE was MM.

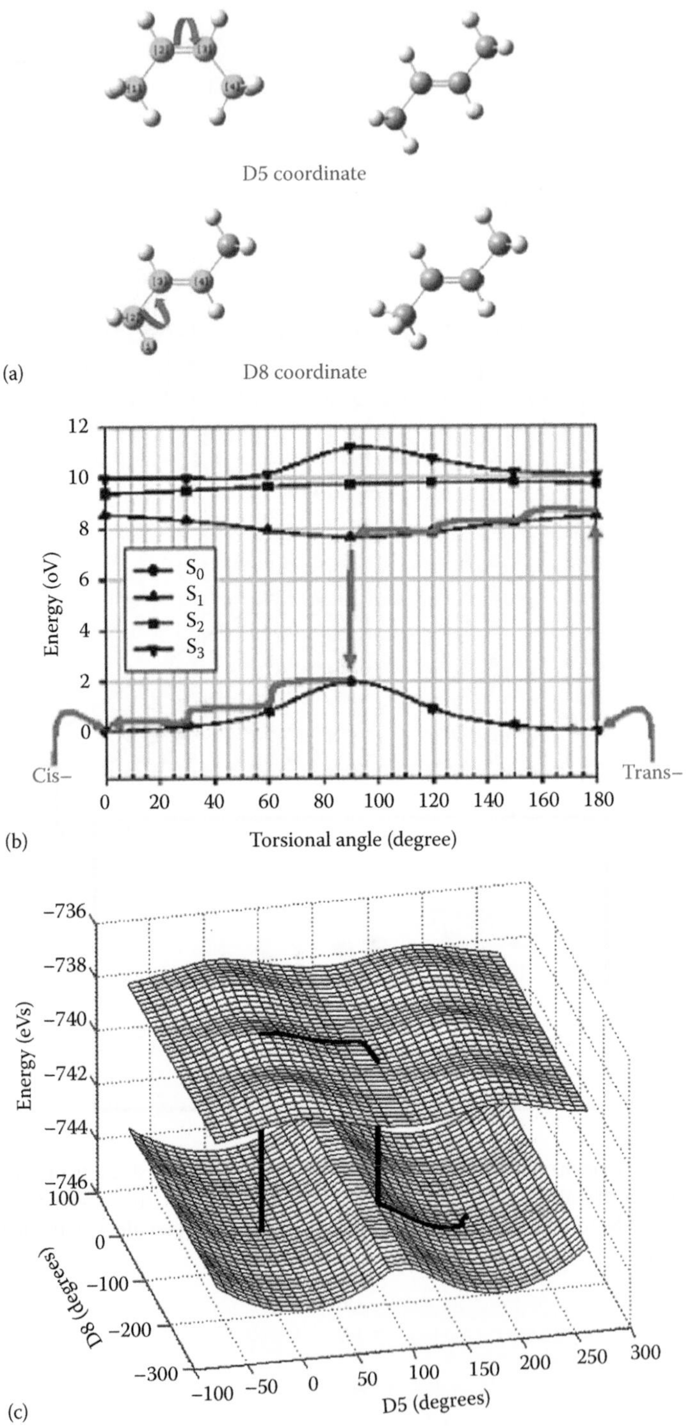

FIGURE 5.12 (a) 2-butene molecule with coordinate rotations for D5 (upper) and D8 (lower) labeled. (b) One-dimensional analysis (i.e. D5 coordinate only) that successfully predicts the trans-to cis-transition of 2-butene. (c) Actual two-dimensional transition path associated with conformational transformation defined by (b) that shows transversals (back and forth) in D8 space.

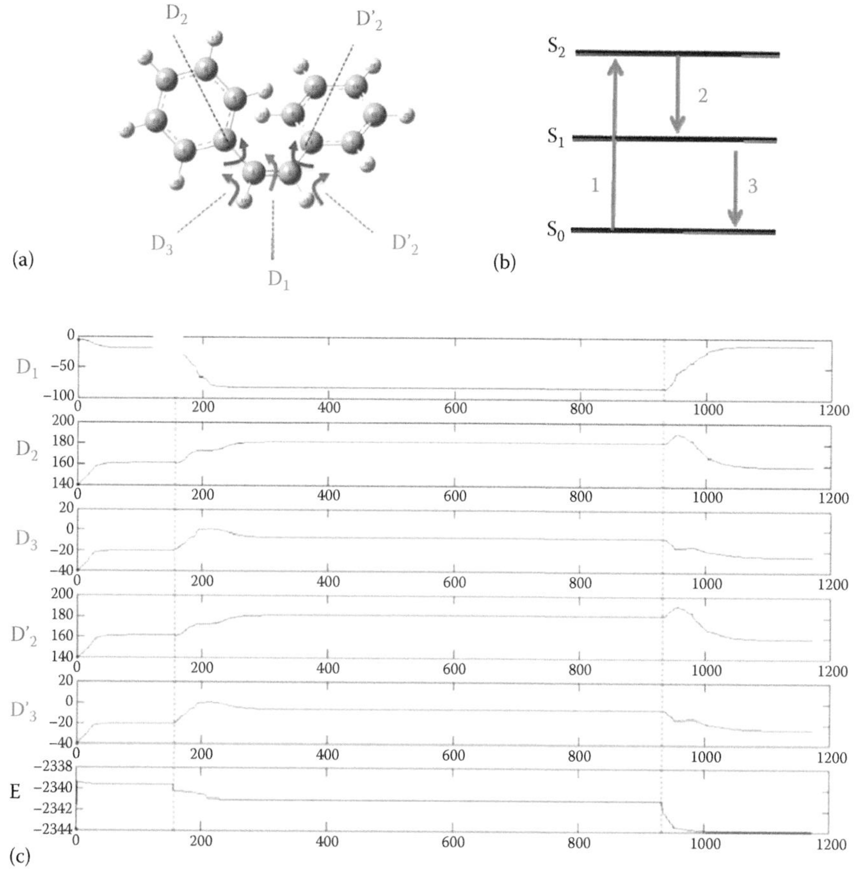

FIGURE 5.13 Five coordinate energy-space trajectory simulation information. (a) Illustration of the five coordinate angles D_1, D_2, D'_2, D_3 and D'_3. (b) Illustration of the excitation/decay series $\{S_0, S_1, S_2\}$. (c) Time evolution of the five coordinate angles (in angular degrees) and molecular energy (in eVs). Note that the red-dashed lines divide the three $\{S_2, S_1, S_0\}$ regions.

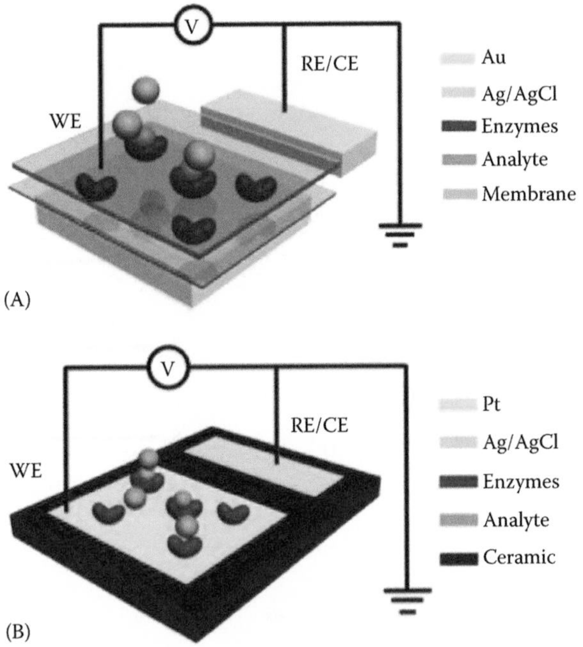

FIGURE 10.3 Schematic illustration of biosensors with (A) a Clark-type oxygen electrode and (B) a screen-printed electrode.

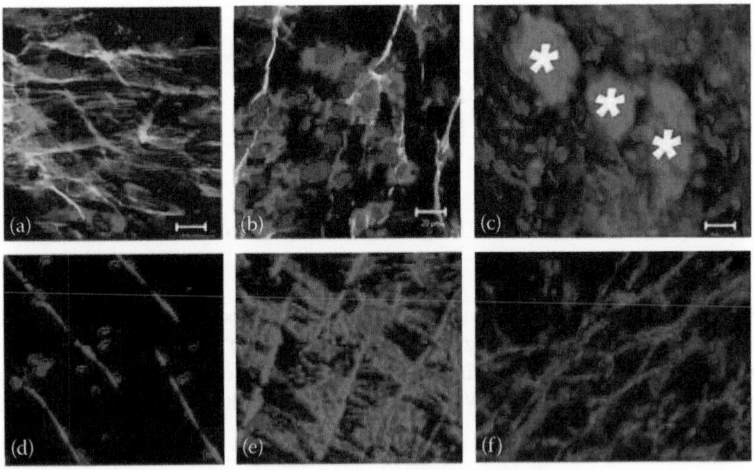

FIGURE 13.1 Expression pattern of osteogenic markers osteopontin (a), osteocalcin (b), and alizarin red (c) in collagen tubes; expression pattern of vasculogenic markers α-SMA of capillary-like structures (red, d), parallel sheets (red, e), and tomato lectin in tubular structures (red, f). DAPI stained blue is for nuclei and phalloidin stained green is for actin (scale bars are 20 μm).

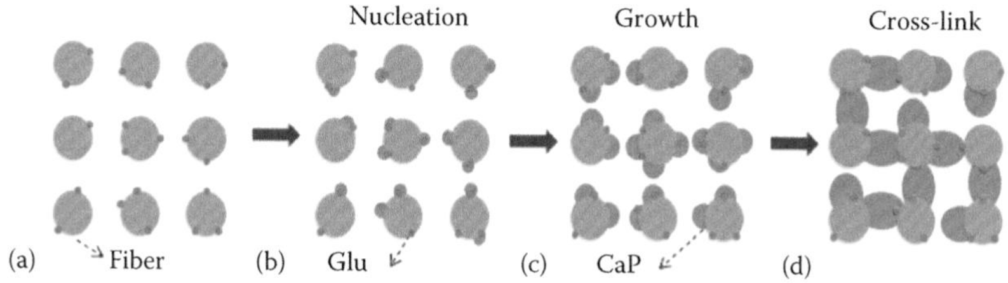

FIGURE 13.5 Schematic diagram demonstrating the effect of CaP deposition on Glu nanofibers on the strength of the microsheets. Fiber cross section, Glu peptide, and CaP crystals are represented by brown, red, and gray circles, respectively. The Glu on the fiber surface (a) initiates the nucleation of CaP crystals (b). The CaP crystals continue to grow with the incubation time (c) and start to merge and fuse to form a network of fibers cross-linked by CaP crystals (d).

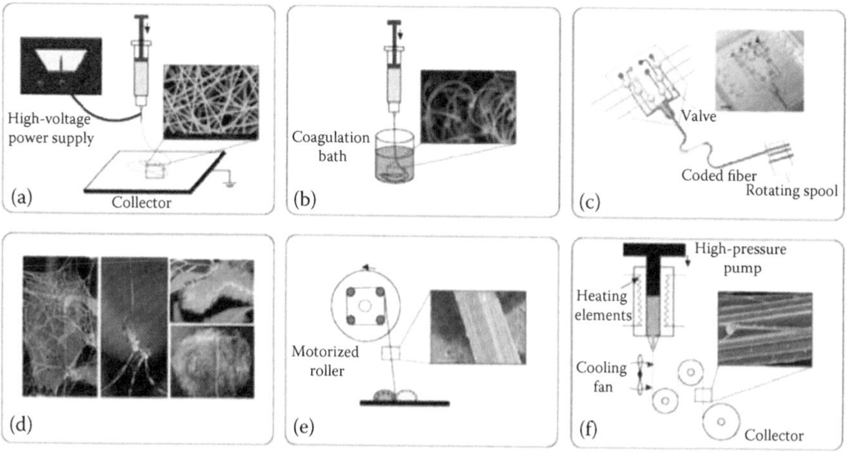

FIGURE 14.1 Novel fiber fabrication techniques. (a) Electrospinning: a charged polymer solution is pulled on a collecting plate in the presence of an applied electrostatic force (From Pham, Q.P., Sharma, U., and Mikos, A.G., *Tissue Engineering,* 12, 1197–1211, 2006.) [7]. (b) Wet spinning: a prepolymer solution is injected into a coagulation bath including either a poor solvent or a nonsolvent to form a polymer and cross-link the fibers (From Jalili, R., Aboutalebi, S.H., Esrafilzadeh, D.E. et al., *Advanced Functional Materials,* 23, 5345–5354, 2013.) [8]. (c) Microfluidic spinning: a prepolymer solution is pushed into a microchannel using coaxial flow and the polymer is cross-linked (From Lee, B.R., Lee, K.H., Kang, E., Kim, D.S., and Lee, S.H., *Biomicrofluidics,* 5, 022208, 2011.) [9]. (d) Biospinning: fibers are naturally produced by various insects such as silkworms and spiders (From Mandal, B.B. and Kundu, S.C., *Acta Biomaterialia,* 6, 360–371, 2010.) [10]. (e) Interfacial complexation: fibers are fabricated at the interface of two oppositely charged polyelectrolyte solutions by means of polyion complex formation (From Yim, E.K.F., Liao, I.C., and Leong, K.W., *Tissue Engineering,* 13, 423–433, 2007.) [11]. (f) Melt spinning: a heated polymer melt is extruded through a spinneret to form fiber strands (From Zhmayev, E., Cho, D., and Joo, Y.L., *Polymer,* 51, 4140–4144, 2010.) [12]. (From Tamayol, A., Akbari, M., Annabi, N., Paul, A., Khademhosseini, A., and Juncker, D., *Biotechnology Advances*, 31, 669–687, 2013. With permission.)

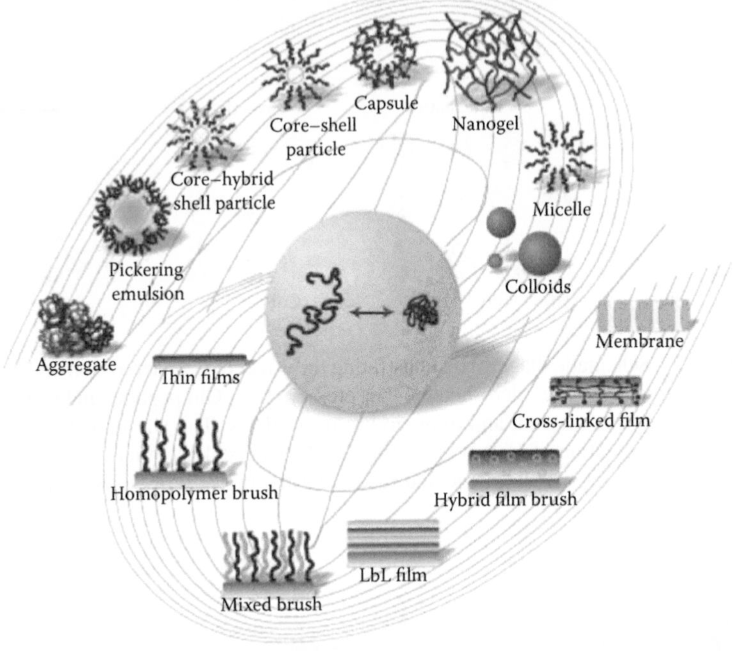

FIGURE 14.6 Galaxy of nanostructured stimuli-responsive polymer materials. (From Stuart, M.A.C., Huck, W.T.S., Genzer, J. et al., *Nature Materials*, 9, 101–113, 2010. With permission.)

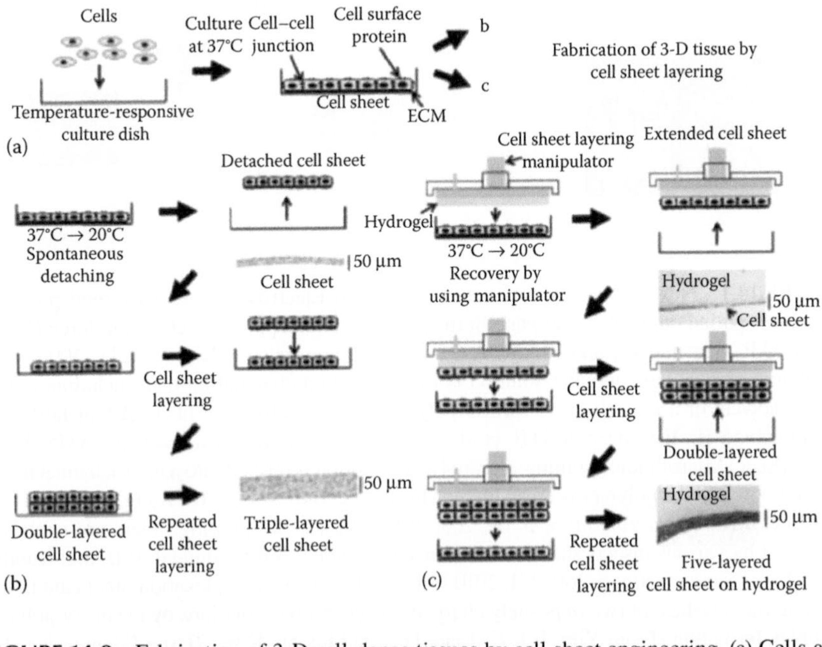

FIGURE 14.8 Fabrication of 3-D cell-dense tissues by cell sheet engineering. (a) Cells are cultivated on a temperature-responsive culture dish until reaching confluence. (b, c) The cell sheet is recovered by changing the temperature, and the 3-D tissues are fabricated by the following alternative methods. Using a simple pipetting method, extremely cell-dense 3-D tissue constructs can be fabricated by layering recovered cell sheets. Using the hydrogel-coated, plunger-like manipulator method, the cell sheet is recovered from the dish by adhering to the hydrogel surface within the manipulator. (From Haraguchi, Y., Shimizu, T., Sasagawa, T. et al., *Nature Protocols*, 7, 850–857, 2012. With permission.)

10 Construction of Enzyme Biosensors Based on a Commercial Glucose Sensor Platform

Yue Cui

CONTENTS

10.1 INTRODUCTION TO GLUCOSE BIOSENSORS

Biosensors are defined as analytical devices incorporating a biological material (e.g., tissue, microorganisms, organelles, cell receptors, enzymes, antibodies, nucleic acids, natural products), a biologically derived material (e.g., recombinant antibodies, engineered proteins, aptamers), or a biomimetic (e.g., synthetic receptors, biomimetic catalysts, combinatorial ligands, imprinted polymers) intimately associated with or integrated within a physicochemical transducer or transducing microsystem, which may be optical, electrochemical, thermometric, piezoelectric, magnetic, or micromechanical,[1] as shown in Figure 10.1. The generated electrical signal is related to the concentration of analytes through the biological reactions. Diabetes is a worldwide public health problem. It is one of the leading causes of death and disability in the world. The diagnosis and management of diabetes mellitus require close monitoring of blood glucose levels, which is the major diagnostic criterion for diabetes and is a useful tool for patient monitoring.[2] To date, the most commercially successful biosensors are amperometric enzyme glucose biosensors for monitoring diabetes, which account for approximately 85% of the current world market.

Glucose oxidase (GOD)[3] is the standard enzyme for glucose biosensors, and it has a higher selectivity for glucose. GOD is a stable enzyme that can withstand extremes

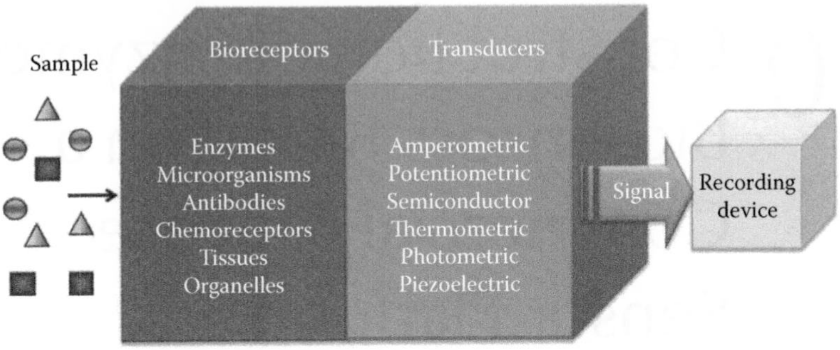

FIGURE 10.1 Schematic illustration of biosensors.

of pH, ionic strength, and temperature compared with many other enzymes, which allows less stringent conditions during the manufacturing process and relatively relaxed storage norms for use. It is also cost-effective and commercially available.[4] Thus, GOD is widely used for the construction of glucose biosensors. As shown in Figure 10.2, glucose is catalyzed by GOD to consume oxygen and produce glucolactone and hydrogen peroxide.

Due to their advantages of being simple, small, cost-effective, and easy to handle, amperometric biosensors have been extensively employed. Commercial glucose biosensors are mainly constructed based on Clark-type electrodes and screen-printed electrodes with the immobilization of GOD. As shown in Figure 10.3A, for a typical Clark-type glucose biosensor, the enzyme matrix of the biosensor is prepared by mixing the enzyme solution and a curing agent.

The enzyme mixture is spread over a Teflon membrane and dried at 4°C overnight. Then, it is covered with a dialysis membrane to sandwich the enzyme layer between the two membranes and is fixed with a membrane holder and an O-ring. The Teflon side facing the electrode receives 15 μL of saturated KCl gel as an electrolyte before the enzyme membrane is placed on top of the Clark-type electrode. The sensor is then screwed into a measuring cell and rehydrated in buffer at room temperature before using the Teflon side. In the presence of glucose, GOD catalyzes the specific oxidation of glucose. Dissolved oxygen acting as an essential material for the enzymatic activity of GOD is consumed proportionally to the concentration

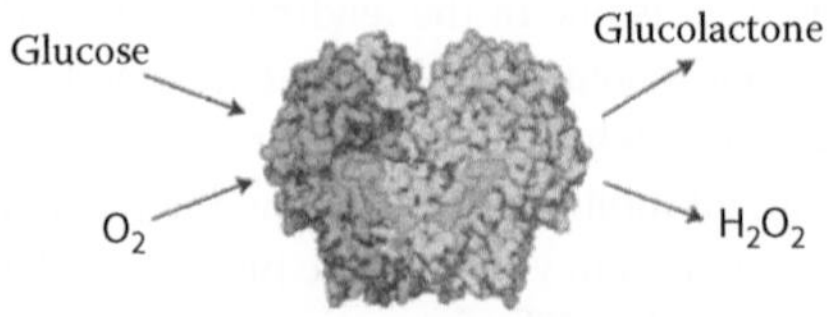

FIGURE 10.2 Enzymatic reactions with GOD for the detection of glucose. (From Wikipedia. Glucose oxidase, last modified December 18, 2013, http://en.wikipedia.org/wiki/Glucose_oxidase.)

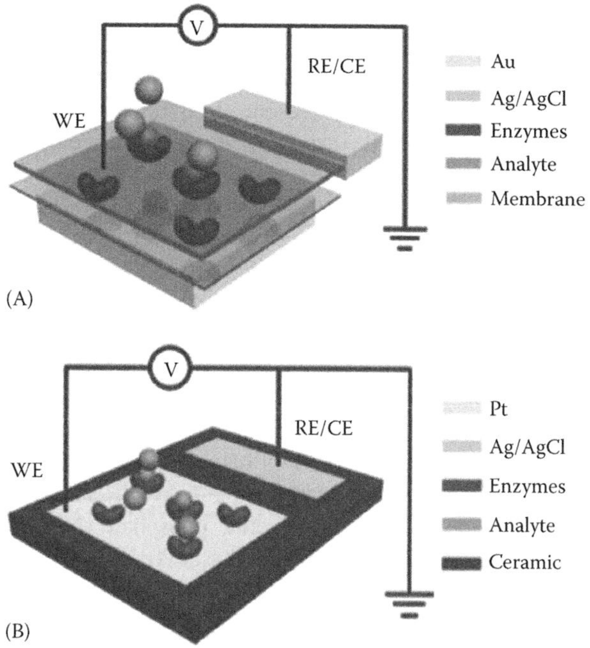

FIGURE 10.3 **(See color insert)** Schematic illustration of biosensors with (A) a Clark-type oxygen electrode and (B) a screen-printed electrode.

of glucose during the measurements. A detectable signal, caused by the consumption of dissolved oxygen by GOD, is monitored by the Clark electrode. Electroactive interferences are eliminated by the Teflon membrane in front of the Clark electrode. As shown in Figure 10.3B, the working electrode of the screen-printed electrode is covered by a mixture containing GOD and a curing agent, followed by drying and storing at 4°C overnight. The enzyme electrode is then screwed into the measuring cell and rehydrated at room temperature to allow the enzyme matrix to swell before use. Experiments were performed at room temperature by applying a specific potential to the screen-printed electrode for the detection of hydrogen peroxide. The current difference (nA) between the stationary currents was recorded, which was proportional to the concentration of glucose.

10.2 BIOSENSOR FOR THE METABOLIC COMPOUND ADENOSINE TRIPHOSPHATE (ATP)

Metabolic compounds are important analytes for medical diagnosis and environmental monitoring, and metabolic biosensors are widely developed based on sequential and competitive enzymatic reactions on Clark-type electrodes and screen-printed electrodes. Here, I describe the biosensor construction for a typical metabolic compound, ATP,[5] which is based on sequential enzymatic reactions and the detection of the substrate consumption and enzymatic product.

ATP, as a mediator of energy exchanges for all living cells, plays an important role in various vital biological processes.[6,7] It is also widely used as an index for biomass determinations in clinical microbiology, food quality control, and environmental analyses owing to its ubiquitous presence in living matter.[8,9]

Due to the importance of ATP, various methods have been developed for its determination, including luminescent, colorimetric, spectrofluorimetric, chromatographic, patch-sniffing, and potentiometric biosensing methods.[10–17] However, these methods result in a long measuring time, a large amount of enzyme consumption, sophisticated procedures that need to be performed by skilled personnel, or complex signal processing that is difficult to adapt for handheld devices.

The importance of amperometric enzyme-based biosensors has increased considerably in recent years, due to their advantages of being highly sensitive, rapid, accurate, economical, and easy to handle for the specific measurement of target analytes in complex matrices such as blood, food products, and environmental samples. Various amperometric ATP biosensors have been developed using different combinations of enzymes or enzymes with mediators.[18–23] Recently, the coimmobilizations of a nicotinamide adenine dinucleotide phosphate ($NAD(P)^+$)-dependent dehydrogenase with p-hydroxybenzoate hydroxylase (HBH) in front of a Clark-type oxygen electrode and on a screen-printed electrode have been investigated for developing a general type of dehydrogenase-based biosensor,[24–26] which shows high-performance characteristics.

Here, the development of two types of ATP biosensors are described, which are based on new combinations of enzymes and electrodes by using the coimmobilizations of HBH, glucose-6-phosphate dehydrogenase (G6PDH), and hexachlorocyclopentadiene (HEX) on a Clark-type oxygen electrode and on a screen-printed electrode. Schematic illustrations for the biosensors' setup and their determination principles are shown in Figure 10.4. HEX transfers the phosphate group from ATP to glucose to form glucose-6-phosphate. G6PDH catalyzes the specific dehydrogenation of glucose-6-phosphate by consuming $NAD(P)^+$. The product, NADPH, initiates the irreversible hydroxylation of p-hydroxybenzoate by HBH to consume dissolved oxygen and generate 3,4-dihydroxybenzoate. During the measurement of

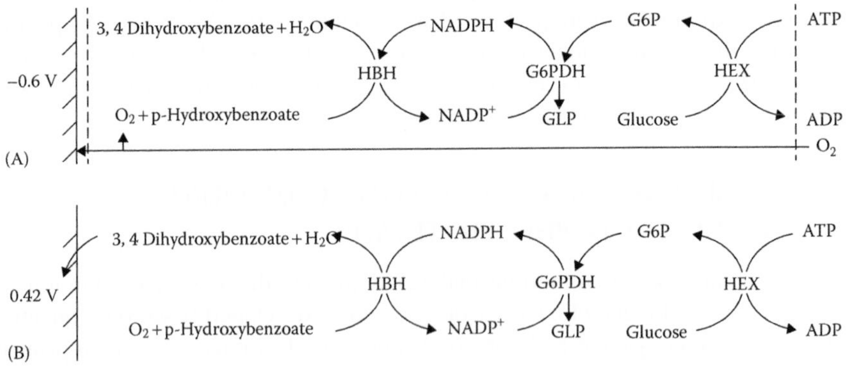

FIGURE 10.4 Schematic illustration of ATP biosensors with (A) a Clark-type electrode and (B) a screen-printed electrode. (From Cui, Y., *IEEE Sensors Journal*, 10, 979–983, 2010.)

ATP, a detectable signal caused by the consumption of oxygen by HBH can be monitored at −0.6 V versus Ag/AgCl by the Clark-type electrode, and another detectable signal caused by the generation of 3,4-dihydroxybenzoate by HBH can be monitored at 0.42 V versus Ag/AgCl by the screen-printed electrode. These electronic signals are monitored and processed with a potentiostat, and the data acquisitions are performed using a computer.

The Clark-type sensor performance for the determination of ATP was characterized. Figure 10.5A shows the current–time curve and the recovery study of the sensor obtained by the addition of various concentrations of ATP. The cathodic current decreased after the addition of ATP due to the consumption of dissolved oxygen through enzymatic reactions by HBH, G6PDH, and HEX, and the reduction in the cathodic current was proportional to the concentration of ATP. The response of this biosensor was rapid (2 s), with a high reproducibility and a short recovery time (1 min). The steady-state background current decreased after the addition of ATP and reached a new stationary state in 20 s, and with a washing step to remove ATP in the buffer solution for recovery, the current increased to the

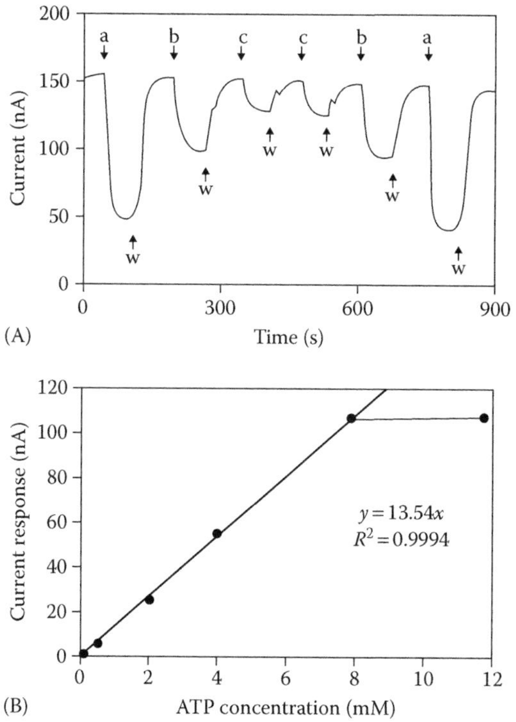

FIGURE 10.5 Characterization of a Clark-type sensor performance. (A) Current–time curve of the sensor to (a) 8 mM, (b) 4 mM, and (c) 2 mM ATP with (w) washing step. (B) Calibration curve for ATP with the sensor ($n = 3$) (sensor: 2 U HBH, 1.2 G6PDH, and 0.8 U HEX on a Clark-type oxygen electrode. Buffer: 100 mM Tris-SO_4 buffer at pH 8.0, 1 mM p-hydroxybenzoate, 0.2 mM $NADP^+$, 2 mM glucose, and 10 mM $MgSO_4$). (From Cui, Y., *IEEE Sensors Journal*, 10, 979–983, 2010.)

initial background current. The total measurement using the sensor took less than 3 min. Figure 10.5B shows the calibration curve for ATP with the sensor. A linear relationship was obtained between the current response and the concentration of ATP from 0.1 to 8 mM with a detection limit of 0.05 mM (slope: 13.5 nA mM^{-1}, $R^2=0.9994$, $n=3$). Also, the Teflon membrane covering the Clark-type electrode was protected from contacting the electroactive interferences and producing unreliable signals. The sensor showed no signal response to the electroactive substances, such as ascorbic acid, uric acid, and 20 L-amino acids (L-alanine, L-valine, L-leucine, L-isoleucine, L-phenylalanine, L-tryptophan, L-methionine, L-glycine, L-glutamate, L-serine, L-threonine, L-cysteine, L-tyrosine, L-asparagine, L-glutamine, L-aspartic acid, L-lysine, L-arginine, L-histidine, L-proline). Therefore, the Clark-type sensor has a high specificity for the determination of ATP.

The screen-printed sensor performance for the determination of ATP was characterized. Figure 10.6A shows the current–time curve and the recovery study of the sensor obtained by the additions of various concentrations of ATP. The anodic current increased after the addition of ATP due to the oxidation of catechol generated

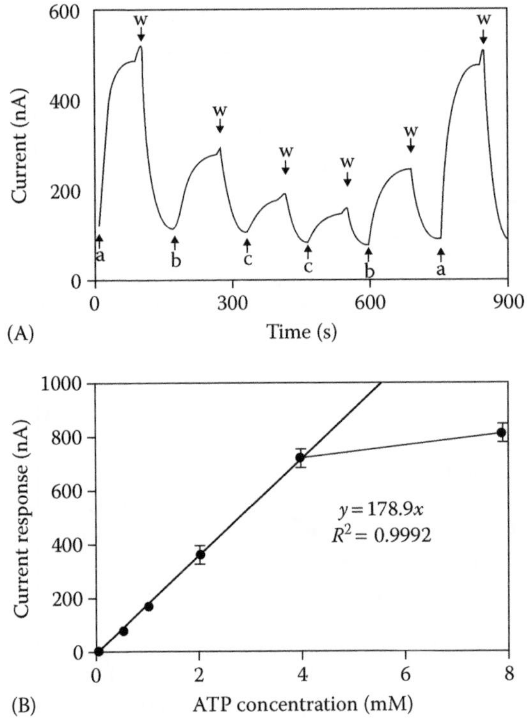

FIGURE 10.6 Characterization of a screen-printed sensor performance. (A) Current–time curve of the sensor to (a) 2 mM, (b) 1 mM, and (c) 0.5 mM ATP with (w) washing step. (B) Calibration curve for ATP with the sensor ($n=3$) (sensor: 2 U HBH, 1.2 G6PDH, and 0.8 U HEX on a Clark-type oxygen electrode. Buffer: 100 mM Tris-SO_4 buffer at pH 8.0, 1 mM p-hydroxybenzoate, 0.2 mM $NADP^+$, 2 mM glucose, and 10 mM $MgSO_4$). (From Cui, Y., *IEEE Sensors Journal*, 10, 979–983, 2010.)

by HBH, G6PDH, and HEX, and the increase in the anodic current was proportional to the concentration of ATP. The response of this biosensor was rapid (2 s), with a high reproducibility and a short recovery time (1 min). The steady-state background current increased after the addition of ATP and reached a new stationary state in around 1 min, and with a washing step to remove ATP in the buffer solution for recovery, the current decreased to the initial background current. The total measurement using the sensor took less than 3 min. Figure 10.6B shows the calibration curve for ATP with the sensor. A linear relationship was obtained between the current response and the concentration of ATP from 5 μM to 4 mM with a detection limit of 4 μM (slope: 178.9 nA mM^{-1}, $R^2=0.9992$, $n=3$). Also, some typical electroactive substances were investigated for their interference effects on the sensor response. The sensor showed almost no signal response.[27,28] The screen-printed sensor could also determine the concentrations of ATP relatively accurately.

Compared with the screen-printed sensor, the Clark-type sensor showed a higher specificity for being free from electroactive interferences, while it also showed a lower sensitivity and a higher detection limit, which were probably mainly due to the smaller electrode area for signal transduction and the diffusion barriers from the dialysis membrane and the Teflon membrane. The determination performance of both types of sensors could be further enhanced by using other kinds of electrodes (e.g., another electrode with a larger diameter) for sensor constructions.

In summary, the development of two types of amperometric trienzyme ATP biosensors is described based on new combinations of enzymes and electrodes by using the coimmobilizations of HBH, G6PDH, and HEX on a Clark-type oxygen electrode and on a screen-printed electrode. The sensors show high-performance characteristics with broad detection ranges, short measuring times, and good specificities. Thus, the methods provide new analytical approaches for the determination of ATP that are rapid, sensitive, accurate and easy to handle.

10.3 BIOSENSOR FOR ENZYME ACTIVITY: PHOSPHOGLUCOMUTASE (PGM)

The determination of enzyme activity is of great importance for various biological and medical applications. Biosensors for enzyme activities are generally constructed based on the detection of a decrease in substrates or an increase in products from the enzymatic reaction. Here, a PGM biosensor will be described as an example to illustrate the construction of a sensor for enzyme activities.[29]

PGM is a ubiquitous enzyme that is expressed in all organisms from bacteria to plants to animals, and it controls a key branch point for carbohydrate metabolism. PGM catalyzes the interconversion of glucose-1-phosphate and glucose-6-phosphate. In this process, the enzyme links various catabolic pathways to yield energy ATP or reducing power NAD(P)H, and several anabolic pathways, which lead to the synthesis of polysaccharides.[30–33]

Due to its importance, measurements for the enzymatic activity of PGM have been widely performed using optical methods with a coupled-enzyme system[32–40] or using a combination of ion/molecule reactions and Fourier transform ion cyclotron resonance (FT-ICR) mass spectrometry.[41] However, these measurements result in a long detection time, a large amount of enzyme consumption, or complex procedures that need to be performed by skilled personnel. The importance of amperometric enzyme-based biosensors has increased considerably in recent years, due to their advantages of being highly sensitive, rapid, accurate, economical, and easy to handle for the specific measurement of target analytes in complex matrices such as blood, food products, and environmental samples. To the best of our knowledge, no report has determined PGM activity using an amperometric biosensor.

Here, an amperometric biosensor for PGM activity with a bienzyme screen-printed biosensor is described. As shown in Figure 10.7, the principle is as follows: PGM (EC 5.4.2.2, from rabbit muscle, Sigma) converts glucose-1-phosphate to glucose-6-phosphate. G6PDH (EC 1.1.1.49, from *Leuconostoc mesenteroides*, Sigma) catalyzes the specific dehydrogenation of glucose-6-phosphate by consuming NAD^+. The product, NADH, initiates irreversible decarboxylation and hydroxylation of salicylate by salicylate hydroxylase (SHL, EC 1.14.13.1, from *Pseudomonas* sp., GDS Technology Inc., USA) in the presence of oxygen to produce catechol, which results in a detectable signal due to its oxidation at the working electrode.

The bienzyme electrode for the measurement of PGM activity is based on the detection of glucose-6-phosphate, the product of the PGM-catalyzed reaction. From the literature, glucose-6-phosphate biosensors have been developed using several methods, including the combination of G6PDH and various mediators, or the combination of phosphatase and GOD.[42–44] Here, the successful coupling of SHL and G6PDH on a screen-printed electrode can also serve as a glucose-6-phosphate biosensor.

The screen-printed electrode was covered with a mixture of SHL, G6PDH, and glutaraldehyde, followed by drying and storing at 4°C overnight. The biosensor was then screwed into a measuring cell, which was filled with buffer solution, and rehydrated for around 1 h at room temperature to allow the enzyme matrix to swell before use. Experiments were performed by applying the specific potential to this type of screen-printed electrode at 0.42 V and magnetically stirring the solution at 300 rpm to obtain a uniform distribution of PGM. The measurement was started by adding PGM into the buffer solution in the measuring cell, and the current velocity (nA min^{-1}) was recorded for plotting a calibration curve. A syringe was used between each measurement for sucking the buffer solution out of the measuring cell to remove PGM in the solution, and thus to remove catechol on the electrode surface.

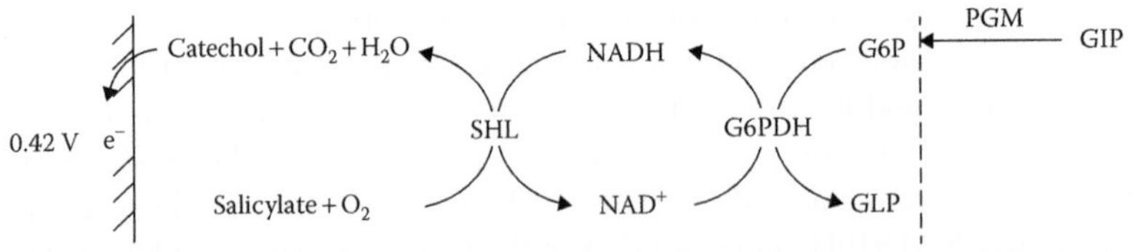

FIGURE 10.7 Schematic illustration of an amperometric screen-printed biosensor for the determination of PGM activity (G1P: glucose-1-phosphate; G6P: glucose-6-phosphate; GLP: D-glucono-1,5-lactone 6-phosphate; PGM: phosphoglucomutase; G6PDH: glucose-6-phosphate dehydrogenase; SHL: salicylate hydroxylase). (From Cui, Y., Barford, J.P., and Renneberg, R., *Analytical Biochemistry*, 354, 162–164, 2006.)

A two-step optimization was needed before the sensor calibration in order to improve the sensor performance for determining PGM activity. The first step was the optimization of the enzyme matrix, including the enzyme loadings and the immobilization agent concentrations. Various loadings of SHL, G6PDH, and glutaraldehyde on screen-printed electrodes were investigated to obtain the maximum current velocity. Based on this optimization, a mixture containing 0.33 U SHL and 1.88 U G6PDH with 1% glutaraldehyde in 0.5 μL of the enzyme matrix was used for further experiments. The second step was the optimization of the working conditions, including the pH value and the substrate and cofactor concentrations in the buffer solution. The effects of glucose-1,6-diphosphate and $MgCl_2$ were also studied as they were the activating cofactor and metal cofactor, respectively, for the PGM-catalyzed reaction. To improve the sensor performance, various loadings of glucose-1-phosphate, salicylate, NAD^+, glucose-1,6-diphosphate, $MgCl_2$, and buffer pH were investigated. In order to obtain the maximum current velocity for the measurement of PGM activity, the substrate and cofactor concentrations in the buffer solution should be sufficient to avoid signal saturation due to their inadequate loadings. The optimized working condition obtained was 100 mM Tris-HCl buffer solution containing 5 mM glucose-1-phosphate, 5 mM salicylate, 5 mM NAD^+, 50 μM glucose-1,6-diphosphate, and 5 mM $MgCl_2$ at pH 8.0.

Figure 10.8 shows the calibration curve for PGM activity using the bienzyme screen-printed sensor. PGM activities were measured by the biosensor with the optimal enzyme matrix and working condition. As the PGM activity was proportional to the rate of production of glucose-6-phosphate, it was further proportional to the rate of production of catechol and the current velocity, thus the current velocity (nA min^{-1}) was used for the determination of PGM activity. As shown in Figure 10.8, a linear relationship was obtained between the current velocity and the PGM activity from 0.05 to 5 U mL^{-1} with a detection limit of 0.02 U mL^{-1} (slope: 76.36 [nA min^{-1}]/[U mL^{-1}], $R^2=0.9988$, $n=3$). The linear detection range and the detection limit are

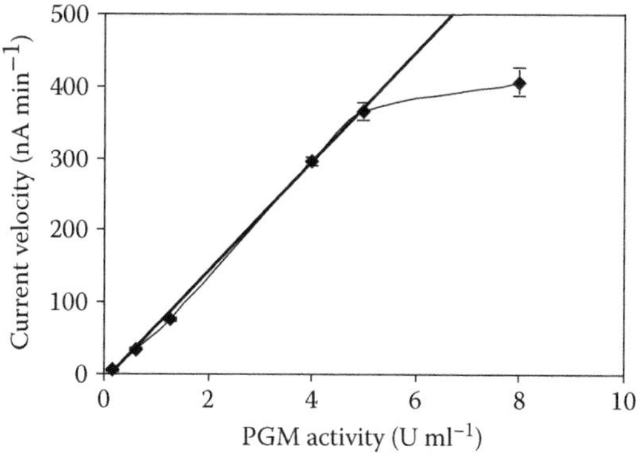

FIGURE 10.8 Calibration curve for PGM activity using an amperometric screen-printed biosensor (sensor: 0.33 U SHL, 1.88 U G6PDH with 1% glutaraldehyde in 0.5 μL of enzyme matrix. Buffer: 100 μM Tris-HCl buffer containing 5 μM glucose-1-phosphate, 5 μM salicylate, 5 mM NAD^+, 50 μM glucose-1,6-diphosphate, and 5 μM $MgCl_2$ at pH 8.0). (From Cui, Y., Barford, J.P., and Renneberg, R., *Analytical Biochemistry*, 354, 162–164, 2006.)

decided by the recording method, and if the record of the current velocity is changed from nA (min^{-1}) to nA $(2\ min)^{-1}$ or nA $(3\ min)^{-1}$, the linear detection range and the detection limit will be changed and improved.

Also, the sensor has a fast measuring time (1 min) and a short recovery time (2 min) with high reproducibility. Hence, the total measurement of PGM activity takes less than 4 min with simple operations, which is more rapid and convenient than conventional methods.

In summary, a biosensor for PGM is described using an amperometric screen-printed biosensor based on the coimmobilization of SHL and G6PDH. The sensor shows high-performance characteristics with a broad detection range (0.05–5 U mL^{-1}) and a rapid measuring time (1 min). Thus, it provides a new analytical approach to the determination of PGM activity that is rapid, sensitive, economical, and easy to handle.

10.4 BIOSENSOR FOR TOXIN: AZIDE

Toxins are harmful for various metabolic activities, and the determination of toxins is important for environmental and medical monitoring. Biosensor constructions are generally based on the inhibition of enzymatic activities on electrodes. Here, azide is used as an example for the construction of biosensors for toxins.[45]

Sodium azide is a rapidly acting toxic chemical that exists as an odorless white solid. It prevents cells from using oxygen, making it very harmful to the organs of the body, especially the heart and the brain. In addition, sodium azide is used daily in the air bags of vehicles with large tonnage, in hospitals and laboratories as a chemical preservative, in agriculture for pest control, and in detonators and other explosives. Moreover, sodium azide is the starting material of heavy sodium azide, pure sodium metal, hydrazoic acid, and a variety of medicines.[46–49]

Due to its importance, various methods have been developed for the determination of azide concentration, including ion chromatography,[50,51] gas chromatography,[52] high-performance liquid chromatography,[53] and the combination of diffusion extraction and spectrophotometry.[54] However, these methods either require expensive equipment or they need to be performed by skilled personnel. As the importance of amperometric biosensors has increased due to their advantages of being highly sensitive, rapid, economical, and easy to handle for measurements, several types of biosensors have been constructed for the determination of azide based on the inhibition of enzymatic activity, including a laccase- or tyrosinase-immobilized mediated carbon electrode,[55,56] and a catalase-immobilized Clark-type oxygen electrode.[57]

Here, two methods to construct biosensors for the amperometric determination of azide are described using a disposable, screen-printed electrode immobilized with catalase or tyrosinase, as shown in Figure 10.9.

The first method for azide determination is based on the inhibition of the enzymatic consumption of H_2O_2 on the catalase-immobilized screen-printed electrode by azide, as shown in Figure 10.9A. H_2O_2 can result in a high current due to its oxidation at the working electrode; however, the immobilized catalase on the electrode initiates the decomposition of H_2O_2 into H_2O and O_2, which consumes H_2O_2 at the electrode surface, and therefore the current response to H_2O_2 is not apparent. The presence of

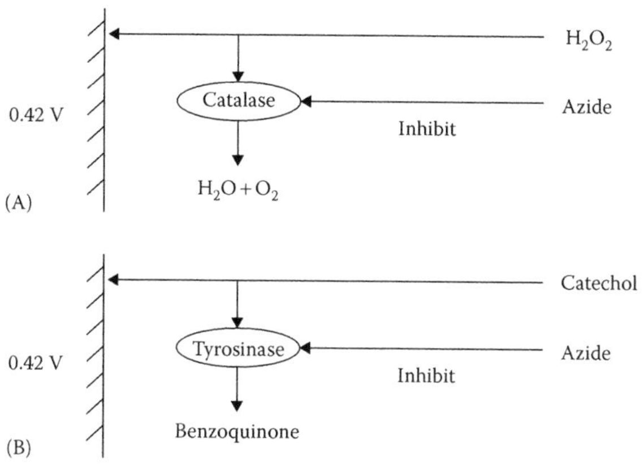

FIGURE 10.9 Schematic illustration for azide determination with (A) a catalase-immobilized screen-printed electrode and (B) a tyrosinase-immobilized screen-printed electrode. (From Cui, Y., Barford, J.P., and Renneberg, R., *Analytical Sciences*, 22, 1279–1281, 2006.)

azide inhibits the enzymatic activity of the catalase, thus the consumption of H_2O_2 by the catalase on the electrode decreases, followed by a significant increase in the current signal, and the current difference after the injection of azide is proportional to the concentration of azide.

The second method for azide determination is based on the inhibition of the enzymatic consumption of catechol on the tyrosinase-immobilized screen-printed electrode by azide, as shown in Figure 10.9B. Catechol can result in a high current due to its oxidation at the working electrode; however, the immobilized tyrosinase on the electrode converts it into benzoquinone in the presence of oxygen, which consumes catechol at the electrode surface, and therefore the current response to catechol using the tyrosinase-immobilized electrode is much smaller than that using a bare electrode. The presence of azide inhibits the enzymatic activity of tyrosinase, thus the consumption of catechol by tyrosinase on the electrode decreases, followed by a significant increase in the current signal, and the current difference after the injection of azide is proportional to the concentration of azide.

A bare screen-printed electrode showed a response of 3480 nA to 1 mM H_2O_2. The catalase-immobilized screen-printed electrode exhibited no apparent current response to 1 mM H_2O_2 due to the enzymatic decomposition of H_2O_2 by the catalase. Figure 10.10A shows a current–time curve with the catalase-immobilized electrode obtained by adding various amounts of azide. As shown in the figure, after the injection of azide, the anodic current increased due to the decrease of H_2O_2 consumption by the catalase, and the increase in the anodic current was proportional to the concentration of azide. The response of the sensor was rapid (1 s) with a high reproducibility and a short recovery time (1 min). The steady background current increased after the addition of azide and reached a new steady state within 30 s. Therefore, the total measurement using the sensor took less than 3 min. Figure 10.10B shows the calibration curve for azide with the catalase-immobilized electrode. A linear relationship was obtained between the current response and the concentration of

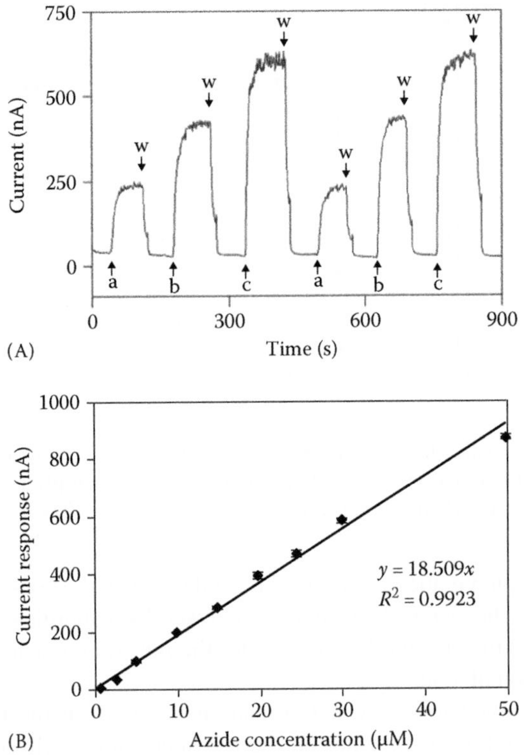

FIGURE 10.10 Determination performance with a catalase-immobilized electrode. (A) Current–time curve to (a) 10 μM, (b) 20 μM, and (c) 30 μM of azide with (w) washing step. (B) Calibration curve for azide ($n = 3$) (sensor: 75 U catalase with 1% glutaraldehyde in 0.5 μL of enzyme matrix. Buffer: 50 mM K-PBS buffer containing 1 mM H_2O_2 at pH 7.0). (From Cui, Y., Barford, J.P., and Renneberg, R., *Analytical Sciences*, 22, 1279–1281, 2006.)

azide from 0.1 to 50 μM with a slope of 18.51 nA μM^{-1} and a correlation coefficient of 0.9923. This method also has a high reproducibility with a relative standard deviation (RSD) of 6.6% for five different catalase-immobilized electrodes by testing their sensitivities. Compared with the biosensor methods reported previously,[55–57] this method shows high-performance characteristics with the most sensitive detection range, a short measuring time, and an easy-to-handle operation.

A bare screen-printed electrode showed a high current response of 2350 nA to 1 mM catechol, and the tyrosinase-immobilized screen-printed electrode exhibited a current response of 350 nA to 1 mM catechol due to the incomplete consumption of catechol by tyrosinase. Figure 10.11A shows a current–time curve with the tyrosinase-immobilized electrode obtained by adding various amounts of azide. As shown in the figure, after the injection of azide, the anodic current increased due to the decrease of catechol consumption by tyrosinase, and the increase in the anodic current was proportional to the concentration of azide. The response of the sensor was rapid (1 s) with a high reproducibility and a short recovery time (1 min).

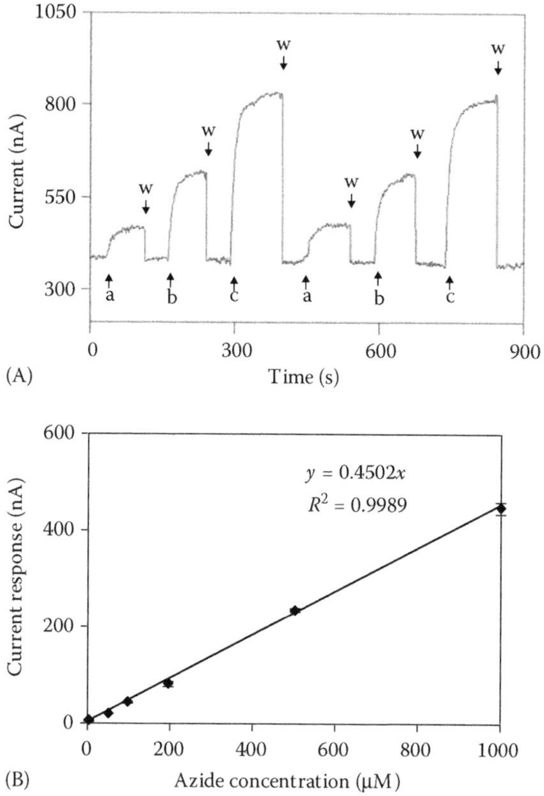

FIGURE 10.11 Determination performance with a tyrosinase-immobilized electrode. (A) Current–time curve to (a) 200 μM, (b) 500 μM, and (c) 1000 μM of azide with (w) washing step. (B) Calibration curve for azide ($n = 3$) (sensor: 75 U tyrosinase with 1% glutaraldehyde in 0.5 μL of enzyme matrix. Buffer: 50 mM K-PBS buffer containing 1 mM catechol at pH 7.0). (From Cui, Y., Barford, J.P., and Renneberg, R., *Analytical Sciences*, 22, 1279–1281, 2006.)

The steady background current increased after the addition of azide and reached a new steady state within 30 s. Therefore, the total measurement using the sensor took less than 3 min. Figure 10.11B shows the calibration curve for azide with the tyrosinase-immobilized electrode. A linear relationship was obtained between the current response and the concentration of azide from 5 to 1000 μM with a slope of 0.45 nA μM^{-1} and a correlation coefficient of 0.9989. This method also has a high reproducibility with an RSD of 4.5% for five different tyrosinase-immobilized electrodes by testing their sensitivities. In this experimental condition, the determination of azide using the tyrosinase-immobilized electrode was not as sensitive as that using a catalase-immobilized electrode, which was probably because the inhibition effect of azide on tyrosinase is not as large as that on catalase.

In summary, the construction of an amperometric azide biosensor is described based on the inhibition of the enzymatic consumption of hydrogen peroxide or catechol with a disposable, screen-printed electrode immobilized with catalase or tyrosinase. Either of these methods provides a new analytical approach to the determination of azide that is rapid, sensitive, economical, and easy to handle. Besides, the two methods also provide new analytical methods for the determination of other toxic substances.

10.5 BIOSENSOR FOR POLYMER: POLY(3-HYDROXYBUTYRATE) (PHB)

Polymers play an important role in various metabolisms and in applications in different fields. The construction of biosensors for polymers is generally based on the digestion of polymers into monomers, and the detection of monomers with biosensors. Here, PHB is used as an example for the construction of a polymer biosensor.[58]

PHB is a common intracellular biodegradable polymer involved in bacterial carbon and energy storage, and it plays an important role in the course of metabolism.[59,60] As one of the most interesting biodegradable materials, it has promising applications in medicine, material science, agriculture, and so on.[61–63] It has also been found in activated sludge samples from conventional wastewater treatment plants.[64]

Due to its importance, various methods have been developed for measuring PHB concentration, including the use of gravimetry,[65] turbidimetry,[66] and spectrophotometry,[67] and the use of chromatography,[68,69] capillary isotachophoresis,[70] and capillary zone electrophoresis[71] with pretreatments. However, these methods result in an unreliable determination, a long measurement time (several hours), or procedures with high temperature or pressure, or they need to be performed by skilled personnel.

The importance of amperometric enzyme-based biosensors has increased considerably in recent years, due to their advantages of being highly sensitive, rapid, reliable, economical, and easy to handle for the specific measurement of target analytes in complex matrices, such as blood, food products, and environmental samples. In this work, we present a simple method for the determination of PHB using a combination of an amperometric enzyme-based biosensor and alkaline hydrolysis, which is the first report concerning the measurement of PHB with a biosensor technique.

As shown in Figure 10.12, PHB is first decomposed to produce its monomer 3-hydroxybutyrate (3-HB) through alkaline hydrolysis, followed by neutralization with acid. The product, 3-HB, formed accordingly, is measured with an enzyme-based 3-HB biosensor. The 3-HB sensor[72] is constructed by immobilizing

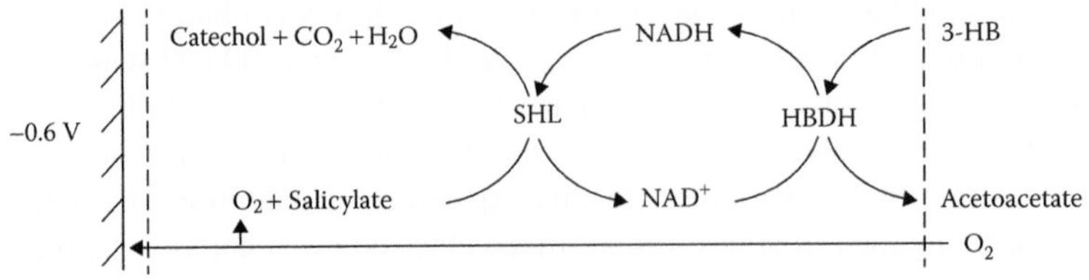

FIGURE 10.12 Schematic illustration for the determination of PHB and an enzyme-based 3-HB biosensor. (From Cui, Y., Barford, J.P., and Renneberg, R., *Analytical Sciences*, 22, 1323–1326, 2006.)

3-hydroxybutyrate dehydrogenase (HBDH, EC 1.1.1.30) and SHL (EC 1.14.13.1) on a Clark-type oxygen electrode. The determination principle of the biosensor is as follows: HBDH catalyzes the specific dehydrogenation of 3-HB in the presence of NAD^+. The product, NADH, initiates the irreversible decarboxylation and hydroxylation of salicylate by SHL. Dissolved oxygen, acting as an essential material for the enzymatic activity of SHL, is consumed proportionally to the concentration of 3-HB during the measurements. A detectable signal, caused by the consumption of the dissolved oxygen by SHL, is monitored at −0.6 V versus Ag/AgCl by the Clark electrode. Both enzymes are entrapped by a poly(carbamoyl) sulfonate (PCS) hydrogel, which is sandwiched between a dialysis membrane and a Teflon membrane. Electroactive interferences are eliminated by the Teflon membrane.

PHB, with a concentration of 21.5 g L^{-1} (250 mM 3-HB unit) in 6 M KOH, was decomposed into its monomer product, 3-HB, through alkaline hydrolysis at 50°C for 30 or 10 min, followed by neutralization with 6 M HCl to a pH of around 8 with a volume ratio of 1.1 for a 30-min hydrolysis or 1.07 for a 10-min hydrolysis. The volume ratios of KOH to HCl, being larger than 1.0, were due to the acid forms of the hydrolyzed products of PHB contained in KOH.

The production of 3-HB from PHB was measured with a 3-HB biosensor. One milliliter of K-PBS (100 mM, pH 7.5) buffer containing 0.5 mM sodium salicylate and 0.5 mM NAD^+ was added into the measuring chamber. After achieving a steady background current (<30 min), the measurement was started at room temperature (22°C) by adding a pretreated PHB solution, or its solution diluted 10 times (0.1–30 μL) into the measuring buffer. The concentration of PHB was determined by the decrease in the dissolved oxygen reduction current. Until a stationary current occurred, the current difference was recorded for plotting the calibration curve. The current response was proportional to the production of 3-HB from PHB, and was further proportional to the concentration of PHB.

Figure 10.13A shows the current response to various concentrations of the pretreated PHB using the enzyme-based 3-HB sensor with hydrolysis in 6 M KOH at 50°C for 30 min. The cathodic current decreased due to the consumption of dissolved oxygen by the SHL and HBDH in the enzyme layer after the addition of the pretreated PHB solution. Also, the reduction in the cathodic current was proportional to the concentration of PHB. The response of the bienzyme 3-HB sensor was rapid (2 s) with high reproducibility and a short recovery time (2 min). The steady-state background current decreased after the addition of the pretreated PHB solution and reached a stationary state within 30 s. The total measuring time using the 3-HB sensor for the pretreated PHB was less than 4 min. Hence, the whole measurement for PHB took only a few minutes, being less than 40 min with a saturated production of 3-HB using a 30-min hydrolysis. If the hydrolysis time is 10 min, the total time is less than 15 min with part production of 3-HB, which mostly depends on the hydrolysis time. These times were much quicker than when using a conventional method.

Figure 10.13B shows the calibration curves for the pretreated PHB using the enzyme-based 3-HB sensor. A linear relationship is observed between the current response and the concentration of PHB from 0.5 to 110 mg L^{-1} (slope: 1.76 nA/[mg L^{-1}], $R^2 = 0.9937$, $n = 3$) with hydrolysis in 6 M KOH at 50°C for 30 min. A sharp saturation at 120 mg L^{-1} was observed due to oxygen depletion by the enzymatic

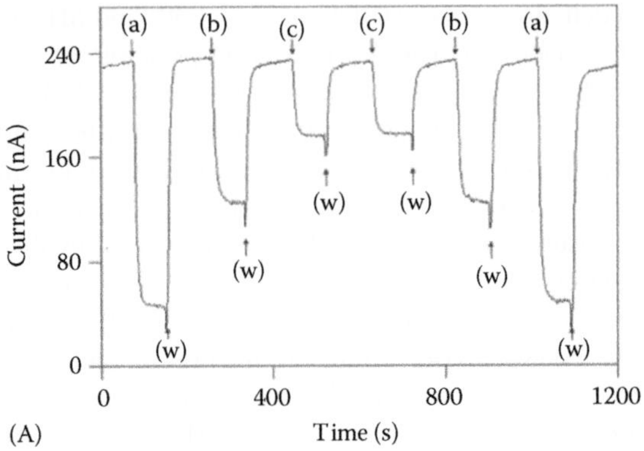

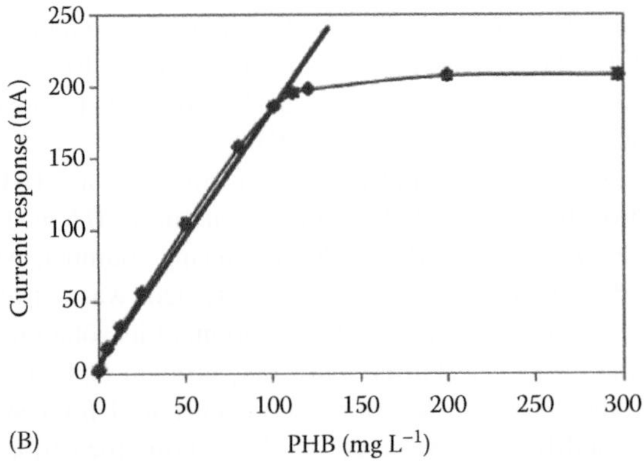

FIGURE 10.13 (A) Current response to (a) 100 mg L^{-1}, (b) 50 mg L^{-1}, and (c) 25 mg L^{-1} of PHB, with (w) washing step. (B) Calibration curve for PHB using the combination of an enzyme-based biosensor and alkaline hydrolysis in 6 M KOH at 50°C for 30 min. (From Cui, Y., Barford, J.P., and Renneberg, R., *Analytical Sciences*, 22, 1323–1326, 2006.)

reaction, as indicated by a net current of zero at the Clark electrode. For calculating the detection limit, 3 M KCl (30 μL), a neutralized solution of 6 M KOH and 6 M HCl with a volume ratio of 1:1, was added as a blank solution into the measuring cell in order to determine the blank signal. The detection limit of the system with a 30-min hydrolysis of PHB, calculated as the mean blank signal plus three times the standard derivation of the mean blank signal, was 0.3 mg L^{-1} of PHB. This method also has a high average reproducibility of 98.6%, which was obtained by repeating the analysis of different concentrations of PHB, ranging from 0 to 300 mg L^{-1}, three times.

The characteristics for the determination of PHB using the combination of an enzyme-based biosensor and alkaline hydrolysis are summarized in Table 10.1. This method shows a fast measurement, which takes several minutes due to the short hydrolysis time and the quick sensor-measuring time. It also has a sensitive detection range and a low detection limit. Also, the determination of PHB using a 30-min hydrolysis has a smaller detection range, and takes a longer measurement time than using a 10-min hydrolysis, while it has a higher sensitivity and a lower detection

TABLE 10.1
Characteristics for the Determination of PHB Using the Combination of an Enzyme-Based Biosensor and Alkaline Hydrolysis in 6 M KOH at 50°C for 30 min or 10 min

Characteristics	With a 30-min Hydrolysis	With a 10-min Hydrolysis
Detection time (min)	<40	<15
Linear range (mg L^{-1})	0.5–110	1.0–160
Sensitivity (nA/[mg L^{-1}])	1.76	1.20
Detection limit (mg L^{-1})	0.3	0.5
Reproducibility (%)	98.6	97.7
Stability	17 days for the half-life of the 3-HB bienzyme sensor[15]	
Specificity	Specific, free from electroactive interferences[15]	

Source: Cui, Y., Barford, J.P., and Renneberg, R., *Analytical Sciences*, 22, 1323–1326, 2006.

limit. This method also shows a high average reproducibility of greater than 97%. From a previous report,[72] the enzyme-based sensor has good stability, which can retain more than 50% of the initial response for 17 days using storage in a buffer solution at 4°C after a measurement. Due to the Teflon membrane, the determination of the pretreated PHB using the enzyme-based biosensor is also free from any electroactive interference. Since this method employs biosensor detection through specific enzymatic reactions and a relatively fast pretreatment, it is more rapid and reliable than conventional methods for the determinations of PHB.

In summary, the determination of PHB is described using a combination of an enzyme-based biosensor and alkaline hydrolysis. It shows high-performance characteristics with a sensitive detection range, a short measurement time, and simple operation. Thus, the method provides a new analytical approach for the determination of PHB that is rapid, sensitive, and easy to handle.

10.6 CONCLUSIONS AND FUTURE PERSPECTIVES

The development of a general approach for the construction of enzyme biosensors could expand opportunities for sensors in both fundamental studies and a variety of device platforms. Glucose biosensors for the detection of diabetes have been widely commercialized. Here, the developments of biosensors for several typical and important analytes based on the commercial glucose biosensor platform are described, including metabolic compounds, toxins, polymers, and enzyme activities. These biosensors are constructed based on the immobilization of an enzyme matrix with sequential enzymatic reactions or competitive enzymatic reactions on a conventional Clark-type electrode or a disposable screen-printed electrode. The characterization shows that the sensors can detect the analytes using an approach that is sensitive, selective, rapid, and easy to handle. These sensors can be easily incorporated into commercial sensor devices and applied to the industrial market. I anticipate that these methods could open up exciting opportunities for the use of biosensors in fundamental bioanalytical research, as well as applications ranging from medical diagnosis to environmental monitoring.

REFERENCES

1. Tang. A. (ed.) *Biosensors and Bioelectronics.* http://www.journals.elsevier.com/09565663/biosensors-and-bioelectronics/aims-and-scope/.
2. Wang, J. Glucose biosensors: 40 Years of advances and challenges. *Electroanalysis* **13**, 983–988 (2001).
3. Wikipedia. Glucose oxidase, last modified December 18, 2013, http://www.en.wikipedia.org/wiki/Glucose_oxidase.
4. Yoo, E. H. and Lee, S. Y. Glucose biosensors: An overview of use in clinical practice. *Sensors* **10**, 4558–4576 (2010).
5. Cui, Y. Amperometric ATP biosensors based on coimmobilizations of p-hydroxybenzoate hydroxylase, glucose-6-phosphate dehydrogenase, and hexokinase on Clark-type and screen-printed electrodes. *IEEE Sensors Journal* **10**, 979–983 (2010).
6. Higgins, C. F., Hiles, I. D. and Salmond, G. P. C. A family of related ATP-binding subunits coupled to many distinct biological processes in bacteria. *Nature* **323**, 448–450 (1986).
7. Stekhoven, F. S. Energy transfer factor A.D (ATP synthetase) as a complex Pi-ATP exchange enzyme and its stimulation by phospholipids. *Biochemical and Biophysical Research Communications* **47**, 7–14 (1972).
8. Jorgensen, P. E., Eriksen, T. and Jensen, B. K. Estimation of viable biomass in wastewater and activated sludge by determination of ATP, oxygen utilization rate and FDA hydrolysis. *Water Research* **26**, 1495–1501 (1992).
9. Pietrzak, E. M. and Denes, A. S. Comparison of luminol chemiluminescence with ATP bioluminescence for the estimation of total bacterial load in pure cultures. *Journal of Rapid Methods and Automation in Microbiology* **4**, 207–218 (1996).
10. Jose, D. A. et al. Colorimetric sensor for ATP in aqueous solution. *Organic Letters* **9**, 1979–1982 (2007).
11. Miao, Y., Liu, J., Hou, F. and Jiang, C. Determination of adenosine disodium triphosphate (ATP) using norfloxacin-Tb^{3+} as a fluorescence probe by spectrofluorimetry. *Journal of Luminescence* **116**, 67–72 (2006).
12. Ronner, P., Friel, E., Czerniawski, K. and FraFnkle, S. Luminometric assays of ATP, phosphocreatine, and creatine for estimation of free ADP and free AMP. *Analytical Biochemistry* **275**, 208–216 (1999).
13. Brown, P. and Dale, N. Spike-independent release of ATP from Xenopus spinal neurons evoked by activation of glutamate receptors. *Journal of Physiology* **540**, 851–860 (2002).
14. Karatzaferi, C., De Haan, A., Offringa, C. and Sargeant, A. J. Improved high-performance liquid chromatographic assay for the determination of "high-energy" phosphates in mammalian skeletal muscle: Application to a single-fibre study in man. *Journal of Chromatography B: Biomedical Sciences and Applications* **730**, 183–191 (1999).
15. Katsu, T. and Yamanaka, K. Potentiometric method for the determination of adenosines 5′-triphosphate. *Analytica Chimica Acta* **276**, 373–376 (1993).
16. Adachi, Y., Sugawara, M., Taniguchi, K. and Umezawa, Y. Na^+/K^+-ATPase-based bilayer lipid membrane sensor for adenosine 5′-triphosphate. *Analytica Chimica Acta* **281**, 577–584 (1993).
17. Gotoh, M., Tamiya, E., Karube, I. and Kagawa, Y. A microsensor for adenosine-5′-triphosphate pH-sensitive field effect transistors. *Analytica Chimica Acta* **187**, 287–291 (1986).

18. Compagnone, D. and Guilbault, G. G. Glucose oxidase/hexokinase electrode for the determination of ATP. *Analytica Chimica Acta* **340**, 109–113 (1997).
19. Cui, Y., Barford, J. P. and Renneberg, R. Amperometric trienzyme ATP biosensors based on the coimmobilization of salicylate hydroxylase, glucose-6-phosphate dehydrogenase, and hexokinase. *Sensors and Actuators, B: Chemical* **132**, 1–4 (2008).
20. Kueng, A., Kranz, C. and Mizaikoff, B. Amperometric ATP biosensor based on polymer entrapped enzymes. *Biosensors and Bioelectronics* **19**, 1301–1307 (2004).
21. Liu, S. and Sun, Y. Co-immobilization of glucose oxidase and hexokinase on silicate hybrid sol-gel membrane for glucose and ATP detections. *Biosensors and Bioelectronics* **22**, 905–911 (2007).
22. Llaudet, E., Hatz, S., Droniou, M. and Dale, N. Microelectrode biosensor for real-time measurement of ATP in biological tissue. *Analytical Chemistry* **77**, 3267–3273 (2005).
23. Yang, X., Johansson, G., Pfeiffer, D. and Scheller, F. Enzyme electrodes for ADP/ATP with enhanced sensitivity due to chemical amplification and intermediate accumulation. *Electroanalysis* **3**, 659–663 (1991).
24. Cui, Y., Barford, J. P. and Renneberg, R. Development of an L-glutamate biosensor using the coimmobilization of L-glutamate dehydrogenase and p-hydroxybenzoate hydroxylase on a Clark-type electrode. *Sensors and Actuators, B: Chemical* **127**, 358–361 (2007).
25. Cui, Y., Barford, J. P. and Renneberg, R. Development of a glucose-6-phosphate biosensor based on coimmobilized p-hydroxybenzoate hydroxylase and glucose-6-phosphate dehydrogenase. *Biosensors and Bioelectronics* **22**, 2754–2758 (2007).
26. Gajovic, N., Warsinke, A., Huang, T., Schulmeister, T. and Scheller, F. W. Characterization and mathematical modeling of a bienzyme electrode for L-malate with cofactor recycling. *Analytical Chemistry* **71**, 4657–4662 (1999).
27. Gribble, F. M. et al. A novel method for measurement of submembrane ATP concentration. *Journal of Biological Chemistry* **275**, 30046–30049 (2000).
28. Chenzhuo, L., Murube, J., Latorre, A. and Martin del Rio, R. The presence of high amounts of amino acid taurine in human tears. *Archivos de la Sociedad Canaria de Oftalmología* **11**, 11–12 (2000).
29. Cui, Y., Barford, J. P. and Renneberg, R. Amperometric determination of phosphoglucomutase activity with a bienzyme screen-printed biosensor. *Analytical Biochemistry* **354**, 162–164 (2006).
30. Ray, W. J. and Peck, E. J. Phosphomutases. *The Enzymes* **6**, 407–477 (1972).
31. Lytovchenko, A., Sweetlove, L., Pauly, M. and Fernie, A. R. The influence of cytosolic phosphoglucomutase on photosynthetic carbohydrate metabolism. *Planta* **215**, 1013–1021 (2002).
32. Akutsu, J. I. et al. Characterization of a thermostable enzyme with phosphomannomutase/phosphoglucomutase activities from the hyperthermophilic archaeon Pyrococcus horikoshii OT_3. *Journal of Biochemistry* **138**, 159–166 (2005).
33. Mesak, L. R. and Dahl, M. K. Purification and enzymatic characterization of PgcM: A β-phosphoglucomutase and glucose-1-phosphate phosphodismutase of Bacillus subtilis. *Archives of Microbiology* **174**, 256–264 (2000).
34. Frazier, D. M., Clemons, E. H. and Kirkman, H. N. Minimizing false positive diagnoses in newborn screening for galactosemia. *Biochemical Medicine and Metabolic Biology* **48**, 199–211 (1992).
35. Inoue, H., Kondo, S., Hinohara, Y., Juni, N. and Yamamoto, D. Enhanced phosphorylation and enzymatic activity of phosphoglucomutase by the Btk29A tyrosine kinase in Drosophila. *Archives of Biochemistry and Biophysics* **413**, 207–212 (2003).
36. Videira, P. A., Cortes, L. L., Fialho, A. M. and Sá-Correia, I. Identification of the pgmG gene, encoding a bifunctional protein with phosphoglucomutase and phosphomannomutase activities, in the gellan gum-producing strain Sphingomonas paucimobilis ATCC 31461. *Applied and Environmental Microbiology* **66**, 2252–2258 (2000).

37. Zhang, G. et al. Catalytic cycling in β-phosphoglucomutase: A kinetic and structural analysis. *Biochemistry* **44**, 9404–9416 (2005).
38. Howard, S. C., Deminoff, S. J. and Herman, P. K. Increased phosphoglucomutase activity suppresses the galactose growth defect associated with elevated levels of Ras signaling in S. cerevisiae. *Current Genetics* **49**, 1–6 (2006).
39. Masuda, C. A., Xavier, M. A., Mattos, K. A., Galina, A. and Montero-Lomeli, M. Phosphoglucomutase is an in vivo lithium target in yeast. *Journal of Biological Chemistry* **276**, 37794–37801 (2001).
40. Sergeeva, L. I. and Vreugdenhil, D. In situ staining of activities of enzymes involved in carbohydrate metabolism in plant tissues. *Journal of Experimental Botany* **53**, 361–370 (2002).
41. Gao, H. and Leary, J. A. Kinetic measurements of phosphoglucomutase by direct analysis of glucose-1-phosphate and glucose-6-phosphate using ion/molecule reactions and Fourier transform ion cyclotron resonance mass spectrometry. *Analytical Biochemistry* **329**, 269–275 (2004).
42. Bassi, A. S., Tang, D. and Bergougnou, M. A. Mediated, amperometric biosensor for glucose-6-phosphate monitoring based on entrapped glucose-6-phosphate dehydrogenase, Mg2+ ions, tetracyanoquinodimethane, and nicotinamide adenine dinucleotide phosphate in carbon paste. *Analytical Biochemistry* **268**, 223–228 (1999).
43. Hung Tzang, C., Yuan, R. and Yang, M. Voltammetric biosensors for the determination of formate and glucose-6-phosphate based on the measurement of dehydrogenase-generated NADH and NADPH. *Biosensors and Bioelectronics* **16**, 211–219 (2001).
44. Mazzei, F., Botrè, F. and Botrè, C. Acid phosphatase/glucose oxidase-based biosensors for the determination of pesticides. *Analytica Chimica Acta* **336**, 67–75 (1996).
45. Cui, Y., Barford, J. P. and Renneberg, R. A disposable, screen-printed electrode for the amperometric determination of azide based on the immobilization with catalase or tyrosinase. *Analytical Sciences* **22**, 1279–1281 (2006).
46. Betterton, E. A. Environmental fate of sodium azide derived from automobile airbags. *Critical Reviews in Environmental Science and Technology* **33**, 423–458 (2003).
47. Heeschen, W. H., Ubben, E. H., Gyodi, P. and Beer, P. Kieler Milchw. *ForschBer* **45**, 109136 (1993).
48. Chang, S. and Lamm, S. H. Human health effects of sodium azide exposure: A literature review and analysis. *International Journal of Toxicology* **22**, 175–186 (2003).
49. Hagenbuch, J. P. Opportunities and limits of the use of azides in industrial production. Implementation of safety measures. *Chimia* **57**, 773–776 (2003).
50. Annable, P. L. and Sly, L. A. Azide determination in protein samples by ion chromatography. *Journal of Chromatography* **546**, 325–334 (1991).
51. Kruszyna, R., Smith, R. P. and Kruszyna, H. Determining sodium azide concentration in blood by ion chromatography. *Journal of Forensic Sciences* **43**, 200–202 (1998).
52. Kage, S., Kudo, K. and Ikeda, N. Determination of azide in blood and urine by gas chromatography-mass spectrometry. *Journal of Analytical Toxicology* **24**, 429–432 (2000).
53. Vácha, J., Tkaczyková, M. and Rejholcová, M. Determination of sodium azide in the presence of proteins by high-performance liquid chromatography. *Journal of Chromatography B: Biomedical Sciences and Applications* **488**, 506–508 (1989).
54. Tsuge, K., Kataoka, M. and Seto, Y. Rapid determination of cyanide and azide in beverages by microdiffusion spectrophotometric method. *Journal of Analytical Toxicology* **25**, 228–236 (2001).
55. Daigle, F., Trudeau, F., Robinson, G., Smyth, M. R. and Leech, D. Mediated reagentless enzyme inhibition electrodes. *Biosensors and Bioelectronics* **13**, 417–425 (1998).
56. Leech, D. Optimisation of a reagentless laccase electrode for the detection of the inhibitor azide. *Analyst* **123**, 1971–1974 (1998).

57. Sezgintürk, M. K., Göktuğ, T. and Dinçkaya, E. A biosensor based on catalase for determination of highly toxic chemical azide in fruit juices. *Biosensors and Bioelectronics* **21**, 684–688 (2005).
58. Cui, Y., Barford, J. P. and Renneberg, R. Determination of poly(3-hydroxybutyrate) using a combination of enzyme-based biosensor and alkaline hydrolysis. *Analytical Sciences* **22**, 1323–1326 (2006).
59. Freier, T. et al. In vitro and in vivo degradation studies for development of a biodegradable patch based on poly(3-hydroxybutyrate). *Biomaterials* **23**, 2649–2657 (2002).
60. Mansfield, D. A., Anderson, A. J. and Naylor, L. A. Regulation of PHB metabolism in Alcaligenes eutrophus. *Canadian Journal of Microbiology* **41**, 44–49 (1995).
61. Koller, M. et al. Biotechnological production of poly(3-hydroxybutyrate) with Wautersia eutropha by application of green grass juice and silage juice as additional complex substrates. *Biocatalysis and Biotransformation* **23**, 329–337 (2005).
62. Pouton, C. W. and Akhtar, S. Biosynthetic polyhydroxyalkanoates and their potential in drug delivery. *Advanced Drug Delivery Reviews* **18**, 133–162 (1996).
63. Saad, B., Neuenschwander, P., Uhlschmid, G. K. and Suter, U. W. New versatile, elastomeric, degradable polymeric materials for medicine. *International Journal of Biological Macromolecules* **25**, 293–301 (1999).
64. Dircks, K., Henze, M., van Loosdrecht, M. C., Mosbaek, H. and Aspegren, H. Storage and degradation of poly-β-hydroxybutyrate in activated sludge under aerobic conditions. *Water Research* **35**, 2277–2285 (2001).
65. Tsuji, H. and Suzuyoshi, K. Environmental degradation of biodegradable polyesters. IV. The effects of pores and surface hydrophilicity on the biodegradation of poly (ε-caprolactone) and poly[(R)-3-hydroxybutyrate] films in controlled seawater. *Journal of Applied Polymer Science* **90**, 587–593 (2003).
66. Murase, T., Suzuki, Y., Doi, Y. and Iwata, T. Nonhydrolytic fragmentation of a poly[(R)-3-hydroxybutyrate] single crystal revealed by use of a mutant of polyhydroxybutyrate depolymerase. *Biomacromolecules* **3**, 312–317 (2002).
67. Yilmaz, M., Soran, H. and Beyatli, Y. Determination of poly-β-hydroxybutyrate (PHB) production by some Bacillus spp. *World Journal of Microbiology and Biotechnology* **21**, 565–566 (2005).
68. Jan, S. et al. Study of parameters affecting poly(3-hydroxybutyrate) quantification by gas chromatography. *Analytical Biochemistry* **225**, 258–263 (1995).
69. Saeki, T., Tsukegi, T., Tsuji, H., Daimon, H. and Fujie, K. Hydrolytic degradation of poly[(R)-3-hydroxybutyric acid] in the melt. *Polymer* **46**, 2157–2162 (2005).
70. Sulo, P., Hudecová, D., Propperová, A. and Bašnák, I. Rapid and simple analysis of poly-β-hydroxybutyrate content by capillary isotachophoresis. *Biotechnology Techniques* **10**, 413–418 (1996).
71. He, J., Chen, S. and Yu, Z. Determination of poly-β-hydroxybutyric acid in Bacillus thuringiensis by capillary zone electrophoresis with indirect ultraviolet absorbance detection. *Journal of Chromatography A* **973**, 197–202 (2002).
72. Kwan, R. C. H. et al. Biosensor for rapid determination of 3-hydroxybutyrate using bienzyme system. *Biosensors and Bioelectronics* **21**, 1101–1106 (2006).

11 Bioanalytical Applications of Piezoelectric Sensors

Şükran Şeker and Y. Murat Elçin

CONTENTS

ABBREVIATIONS

AW	Acoustic wave
BAW	Bulk acoustic wave
F	Frequency
FPW	Flexure plate wave
GOx	Glucose oxidase
LW	Love wave
PLGA	Poly(lactic-co-glycolic acid)
PS	Polystyrene

PZ	Piezoelectric
QCM	Quartz crystal microbalance
QCR	Quartz crystal resonator
R	Resistance
SAM	Self-assembled monolayer
SAW	Surface acoustic wave
SEM	Scanning electron microscopy
SH-APW	Shear horizontal acoustic plate wave
TSM	Thickness shear mode

11.1 INTRODUCTION

A biosensor is an analytical device that determines the specific interactions between biological molecules. Biosensors are commonly used in various fields such as the food and environmental industries, medicine, and drug discovery because of their advantages, such as label-free detection of molecules, fast response time, minimum sample pretreatment, and high sample throughput. Biosensors that convert a biological response into an electrical signal are integrated by a physicochemical transducer in various forms such as optical, electrochemical, thermometric, magnetic, or piezoelectric (PZ) [1]. Among the current biosensor systems, PZ biosensors play an important role in biochemical sensing. The quartz crystal microbalance (QCM) as a PZ sensor is fundamentally a mass sensing device able to measure very small amounts of mass changes on a quartz crystal resonator (QCR) in real time. It operates on the principle that the resonance frequency of quartz crystal changes with the amount of mass deposited. One of the major advantages of this technique is that it allows for label-free detection of molecules.

The QCM, often called the QCR, is one of the most common acoustic wave (AW) sensors operating with mechanical AWs as the transduction mechanism [2]. It works with a thickness shear mode (TSM) and has become a mature, commercially available, robust, and affordable technology [3]. The sensor's transducer element is based upon a solid plate of PZ material, usually a quartz crystal, which can generate AWs in the substrate in order to detect small mass uptakes in the nanogram level. Typically, in QCM applications, a thin piece of quartz is compressed between two metal plates used as electrodes that have evaporated onto both sides.

It is apparent that the performance of a mechanically resonating sensor is largely influenced by the properties of the sensory coating or a film interacting with target molecules. Thus, the surface coating and modification method is one of the most important factors for the improvement of the sensitivity, selectivity, and time response of a sensor system.

In the past decade, the QCM has found a wide range of applications as chemical and biological sensors in various areas from food to environmental and clinical analysis due to its specific attributes such as its small size, high sensitivity, high specificity, cost-effectiveness, short response time, reproducibility, requisition for small sample volume, and its easy application. The PZ biosensor is based on the measurement of mass changes on a quartz crystal surface caused by the specific adsorption of biomolecules. QCM techniques have been widely applied to investigate the interaction of biomolecules; thus, their types are categorized according to the biorecognition component. QCM immunosensors have been used to detect antigen–antibody binding on the quartz crystal [4]. An enzyme-based QCM measures the product of the substrate conversion by means of the enzyme functioning in the enzymatic reaction [5]. Nucleic acid–based QCM biosensors are used to measure the hybridization reaction between a probe oligonucleotide on the quartz and a complementary single-stranded nucleotide in a sample solution [6]. QCM cell biosensors are used to detect the adhesion and spreading of cells on the crystal surface [7]. Moreover, cell–drug interaction in the drug discovery field [8] and cell–materials interaction in biomaterials studies can be measured by means of the QCM [9].

This chapter aims to give an overview of the fundamental aspects of resonant sensors and to describe some of the quartz crystal surface modification methods. In addition, this chapter further specifies the application of the QCM technique to various measurements involving enzymes, proteins, nucleic acids, and cell biosensors in the fields of drug discovery and biomaterials research.

11.2 PIEZOELECTRIC RESONATORS

11.2.1 Fundamentals of Piezoelectricity and Quartz Crystals

PZ sensing is fundamentally based on the PZ effect as described by Jacques and Pierre Curie in 1880. They found that when subjected to mechanical stress, a voltage occurred on the surface of certain types of crystals, including quartz, tourmaline, and Rochelle salt. A year after their discovery, the opposite effect, the so-called inverse PZ effect, was predicted from fundamental thermodynamic principles by Lippmann [10]. According to these phenomena, the application of an electrical field on a PZ material causes mechanical deformation.

Figure 11.1 shows a schematic illustration of the PZ effect. In a PZ material, the positive and negative charges are randomly distributed in each part of the surface. When some pressure on the material is applied, positive and negative charges in the molecules are separated from each other. This polarization generates an electrical field and can be used to transform the mechanical energy used in the material deformation into electrical energy [11].

A large number of crystals exhibit the PZ effect, but the electrical, thermal, mechanical, and chemical properties of quartz make it the most common crystal type used in analytical applications [12]. Quartz crystal is a SiO_2 monocrystal with a zinc-blend structure type, present in nature in various forms, including α-quartz and β-quartz. The α-quartz form is used for PZ applications since this form is insoluble in water and resistant to high temperatures.

The resonant frequency of quartz crystal vibration depends on the physical properties of the crystal (e.g., size, cut, density, and shear modulus). The quartz crystal is a precisely cut slab from a natural or synthetic crystal. Different types of quartz cuts have different properties and vibration modes. AT- and BT-cut crystals have commonly been used as PZ sensors. AT-cut quartz crystals are used as QCMs due to their near-zero frequency changes at the room temperature range.

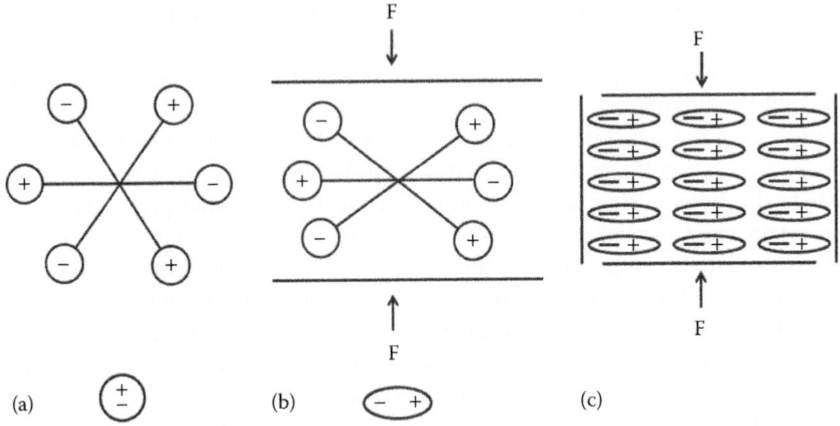

FIGURE 11.1 Simple molecular model explaining the PZ effect: (a) unperturbed molecule; (b) molecule subjected to an external force; and (c) the polarizing effect on the material surfaces. (Adapted and redrawn from Springer Science+Business Media: *Piezoelectric Transducers and Applications*, 1st edn., 2004, Arnau, A.)

Two crystal orientations oscillate exclusively in the TSM resonators, AT- and BT-cut crystals [13]; AT-cut crystals are used as PZ sensors for bioanalytical applications. The AT-cut is made at a +35°10′ angle from the z-axis [14] (Figure 11.2). The advantage of the AT-cut quartz crystal is that it has a temperature coefficient of nearly zero at around room temperature.

In QCM sensor applications, the quartz crystal consists of a thin piece of AT-cut quartz crystal compressed between two metal plates used as electrodes on both sides, as can be seen in Figure 11.3. When the electrodes are exposed to an alternating current (ac), the quartz crystal starts to oscillate at its resonance frequency due to the PZ effect.

The QCM technique was initially introduced to monitor the adsorption of mass on a quartz surface from chemical species in the gas phase [15]. More recently, after the integration of the quartz crystal into liquid conditions, the solution-based QCM became a medium in bioanalytical applications to measure adsorption at solution–surface interfaces. A QCM-flow injection analysis (QCM-FIA) system has the advantages that it works continuously and it monitors the binding of the analyte on the crystal surface. A typical experimental apparatus setup of QCM-FIA is shown in Figure 11.4.

11.2.2 Thickness Shear Mode Resonators (Quartz Crystal Microbalances)

There are several types of AW devices (Figure 11.5), depending on their wave propagation modes [14]. For example, TSM or bulk AW (BAW), flexure plate wave (FPW), surface AW (SAW), Love wave (LW), and shear horizontal AW mode resonators are the most commonly used in sensor systems. BAWs using QCMs travel through the interior of the substrate. By contrast, SAWs propagate on the surface of the crystal.

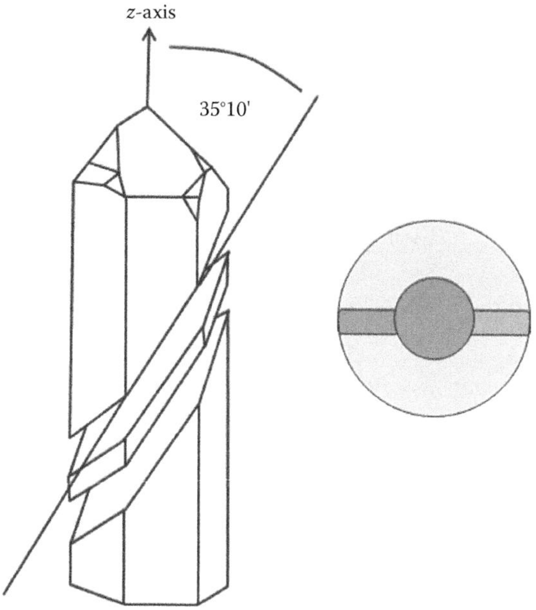

FIGURE 11.2 AT-cut quartz crystal. A quartz plate is cut at an angle of 35°10′ with respect to the optical *z*-axis. Deviating just 5′ from the proper cut angle leads to a temperature coefficient that is different from zero (between 0°C and 50°C). (Adapted and redrawn from Janshoff, A. et al., *Angewandte Chemie International Edition English*, 39, 4004–4032, 2000.)

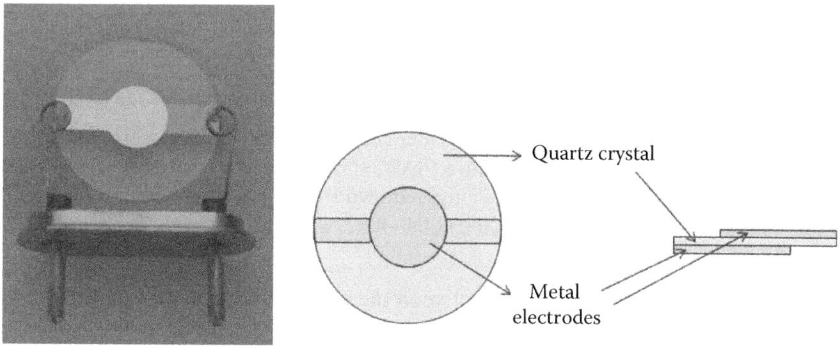

FIGURE 11.3 AT-cut quartz crystal used in QCM applications. The QCM consists of a thin piece of quartz crystal compressed between two metal plates coated on both sides, used as electrodes.

The QCM is a TSM-type resonator consisting of a thin disc of an AT-cut quartz with metal electrodes on both sides. The quartz crystal has a number of different resonator types depending on the cut angle determining the mode of mechanical vibration. An AT-cut crystal operates in the TSM mode and is used for QCM systems. The oscillation in this mode creates a displacement parallel to the surface of

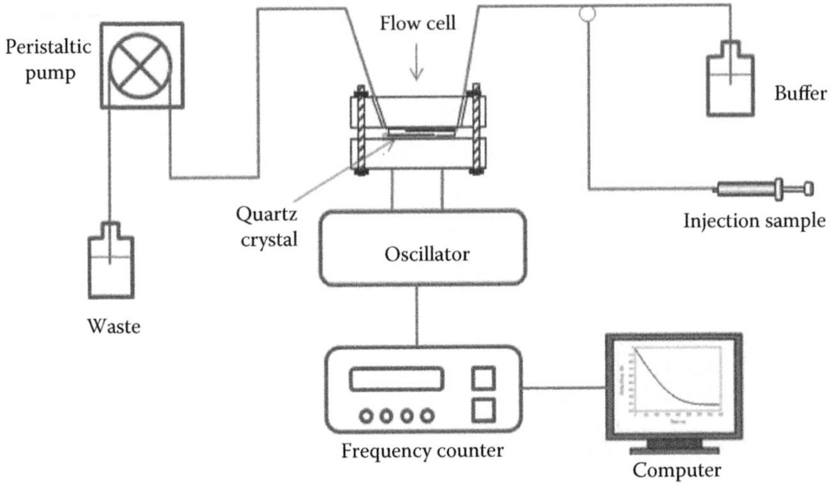

FIGURE 11.4 Typical experimental apparatus setup of the QCM-FIA system.

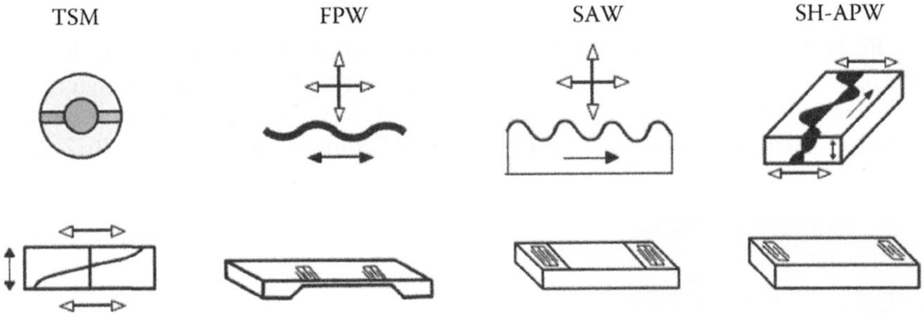

FIGURE 11.5 Schematic illustration of the four common kinds of AW sensors. TSM, thickness shear mode; FPW, flexure plate wave; SAW, surface acoustic wave; SH-APW, shear horizontal acoustic plate wave. (Adapted and redrawn from Janshoff, A. et al., *Angewandte Chemie International Edition English*, 39, 4004–4032, 2000.)

the quartz wafer. A voltage applied between the electrodes causes a shear deformation of the quartz crystal. The TSM oscillation responds very sensitively to any mass changes on the crystal surface [16].

11.3 MASS LOADING EQUATIONS

The deposition of a mass on a thin quartz crystal surface induces a decrease in the resonant frequency for a rigid substance. In 1959, Sauerbrey formulated the relationship between the change of the resonant frequency of a quartz crystal and the adsorption of a mass on the surface of a PZ resonator [17]. The application of an alternating voltage potential across the quartz crystal surface causes the crystal to oscillate at a characteristic resonant frequency. According to this equation, the mass of a thin layer accumulated on the surface of a crystal can be calculated by measuring the changes in the resonant frequency:

$$\Delta F = -2F_0^2\,\Delta M \Big/ \left[A\left(\mu_q \rho_q\right)^{1/2}\right] \qquad (11.1)$$

where:

ΔF is the change in frequency in hertz
F_0 is the initial resonant frequency of the quartz crystal
ΔM is the mass change in grams per square centimeter
A is the area of the crystal in square centimeters
ρ_q is the density of quartz (2.648 g cm^{-3})
μ_q is the shear modulus of quartz (2.947×10^{11} dyn cm^{-2})

The Sauerbrey equation is valid for oscillation in air and only applies to rigid masses attached to the crystal surface. Kanazawa and Gordon showed that when a crystal is dipped into a solution, the frequency shift also depends on the density and the viscosity of the liquid in contact with the QCM at one surface [18]. In this case, the viscosity and density of the liquid affect the propagation of the shear wave that radiates from the resonator into the liquid media. Kanazawa's equation (Equation 11.2) was developed for the QCM measurements in the liquid phase:

$$\Delta F = -F_0^{3/2}\left[\left(\rho_L \eta_L\right)/\left(\pi \rho_q \mu_q\right)\right]^{1/2} \qquad (11.2)$$

where:

ΔF is the measured frequency shift in this nongravimetric regime
ρ_L is the density of the liquid in contact with the crystal
η_L is the viscosity of the liquid in contact with the crystal

11.4 METHODS OF QUARTZ CRYSTAL SURFACE MODIFICATIONS

The quality and sensitivity of the resonance frequency of crystal affect its surface characteristics, such as its coating material, roughness, and hydrophobicity. Therefore, modification of the sensing surface is an important process in QCM biosensor studies. A variety of methods exist for surface coating or modification. Thin-film deposition methods such as self-assembled monolayers (SAMs), electrochemical deposition, spin coating, and electrospinning have been commonly used to improve the functionalization of quartz crystal surfaces.

11.4.1 Self-Assembled Monolayers

Functionalized alkanethiolate SAMs are important in preparing biosensor surfaces. Modifying a quartz crystal surface with SAMs generates a suitable recognition layer with a specific property or function, for the use of a single sensor sensitive to a single compound. There are generally two kinds of methods depending on self-assembly for use in biosensor studies. The most commonly used are the gold-alkylthiolate monolayers and the alkylsilane monolayers. The gold-alkylthiolate monolayer was first produced by Nuzzo and Allara in 1983 [19], demonstrating that alkanethiolates could be ordered on gold by the adsorption of di-*n*-alkyl disulfides from dilute solutions.

In order to obtain well-ordered, defect-free SAMs, the quartz crystals are immersed in low concentrations of thiol solution (typically 1–2 mM) in ethanol overnight at room temperature, forming a monolayer bearing many active tails (–COOH, $-NH_2$, or –OH), with improved analyte attachment features (Figure 11.6). The most commonly used solvent in the preparation of SAMs is ethanol. One of the advantages of the gold-alkylthiolate monolayer is that it is stable when exposed to air and aqueous or ethanolic solutions for several months. The reason for this could be ascribed to the fact that SAMs on gold surfaces adsorb very strongly due to the formation of a covalent bond between the gold and the sulfur atoms [20,21].

The SAMs of functionalized alkanethiols on gold surfaces are used in various applications, such as molecular recognition, selective immobilization of enzymes to surfaces, corrosion protection, and patterned surfaces in micrometer scale. The basic principle is that the thiol molecules adsorb readily from a solution onto the gold

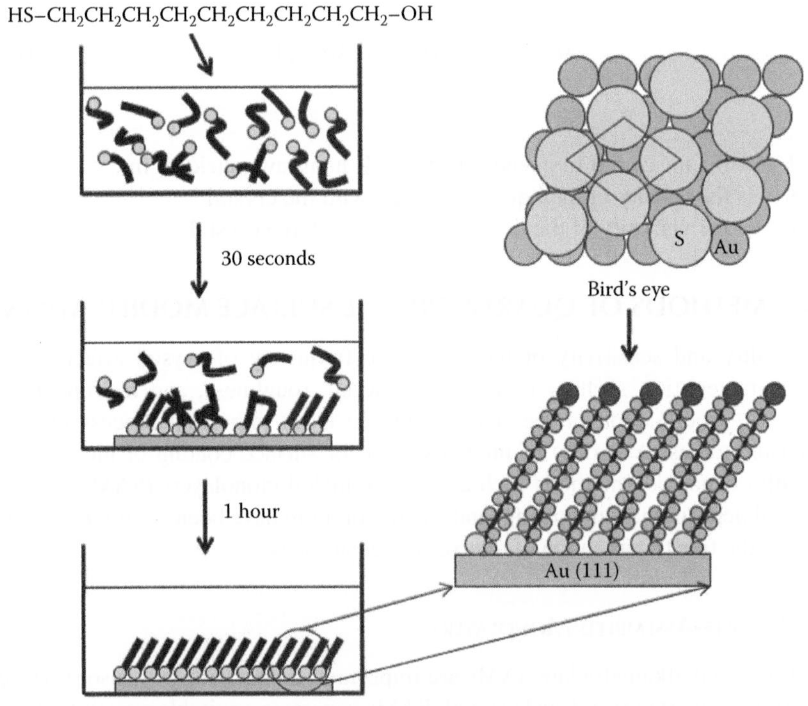

FIGURE 11.6 Schematic illustration of the preparation of SAMs. (Adapted and redrawn from Castner, D.G. and Ratner, B.D., *Surface Science,* 500, 28–60, 2002 [111].)

surface, forming an ordered monolayer with the tail group. Different thiol molecules or a mixture of thiol molecules can create SAM surfaces of the desired chemical surface functionality and size. By performing some activation reactions, the tail groups can be chemically functionalized following the assembly of the SAM. Additionally, a thiol-derivatized probe can be covalently linked to a gold electrode of a QCM by thiol groups [22].

11.4.2 Electrochemical Deposition

The electropolymerization technique has been used to form thin polymeric films on electrode surfaces. This electrochemical method has many advantages over other coating methods for creating biosensors [23]. One major advantage is that the reproducible and precise formation of a polymeric coating over surfaces of a desired size and geometry provides a polymeric coating with an electrochemically controlled thickness.

The electrochemical QCM (EQCM) system is a very convenient and powerful tool for creating electropolymerized thin films on the electrode surface of a quartz crystal (the working electrode) using cyclic voltammetry (CV); thus, the investigation of mass and viscoelastic properties is possible [24].

The electropolymerization of a wide range of different monomer types, such as pyrrole and substituted pyrrole, a large 32-unit ferrocenyl dendrimer, reversible fullerene C_{60} derivatives, phenols, and biomimetic tyrosine derivatives, has been performed [25].

11.4.3 Electrospinning and Spin Coating

The electrospinning [26] technique has become a popular method with the potential to produce nanofibrous nonwoven surfaces for a variety of applications, such as the modification of sensor surfaces [27–29]. Nanofibers fabricated via electrospinning have a surface area of approximately 10–100 times larger than that of continuous films, further increasing the adsorption rate and also the sensor sensitivity. The most important advantage of using nanofibers for coating is that they increase the surface area by their highly porous topography for adhesion. In principle, using a syringe pump, a polymer solution alone or with its contents is transferred along a glass pipe to a needle where a positive charge is applied. The charged polymer solution moves toward the grounded target, the solvent evaporates, attaining a three-dimensional nanofiber deposit [28] (Figure 11.7). Electrospun fibers with a controllable membrane thickness, fine structures, a diversity of materials, and a large specific surface can be prepared by this technique.

Recently, the electrospinning method has been used to produce polymeric nanofibers for sensing applications [5,30] (Figure 11.8). The fibrous membranes obtained by the electrospinning technique have a strong potential application for sensor systems since the fibrous membranes have a larger surface area than that of continuous films.

Spin coating is an efficient, relatively simple, and low-cost way to produce a thin, uniform polymer coating on a planar substrate [31]. In the spin-coating

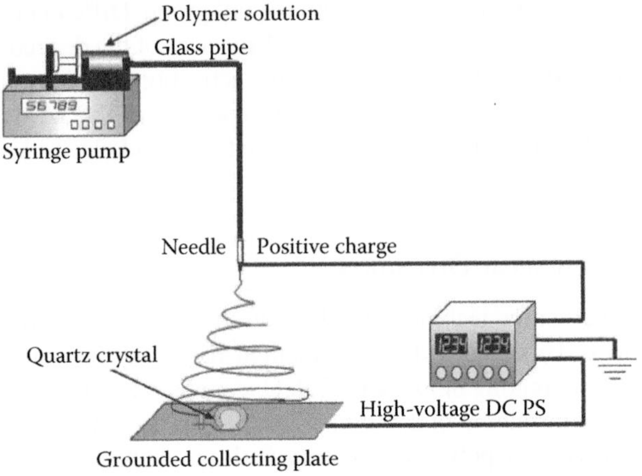

FIGURE 11.7 Schematic illustration of the electrospinning process used to coat quartz crystals with polymer solutions. (From Şeker, Ş. et al., *IEEE Sensors Journal*, 10, 1342–1348, 2010. With permission.)

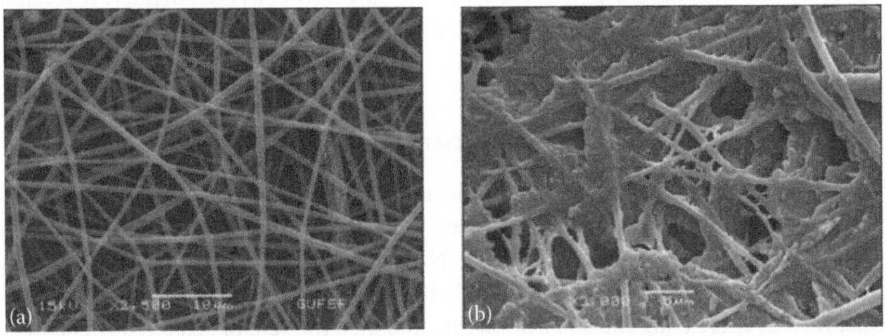

FIGURE 11.8 Representative scanning electron microscope (SEM) images of (a) newly prepared and (b) enzyme (glucose oxidase) immobilized electrospun polymer coatings on coverslips used for the enzymatic oxidation of β-D-glucose. The images demonstrate a non-woven mesh with randomly oriented nanofibers. Scale bars: (a) 10 μm and (b) 5 μm. (From Şeker, Ş. et al., *IEEE Sensors Journal*, 10, 1342–1348, 2010. With permission.)

process, the solution is first deposited onto a substrate, which is then accelerated rapidly to the desired spin speed (Figure 11.9). The centrifugal force spreads the solution, resulting in the formation of a homogeneous film on the substrate surface. The coating thickness depends on the polymer concentration, the solvent, and the spin-coating speed. Using the spin-coating technique, it is possible to reduce the surface roughness of commercially produced quartz, in order to protect the electrode from oxidation [32].

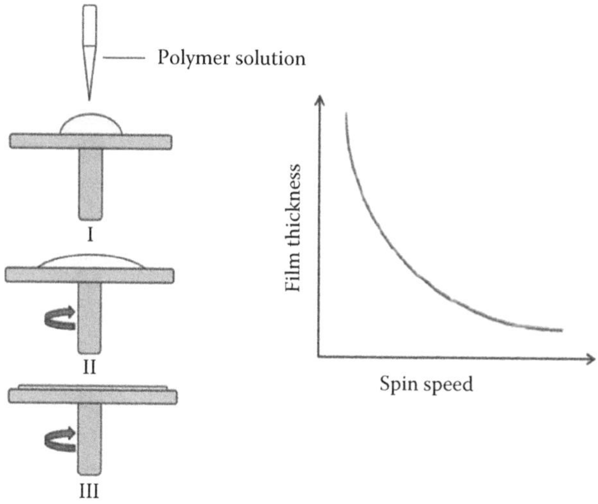

FIGURE 11.9 Schematic illustration of the spin-coating process.

11.5 BIOLOGICAL APPLICATIONS OF QCM

11.5.1 Enzyme Biosensors

Enzyme biosensors based on highly specific enzymatic reactions have been widely exploited (in clinical and food analysis). In general, QCM enzyme biosensors are used for the measurement of the mass deposition of the product molecule from the enzymatic reaction.

Several QCM enzyme biosensors using immobilized urease [33] and glucose oxidase (GOx) [34] have been studied. Wei and Shih [33] have developed a fullerene-cryptand-coated PZ urea sensor by measuring the ammonium ion, a product of the catalytic hydrolysis of urea by urease. They have found that the fullerene C_{60}-cryptand22 PZ crystal detection system exhibited good sensitivity and selectivity for urea with respect to other biological species in aqueous solutions. In the last case, a PZ sensor coated with a nanofibrous layer of poly(lactic-co-glycolic acid) (PLGA) containing saturated fullerene C_{60} was developed to detect gluconic acid, the oxidation product of β-D-glucose by GOx [5] (Figure 11.10). Fullerene-C_{60} containing nanofibrous poly(DL-lactide-co-glycolide) coatings with a thickness of ~625 nm was prepared by electrospinning on PZ quartz crystals. The sensor was able to monitor D-gluconic acid in real time, in the glucose concentration range between 1.4 and 14.0 mM quite linearly.

In another enzyme biosensor system, substrates are immobilized onto PZ crystal surfaces [35–37]. The frequency shift caused by the enzymatic hydrolysis reaction on the crystal surface is determined by the QCM system. For example, Hu et al. (2009) investigated the interactions between the cellulase enzyme and a cellulose substrate using the QCM system. The substrates consisted of thin films of cellulose that were spin-coated onto a polyvinylamine (PVAm) precoated quartz crystal

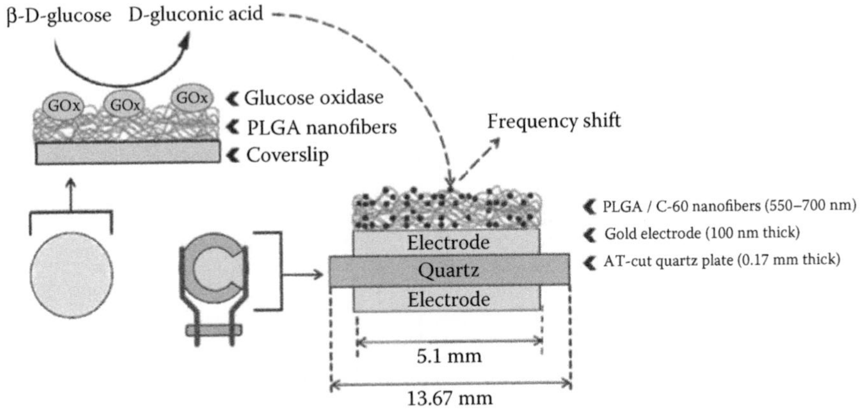

FIGURE 11.10 Schematic illustration of the real-time monitoring of gluconic acid using a QCM. (From Şeker, Ş. et al., *IEEE Sensors Journal*, 10, 1342–1348, 2010. With permission.)

surface. The frequency and dissipation shifts resulted in substrate hydrolysis on the crystal surface at various temperatures and the enzyme concentrations were monitored *in situ* and in real time by the QCM system [35].

In another study, the functional characterization of starch-degrading enzymes (α-amylase and glucoamylase) was performed using the QCM with dissipation monitoring (QCM-D) system. Starch that is produced by plants as a result of photosynthesis was immobilized onto gold-coated quartz surfaces. Two starch-degrading enzymes with different action mechanisms were injected separately into the flow cell. An increase in the oscillation frequency and a decrease in dissipation were observed due to the enzymatic hydrolysis of starch, resulting in a mass decrease on the PZ crystal surface [36].

11.5.2 QCM DNA Biosensors

QCM DNA biosensors (genosensors) are analytical devices consisting of an immobilized sequence-specific probe to detect the complimentary sequence by hybridization reaction. The analysis of specific DNA sequences in clinical, food, and environmental samples facilitates the detection and identification of infectious diseases from biological species or living systems such as viruses and bacteria in real time, without the use of labels such as radioisotopes, enzymes, and fluorophores. This has been presented as an alternative to traditional methods of detecting specific DNA sequences, in which labeled probes are required [6]. The basis of operation for a PZ DNA biosensor is the monitoring of the decrease in oscillation frequency due to a hybridization reaction on the quartz surface. The hybridization reaction is detected following the frequency change, resulting from the interaction between the single-stranded oligonucleotides (probe) immobilized on the quartz crystal and the target complementary strand in the solution (Figure 11.11). Many types of nucleic acid–based biosensors have already found use in several analytical fields, for instance,

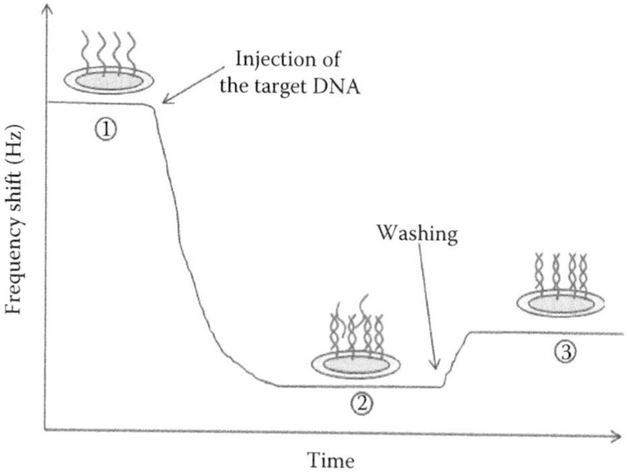

FIGURE 11.11 Frequency shift during a hybridization reaction on the surface of the quartz crystal. (1) DNA probe on the crystal surface. (2) After the injection of the target DNA, hybridization reaction is carried out, which causes a decrease in the frequency. (3) The crystal is washed to remove the unbound oligonucleotide.

gene mutation [38], the detection of genetically modified organisms [22], bacteria [39], and viruses [40], and toxicology studies [41].

The probe is usually a short synthetic oligonucleotide that is immobilized onto the quartz crystal. In general, the probe sequence is 18–25 nucleotides in size; longer capture oligonucleotides often exhibit particularly unfavorable hybridization specificity, yielding to intramolecular hydrogen bonding and the consequent formation of nonreactive hairpin structures [42].

The immobilization of a probe DNA on the quartz crystal surface is a fundamental step in the development of a DNA biosensor since the affinity and specificity of the biosensor can be greatly improved by choosing the proper immobilization procedure. The probe DNA should be attached to the crystal surface without losing its native conformation and activity. There are a variety of immobilization procedures for designing a QCM DNA biosensor to absorb single-stranded DNA. A thiol- or disulfide-labeled DNA probe can be adsorbed chemically onto a gold surface of quartz crystal [43]. An amine-labeled DNA probe can be bound covalently on a gold surface covered with mercaptopropanoic acid by the self-assembly technique [44]. A DNA probe labeled with biotin can be immobilized on the gold surface of a quartz crystal coated with an avidin SAM via the formation of a biotin–avidin complex [45]. The 5′-phosphate residues of the probe strand can be easily modified by thiol [46] or biotin [47]. The immobilization of oligonucleotide probes onto a quartz crystal surface is most commonly achieved using the biotin–avidin interaction. In this method, the probe DNA biotinylated at the 5′ end is immobilized onto streptavidin-coated gold electrodes. The thiolated probe can be directly immobilized onto the gold surface of quartz crystal by SAM formation. The immobilization of thiol-modified probes is easily performed in one step. Additionally, an amine-labeled DNA probe can be immobilized onto a surface modified with a silane derivative [48].

Apart from DNA hybridization studies, DNA biosensors are used for the investigation of DNA interaction with various substances, such as drugs, nucleic acid mimics, and proteins [49]. Pope et al. [41] have shown the interaction of two low molecular weight drugs binding to short duplex sequences in an intercalative (nogalamycin) and groove-binding manner (berenil). They reported that the QCM technique may be exploited to monitor changes in the energy dissipation of the oscillated film except for detecting mass changes.

11.5.3 QCM Immunosensors

Nowadays, QCM immunosensors are the most widely used analytical tool for detecting antibody–antigen reactions. The method is based on the detection of the frequency shift, resulting from the highly specific interaction of the antibody immobilized on the quartz crystal surface and the antigen inside the solution. The QCM immunosensor has many potential applications, ranging from clinical diagnosis and food control to environmental analysis for the detection of bacteria or organic compounds by using an antigen–antibody reaction (Table 11.1). The advantage of the QCM immunosensor is that it allows for the direct measurement of immunointeraction without

TABLE 11.1
Immunosensor Applications of Piezoelectric Quartz Crystal Microbalance

Target	Immobilization Method	Detection Limit
Microorganism Detection		
Escherichia coli [52]	SAM, protein A	1.0×10^3 CFU mL^{-1}
Pseudomonas aeruginosa [53]	SAM via sulfo-LC-SPDP	1.3×10^7–1.3×10^8 cells mL^{-1}
Listeria monocytogenes [4]	SAM	1.0×10^7 cells mL^{-1}
Staphylococcal enterotoxins [54]	PEI, SAM, protein A	2.7–12.1 µg mL^{-1}
Salmonella enteritidis [55]	SAM via MPA	1.0×10^5 cells mL^{-1}
Herpes viruses [56]	Protein A	5.0×10^4 cells mL^{-1}
Vibrio harveyi [57]	SAM	10^3–10^7 CFU mL^{-1}
Mycobacterium tuberculosis [58]	Protein A	8.7×10^4–8.7×10^7 cells mL^{-1}
Protein Detection		
Ferritin [59]	SAM	2.4 nmol L^{-1}
α-Fetoprotein [60]	SAM	1.5 nmol L^{-1}
Albumin [61]	SAM/protein A	1–5 µg mL^{-1}
Human chorionic gonadotropin [62]	SAM via sulfo-LC-SPD	2.5–500 mIU mL^{-1}
CRP [63]	SAM via sulfo-LC-SPD	0.130 ng mL^{-1}
IgG [64]	CMD	46 ng mL^{-1}
COMP [65]	SAM	1–200 ng mL^{-1}

Notes: SAM, self-assembled monolayer; MPA, 3-mercaptopropionic acid; sulfo-LC-SPDP, sulfosuccinimidyl 6-[3-(2-pyridyldithio)propionamido] hexanoate; CRP, C-reactive protein; IgG, immunoglobulin G; CMD, carboxymethyl dextran; COMP, cartilage oligomeric matrix protein.

using any labels or additional chemicals. By means of this interaction, the QCM immunosensors can qualitatively or quantitatively detect an antigen or antibody on quartz crystal, which is caused by a change in the resonant frequency.

Protein A is a widely used component for immobilizing an antibody, such as for immunosensor development. Gao et al. [50] developed a Staphylococcal enterotoxin C_2 (SEC_2) immunosensor, which was coated with different immobilization methods of SEC_2 antibody on a gold electrode of the PZ crystal. In this study, the electrode coated with protein A demonstrated the best result for SEC_2 detection. Covalent immobilization of organic polymer layers is also used for immunosensor development. Polyethylenemine (PEI) cross-linked with glutaraldehyde is the most commonly used polymer in immunosensor development studies. Tsai and Lin [51] showed that the amount and the reaction activity of a bound antibody on PEI film were better than those on SAM. The interaction between thiols and the gold surface of PZ crystals has been used for the immobilization of antibodies. A monolayer of thiol on the gold surface of crystal is facilitated by covalent binding due to the high affinity between the gold and sulfur atoms. Vaughan et al. [4] have developed an immunosensor to detect *Listeria monocytogenes* against an antibody immobilized on an SAM of thiosalicylic acid. They have shown that the sensor could detect *L. monocytogenes* cells in real time in a solution of 1.0×10^7 cells mL^{-1}.

11.5.4 QCM Mammalian Cell Biosensors

11.5.4.1 Monitoring of Cell Adhesion and Spreading to Surfaces

The detection of cell attachment to various substrates is essential for determining mammalian cell behavior toward biomaterials and substances that are of biological importance. Developing medical materials either favoring or preventing cell adhesion may be crucial, according to the desired condition [66]. One of the major shortcomings in biomaterials research and development studies is the lack of suitable tools to detect cell–material interactions under *in vitro* culture conditions in real time [67]. In order to detect cell attachment to biomaterials, a variety of techniques have been used [68], that is, counting of labeled cells directly, measurement of the cellular zones and the cell density using optical techniques, and fluorescence density measurements to analyze cell adhesion strength. Following cell adhesion, the removal of cells from the substrate surface and the staining or fixation stages are regarded as disadvantages of such techniques [66]. The QCM system, however, is a useful technique to detect cell adhesion without removing cells from the surface and without using any labels [69].

PZ biosensors incorporating living cells as the active sensing element provide useful information regarding the properties of cultured cells, such as attachment, proliferation, cell–substrate interaction, and cell–drug interaction under different conditions. The behavior of cells toward materials is important in determining the compatibility of biomaterial surfaces (Figure 11.12).

The QCM mammalian cell biosensor is a less intensely studied area than other QCM biosensor applications due to its added difficulties [70]. Sterile culture conditions are required in these experiments. The cells must be attached to the quartz surface under cell culture conditions (e.g., 5% CO_2/95% air and 37°C) to create QCM

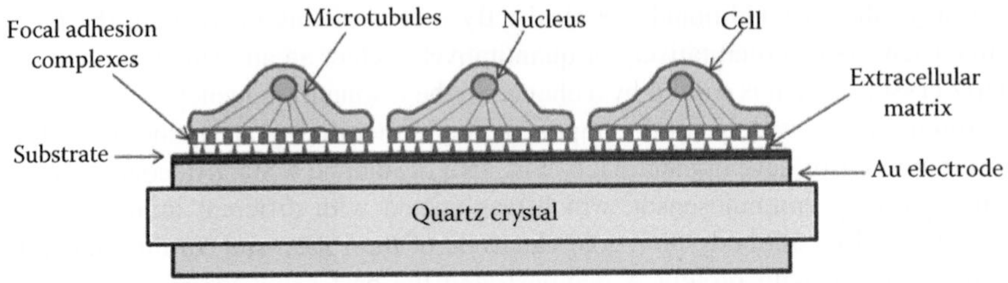

FIGURE 11.12 Schematic representation of the QCM cell biosensor. (Adapted and redrawn from Marx, K.A. et al., *Analytical Biochemistry*, 361, 77–92, 2007.)

cell biosensors. The quartz crystal surface should attain a certain level of hydrophilicity for cell attachment. Therefore, some modifications or coating techniques have been applied to improve the hydrophilic character of the hydrophobic gold surface of quartz crystal. In many studies, the gold surface of quartz crystal is treated chemically to render its hydrophilicity [71].

A number of studies have focused on the process of cellular attachment as this behavior is important for biomaterial development. In QCM cell biosensors, endothelial cells (ECs) [72], osteoblasts [73], MDCK I and II cells [74], 3T3 cells [74], CHO cells [66], Neuro-2A cells [75], McCoy human fibroblast cells [9], MC3T3-E1 cells [69], MCF-7 cells [76], and Vero cells [77] have been used as the cell source. These studies have indicated that mammalian cell attachment and spreading on a quartz surface can be monitored by the QCM sensor. Overall, the results have shown that adherent cells on the surfaces cause a decrease in frequency. However, most studies have suggested that the cell layer formed on the crystal surface does not act as a rigid mass, and in this case, the Sauerbrey equation would not be valid.

In the past decade, QCM-based mammalian cell biosensors have been successfully developed to monitor living cell proliferation and cell attachment to a quartz electrode surface. For instance, Khraiche et al. [75] demonstrated the use of an acoustic sensor to measure the adhesion of neuroblastoma cells (Neuro-2A) on uncoated gold electrodes and on poly-L-lysine (PLL) coatings. They have shown that the acoustic sensor has sufficient sensitivity to monitor neuronal adhesion on the PLL coating in real time [75]. Fohlerová et al. [78] investigated the attachment of rat epithelial cells (WB F344) and lung melanoma cells (B16F10) to a QCM electrode coated with extracellular matrix proteins, vitronectin, and laminin. Their results demonstrated that the QCM cell sensor was suitable for the evaluation of different cell adhesion processes. Redepenning et al. [73] investigated the attachment and spreading of osteoblasts to a quartz crystal surface, by recording the shift of the resonance frequency responding proportionally to the surface cell coverage. Fredriksson et al. [66] have shown that cell adhesion on polystyrene (PS) surfaces was dependent on wettability. Wegener et al. [79] revealed that different mammalian cell types generated individual responses to QCM when in contact with the quartz crystal surface. Marx et al. [80] studied the changes in the cellular viscoelastic properties of cells based on frequency and resistance shifts measured by the QCM cell biosensor with living ECs or human breast cancer cells (MCF-7) on a gold electrode surface. Guillou-Buffello et al. [9] disclosed that PMMA-based bioactive polymers exhibiting either carboxylate or sulfonate functional groups or both, inhibited the adhesion of the McCoy fibroblastic cells when compared with nonfunctionalized PMMA and PMMA-based copolymers, which comprised only one of these functional groups (Table 11.2).

11.5.4.2 Monitoring of Cellular Responses to Soluble Factors

QCM cell biosensors have been used for the monitoring of cellular responses to soluble factors that alter the viscoelastic properties of the attached cells. In these studies, cells are attached to the surface of quartz crystals and are exposed to various soluble factors. By measuring the resonance frequency and resistance shifts of quartz crystals, important biological information, such as cytotoxicity, can be obtained. The cell-based QCM technology approach offers several advantages over other methods currently in use. The QCM cell biosensor does not require cell removal from the surface or a labeling step for monitoring cell response. There have been many studies on real-time detection of cellular responses to various soluble factors (Table 11.3).

TABLE 11.2
QCM Cell Biosensor Studies Investigating Cell Adhesion and Spreading onto Various Surfaces

Cell Type	Surface Type
African green monkey kidney (Vero) cells [77]	Gold surface
Human dermal fibroblasts cells [81]	PCL, PEG/PCL, CS/PCL or HA/CS/PCL–coated surface
Malignant human mammary epithelial cells [82]	Gold surface
Human mammary cancer cells [83]	Rough and mirror-finished gold surface
Mouse fibroblasts cell line [84]	Gold surface, PMSH-coated gold surface
Neonatal rat calvaria osteoblasts [73]	Gold surface
Human breast cancer MCF-7 cells [76]	Polyporphyrin-modified gold surface
Canine MDCK-I, MDCK-II epithelial cells [85]	Gold surface
Canine MDCK-I, MDCK-II epithelial, porcine choroid plexus epithelial, murine Swiss 3T3 fibroblast, and bovine aortic endothelial cells [79]	Gold surface
Human skin fibroblast cells [86]	Gold surface adsorbed fibronectin
Canine MDCK-II, rat epithelial-like NRK-52E, and murine fibroblast NIH-3T3 cells [87]	Gold surface adsorbed gelatin
Human ovarian cancer OV-MZ-6 cells [88]	Gold surface adsorbed fibronectin, vitronectin, laminin
Bovine aortic endothelial cells [89]	Gold surface adsorbed gelatine and poly-D-lysine
Human hepatocellular carcinoma HepG2 cells [90]	Poly(ethyleneimine), poly(sodium-4-styrenesulfonate), poly(allylamine-hydrochloride)
Human bone mesenchymal cells [91]	Hydroxyapatite, gold coated with osteopontin
Murine NIH-3T3 fibroblast cells [67]	Tantalum and oxidized polystyrene coated with albumin, fibronectin, and serum proteins

Notes: PEG, polyethylene glycol; PCL, poly (ε-caprolactone); CS, chitosan; HA, hyaluronic acid; PMSH, (poly(MPC-co-2-(methacryloyloxy)ethylthiol)).

TABLE 11.3
QCM Cell Biosensor Studies Investigating Cellular Responses to Various Soluble Factors

Cell Type	Surface Type	Soluble Factors
DH82 macrophage cell line [92]	Gold surface	Zymosan A, polystyrene beads, SWCNT
Human colorectal cancer cell lines [93]	Polystyrene-coated surface	Lectins
Human hepatoma cell line [94]	Collagen-coated surface	GA, t-BHP
Human endothelial cells, canine macrophages [95]	Gold surface	NaN_3
Human breast cancer cells [96]	Gold surface	Doxorubicin-loaded magnetic silica composite
Bovine aortic endothelial cells [97]	PLL-coated surface	TX-100, bacterial LPS
Rat liver epithelial cells, lung melanoma cells [98]	Polystyrene- or fibronectin-coated surface	α-TAM
Bovine aortic endothelial cells [72]	Gold surface	Fibroblast growth factor
Human breast cancer MCF-7 and MD-MBA-231 cells [8]	Gold surface	Paclitaxel, docetaxel
Bovine aortic endothelial cells [99]	Gold surface	Nocodazole
African green monkey kidney TC7, human ovarian cancer HeLa, human lung carcinoma A549 cells [100]	Gold surface	Cytochalasin D, latrunculin B, jaspakinolide, nocodazole, paclitaxel
Bovine aortic endothelial and human breast cancer MCF-7 and MD-MBA-231 cells [101]	Gold surface	Nocodazole, taxol
Human hepatocellular carcinoma HepG2 cells [102]	Gold surface	Gold nanoparticles and paclitaxel
Human oral epithelial H376 cells [103]	Polystyrene-coated surface	Fluorescent polystyrene microspheres
Human hepatic normal (L-02) and cancer (Bel7402) cells [104]	Gold surface	Concanavalin A, wheat germ agglutinin
Human hepatic cancer Bel7402 cells [105]	Gold surface	Adriamycin, selenium nanoparticles
Human B-lymphoblastoid LG2 cells [106]	Gold surface adsorbed protein G	Cytochalasin D
Human skin fibroblast cells [86]	Gold surface	Cytochalasin D
Human osteoblast-like MG-63 cells [107]	Gold surface	Selenium-ferroferric oxide nanocomposites
Canine MDCK-I, MDCK-II epithelial, porcine choroid plexus epithelial, murine Swiss 3T3 fibroblast, and bovine aortic endothelial cells [79]	Gold surface	Antiadhesive peptides (GRGDS, SDGRG), cytochalasin D

Notes: SWCNT, single-walled carbon nanotubes; GA, glutaraldehyde; t-BHP, t-butylhydroperoxide; NaN_3, sodium azide; TX-100, Triton-X 100; LPS, lipopolysaccharide; PLL, poly-L-lysine; α-TAM, α-tocopherol amidomalate.

QCM mammalian cell biosensors have also been used to investigate microtubule dynamics for drug discovery research. Microtubules are long tubular polymers composed of dimeric subunits consisting of α- and β-tubulin proteins, which form the tubulin heterodimer. In microtubule formation (Figure 11.13), α- and β-tubulin heterodimers polymerize to form a short microtubule nucleus. The second step is the elongation of the microtubule at both ends ([+] or [−] end) to form a cylinder [108].

Microtubules are involved in the process of cell division and mitosis, in cell signaling, in the development and maintenance of cell shape, and in intercellular transport. Their function in mitosis and cell division makes microtubules important agents in cancer chemotherapy [109].

A number of chemically diverse compounds bind to tubulin in the microtubules (Figure 11.14). These compounds with diverse structures disrupt the polymerization

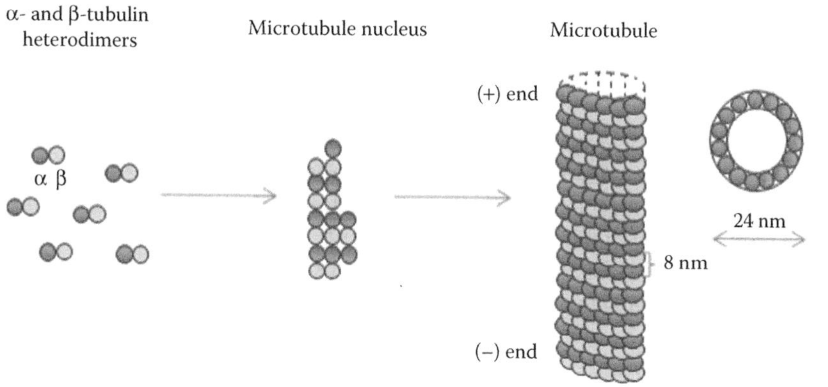

FIGURE 11.13 Polymerization of microtubules. (Adapted and redrawn from Jordan, M.A. and Wilson, L., *Nature Reviews Cancer*, 4, 253–265, 2004.)

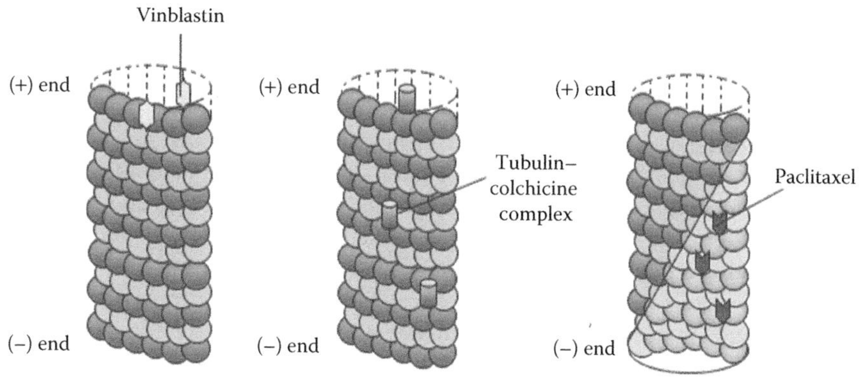

FIGURE 11.14 Antimicrotubule agents binding to microtubules at diverse sites. (Adapted and redrawn from Jordan, M.A. and Wilson, L., *Nature Reviews Cancer*, 4, 253–265, 2004.)

dynamics of the microtubules and cause inhibition of cell proliferation. These events make it possible to discover and develop important anticancer agents for the treatment of cancer in the clinical and preclinical stages. Microtubule-binding antimitotic drugs are classified into two major groups [108]. The first group causes destabilization of the microtubule, inhibiting microtubule polymerization at high drug concentrations, namely, the *Vinca* alkaloids (vinblastine, vincristine, vinorelbine, vindesine, and vinflunine), cryptophycins, halichondrins, estramustine, colchicine, and combretastatins. The second group of compounds is called the microtubule-stabilizing agents; these compounds induce microtubule polymerization and stabilize microtubules at high concentrations. These drugs include paclitaxel, docetaxel, epothilones, discodermolide, eleutherobins, sarcodictyins, laulimalide, and rhazinalam. The antimitotic drugs are able to bind to diverse sites on tubulin and at different positions in the microtubule and show different effects on microtubule dynamics through different chemical mechanisms. For example, vinblastine binds to the microtubule (+) end, suppressing microtubule dynamics. Colchicine binds to tubulin dimers and copolymerizes into the microtubule lattice, suppressing microtubule dynamics. Paclitaxel binds to the interior region of the microtubule, thus suppressing microtubule dynamics [108].

Cancer cells can develop resistance to microtubule-binding compounds following long-term treatment. Acquired resistance to microtubule-binding drugs loses their function on cancer cells. Thus, there is a need to develop more antimitotic drugs that can be used for the treatment of cancer [109,110]; the discovery of novel microtubule-binding drugs is considered invaluable for effective cancer treatment.

In the pharmaceutical industry, drug discovery studies are searching for novel methods to determine pharmaceutical compounds and drugs, using a PZ crystal coated with a cell layer. There are a few studies based on the QCM drug biosensor for detecting the effect of drugs on cells. A detailed evaluation of cell–drug interactions was performed by Marx et al. [99] using bovine aortic ECs. They investigated the behavior of ECs on a QCM biosensor by adding different nocodazole concentrations (in the range of 0.11–15 μM) to cell growth media and determining nocodazole doses based on the frequency and resistance shift effects. Following this study, the effects of taxol and nocodazole on two different cell types, namely, ECs and the metastatic human mammary cancer cell line (MDA-MB-231), were investigated by PZ whole-cell biosensors [101]. In this particular study, the response of cells to varying concentrations of microtubule-binding drugs—taxol and nocodazole—was detected by measuring the changes in the frequency and resistance of the quartz crystal [101]. Braunhut et al. [8] investigated the response of two human breast cancer cell lines (MCF-7 and MDA-MB-231) to docetaxel and paclitaxel drugs. The results accurately predicted that docetaxel was more effective than paclitaxel, and MCF-7 cells were more resistant to MDA-MB-231 cells than taxanes were.

These results indicate that the QCM mammalian cell biosensor can be used for the detection of cytoskeletal alterations in microtubules as well as changes in the shape, attachment, and viscoelastic properties of cells. To sum up, the QCM cell biosensor may be useful in the discovery of new antimitotic drugs that affect cellular attachment in real time.

ACKNOWLEDGMENT

YME acknowledges the support of the Turkish Academy of Sciences, TÜBA (Ankara, Turkey).

REFERENCES

1. Cooper J. M. and A. E. G. Cass. 2004. *Biosensors: A Practical Approach*. Oxford University Press, Oxford.
2. Lucklum R. and P. Hauptmann. 2000. The quartz crystal microbalance: Mass sensitivity, viscoelasticity and acoustic amplification. *Sensors and Actuators B* 70: 30–36.
3. Rocha-Gaso M. I., C. March-Iborra, A. Montoya-Baides, and A. Arnau-Vives. 2009. Surface generated acoustic wave biosensors for the detection of pathogens: A review. *Sensors* 9: 5740–5769.
4. Vaughan R. D., C. K. O'Sullivan, and G. G. Guilbault. 2001. Development of a quartz crystal microbalance (QCM) immunosensor for the detection of *Listeria monocytogenes*. *Enzyme and Microbial Technology* 29: 635–638.
5. Şeker Ş., Y. E. Arslan, and Y. M. Elçin. 2010. Electrospun nanofibrous PLGA/Fullerene-C_{60} coated quartz crystal microbalance for real-time gluconic acid monitoring. *IEEE Sensors Journal* 10: 1342–1348.
6. Minunni M., S. Tombelli, R. Scielzi, I. Mannelli, M. Macsini, and C. Gaudiano. 2003. Detection of β-Thalassemia by a DNA piezoelectric biosensor coupled with polymerase chain reaction. *Analytica Chimica Acta* 481: 55–64.
7. Jia X., L. Tan, Q. Xie, Y. Zhang, and S. Yao. 2008. Quartz crystal microbalance and electrochemical cytosensing on a chitosan/multiwalled carbon nanotubes/Au electrode. *Sensors and Actuators B* 134: 273–280.
8. Braunhut S. J., D. McIntosh, E. Vorotnikova, T. Zhou, and K. A. Marx. 2005. Detection of apoptosis and drug resistance of human breast cancer cells to taxane treatments using quartz crystal microbalance biosensor technology. *Assay and Drug Development Technologies* 3: 77–88.
9. Guillou-Buffello D. L., G. Helary, M. Gindre, G. Pavon-Djavid, P. Laugier, and V. Migonney. 2005. Monitoring cell adhesion processes on bioactive polymers with the quartz crystal resonator technique. *Biomaterials* 26: 4197–4205.
10. Lippmann G. 1881. Principe de conservation de l'électricité. *Annales de Physique et de Chimie*, 5ª Serie 24: 145–178.
11. Arnau A. 2004. *Piezoelectric Transducers and Applications*, 1st edn. Springer Verlag, Berlin.
12. Deakin M. R. and D. A. Buttry. 1989. Electrochemical applications of the quartz crystal microbalance. *Analytical Chemistry* 61: 1147A–1154A.
13. Lu C. and A. W. Czanderna. 1984. *Methods and Phenomena 7: Application of Piezoelectric Quartz Crystal Microbalance*. Elsevier, New York.
14. Janshoff A., H. J. Galla, and C. Steinem. 2000. Piezoelectric mass-sensing devices as biosensors: An alternative to optical biosensors? *Angewandte Chemie International Edition English* 39: 4004–4032.
15. King W. H. 1964. Piezoelectric sorption detector. *Analytical Chemistry* 36: 1735–1739.
16. Guilbault G. G. and J. M. Jordan. 1988. Analytical uses of piezoelectric crystals: A review. *CRC Critical Reviews in Analytical Chemistry* 19: 1–28.
17. Sauerbrey G. 1959. Verwendung von Schwingquarzen zur Wägung dünner Schichten und zur Mikrowägung. *Zeitschrift Physik* 155: 206–212.
18. Kanazawa K. K. and J. G. Gordon. 1985. Frequency of a quartz microbalance in contact with liquid. *Analytical Chemistry* 57: 1770–1771.

19. Nuzzo R. G. and D. L. Allara. 1983. Adsorption of bifunctional organic disulfides on gold surfaces. *Journal of the American Chemical Society* 105: 4481–4483.
20. Nuzzo R. G., B. R. Zegarski and L. H. Dubois. 1987. Fundamental studies of the chemisorption of organosulfur compounds on Au(111). Implications for molecular self-assembly on gold surfaces. *Journal of the American Chemical Society* 109: 733–740.
21. Finklea H. O., S. Avery, M. Lynch, and T. Furtsch. 1987. Blocking oriented monolayers of alkyl mercaptans on gold electrodes. *Langmuir* 3: 409–413.
22. Mannelli I., M. Minunni, S. Tombelli, and M. Mascini. 2003. Quartz crystal microbalance (QCM) affinity biosensor for genetically modified organisms (GMOs) detection. *Biosensors and Bioelectronics* 18: 129–140.
23. Cosnier S. 2003. Biosensors based on electropolymerized films: New trends. *Analytical and Bioanalytical Chemistry* 377: 507–520.
24. Marx K. A., T. Zhou, D. McIntosh, and S. J. Braunhut. 2009. Electropolymerized tyrosine-based thin films: Selective cell binding via peptide recognition to novel electropolymerized biomimetic tyrosine RGDY films. *Analytical Biochemistry* 384: 86–95.
25. Marx K. A. 2007. The quartz crystal microbalance and the electrochemical QCM: Applications to studies of thin polymer films, electron transfer systems, biological macromolecules, biosensors, and cells. *Springer Series on Chemical Sensors and Biosensors* 5: 371–424.
26. Taylor G. I. 1964. Disintegration of water drops in an electric field. *Proceedings of the Royal Society A: Mathematical, Physical and Engineering Sciences* 280: 383–397.
27. Ding B., J. Kima, Y. Miyazaki, and S. Shiratori. 2004. Electrospun nanofibrous membranes coated quartz crystal microbalance as gas sensor for NH_3 detection. *Sensors and Actuators B* 101: 373–380.
28. Ramaseshan R., S. Sundarrajan, and R. Jose. 2007. Nanostructured ceramics by electrospinning. *Journal of Applied Physics* 102(11): 111101–111118.
29. Frey M. W., A. J. Baeumner, D. Li, and P. Kakad. 2009. Electrospun nanofiber-based biosensor assemblies. U.S. Patent 7485591, filed May 4, 2006, and issued February 3, 2009.
30. Ding B., M. Yamazaki, and S. Shiratori. 2005. Electrospun fibrous polyacrylic acid membrane-based gas sensors. *Sensors and Actuators B* 106: 477–483.
31. Hall D. B., P. Underhill, and J. M. Torkelso. 1998. Spin coating of thin and ultrathin polymer films. *Polymer Engineering and Science* 38: 2039–2045.
32. Sakti S. P., S. Rösler, R. Lucklum, P. Hauptmann, F. Bühling, and S. Ansorge. 1999. Thick polystyrene-coated quartz crystal microbalance as a basis of a cost effective immunosensor. *Sensors and Actuators* 76: 98–102.
33. Wei L. F. and J. S. Shih. 2001. Fullerene-cryptand coated piezoelectric crystal urea sensor based on urease. *Analytica Chimica Acta* 437: 77–85.
34. Reddy S. M., J. P. Jones, T. J. Lewis, and P. M. Vadgama. 1998. Development of an oxidase-based glucose sensor using thickness-shear-mode quartz crystals. *Analytica Chimica Acta* 363: 203–213.
35. Hu G., J. A. Heitmann Jr, and O. J. Rojas. 2009. In situ monitoring of cellulase activity by microgravimetry with a quartz crystal microbalance. *Journal of Physical Chemistry B* 113: 14761–14768.
36. Bouchet-Spinelli A., L. Coche-Guérente, S. Armand, F. Lenouvel, P. Labbé, and S. Fort. 2013. Functional characterization of starch-degrading enzymes using quartz crystal microbalance with dissipation monitoring (QCM-D). *Sensors and Actuators B* 176: 1038–1043.
37. Josefsson P., G. Henriksson, and L. Wågberg. 2008. The physical action of cellulases revealed by a quartz crystal microbalance study using ultrathin cellulose films and pure cellulases. *Biomacromolecules* 9: 249–254.

38. Dell'Atti D., S. Tombelli, M. Minunni, and M. Mascini. 2006. Detection of clinically relevant point mutations by a novel piezoelectric biosensor. *Biosensors and Bioelectronics* 21: 1876–1879.
39. Mao X., L. Yang, X. L. Su, and Y. Li. 2006. A nanoparticle amplification based quartz crystal microbalance DNA sensor for detection of *Escherichia coli* O157:H7. *Biosensors and Bioelectronics* 21: 1178–1185.
40. Skládal P., C. dos Santos Riccardi, H. Yamanaka, and P. I. da Costa. 2004. Piezoelectric biosensors for real-time monitoring of hybridization and detection of hepatitis C virus. *Journal of Virological Methods* 117: 145–151.
41. Pope L. H., S. Allen, M. C. Davies, C. J. Roberts, S. J. B. Tendler, and P. M. Williams. 2001. Probing DNA duplex formation and DNA–drug interactions by the quartz microbalance technique. *Langmuir* 17: 8300–8304.
42. Lucarelli F., S. Tombelli, M. Minunni, G. Marrazza, and M. Mascini. 2008. Electrochemical and piezoelectric DNA biosensors for hybridisation detection. *Analytica Chimica Acta* 609: 139–159.
43. Lazerges M., H. Perrot, N. Zeghib, E. Antoine, and C. Compere. 2006. In situ QCM DNA-biosensor probe modification. *Sensors and Actuators B* 120: 329–337.
44. Hong S. R., H. D. Jeong, and S. Hong. 2010. QCM DNA biosensor for the diagnosis of a fish pathogenic virus VHSV. *Talanta* 82: 899–903.
45. Wang D., G. Chen, H. Wang, et al. 2013. A reusable quartz crystal microbalance biosensor for highly specific detection of single-base DNA mutation. *Biosensors and Bioelectronics* 48: 276–280.
46. Tombelli S., M. Minunni, and M. Mascini. 2005. Piezoelectric biosensors: Strategies for coupling nucleic acids to piezoelectric devices. *Methods* 37: 48–56.
47. Tombelli S., M. Mascini, and A. P. F. Turner. 2002. Improved procedures for immobilisation of oligonucleotides on goldcoated piezoelectric quartz crystals. *Biosensors and Bioelectronics* 17: 929–936.
48. Hang T. C. and A. Guiseppi-Elie. 2004. Frequency dependent and surface characterization of DNA immobilization and hybridization. *Biosensors and Bioelectronics* 19: 1537–1548.
49. Papadakis G., A. Tsortos, F. Bender, E. E. Ferapontova, and E. Gizeli. 2012. Direct detection of DNA conformation in hybridization processes. *Analytical Chemistry* 84(4): 1854–1861.
50. Gao Z., F. Chao, Z. Chao, and G. Li. 2000. Detection of staphylococcal enterotoxin C_2 employing a piezoelectric crystal immunosensor. *Sensors and Actuators B* 66: 193–196.
51. Tsai W. C. and I. C. Lin. 2005. Development of a piezoelectric immunosensor for the detection of α-fetoprotein. *Sensors and Actuators B* 106: 455–460.
52. Su X. L. and Y. Li. 2004. Self-assembled monolayer-based piezoelectric immunosensor for rapid detection of *Escherichia coli* O157:H7. *Biosensors and Bioelectronics* 19: 563–574.
53. Kim N., I. S. Park, and D. K. Kim. 2004. Characteristics of a label-free piezoelectric immunosensor detecting *Pseudomonas aeruginosa*. *Sensors and Actuators B* 100: 432–438.
54. Lin H. C. and W. C. Tsai. 2003. Piezoelectric crystal immunosensor for the detection of staphylococcal enterotoxin B. *Biosensors and Bioelectronics* 18: 1479–1483.
55. Si S. H., X. Li, Y. S. Fung, and D. R. Zhu. 2001. Rapid detection of *Salmonella enteritidis* by piezoelectric immunosensor. *Microchemical Journal* 68: 21–27.
56. Konig B. and M. Gratzel. 1994. A novel immunosensor for herpes viruses. *Analytical Chemistry* 66: 341–344.
57. Buchatip S., C. Ananthanawat, P. Sithigorngul, P. Sangvanich, S. Rengpipat, and V. P. Hoven. 2010. Detection of the shrimp pathogenic bacteria, *Vibrio harveyi*, by a quartz crystal microbalance-specific antibody based sensor. *Sensors and Actuators B* 145(1): 259–264.

58. Hiatt L. A. and D. E. Cliffel. 2012. Real-time recognition of *Mycobacterium tuberculosis* and lipoarabinomannan using the quartz crystal microbalance. *Sensors and Actuators B* 174: 245–252.
59. Chou S. F., W. L. Hsu, J. M. Hwang, and C. Y. Chen. 2002. Development of an immunosensor for human ferritin, a nonspecific tumor marker, based on a quartz crystal microbalance. *Analytica Chimica Acta* 453: 181–189.
60. Chou S. F., W. L. Hsu, J. M. Hwang, and C. Y. Chen. 2002. Determination of α-fetoprotein in human serum by a quartz crystal microbalance based immunosensor. *Clinical Chemistry* 48: 913–918.
61. Navrátilová I., P. Skládal, and V. Viklický. 2001. Development of piezoelectric immunosensors for measurement of albuminuria. *Talanta* 55: 831–839.
62. Zhang B., Q. Mao, X. Zhang, et al. 2004. Novel piezoelectric quartz micro-array immunosensor based on self-assembled monolayer for determination of human chorionic gonadotropin. *Biosensors and Bioelectronics* 19: 711–720.
63. Kim N., D. K. Kim, and Y. J. Cho. 2009. Development of indirect-competitive quartz crystal microbalance immunosensor for C-reactive protein. *Sensors and Actuators B* 143: 444–448.
64. Crosson C. and C. Rossi. 2013. Quartz crystal microbalance immunosensor for the quantification of immunoglobulin G in bovine milk. *Biosensors and Bioelectronics* 42: 453–459.
65. Wang S. H., C. Y. Shen, T. C. Weng, et al. 2010. Detection of cartilage oligomeric matrix protein using a quartz crystal microbalance. *Sensors* 10: 11633–11643.
66. Fredriksson C., S. Khilman, B. Kasemo, and D. M. Steel. 1998. In vitro real-time characterization of cell attachment and spreading. *Journal of Materials Science: Materials in Medicine* 9: 785–788.
67. Lord M. S., C. Modin, M. Foss, et al. 2006. Monitoring cell adhesion on tantalum and oxidised polystyrene using a quartz crystal microbalance with dissipation. *Biomaterials* 27: 4529–4537.
68. Anselme K. 2000. Osteoblast adhesion on biomaterials. *Biomaterials* 21: 667–681.
69. Modin C., A. L. Stranne, M. Foss, et al. 2006. QCM-D studies of attachment and differential spreading of pre-osteoblastic cells on Ta and Cr surfaces. *Biomaterials* 27: 1346–1354.
70. Marx K. A. 2003. Quartz crystal microbalance: A useful tool for studying thin polymer films and complex biomolecular systems at the solution-surface interface. *Biomacromolecules* 4(5): 1099–1120.
71. Marx K. A., T. Zhou, M. Warren, and S. J. Braunhut. 2003. Quartz crystal microbalance study of endothelial cell number dependent differences in initial adhesion and steady-state behavior: Evidence for cell–cell cooperativity in initial adhesion and spreading. *Biotechnology Progress* 19: 987–999.
72. Zhou T., K. A. Marx, M. Warren, H. Schulze, and S. J. Braunhut. 2000. The quartz crystal microbalance as a continuous monitoring tool for the study of endothelial cell surface attachment and growth. *Biotechnology Progress* 16: 268–277.
73. Redepenning J., T. K. Schlesinger, E. J. Mechalke, D. A. Puleo, and R. Bizios. 1993. Osteoblast attachment monitored with a quartz crystal microbalance. *Analytical Chemistry* 65: 3378–3381.
74. Wegener J., A. Janshoff, and H. J. Galla. 1998. Cell adhesion monitoring using a quartz crystal microbalance: Comparative analysis of different mammalian cell lines. *European Biophysics Journal* 28: 26–37.
75. Khraiche M. L., A. Zhou, and J. Muthuswamy. 2005. Acoustic sensor for monitoring adhesion of Neuro-2A cells in real-time. *Journal of Neuroscience Methods* 144: 1–10.

76. Guo M., J. Chen, Y. Zhang, K. Chen, C. Pan, and S. Yao. 2008. Enhanced adhesion/spreading and proliferation of mammalian cells on electropolymerized porphyrin film for biosensing applications. *Biosensors and Bioelectronics* 23: 865–871.
77. Gryte D. M., M. D. Ward, and W. S. Hu. 1993. Real-time measurement of anchorage-dependent cell adhesion using a quartz crystal microbalance. *Biotechnology Progress* 9: 105–108.
78. Fohlerová Z., P. Skládal, and J. Turánek. 2007. Adhesion of eukaryotic cell lines on the gold surface modified with extracellular matrix proteins monitored by the piezoelectric sensor. *Biosensors and Bioelectronics* 22: 1896–1901.
79. Wegener J., J. Seebach, A. Janshoff, and H. J. Galla. 2000. Analysis of the composite response of shear wave resonators to the attachment of mammalian cells. *Biophysical Journal* 78: 2821–2833.
80. Marx K. A., T. Zhou, A. Montrone, D. McIntosh, and S. J. Braunhut. 2005. Quartz crystal microbalance biosensor study of endothelial cells and their extracellular matrix following cell removal: Evidence for transient cellular stress and viscoelastic changes during detachment and the elastic behavior of the pure matrix. *Analytical Biochemistry* 343: 23–34.
81. Chung T. W., Y. C. Tyan, R. H. Lee, and C. W. Ho. 2012. Determining early adhesion of cells on polysaccharides/PCL surfaces by a quartz crystal microbalance. *Journal of Materials Science: Materials in Medicine* 23: 3067–3073.
82. Zhou T., K. A. Marx, A. H. Dewilde, D. McIntosh, and S. J. Braunhut. 2012. Dynamic cell adhesion and viscoelastic signatures distinguish normal from malignant human mammary cells using quartz crystal microbalance. *Analytical Biochemistry* 421: 164–171.
83. Tan L., Q. Xie, X. Jia, M. Guo, Y. Zhang, H. Tang, and S. Yao. 2009. Dynamic measurement of the surface stress induced by the attachment and growth of cells on Au electrode with a quartz crystal microbalance. *Biosensors and Bioelectronics* 24: 1603–1609.
84. Watarai E., R. Matsuno, T. Konno, K. Ishihara, and M. Takai. 2012. QCM-D analysis of material–cell interactions targeting a single cell during initial cell attachment. *Sensors and Actuators B* 171–172: 1297–1302.
85. Janshoff A., J. Wegener, M. Sieber, and H. J. Galla. 1996. Double-mode impedance analysis of epithelial cell monolayers cultured on shear wave resonators. *European Biophysics Journal* 25(2): 93–103.
86. Li F., J. H. Wang, and Q. M. Wang. 2007. Monitoring cell adhesion by using thickness shear mode acoustic wave sensors. *Biosensors and Bioelectronics* 23(1): 42–50.
87. Heitmann V. and J. Wegener. 2007. Monitoring cell adhesion by piezoresonators: Impact of increasing oscillation amplitudes. *Analytical Chemistry* 79(9): 3392–3400.
88. Li J., C. Thielemann, U. Reuning, and D. Johannsmann. 2005. Monitoring of integrin-mediated adhesion of human ovarian cancer cells to model protein surfaces by quartz crystal resonators: Evaluation in the impedance analysis mode. *Biosensors and Bioelectronics* 20(7): 1333–1340.
89. Hong S., E. Ergezen, R. Lec, and K. A. Barbee. 2006. Real-time analysis of cell–surface adhesive interactions using thickness shear mode resonator. *Biomaterials* 27(34): 5813–5820.
90. Saravia V. and J. L. Toca-Herrera. 2009. Substrate influence on cell shape and cell mechanics: HepG2 cells spread on positively charged surfaces. *Microscopy Research and Technique* 72(12): 957–964.
91. Jensen T., A. Dolatshahi-Pirouz, M. Foss, et al. 2010. Interaction of human mesenchymal stem cells with osteopontin coated hydroxyapatite surfaces. *Colloid Surface B* 75(1): 186–193.

92. Wang G., A. H. Dewilde, J. Zhang, et al. 2011. A living cell quartz crystal microbalance biosensor for continuous monitoring of cytotoxic responses of macrophages to single-walled carbon nanotubes. *Particle and Fibre Toxicology* 8(4): 1–17.
93. Peiris D., A. Markiv, G. P. Curley, and M. V. Dwek. 2012. A novel approach to determining the affinity of protein–carbohydrate interactions employing adherent cancer cells grown on a biosensor surface. *Biosensors and Bioelectronics* 35: 160–166.
94. Kang H. W. and H. Muramatsu. 2009. Monitoring of cultured cell activity by the quartz crystal and the micro CCD camera under chemical stressors. *Biosensors and Bioelectronics* 24: 1318–1323.
95. Dewilde H., G. Wang, J. Zhang, K. A. Marx, J. M. Therrien, and S. J. Braunhut. 2013. Quartz crystal microbalance measurements of mitochondrial depolarization predicting chemically induced toxicity of vascular cells and macrophages. *Analytical Biochemistry* 439: 50–61.
96. Zhou Y., Y. Zeng, S. Huang, et al. 2012. Quartz crystal microbalance monitoring of intervention of doxorubicin-loaded core–shell magnetic silica nanospheres on human breast cancer cells (MCF-7). *Sensors and Actuators B* 173: 433–440.
97. Fatisson J., F. Azari, and N. Tufenkji. 2011. Real-time QCM-D monitoring of cellular responses to different cytomorphic agents. *Biosensors and Bioelectronics* 26: 3207–3212.
98. Fohlerová Z., J. Turánek, and P. Skládal. 2012. The cell adhesion and cytotoxicity effects of the derivate of vitamin E compared for two cell lines using a piezoelectric biosensor. *Sensors and Actuators B* 174: 153–157.
99. Marx K. A., T. Zhou, A. Montrone, H. Schulze, and S. J. Braunhut. 2001. Quartz crystal microbalance cell biosensor: Detection of microtubule alterations in living cells at nM nocodazole concentrations. *Biosensors and Bioelectronics* 16: 773–782.
100. Galli M. C., M. Collaud Coen, T. Greber, U. F. Greber, and L. Schlapbach. 2003. Cell spreading on quartz crystal microbalance elicits positive frequency shifts indicative of viscosity changes. *Analytical and Bioanalytical Chemistry* 377(3): 578–586.
101. Marx K. A., T. Zhou, A. Montrone, D. McIntosh, and S. J. Braunhut. 2007. Comparative study of the cytoskeleton binding drugs nocodazole and taxol with a mammalian cell quartz crystal microbalance biosensor: Different dynamic responses and energy dissipation effects. *Analytical Biochemistry* 361: 77–92.
102. Wei X. L., Z. H. Mo, B. Li, and J. M. Wei. 2007. Disruption of HepG2 cell adhesion by gold nanoparticle and Paclitaxel disclosed by in situ QCM measurement. *Colloid Surface B* 59(1): 100–104.
103. Elsom J., M. I. Lethem, G. D. Rees, and A. C. Hunter. 2008. Novel quartz crystal microbalance based biosensor for detection of oral epithelial cell-microparticle interaction in real-time. *Biosensors and Bioelectronics* 23(8): 1259–1265.
104. Tan L., X. Jia, X. F. Jiang, et al. 2008. Real-time monitoring of the cell agglutination process with a quartz crystal microbalance. *Analytical Biochemistry* 383(1): 130–136.
105. Tan L., X. Jia, X. Jiang, et al. 2009. In vitro study on the individual and synergistic cytotoxicity of adriamycin and selenium nanoparticles against Bel7402 cells with a quartz crystal microbalance. *Biosensors and Bioelectronics* 24: 2268–2272.
106. Saitakis M., A. Tsortos, and E. Gizeli. 2010. Probing the interaction of a membrane receptor with a surface-attached ligand using whole cells on acoustic biosensors. *Biosensors and Bioelectronics* 25(7): 1688–1693.
107. Zhou Y. P., X. E. Jia, L. Tan, Q. J. Xie, L. H. Lei, and S. Z. Yao. 2010. Magnetically enhanced cytotoxicity of paramagnetic seleniumferroferric oxide nanocomposites on human osteoblast-like MG-63 cells. *Biosensors and Bioelectronics* 25(5): 1116–1121.
108. Jordan M. A. and L. Wilson. 2004. Microtubules as a target for anticancer drugs. *Nature Reviews Cancer* 4: 253–265.

109. Kavallaris M. 2010. Microtubules and resistance to tubulin-binding agents. *Nature Reviews Cancer* 10: 194–204.
110. Drukman S. and M. Kavallaris. 2002. Microtubule alterations and resistance to tubulin-binding agents. *International Journal of Oncology* 21: 621–628.
111. Castner D. G. and B. D. Ratner. 2002. Biomedical surface science: Foundations to frontiers. *Surface Science* 500: 28–60.

Part 3

Biological Applications

12 Monitoring, Controlling, and Improving Engineered Tissues

Nanoscale Technologies and Devices for Tissue Engineering

Irina Pascu, Hayriye Ozcelik, Albana Ndreu-Halili, Yurong Liu, and Nihal Engin Vrana

CONTENTS

12.1 INTRODUCTION

Tissue engineering aims to produce functional artificial tissues, which can be used as either substitutes or models for drug tests and disease studies. The initial paradigm was to develop three-dimensional (3-D) structures from preferably biodegradable materials, and seed these structures with the prevalent cells of the target organ. These conventional engineered tissues generally have two main parts: the cells and the scaffold [1]. The cells provide the biological functionality while the scaffold provides a platform for cell growth. Over time, cells are able to remodel the initial scaffold into the extracellular matrix (ECM) structure of the target tissue. To this end, under the flow stress/strain conditions relevant to the target organ, the construct is matured *in vitro*. When the structure is deemed ready, it is implanted into the target site for integration [2]. The maturation period can be as short as days to months, depending on the technique utilized.

This methodology was reasonably successful in some tissues, resulting in clinically usable artificial tissues, thus opening the way for further investigation. Over the past few decades, implantable tissues have been created and are already in use in humans or are undergoing clinical trials (e.g., skin, cartilage, blood vessels, heart valves, and bladder) [3,4]. However, for more complex tissues and whole organs, their functionality depends on many parameters. Especially for metabolically active tissues, the correct cell phenotype is crucial [5]. In other tissues, some hard-to-attain physical properties are necessary for proper tissue function, such as long-term transparency in the cornea. In other cases, the synchronization of cells, such as in the cardiac tissue, or the large-scale control over cell–cell contacts, such as in the case of nerve tissue, are required. Also, the availability of cells is a general problem, especially considering the fact that several cell types, such as cardiomyocytes and neurons, have no or very limited regenerative and proliferative capacity [6].

The problems encountered with cell sources have been partially overcome by the availability of stem cell sources. Pluripotent stem cells, with their ability to differentiate many tissue types and to proliferate rapidly, provide a venue to obtain the large amount of cells that is necessary for macroscale tissue engineering applications. However, when stem cells are used, the directed differentiation of cell populations into different cell types simultaneously in microenvironments is a necessity for the development of more complex organs and tissues [7]. Also, the addition of the differentiation step into the production line of tissue engineering brought in the requirement for closer monitoring of the structure as tumor formation and cell dedifferentiation had become real concerns. Moreover, the large-scale processing of stem cells under reproducible conditions is not a trivial problem [8]. Regardless, there have already been clinical applications of stem cell–based engineered tissues, such as in the case of the bladder and trachea [9].

Tissue formation is a complex process. During development, tissue morphogenesis is controlled by a combination of soluble factors, cell–cell interactions, and cell–ECM interactions. This implies that cell–surface interactions occurring at the interface between man-made substrates and tissue would have a substantial influence on cell function. Nanotechnology allows for the construction of devices that interact at the subcellular level and provides a high level of control over cell behavior.

In order to control cell behavior, the design of tissue engineering systems has become increasingly complex [10]. Yet, many fundamental problems remain to be solved such as (i) the production of tissues at macroscales (at cubic centimeter scale and higher); (ii) the control of the host response; (iii) the incorporation of several cell types with correct phenotypes and physiologically relevant heterotypic interactions; and (iv) vascularization. Most of these problems cannot be solved just with improved scaffold designs and would require the incorporation of other functionalities into the engineered tissues. The initial complexity of the engineered tissue is not enough for its correct remodeling *in vivo* as the remodeling process has many components with different effects on the scaffold, which can negate the projected effects of the designed features (such as excessive inflammation).

One of the biggest challenges in tissue engineering is the management and control of the *in vivo* healing period. The design of tissue engineering scaffolds and products, generally optimized for the properties of the target tissue, are not well suited to deal with the initial immune reaction to implantation. In the absence of stimuli response in the design of the engineered tissue, it is not possible to control the switch between inflammation and healing. This is a commonly encountered problem in drug release too [11], and miniaturization can only be done to a certain limit as small structures also have small volumes, which diminishes their ability to carry relevant volumes of bioactive agents. It is possible to control the engineered tissue microenvironment using biochemical agents such as growth factors and cytokines, but these molecules have short half-lives even in delivery systems. Nanoparticle loading of tissue-engineered (TE) scaffolds for the sequential delivery of growth factors has been a well-established method [12]; however, in reality, an even bigger number of growth factors need to be delivered in an orchestrated manner. For this, micro/nanotechnologies that integrate engineered tissues with controlled delivery systems are necessary. Given the volume limitations, this can be done either with devices with considerable volume or separate miniaturized systems that can be accurately controlled by external signals, such as photoactivation, ultrasound, or high-frequency radio waves, or with direct telemetric control via on and off switches. By incorporating stimuli-responsive microvalves and micropumps, delivery systems with larger loads can be connected to the engineered tissue to obtain "on-demand" delivery *in vivo*. Another possibility is to use cell chambers connected to the engineered tissue to deliver the necessary factors. Nanoparticles such as quantum dots (QDs) can be utilized for both cell tracking and delivery. Nanowires and carbon nanotubes (CNTs) would render the scaffold responsive to electrical stimuli, which is important for the development of cardiac, musculoskeletal, and nerve tissue [13]. All these methods require the implementation of nanoscale manufacturing techniques into the scaffold design.

Nanoscale manufacturing is a multidisciplinary area that involves mechanics, electrical engineering, physics, chemistry, biology, and biomedical engineering. Nanoscale technologies are emerging as important and powerful tools for tissue engineering and drug discovery purposes. The nanofabrication processes can be divided into two well-known approaches: "top-down" and "bottom-up." Bottom-up production methods also provide the basis of modular tissue engineering approaches as there is a strong biological basis for them since many tissues are comprised of repeating functional units (e.g., the lobules in the liver) [14].

A number of these nanotechnologies can be used to produce and manufacture biomimetic scaffolds with varying morphologies and properties [15]. Also, nanoscale manufacturing offers exciting possibilities to address the limitations of tissue engineering (e.g., the inability to create vascularized tissue structures [16], inappropriate mechanical strength, the requirement for large sample volumes, and the lack of suitable and functional cell that can overcome the immunological response of the host). Further, nanotechnologies can be used to control the microenvironment (e.g., cell–cell, cell–matrix, and cell–soluble factor interactions) in a reproducible way and with improved temporal and spatial resolution when compared with macro- and microscale technologies [17].

This chapter will give an overview of the utilization of nanotechnologies in the development of engineered tissues. First, the incorporation of nanoscale features into TE scaffolds for improved functionality will be described. Then, the nanoscale techniques to measure cell properties at the single-cell level and modulate cell response at the single-cell level will be covered. The utilization of nanoscale materials for *in vivo* and intracellular visualization will be explained, and, finally, developments in "smart scaffolds" with the implementation of nanoelectronics and biosensors will be reviewed to explain their possible applications in tissue engineering.

12.2 INCORPORATION OF NANOSCALE FEATURES

Native biological tissue is essentially composed of cells and ECMs. Cells receive signals from the surrounding environment and express the appropriate genes to maintain tissue functions. The ECM acts as a supporting scaffold for cells and also as a reservoir to store cell-responsive cues, such as chemical and topographical cues. Some features of both cells and ECMs are at the nanoscale (Figure 12.1).

While it is well known that scaffold features at the scales of individual cells (1–100 μm) can influence cell growth and function, when designing a scaffold/artificial tissue all size scales need to be considered. Scaffolds must provide a full tissue-specific ECM, the substratum in which cells live *in vivo*, and a microenvironment to maintain and regulate cell behavior and function [17], as the ECM plays a vital role in storing, activating, and releasing bioactive agents such as growth factors, while helping the intercellular and cell-soluble biochemicals' interactions [18]. The ECM has considerable topographic detail down to the nanometer scale (e.g., the 66 nm repeat banding of collagen fibers) [19]. Cells' interior structures such as actins are 5–9 nm in diameter and microtubes are 25 nm in diameter. As for the ECM elements, collagen has 1.5 nm triple helices in its structure, which form 10–300 nm fibers, whereas fibronectin has 10–1000 nm fibrils with 2–3 nm arms. As a result, cellular contact with the ECM is also at the nanoscale (focal contacts are generally ~1.5 μm × 250 nm).

Cells are in contact with the ECM forming adhesions to the fibers via integrins, which are transmembrane proteins that interact with specific amino acid sequences found within the macromolecules that make up the ECM [20,21]. The texture, structural organization, and content of an ECM at different length scales significantly

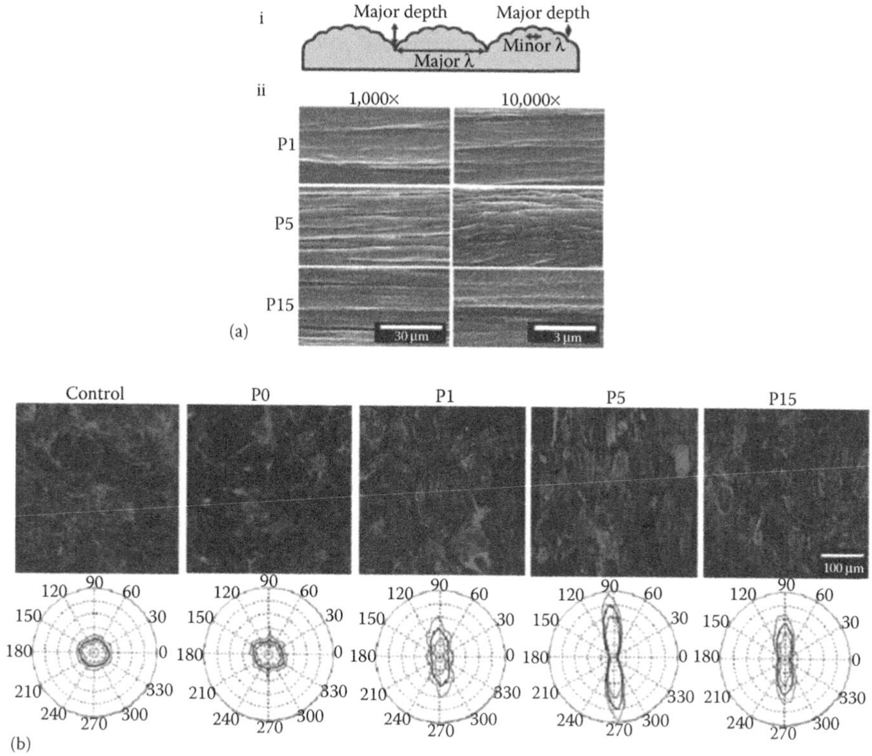

FIGURE 12.1 Application of nanoscale wrinkles on fiber surfaces significantly changes the cell behavior. (a) Surface wrinkles ranged from 150 nm to 1 μm. (b) The presence of wrinkles causes preferential orientation of the cells and the effect can be applied at centimeter scales, which is hard to achieve with conventional patterning methods. (From Chen, A., Lieu, D.K., Freschauf, L., Lew, V., Sharma, H., Wang, J., Nguyen, D. et al., *Advanced Materials*, 23, 5785–5791, 2011.)

affect the cellular response to scaffolds; therefore utilization of contact guidance is an important part of tissue engineering efforts, especially for organs with a high level of anisotropic organization. However, it is generally difficult to achieve such hierarchical organization at larger length scales, even though methods such as the introduction of surface wrinkles can provide nanoscale-aligned topographies at centimeter scales [22] (Figure 12.1).

Individual cells also have a hierarchical structure. Typically, cells are tens of microns in diameter/length. When cells migrate, they extend pseudopodia (filipodia) from the main body of the cell. Important observations have shown the existence of pseudopodia on the order of 500 nm extending from the body of a cell to explore and probe the surrounding environment [23]. This exploration by the cells appears to happen randomly, with pseudopodia sweeping the surface until a point of stable contact is made. Cell movement then occurs in the direction of the contact through retraction of the attached pseudopodium [24]. This apparent exploration of the surrounding environment by cells is the key to the development of nano-featured scaffolds and drug delivery devices that provide information to the cells as they migrate across its surface. By providing the necessary topographical cues, cells and also biological fluids can be controlled in the microenvironment of the scaffold, thus enabling tissue engineers to exert more control over the scaffold remodeling. This can be achieved by the development of biomimetic nanostructures that are physically very close to the topographical features of the natural ECM (Figure 12.2). Some nanotechnology methods, such as electrospinning, self-assembly, and phase separation, have been developed to produce nanostructures within TE products, such as nanofibers and nanoparticles.

12.2.1 Self-Assembly as a Tool in Scaffold Production

One of the possible methods to achieve controlled biomimicry is the utilization of nature's building blocks, such as amino acids and DNA, to develop synthetic materials with precisely controlled physical properties and self-assembly capacities. Self-assembly is a bottom-up technique that mimics the process of forming important biomacromolecules in native tissue. The 3-D conformations of proteins and nucleic acids are among the well-known nanolevel structures built by the self-assembly approach in our body. Although it is very challenging to reconstitute the complex biological structures of these molecules, considerable advances have been made to mimic some functional substructures. Artificial proteins can be used (a) to study the *in vivo* folding and catalytic properties of polypeptides; (b) as sources of specific amino acids for patients with genetic defects in their amino acid metabolism; and (c) to produce proteins containing repeated amino acids, which are sources of useful oligopeptides. Engineered polypeptides have emerged as attractive materials to construct artificial ECMs in tissue engineering [25]. These materials offer advantages over conventional synthetic materials in mimicking the essential characteristics of complex and dynamic native ECMs. The structures and functions of these proteins

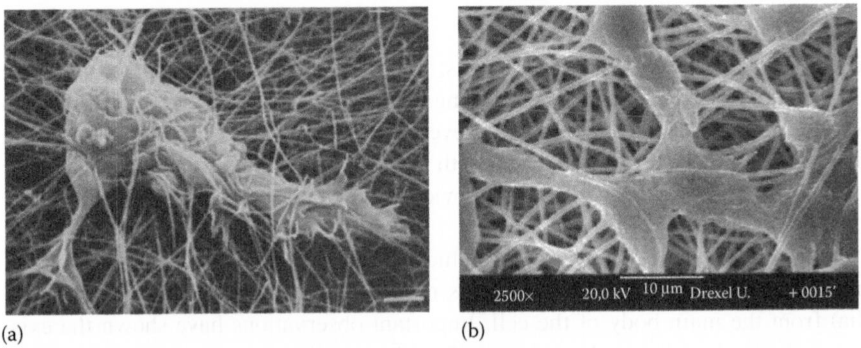

FIGURE 12.2 Biomimetic microenvironment creation by nanobiomaterials: (a) cells in natural collagen matrix (From Heath, J.P. and Peachey, L.D., *Cell Motility and the Cytoskeleton*, 14, 382–392, 1989.); (b) cells in electrospun fibers. (From Li, W.J., Laurencin, C.T., Caterson, E.J., Tuan, R.S., and Ko, F.K., *Journal of Biomedical Materials Research*, 60, 613–621, 2002.)

and their domains, as well as those of novel designed polypeptide domains with self-assembly and molecular recognition abilities, can be combined in engineered polypeptides in a modular manner to yield multifunctional, bioactive materials to mimic native ECMs and optimize tissue engineering outcomes. Engineered polypeptides can be synthesized both chemically and biosynthetically [15].

The self-assembly approach generally employs weak interactions, such as ionic bonds, hydrogen bonds, and van der Waals' interactions, to link molecules [15]. These interactions are generally not as stable as covalent bonding to support the mechanical strength of TE products; however, on the other hand, they allow the self-assembled structures to respond quickly to surrounding conditions, such as the pH or temperature, which makes these self-assembling polypeptides promising candidates as vehicles for the addition of drug delivery capacity and local functionalities to tissue engineering scaffolds. For example, using pH change–induced self-assembly, the development of supramolecular nanofibers made of a peptide-amphiphile (PA) with the cell ligand RGD (Arg-Gly-Asp) was reported. The RGD-modified PA showed an increase in alkaline phosphatase activity and osteocalcin content by osteoblasts [26]. Osteogenic helical rosette nanotubes are other novel nanomaterials whose development was reported recently. They were produced by the self-assembly of DNA base pairs (guanine and cytosine) in aqueous solutions. Their novelty consists in their tailored amino acid side chains (RGD and lysine—promoters of osteoblast adhesion and fibroblast adhesion inhibitors) and their ability to provide a mineralization template for biomimetic nanotube/HA structures [27].

Protein microarray technologies are beginning to advance the field of proteomics by providing miniaturized platforms to probe the interactions and functions of proteins. Despite the growing success of protein microarrays, the central challenge remains to develop simple and general techniques to immobilize functional proteins onto solid supports. Researchers have been able to create an artificial polypeptide scaffold that can be used to immobilize recombinant proteins on substrates. The polypeptide contains separate surface anchor and protein capture domains, and it uses an artificial amino acid to covalently cross-link the polypeptide to surfaces [28]. The chemical synthesis of DNA provides a means for constructing or modifying natural genes. Researchers have been able to present a study where they have expressed a gene encoding an artificial polypeptide enriched with the essential amino acids, tryptophan and lysine. Their results show that a chemically synthesized DNA might be a useful means for producing a novel protein while further studies are necessary to increase the production level [29].

12.2.2 Bioprinting

Another level of self-assembly that is relevant to tissue engineering is the self-assembly of different cell types in tissues, especially during the developmental stages. One approach used to engineer tissues is controlling the cellular self-assembly process by restricting the geometry of cell aggregation [30]. In this way, engineers can create modular tissues with a controlled microarchitecture. This control can be obtained in many ways, such as seeding cells in microwells [31] or microchannels [6], micromolding cells in hydrogels [32,33], or culturing cells in sheets [33]. Researchers have been using micromolds as templates and have been able to create large, intricate modular structures such as linear channels to form more tissue-like structures [14]. While the technique offers the possibility of creating modular tissues of a specific shape and geometry, its limitations are evidenced when some cell types are unable to produce sufficient ECM, migrate, or form cell–cell junctions.

The tissue liquidity concept, in which embryonic tissues are treated as viscous, incompressible liquids with a specific viscosity (γ) and interfacial tension properties (η), provides a means to understand the self-assembly ability of embryonic cells due to their differential adhesion capacities [34]. This property has been used to bioprint complex structures based on cell beads that can self-assemble via merging *in vitro* (Figure 12.3).

Organ printing creates two-dimensional (2-D) arrays of cells and ECM/polymer, which can be built layer by layer into tissues. The advantage of such a technique is the high level of control over the cell and ECM placement and alignment

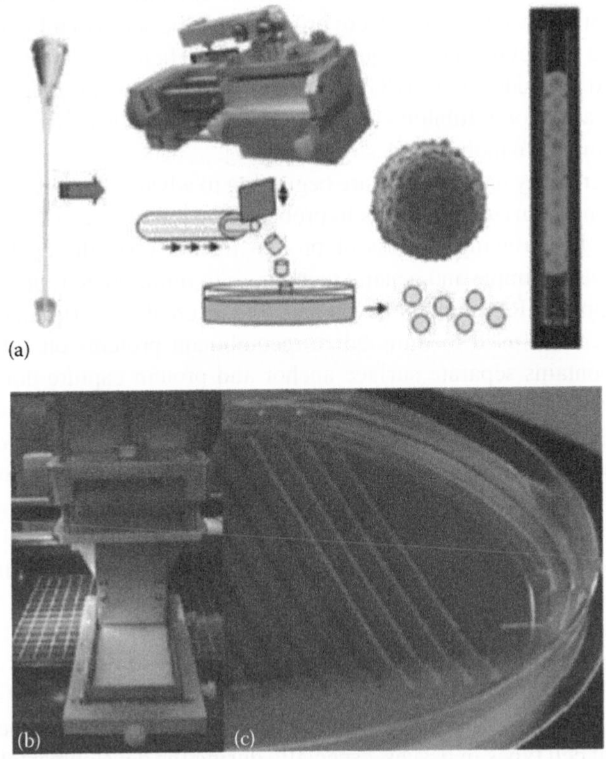

FIGURE 12.3 Control of the self-assembly of cell aggregates (a) using a "biopaper," which is nondegradable (agarose) by cellular activities. This leads to higher control of the aggregation process and highly adjustable artificial tissue shapes that can be printed (b,c). (From Karoly, J., Cyrille, N., Francoise, M., Keith, M., Gordana, V.-N., and Gabor, F., *Biofabrication*, 2, 022001, 2010.)

to create engineered tissues with a wide array of properties and geometries. Further, it is possible to print subsequent layers once the printed layers have sufficient culture time, creating 3-D arrays of cells and tissues [14]. Another way of printing that would incorporate cells within nanoscale structures is the simultaneous ink-jet printing of cells and the electrospinning of nanofibers. Electrospinning is probably the most scalable, versatile, and simple technology to produce nanofibers [35]. The principle is to apply a high electric voltage to draw polymer solutions or melts, and deposit them onto a target collector. Usually, the electrospun nanofibers are tightly packed, which raises the challenge of cell infiltration. Xu et al. combined both techniques to produce a hybrid cartilage tissue construct [36]. An electrospinning setup deposited a layer of poly(e-caprolactone) (PCL) nanofibers onto the collector. Then ink-jet printing was programmed to print chondrocytes onto electrospun layers. Repeating the electrospinning and ink-jet printing steps resulted in a multilayer 3-D construct. The separation of successive layers was prevented by suspending chondrocytes in fibrin–collagen gels, which also acted as a glue to bind layers of nanofibers together. One of the advantages of bioprinting systems in tissue engineering is their ability to provide architectures for the vascularization of the engineered tissue by controlling the liquid flow at micro- and nanolevels. The use of microfluidics in developing modular tissue engineering scaffold building blocks or incorporating microtissues in lab-on-a-chip systems for analysis and diagnostic systems has been widely studied for the last decade [37] and the addition of such systems within the tissue engineering scaffold for local control of the liquid flow can be a remedy for the vascularization-related problems in tissue engineering.

12.3 TEMPORAL ASPECTS OF TISSUE ENGINEERING

Tissue engineering is by definition a dynamic process. An engineered tissue not only needs to respond to the requirements of its cellular component, but it also it needs to be permissive to the temporal remodeling process *in vivo.* Two such events that need to be controlled *in vivo* are the immune response and the vascularization of the implanted structures.

One possible method to achieve this control is the "*in situ* tissue engineering" approach [38]. *In situ* tissue engineering methods work through the local modulation of the microenvironment via physical or biochemical signals to trigger favorable responses to the implanted material for integration. However, it should be noted that most of the possible recipients of TE products would be aged or might have metabolic impairments. This necessitates methods that are more than just a stimulation of the local cell populations. For example, transient delivery of telomerase (hTERT) to increase the local proliferation capacity in a controlled manner can improve *in situ* tissue engineering efforts [39]. However, due to the links between telomerase activity and malignancy, this line of work is moving slowly. A high level of local control of transfected cells or their stimuli-driven senescence or apoptosis, would provide a new venue for cellular expansion. However, until that point, keeping the seeded primary or stem cells alive in scaffolds is still a priority.

12.3.1 Prevascularization/Microfluidic Control of Liquid Flow within Scaffolds

Vascularization has been recognized as one of the most important obstacles for large-scale tissue engineering products [40]. The early TE constructs were unsuccessful when implanted because they were unable to sustain themselves without vascularization. The lack of a microvasculature in TE scaffolds is a severe problem as cells need access to a vasculature for their nutrient needs and waste removal processes [16]. As the scaffold thickens, there is an effective barrier to having a diffusion-based system. The diffusion process takes longer than the needs of the cells, which causes cell death. Any volume above several cubic millimeters is prone to develop a necrotic core and any realistic application of tissue engineering requires tissues of larger volumes (at least in the cubic centimeter range).

Currently, approaches to solving the vascularization problem are either biochemical, architectural, or cell based [41]. In the biochemical route, vascularization-promoting signals are incorporated into the scaffold to promote vascularization *in vivo*. In the architectural route, capillary-like structures are introduced into the scaffold structures via microfabrication techniques. Another possible method is the prevascularization of the scaffold with blood vessels *in vitro* prior to implantation via the incorporation of vascular endothelial cells (ECs) or endothelial progenitor cells. The main advantage of prevascularization is that it is incorporated with the host vasculature significantly faster than the angiogenic factor delivery systems are [42].

Prevascularization can be obtained in isotropic scaffolds by culturing ECs, preferentially in a coculture with smooth muscle cells or fibroblasts in the presence of vasculogenic signals. The level of information obtained by modeling studies has provided the necessary design criteria [43] for a microvascular network, such as structures within scaffolds that would prevent the formation of necrotic cores. For *in vitro* vascularization, the selection of the materials is important, as the *in vivo* addition of hyaluronic acid to collagen gels improves EC sprouting [44]. Other natural biomaterials such as fibrin with known angiogenic properties *in vivo* can be used to facilitate prevascularization. Important concerns are the complete endothelialization of these structures and the creation of a near physiological shear stress environment within the scaffold to ensure proper endothelial phenotype. The coculturing of ECs with several different cell types showed the importance of cell–cell interactions in the development of complex features both *in vivo* and *in vitro* [41].

Most cells are always in contact with other cells in the body to achieve their full functionality [16]. For example, there are supporting cells in the environment of stem cells [45]. Recent advances in coculture systems have tried to capture this phenomenon for regenerative medicine uses. The effects of the presence of smooth muscle cells, stem cells, and osteoblasts on capillary sprout formation *in vitro* have been well documented [40,46]. Recently, it has been shown that the coculturing of endothelial colony-forming cells and mesenchymal stem cells can form vascular networks within photo-cross-linkable hydrogels [47]. These prevascularized hydrogels can easily anastomose with the host vasculature when implanted into mice (Figure 12.4).

For example, to ensure proper blood perfusion and connection with surrounding tissues, a random packing system with microgels can provide necessary flow

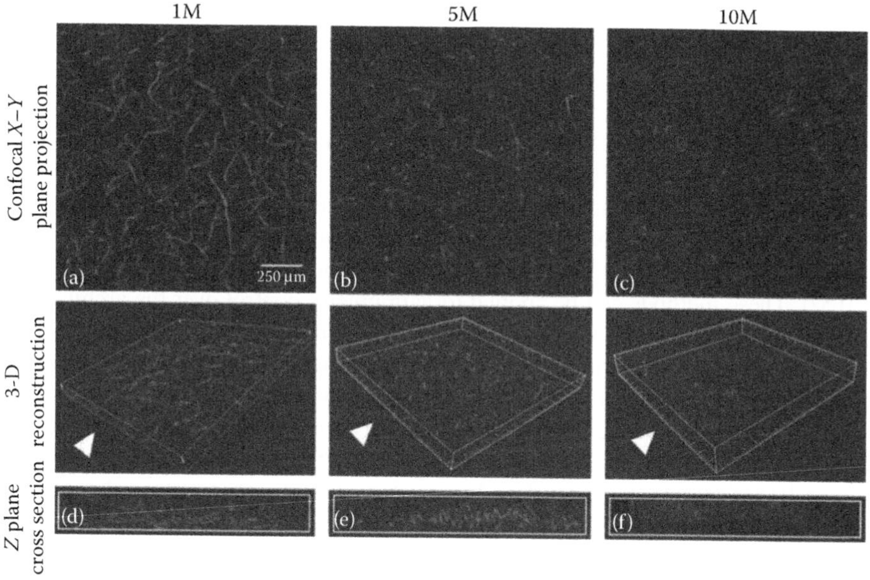

FIGURE 12.4 Effect of methacrylation degree of gelatin on the capillary sprout-like formations obtained from endothelial colony-forming cells and mesenchymal stem cells. In 6 days, a well-developed network of capillary-like structures can be obtained in low methacrylation degree samples (a,d) compared to higher degrees of methacrylation (b,e: 5M; c,f: 10M). The implantation of these structures in nude mice resulted in perfused vascular structures evidenced by the presence of murine erythrocytes in the lumen. (Adapted from Chen, Y.-C., Lin, R.-Z., Qi, H., Yang, Y., Bae, H., Melero-Martin, J.M., and Khademhosseini, A., *Advanced Functional Materials*, 22, 2027–2039, 2012.)

routes. This system was developed by assembling cell-laden collagen units together within perfused tubing to obtain tortuous channels of perfusable modular tissues [48]. While this technique can be used to obtain tissues whose primary function is filtration, such as the liver or the kidney, there are potential drawbacks when considering the lack of mechanical integrity in the absence of an enclosure. Further, the viability of the secondary cellular types may be influenced by the time that elapses between two subsequent cell layers [14]. When the used tissue modules are fragile or difficult to manipulate, directed assembly is another approach that is used to create higher-order tissue structures, in particular using ultraviolet (UV) cross-linked hydrogel modular microtissues [49]. This technique introduces novel potential methods for microfluidics or an automated technique for creating modular microtissues.

However, control over the location and the direction of the induced vascular networks is also important. Control of the fluid flow for determining the microenvironment of the cells is crucial. In the lab-on-a-chip field, utilizing the advantages related to microfluidics introduced tools that enabled better determination of biochemical and biophysical events within the cellular microenvironment [50]. The main advantages of microfluidic systems are the higher precision of fluid movement and the improved control over the concentration of important bioactive agents within small volumes. If these advantages can be imparted to the 3-D culture systems used in drug screening and toxicology tests and in the development of tissue engineering scaffolds, control of the fluid flow around cells similar to that exerted by capillaries can be achieved.

To this end, a biodegradable microfluidic system can be implemented within the scaffold structure. Since the initial work on printing capillary bed-like structures on polydimethylsiloxane (PDMS) [51], microfluidic structures based on several commonly used biodegradable biomaterials such as poly(lactic-co-glycolic) acid (PLGA), poly(glycerol sebacate) (PGS), and silk have been developed, even though such systems generally necessitate fabrication protocols specific to the material [10,52]. These methods can be applied to gain precise control over the architecture of prevascularization and to provide tubular structures for the growth of ECs [53]. By utilizing direct ink writing methods, omnidirectional, highly complex 3-D vascular networks can be obtained [54]. Another area where the microfluidic systems have been highly effective is in the development of model systems. For example, a 12-chamber microfluidic structure with the ability to exert compressive stress on stem cells has recently been developed to compare the osteogenic potential of human mesenchymal stem cells (hMSCs) with human adipose–derived stem cells [55]. This is especially important for those structures in the body that are formed of repetitive units, which can be produced simultaneously and put together in a more permissive environment with the necessary perfusion. The formation of nano/microfibers in microfluidic devices provides systems where fibrillar structures can be produced modularly.

Developments in the nanofluidics area have not yet been widely applied to the tissue engineering field; however, this scale not only provides higher precision but it also helps in more basic discoveries and the determination of the effects of changing material properties, such as viscosity and dielectric constant, on processes [56]. Moreover, low fluid volume analysis will be helpful in single-cell analysis, which will be crucial for metabolomics and proteomics analysis, which in turn will provide more in-depth information about the cellular response to a variety of signals used in tissue engineering research. Even though, normally, tissue engineering is a macroscale field with the aim to produce high-volume tissues, most of the current advances require single-cell or cell colony level observations that can be converted to macroscale events. Single-cell level analysis, or single-molecule level detection within a TE construct *in vitro* would provide the necessary information to control cellular behavior *in vivo*.

12.3.2 Modifications at Single-Cell Level

12.3.2.1 Cell Biomechanics: A Single-Cell-Based View

In the field of tissue engineering, the initial concern was the match between the mechanical properties of the scaffolds with the target tissue and the evaluation of the changes in the mechanical properties due to cellular activities and degradation [57]. However, technological advances over the last few decades in the fields of micro- and nanotechnology and in visualization techniques have increased the number of studies on the mechanical properties of cells. Thus, scientists started to measure forces and displacements at piconewton and nanometer scales, respectively. Hence, the data obtained from such studies help to understand the behavior of the cells in tissues, the development of diagnostic methods for cell identification, and the development of therapeutic approaches (Figure 12.5). The mechanical properties of cells are especially important in the context of tissue engineering, as mechanotransduction is an important determinant of cell response to biomaterials [58].

The measurement of the mechanical properties of individual cells has received much attention in recent years. Today, scientists have a rich "biomechanical toolbox" to study cell biomechanics. It includes magnetic beads [59], optical tweezers [60], optical stretchers with microfluidics [61], magnetic twisting and pulling cytometry [62], micropipette aspiration [63], optical stretchers [64], and atomic/molecular force probes [65].

Micropipette aspiration, which is a historically prevalent tool, has been used extensively to evaluate the mechanical behavior of cells and nuclei [66]. Nash et al. were the first to use micropipette aspiration to study the deformabilities of red blood cells (RBCs) infected with *Plasmodium falciparum* at different stages of infection [67]. They reported a decrease in the stiffness of RBCs infected by the parasite-exported proteins. In a comparison between malignantly transformed fibroblasts and their normal counterparts, a 50% reduction in elasticity and a 30% reduction in the viscous response were found [63]. The rigidity of the cell membrane and cytoskeleton can be calculated from the applied pressure and the resultant aspiration length.

Atomic force microscopy (AFM) functions as a powerful tool due to its high-resolution topographical imaging ability and the presence of a force sensor with piconewton sensitivity. AFM can be used to probe an individual cell's surface at specific locations to measure its localized elasticity, as determined by the force necessary to indent the cell a specified distance. Thus, the local stiffness of a cell is measured and this stiffness is mapped by indentation with a microscale cantilever. Experiments performed using the two methods revealed that the local stiffness of fibroblasts was not homogeneous on the cell surface but largely varied from point

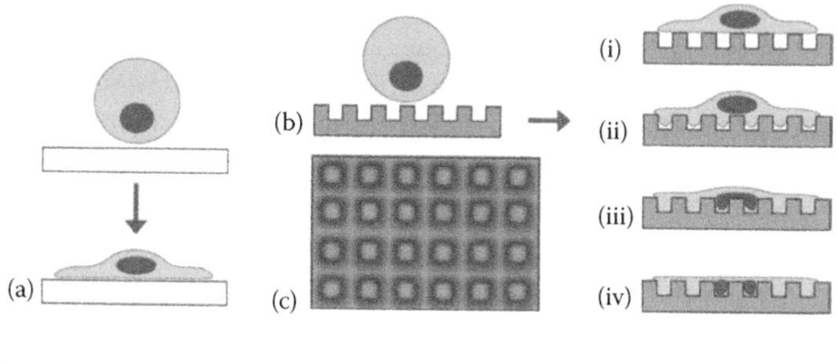

FIGURE 12.5 Effect of nano- and microscale surface patterns on cells with different mechanical properties. Cell spreading on a smooth surface (a). Cancer cells, which have a higher capacity to deform, can have their nuclei deformed in the presence of patterns (b,c), whereas healthy cells only align themselves in the direction of patterns without a change in the shape of their nuclei. (From Davidson, P.M., Özçelik, H., Hasirci, V., Reiter, G., and Anselme, K., *Advanced Materials*, 21, 3586–3590, 2009.)

to point [68]. Tissue engineering scaffolds provide a strong tool to understand the underlying reasons for this heterogeneity by the addition of nanoscale features, and at the same time this knowledge provides certain design criteria for the scaffold to control cell migration.

AFM measurements are potentially useful for detecting the differences between different cellular states, such as disease states, because *in vitro* studies have shown that cancer cells have significantly lower stiffness than normal cells [65]. In an earlier study, Lekka et al. measured the elastic properties of two similar lines of normal cells (Hu609) and cancerous ones (T24) [69]. The results for a normal epithelial cell line from a nonmalignant ureter (Hu609) were considerably higher than the data obtained for cancerous cell line T24. By using the AFM-based microrheology technique, different mechanical properties between the lamellipodia of malignantly transformed fibroblasts (H-ras transformed and SV-T2 fibroblasts) and normal fibroblasts (BALB 3T3 fibroblasts) were shown. In addition, the analysis of time-lapse phase-contrast images showed that the decrease in the elastic constant, K, for malignantly transformed fibroblasts was correlated with the enhanced motility of the lamellipodium. Similarly, in another study it was shown that live metastatic cancer cells taken from the lung, chest, and abdominal cavities of a patient were nearly four times less stiff than the benign cells that lined the respective cavities. Even metastatically competent cells can be distinguished from nonmetastatic cancer cells [70]. Similarly, Remmerbach et al. compared primary oral cells obtained by tissue biopsy from the oral mucosa of four voluntary donors with the oral squamous cell carcinoma cells of five cancer patients [61]. The cancer cells were on average 3.5 times more compliant than those of the healthy donors ($D_{normal} = (4.43 \pm 0.68)10^{-3}$ Pa^{-1}; $D_{cancer} = (15.8 \pm 1.5)10^{-3}$ Pa^{-1}; $P < 0.01$).

The direct application of the information obtained from the comparison of cancer versus noncancerous cells in the tissue engineering area is the use of AFM-based techniques to characterize the mechanical markers of stem cells. Stem cells differ from differentiated cells in their biochemical, structural, and mechanical properties [71]. During differentiation, it has been observed that the stiffness of the nuclei of stem cells increases. The major reasons for this stiffening are the changes in chromatin conformation and nuclear protein expression, and also DNA and histone modification. In two studies, this technique has been used to detect changes in the cytoskeleton during differentiation across different regions of hMSCs [72].

The elasticity of cells is determined by their cytoskeleton. Actin microfilaments, intermediate filaments, and microtubules are the major constituents of the cytoskeleton in eukaryotic cells. Changes in the amount of cytoskeletal proteins and their associated networks are reflected in the cellular function. Cancer is only one example of a disease in which the altered cytoskeleton is diagnostic. There are many other examples that show a strong connection between disease and the cytoskeletal status: circulatory problems [73]; genetic disorders of intermediate filaments and their cytoskeletal networks [74]; various blood diseases, including sickle-cell anemia, hereditary spherocytosis, or immune hemolytic anemia; and certain autoimmune diseases [75]. TE structures allow the modeling of such disease states, which provides a robust tool to understand the effect of cell mechanical properties on the progression of the disease state.

There are some major differences between the optical stretcher and other techniques. The broad and continuous distribution of stress over the cell surface can be easily achieved. The lack of any mechanical contact with the cell, and the possibility to measure suspended, nonattached cells such as white blood cells, are the other properties of the optical stretcher. In addition to its speed, it has the added advantage of being noninvasive; the cells need no marker. It is possible to perform detailed microrheological measurements on single suspended cells with the optical stretcher and to identify the origin of their specific viscoelastic nature. This is particularly useful for the measurement of changes in the viscoelastic properties of immune cells in the presence of nanoparticles. Optical stretchers also allow a glimpse of the microscopic origin of the mechanical properties of cells. In addition, with the incorporation of a microfluidic delivery of cells, cellular deformability can be measured as an inherent cell marker. A substantial difference has been seen between the response of RBCs and polymorphonuclear cells for a given optically induced stress. In another study, a noncontact, microfluidic optical stretcher was used to study cell mechanics, isolated from other parameters, in the context of tissue infiltration by acute promyelocytic leukemia (APL) cells, which occurs during differentiation therapy with retinoic acid. The compliance measurements of APL cells reveal a significant softening during differentiation [76].

One emerging trend in biomaterials research is the design of functional materials that can modulate cellular responses via the incorporation of instructive signals into the materials. In such systems, mechanotransduction is also an important determinant. For example, Davila et al. have developed nanoscale, cytoresponsive systems based on polyelectrolyte multilayers [77]. RGD sequence grafted polyelectrolyte chains that can only be exposed in the presence of mechanical stretch can be used to control selective cell adhesion (Figure 12.6). By ensuring that the cells can only attach in the presence of a certain level of stretch, an accurate temporal control over the cell–biomaterial interface can be obtained for mechanically active tissues.

The next step in this line of tissue engineering research is the concomitant use of contact guidance and mechanotransduction. Patterned substrates have already been used to study cytoskeletal dynamics and cell motility with a single-cell resolution. Studies have also been carried out to measure the cellular traction force on cells seeded onto elastomeric micropost arrays [78,79] using the finite elements method. For example, Fu et al. prepared a series of micropost arrays with a rigidity varying between 1.31 and 1556 nN μm^{-1} and with different post heights (0.97, 6.1, 12.9 μm) to investigate the regulatory role of micropost rigidity on stem cell lineage commitment [80]. They plated hMSCs on micropost arrays and incubated them in either a growth medium or a bipotential differentiation medium, which can promote both osteogenic and adipogenic differentiation. After a 2-week induction, osteogenic differentiation was observed on rigid micropost arrays, whereas adipogenic differentiation took place on soft microposts. Stem cells cultured in growth medium did not differentiate either bone or adipose tissues. Micropost rigidity was characterized by the displacement of the center node on the top surface due to an applied horizontal traction force. Vogel and coworkers also quantified the cellular force by using the same method with an additional parameter: the contribution of substrate warping to pillar deflection [81].

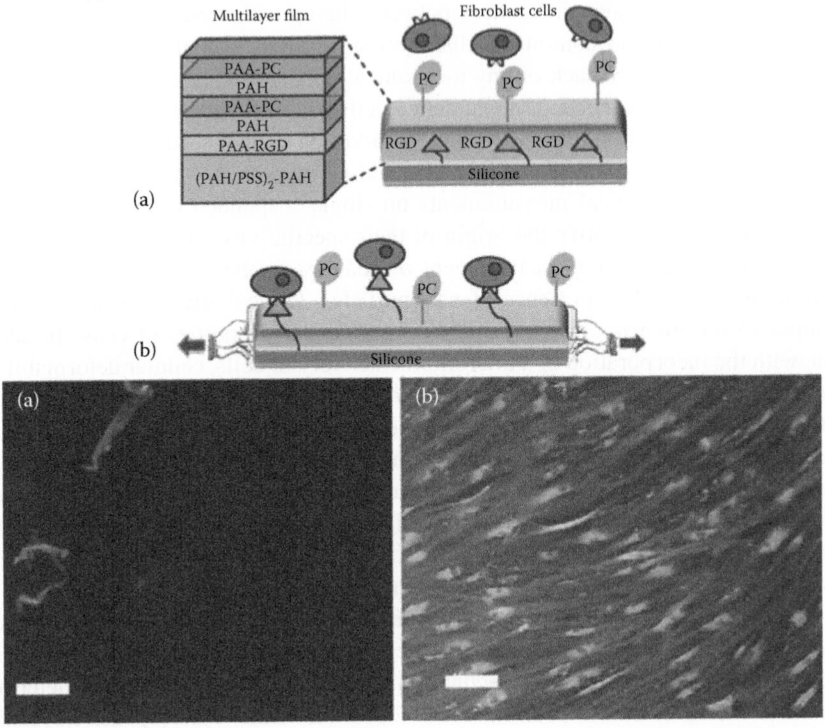

FIGURE 12.6 Cytoresponsive nanocoatings. Cells can only attach to the surface under stretch, which exposes RGD-grafted layers of polyelectrolyte multilayers (top: a,b). By having an antifouling phosphorylcholine, the grafted layer cell attachment can be temporally controlled as a function of stretch (bottom: a, unstretched; b, stretched). (From Davila, J., Chassepot, A., Longo, J., Boulmedais, F., Reisch, A., Frisch, B., Meyer, F. et al., *Journal of the American Chemical Society,* 134, 83–86, 2012.)

Another line of research that uses the single-cell level is the membrane modification of cells. This way it is possible to add structures such as molecular backpacks to cells, which can control their behavior.

12.3.2.2 Cloths for Cells: Nanoscale-Controlled Interactions with Single Cells

To influence cell behavior at the single-cell level, either the cell membrane or the cytosol should be targeted to introduce the necessary structures. Using stimuli-responsive particles that can selectively interact with a specific type of cell, a high level of control can be exerted on cells. For example, by selectively incorporating a CD31 antibody on the surface of a particle, it is possible to produce particles that can bind to ECs selectively through one side. Such Janus particles with preferential differences in their properties in two hemispheres can not only become an important part of growth factor delivery in tissue engineering constructs, but they can also be used for direct control of cell behavior. Moreover, their distinct surface properties can be exploited in such a way that they can be used as building blocks for materials with completely different bulk and surface properties [82]. These particles can also be used as molecular "backpacks" that can aggregate cells into desired configurations [83]. These backpacks have already been used for the aggregation of T cells and B cells by controlling the affinity of the particles to the cell surface. Also, by having backpacks with different loads of drugs, growth factors, or other differentiation-inducing molecules, it is possible to temporally control the processes within the aggregate without any external stimuli.

Conventional magnetic nanoparticles are being used for cell sorting. Several commercially available magnetic particles provide ways to separate various cells from blood, bone marrow, and adipose tissue (Miltenyi, Stem Cells, R&D, Beckman Coulter, and BD Biosciences). Generally, specific cells are categorized by specific surface antigens. On binding magnetic particles to antigens and exposing cells to a magnetic field, the cells of interest can be retained and isolated. A positive selection can be applied to isolate wanted cells, such as rare cancer cells, from peripheral blood and bone marrow. A negative selection can be used to deplete unwanted cells, such as the removal of T cells, for mismatched bone marrow transplantation. Both positive and negative selection strategies can also be combined to isolate more specific cell populations, such as the isolation of CD34+/CD31 stem cells from adipose tissue, which introduces a new cell therapy area of directly implanting freshly isolated stem cells without expensive and time-consuming *in vitro* expansion [84].

Magnetic nanoparticles have also been utilized in the cytosolic pathway. Nanoparticles with a magnetic core offer the feature of reacting to magnetic forces. The core is usually coated with a polymeric layer to minimize hydrophobic interactions, enhancing colloid dispersion and biocompatibility. Upon introduction into the cells to be cultured, they permit the cells' defined positioning by the appropriate use of magnets, creating more complex tissue structures than those that are achieved by conventional culture methods. An emerging tissue engineering strategy, namely, magnetic force–based tissue engineering (Mag-TE), employs cells that have been magnetically labeled with magnetite cationic liposomes (MCLs). In the Mag-TE approach, a magnet is applied under the culture plate, attracting and accumulating magnetically labeled cells. In addition, Mag-TE allows the *in vitro* fabrication and harvesting of cell sheets and heterotypic, layered cocultures containing different cell lines, providing a proof-of-principle for the applicability of this approach for generating complex heterogeneous tissues [85,86] (Figure 12.7).

12.4 NANOSCALE VISUALIZATION IN TISSUE ENGINEERING

12.4.1 Cell Tracking and Intracellular Labeling

Even though during the production stages there is high-level control over the behavior of cells, once implanted, engineered tissue–host interaction becomes more determinant. In tissue engineering applications, it is important to distinguish whether the cells involved in the repair process come from the surrounding tissue or the engineered tissue. To this end, cell-tracking dyes are utilized. Yang and coworkers [87] realized *in vivo* cellular tracking of PKH26-labeled bone marrow stem cells (BMSCs) seeded on cartilage ECM-derived scaffolds, prepared with the aim of

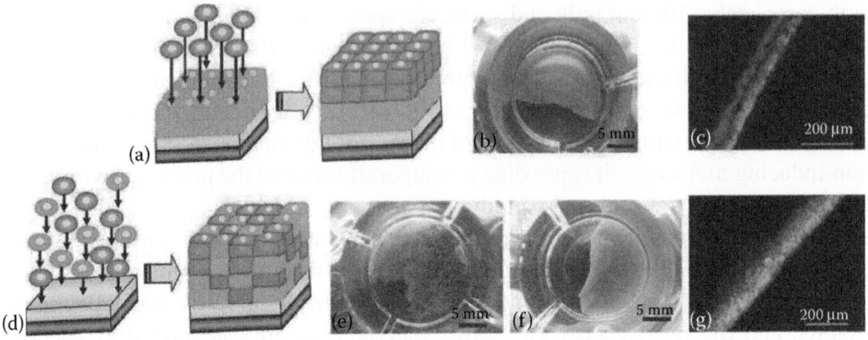

FIGURE 12.7 Control of cellular position with magnetic nanoparticle–loaded cells. Homotypically (b,c) and heterotypically (f,g) layered cell sheets can be produced in the presence of magnets. Magnetic particles promote the formation of the cell sheets because in the absence of the magnetic field a cell sheet does not form in the same amount of time (e). (From Ito, A., Jitsunobu, H., Kawabe, Y., and Kamihira, M., *Journal of Bioscience and Bioengineering,* 104, 371–378, 2007.)

ectopic chondrogenesis assessment. A cartilage-like tissue was observed 4 weeks after implantation in mice. Since the cells in the samples showed red fluorescence due to the presence of PKH26, the researchers claimed that cells found in the newly formed cartilage tissue had originated from the implanted scaffold. In other studies, PHK26 has been shown to have a labeling efficiency of more than 90%. When compared with 5-chloromethylflurorescein diacetate (CMFDA), a green cell tracker that is known to integrate into the cytoplasm of live cells, PKH26-labeled chondrocytes were shown to retain their fluorescence for a longer period (63 days compared with 14 days), even though the duration of fluorescence has been demonstrated to be cell-type dependent [88].

To further improve the term of possible *in vivo* cellular tracking, nanostructured QDs [89] can be utilized where fluorescent nanoparticles are delivered into the cell cytoplasm by using a custom targeting peptide. As they enter the cells, stable fluorescence through some generations can be traced and the technique can be used for long-term studies of both live cells and tissues. Since conventional observation techniques have the limitation of photobleaching, semiconductor QDs have attracted the interest of many researchers due to being photostable and their emissive properties being tunable.

Another possibility in using QDs is their delivery into specific organelles to observe intracellular events. Therefore, researchers have tried many strategies in recent years. Derfus et al. used a combination of epifluorescence microscopy and flow cytometry to obtain both quantitative and qualitative data for different QD delivery strategies [90]. First, they coated QDs with polyethylene glycol (PEG), to prevent their cellular uptake through endocytosis, and studied the biochemical properties of these QDs by combining them with cationic liposomes, dendrimers, and translocation peptides. Moreover, three different delivery methods (transfection, electroporation, and microinjection) were used to study the physical properties of

QDs. Furthermore, by using specific, known sequences, such as the 23mer nuclear localization sequence (NLS) peptide and the 28mer mitochondrial localization sequence (MLS) peptide, they were able to deliver QDs specifically to the nucleus and mitochondria. The best results were obtained with cationic liposomes using the microinjections method [90] (Figure 12.8). Such high-precision localization of QDs is not only effective for intracellular visualization, but it can also be used for delivery purposes [91].

12.4.1.1 Bioluminescence

Bioluminescence imaging is another method that is used for the visualization of cellular functions, such as viability, gene expression, proliferation, differentiation, and so on. It utilizes the emission of native light from a living organism that is bioluminescent [92]. This imaging technique is based on the transfection of the cells to be studied with a bioluminescent reporter (e.g., firefly luciferase, renilla luciferase, or bacterial luciferase), which, upon addition of the required substrate, gives rise to light emission at specific wavelengths. Since, by using this method, the cell-seeded scaffolds do not have to be destroyed, bioluminescence imaging has recently started to be used in tissue engineering applications both *in vitro* and *in vivo* [93,94]. *In vitro* studies are more related to monitoring cells in bioreactor systems. For instance, recently, Liu et al. [95] constructed a 3-D perfusion bioreactor suitable

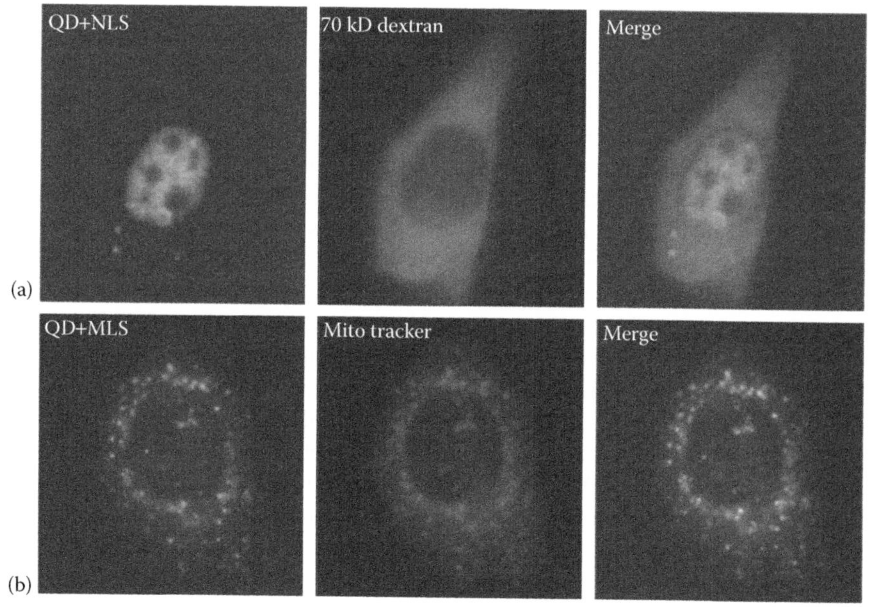

FIGURE 12.8 Intracellular localization of QDs. Using the nuclear localization sequence (NLS) (a) and mitochondrial localization sequence (MLS) (b), quantum dots can be directed with high efficiency to mitochondria and nucleus (From Derfus, A.M., Chan, W.C.W., and Bhatia, S.N., *Advanced Materials,* 16, 961–966, 2004.), providing multichannel, real-time imaging and also delivery functionalities, which can be used in tissue engineering applications.

for studying cell cultures and TE constructs (MC3T3-E1 cells transfected with pRL-SV40 reporter gene) [95]. The researchers studied the expression of BMP-2 in chitosan-based scaffolds seeded again with MC3T3-E1 cells, which were transfected with the BMP-2 luciferase reporter, with the aim of constructing a TE bone graft. In an *in vivo* study on mice for a period of 6 weeks, the expression of the collagen promoter for chondrogenic differentiation was observed by using double bioluminescent labeling (firefly and renila luciferase gene reporters tested on both a bone marrow stromal cell line and human adipose tissue–derived mesenchymal stem cells). The authors reported a good correlation between the bioluminescent images and their previously published *in vitro* results [96].

12.4.1.2 Imaging Flow Cytometry

The addition of visualization capacities to existing analysis techniques is another area of development that will benefit tissue engineering. Flow cytometry is a laser-based biophysical technology with a wide range of applications in basic research and clinical practice [97]. Its main advantages are its high quantitative fluorescence sensitivity and high speed of data acquisition (with an analytical throughput of 5000 cells/s). The principle of flow cytometry is that detection is based on numerical measurements of fluorescence sensitivity and the forward and side scatter degrees of laser light, where the forward scattering gives information on the size of the cell and the side scattering is more related to the cellular complexity (granularity). However, it should be mentioned that this technique does not give information about the localization of subcellular fluorescence [98]. However, currently, there are various techniques of cell imaging in flow, such as flying spot scanning, mirror tracking, and strobed illumination. A commercially available system for imaging flow cytometry is ImageStream®, which has the advantage of observing cells in flow with a high resolution. Imaging flow cytometry includes a combination of conventional flow cytometry with that of a digital microscope [99]. This provides highly detailed information about single-cell–nanobiomaterial interactions, such as the results reported by Marangon et al. on the intercellular translocation of functionalized multiwalled CNTs (f-MWCNTs) [100]. By using an advanced flow cytometry imaging method, Marangon et al. were able to present the location and quantification of CNTs on two different types of cells, namely, human vein ECs and human monocyte-derived macrophages (HMM). The technique they used is based on the combination of bright-field, dark-field, and fluorescent channels. The authors demonstrated for the first time the transfer of f-MWCNTs between heterotypic (HMM to EC or vice versa) and homotypic (EC to EC or HMM to HMM) cell types and they also revealed that this transfer was made possible by means of vesicles released by cells. As expected, an increase in the intercellular translocation occurred with an increase in the dose of CNTs.

12.4.2 Nonlinear Microscopy

Cell–ECM and cell–cell interactions create 3-D dynamic microenvironments that maintain the specificity and homeostasis of an engineered tissue. However, moving from cell monolayers to thick 3-D structures with a variety of material properties requires the development of standard protocols and quantitative analysis methods, which include suitable 3-D imaging techniques [101].

Currently, confocal fluorescence microscopes create thin optical sections, making it possible to build 3-D images. The data gathered from a series of optical sections imaged at short and regular intervals along the optical axis are used to create a 3-D reconstruction of moderately thick specimens. However, photobleaching and the phototoxic effects of excitation limit the efficiency of this type of imaging. A further shortcoming of confocal fluorescence microscopy is its limited penetration depth, especially while using high numerical aperture lenses [102].

Biological tissues tend to scatter light strongly, making high-resolution deep imaging impossible for traditional—including confocal—fluorescence microscopy [103]. Compared with confocal microscopy, two-photon fluorescence microscopy (also referred to as nonlinear, multiphoton microscopy) offers the advantages of a twofold penetration depth and less photodamage [104]. This is mainly because even multiple-scattered signal photons can be assigned to their origin as a result of localized, nonlinear signal generation, using longer excitation wavelengths than traditional one-photon excitation. Two-photon laser scanning microscopy has already been used without sacrifice, sectioning, or fixation to study brain, kidney, lymphoid, and skin tissue as well as the development of plaques in the progression of Alzheimer's disease to depths of nearly 1 mm.

Since two-photon excitation allows tracking for long-time biological events in the cells and tissues of living systems, within subfemtoliter volumes, it may find applications in areas such as biotechnology, neurobiology, embryology, tissue engineering, and materials science. Furthermore, the marriage of conventional chemical nonlinear optical (NLO) spectroscopic methods with biological microscopy has supported the development of several investigation techniques such as second-harmonic generation (SHG), third-harmonic generation (THG), and coherent anti-Stokes Raman scattering (CARS).

SHG is a second-order coherent process in which two lower-energy photons are up-converted to exactly twice the incident frequency (half the wavelength) of an excitation laser [105]. Biological SHG imaging was first reported in 1986, when Freund investigated the polarity of collagen fibers in rat tail tendon at a resolution of ~50 μm [106]. The majority of the recent reports on SHG microscopic imaging of an ECM focus on visualizing collagen fibers in a variety of connective tissues and internal organs [107]. For example, in a recent study, Bilgin et al. grew hMSC on 3-D collagen I gels [108]. They then imaged 3-D gels using SHG to model cell–cell interaction and to calculate the gradient vectors and the metrics. SHG can be used for disease diagnostic purposes. The collagen in an ECM is highly altered in cancer, connective tissue diseases, autoimmune disorders, and cardiovascular diseases. Using SHG imaging microscopy, Campagnola showed significant differences in collagen morphology when comparing a normal human ovary with a malignant one [105].

In another application, the quantitative use of two-photon excited fluorescence (TPEF) and SHG as a noninvasive means of monitoring the differentiation of hMSCs using entirely endogenous sources of contrast was presented [109]. SHG images revealed the deposition of fibrous collagens by cells undergoing osteogenic differentiation. An analysis of the SHG images indicated the accumulation of Type

I collagen in the hypoxic group (HO) (5% oxygen), significantly earlier than in the normoxic group (NO) (20% oxygen). Specifically, the SHG pixel density from HO cells was significantly increased at each time point (compared with the previous day), while no significant increase was observed in NO cells until Day 12. No collagen deposition was observed in propagation and adipogenic differentiation groups.

In recent years, another nonlinear molecular imaging technique, CARS microscopy, has also been developed. CARS was first reported by Maker and Terhune at the Ford Motor Company in 1965, and was named CARS in 1974 by Byer and colleagues. This imaging technique is based on the vibrational signatures of molecules. CARS microscopy provides chemically selective information by tuning into the characteristic vibrational resonances in samples without the use of labels or the complication of photobleaching. The most important advantage of this method is that the sample remains almost unaffected. Developments over the past several years have enabled the application of CARS microscopy to the chemical, materials, biological, and medical fields [110].

Using CARS to image cells with chemical selectivity, the cellular organelles of mammalian cells were imaged. For example, rapid intracellular transport in mammalian cells is driven by molecular motors [111] and tracking of lipid droplets (LDs) [112]. The vibrational selectivity of CARS microscopy makes it a powerful tool to visualize the transport, delivery, and localization of metabolites. In an early experiment, CARS was used to follow the differentiation of cells into adipocytes, following the growth of LDs in cells over the course of several days [113]. Like many previously mentioned techniques, CARS has also been shown to have the potential to distinguish between healthy and disease states in histopathological examinations. Tumors often show distinct chemical differences from healthy tissue. For instance, a large astrocytoma has been observed in a lipid band CARS image due to the lipid-poor nature of the tumor. In the tissue engineering field, SHG and CARS have been used simultaneously to monitor the growth and ECM secretion behavior of vascular smooth muscle cells in nanofibrillar bacterial cellulose scaffolds [114]. The ability to differentiate cells, nanofibrillar scaffolds, and microfibrillar ECM secretions in real time without labeling is a very strong tool for the temporal monitoring of tissue engineering scaffolds (Figure 12.9).

12.4.3 Fiber-Optic-Based Imaging

Most of the available imaging methods have the limitation that the sample to be visualized needs to be destroyed; therefore, only “some moments” of tissue development can be imaged at different intervals. However, TE scaffolds, especially those grown in bioreactors under various controlled preconditions, require nondestructive visualization techniques. Conventional confocal or nonlinear microscopy can do this, but two important limitations of this technique remain, namely, the short working distance and the limited imaging depth. For example, a group of researchers used confocal microscopy to observe green fluorescent protein (GFP)-tagged fibroblasts seeded on silk fibroin scaffolds and the system was investigated in a bioreactor under mechanical stress. The highest penetration depth achieved was 162 μm [115]. Similar to this study, imaging at a depth of 135 μm was possible with an NLO microscope, as

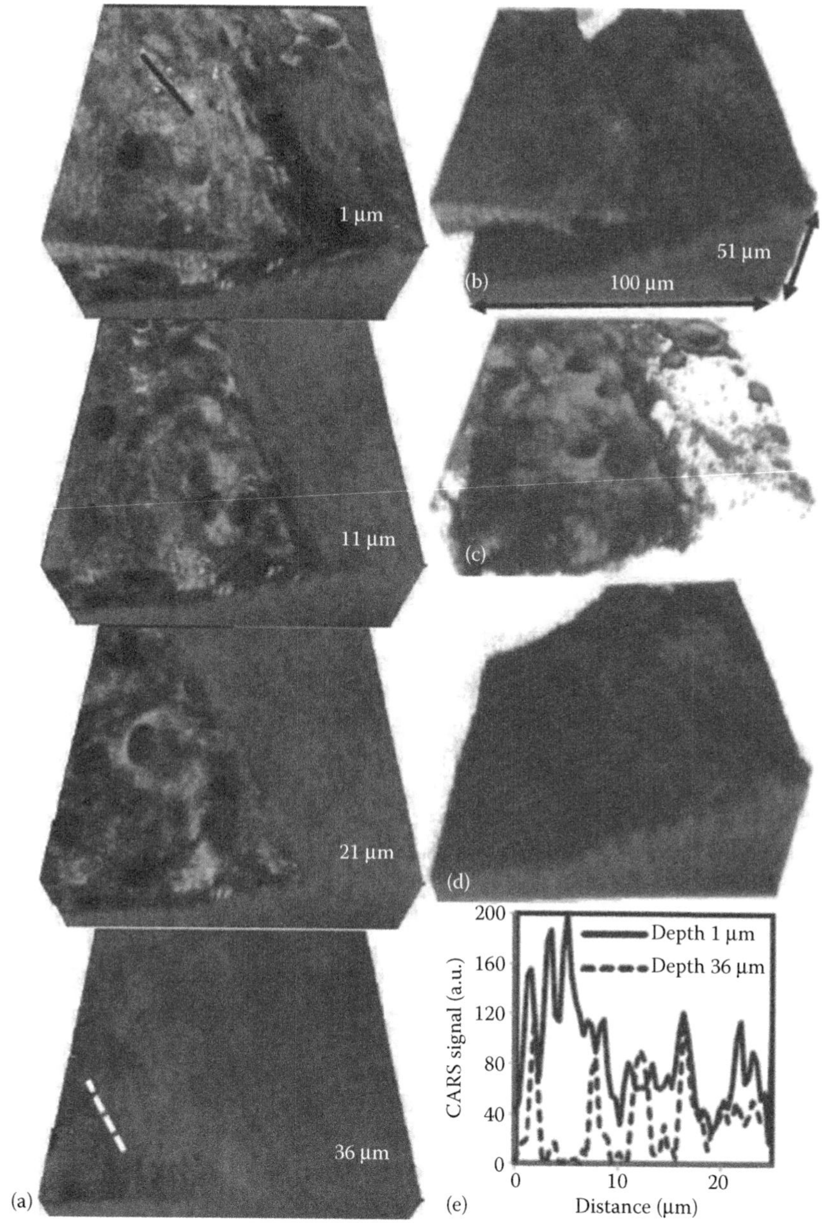

FIGURE 12.9 Simultaneous CARS and SHG imaging of cell-seeded nanofibrous tissue engineering scaffolds (a). High precision monitoring of the distribution of smooth muscle cells (observed by the CARS signal (c)) within a nanofibrillar bacterial cellulose scaffold (observed with an SHG signal (b,d)) is possible. The SHG signal produced by microfibrillar collagen can also be distinguished from the signal coming from the nanofibrous scaffold. (From Brackmann, C., Dahlberg, J.O., Vrana, N.E., Lally, C., Gatenholm, P., and Enejder, A., *Journal of Biophotonics*, 5, 404–414, 2012.)

mentioned by Niklason et al. [116]. Both of these studies showed that the resolution could not exceed 200 μm. Therefore, in the last decades, researchers have focused more on developing sample nondestructive techniques in order to image tissue development both *in vitro* and *in vivo*. Hoffmann and colleagues have recently developed a novel fiber-optic-based (FOB) imaging method that is capable of visualizing a thick and optically opaque electrospun nanofibrous scaffold [117]. The imaging principle lies in the fact that some thin and microimaging channels (MIC) were embedded into this fibrous scaffold and the cells seeded on the scaffold were genetically modified with a GFP-expressing vector. Therefore, the imaging process was based on the local excitation of these fluorescent cells and, based on the position of the excitation light, a fluorescent mapping was obtained. The working distance was increased up to 8 cm, the resolution achieved was 20–30 μm, and the imaging depth was increased to ~500 μm. In this study, it is emphasized that this method is more suitable for the evaluation of TE vascular grafts, but it is aimed at finding applications for constructs that contain epithelial tissues as well as other TE scaffolds.

However, the techniques described above rely on high-end equipment, which is not widely available. Aside from developments of means of visualization, there have been recent developments in the mode of visualization too. One disadvantage of conventional microscopy is the size and sophisticated design of microscopes, which restrict the ability to monitor 3-D structures within settings such as bioreactors and incubators, or in resource-poor conditions. A technique that is becoming comparable to traditional microscope imaging is lens-free on-chip imaging. This is a technique based on the use of a digital optoelectronic sensor array, which samples the light transmitted through the specimen directly; thus, there is no need for a lens between the sample and the sensor planes. The main advantages of this type of imaging technique over the conventional lens-based ones are its compactness, lighter weight, simpler design, lower price, ability to have larger field of view (FOV) and depth of field for 3-D screening of big volumes, and decoupling of the FOV from the resolution. Due to these advantages, it has started to find wide applications in medical areas even though some challenges still need to be solved such as its limited spatial resolution and that it is not yet able to image thick and fluorescent samples [118]. As a step further in this technique, Isikman et al. developed a simple, partially coherent lens-free tomographic microscope that can produce 3-D spatial resolution ($<1 \times <1 \times <3$ μm) of a sample with a volume of 15 mm^3 [119]. In this case, the sample is observed at different angles and a partially coherent quasi-monochromatic light situated approximately 70 mm away from the sensor array is used, which illuminates objects placed on a chip. The wider availability of lens-free imaging systems would improve the real-time imaging of engineered tissues during their maturation.

12.5 BIOSENSORS IN TISSUE ENGINEERING: INCORPORATION OF DEVICES INTO TISSUE ENGINEERING SCAFFOLDS

12.5.1 On-Demand Release

The majority of the most challenging tissues to regenerate have the property of excitability. Control over this property is an essential part of future tissue engineering efforts, which necessitates the integration of parts into the tissue engineering scaffolds with excitation, stimulation, and interrogation capabilities. The addition of such systems would also pave the way for on-demand drug and growth factor delivery within scaffolds and self-sustaining delivery systems [120].

Nanomembranes were the first implanted controlling system of cell activities [121]. They passively control the release of molecules, where the ratio of the average pore size of the membrane to the hydrodynamic diameter of the solute determines the rate of release. Membrane systems have been widely used for pancreatic cell encapsulation, where the movement of molecules and cells is restricted to prevent an immune response, by having pores small enough to allow the transfer of glucose and insulin but not big enough for immunoglobulins. A further step of these devices would be the addition of a glucose biosensor connected to a release system, which would release necessary chemicals for the induction of insulin release by the encapsulated pancreatic cells. A glucose-sensitive valve has been used to control the release of insulin based on the swelling behavior of glucose-sensitive hydrogels. This biosensor, drug delivery, and drug combination need is simpler than a TE construct where the remodeling of the system is not necessary. Other systems include more intricate structures such as polymeric pumps to deliver growth factors. Another possible method is the utilization of multilayers of drug-loaded structures, which release their loads either through passive diffusion or degradation or under thermal, electrical, or mechanical stimuli. However, a biosensor or a drug delivery system in a tissue engineering construct will have a very dynamic environment and if it is to be responsive, complex algorithms that can provide the right dosing will be necessary [122].

Microelectromechanical systems (MEMS) and nanoelectromechanical systems (NEMS) initially offer options for drug delivery for dosing-related problems. The device-based drug delivery within TE systems ensures the correct concentration, prevents dilution, and limits the possible systemic effects that can occur in intravenous delivery methods. Their further use might be of benefit for the administration of multidrugs in a temporally designed manner. An important example of this is the delivery of subsequent levels of growth factors, which requires precise dosing, and also the long-term protection of these highly labile entities within a reservoir. Initially, such goals were achieved by loading drugs in complex matrices and the control of their release was achieved by more physical constraints. For example, one such product was developed by Biomimetics Inc., where platelet-derived growth factor BB (PDGF-BB) was released from a tricalcium phosphate (TCP)-based matrix for healing bone defects [123]. However, for effective bone healing, the activity of more than one growth factor is necessary.

Moreover, the release behavior generally needs to be synchronized with the situation in the surroundings of the reservoir, which necessitates feedback systems to control the release. These needs require more sophisticated systems to be included in tissue engineering scaffolds. MEMS and NEMS structures first improved the drug delivery systems; the next step in their utilization is drug screening in combination with the TE model tissues. It is important to control the rate of administration, either in a pulsatile, discrete, or continuous manner. These devices can be either nanoscale local sensors or delivery devices that can have either inner sensor–based controls or they can be controlled from outside.

12.5.2 Nanoelectronics

The presence of stimulating or sensing structures can be likened to the prevalent presence of the nervous system within the tissues as a commanding structure. Also similar to the neural stimulation of hormonal secretions, such nanoscale devices can be used for stimulating the TE components, which can be used for local level or macroscale stimulation with higher control over the length of stimulation. A further iteration of this would be the integration of sensing and stimulation, which would make scaffolds very responsive to minute changes in the environment and provide a much better chance of attaining functionality and responsiveness.

One of the most widely used nanostructures for electrical responsiveness is CNTs. In addition to the advantage of being in nanoscale, CNTs or carbon nanofibers have been shown to have superior mechanical and electrical properties. Depending on the process used in their production, CNTs are classified as single-walled (SW) CNTs and MWCNTs. Various applications of CNTs have been reported especially in the last decades. They have been used alone or in combination with a variety of materials, such as metals [124], ceramics [125], inorganic materials, and polymers [126], in order to obtain materials with enhanced physicochemical properties. They have been used in various fields including sensors and electrochemical and electronic devices. Regarding tissue engineering, they have mostly been used in bone cell proliferation and bone formation and the regeneration of damaged neural tissues. Nowadays, a combination of CNTs with other materials is in very common use. According to Supronowicz et al. [126], blending CNTs with poly(L-lactic acid) (PLA) introduced electrical conductivity to the nonconductor PLA, since after 2 and 21 days of electrical stimulation, a higher (30%) calcium content and an increase in bone cell proliferation were observed. Similar to this, electrical conductivity in Bioglass®-based foams was introduced by coating these scaffolds with MWCNTs by means of electrophoretic deposition. These structures showed a higher level of interaction between the bone cells and the nanofibrous 3-D scaffolds [127].

Aside from imparting electrical properties, CNTs have been shown to be beneficial for cell attachment and proliferation. Osteoblast adhesion and improved cellular functions in the presence of CNTs have been observed. Also, it has been suggested that the characteristic tubular shape of the CNTs and the different reactivity of the side walls versus the ends would represent an advantage in drug delivery systems, since two different functional groups in each location could be created [128].

Due to their electrical properties, CNTs have been widely studied for neural tissue engineering applications. Mattson and colleagues found that neurons can grow on MWCNTs [129]. A group of researchers, with the aim of neural tissue engineering, produced biohybrid SWCNTs/agarose fibers using a wet-spinning process. In order to increase cellular attachment, these fibers were functionalized with laminin (LN) or bovine serum albumin (BSA). Moreover, preliminary *in vivo* studies in the cerebral cortex of rats revealed promising results in the case of peptide-functionalized structures [130].

Another interesting application of CNTs is their combination with neural stem cells to deliver these cells to the correct site and differentiate them into neuronal cell types (neurons and astrocytes) with the aim of repairing damaged neural tissue. For instance, Jan and Kotov used layer-by-layer assembled CNTs/polyelectrolyte composites to differentiate embryonic mouse neural stem cells into neurons [131].

The size of the electrical structure–cell interface determines the sensitivity of the signal produced. When the size of the sensor is in the micron or millimeter range, it is much less sensitive. This leads to a less spatial resolution of either detection or stimulation. Nanodevices provide a higher signal-to-noise ratio with higher sensitivities than other available systems. Here, the use of nanowires or nanotubes can dramatically improve the sensitivity and precision of such structures. This is especially important for molecular detection studies as the surface–volume ratio at the nanoscale is significantly higher than at the microscale [56].

Nanowires of gold can be distributed within porous scaffolds to obtain excitable structures, nanotubes, graphenes, and 3-D structures from lower-order building blocks. These nanowire structures are also shown to induce less immunological responses [132]. Group IV materials such as silicon, graphenes, coordination polymers, and conductive polymers are other candidates for imparting electrical properties to scaffolds. Conductive polymers have been utilized to enhance the electrical stimulation capacity of scaffolds. Recently, nanoelectronic biomaterials based on silicon nanowire field-effect transistors (FET) have been developed, which can be used to monitor the electrical activity of the cells within scaffolds [133]. Incorporation of materials such as QDs, CNTs, and mesoporous silicon nanoparticles can also be used for monitoring purposes, such as the controlled delivery of magnetic resonance imaging (MRI) contrast agents for imaging engineered tissue (Figure 12.10).

An *in vivo* biosensor is an exciting technology and its incorporation into tissue engineering scaffolds holds great promise. However, biosensors are most sensitive to immune responses such as fibrotic encapsulation, foreign-body response, and biofouling [134]. To prevent these events, several methodologies have been developed such as antifouling of biosensors with PEG and polyelectrolyte multilayers and biomimetic phospholipid coatings. Electrochemical sensors are generally based on enzymes such as oxidases and the conversion of an electrochemical signal to an electrical signal. One such application would be the detection of levels of the important neurotransmitter L-glutamate via glutamate oxidase. Another possible route is the direct detection of action potentials by nanoscale and microscale biosensors; via multielectrode arrays, more complex signals can be measured. Mechanical loading–based effects can be distinguished by using piezoelectric effects. Piezoelectric materials change their electrical properties with respect to the mechanical loading. This change in resistance is one of the piezoelectric effects that can be used to measure absorbed materials, which can be used in vascular or tracheal implants to monitor stenosis or for the detection of necrosis or foreign-body response. However, nonspecific adsorption is a big problem and can create noise problems. To this end, precoating of the implant surface is generally undertaken. Microcantilevers, in a system akin to the cantilevers of AFM, have also been used to detect analyte levels and to trigger the delivery of antimicrobial agents. This system works through the direct binding of a bacterial molecule to its substrate, which in turn triggers an enzyme to cleave its substrate, causing the release of antimicrobial agents. Such complex systems are necessities in the field of dentistry and orthopedic surgeries as postsurgery

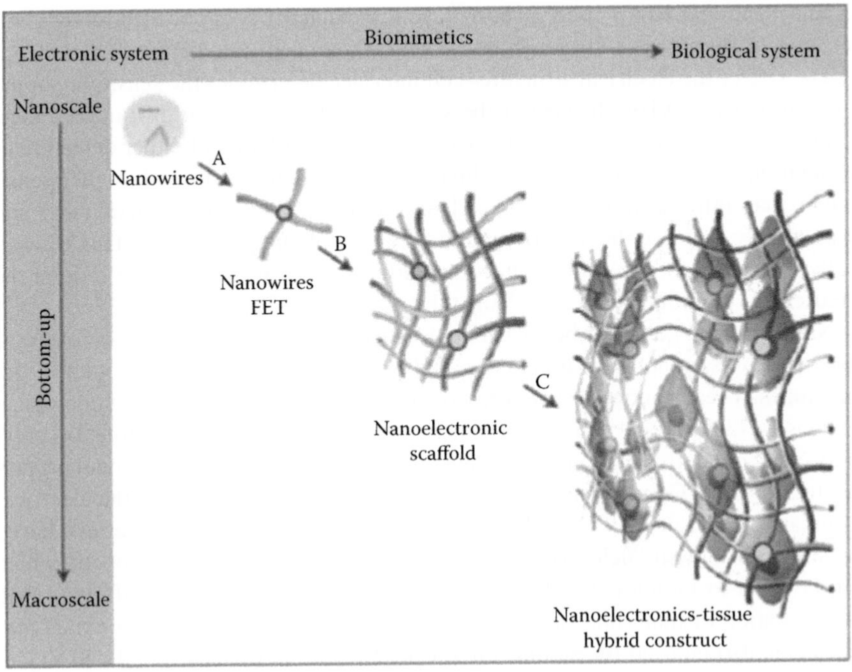

FIGURE 12.10 3-D nanoelectronic scaffolds. Porous nanoelectronic scaffolds based on nanowires and their hybrids with other materials (collagen, alginate, PLGA) with sensory capacities. A nanoelectronic structure's macroporosity provides the necessary volume for cellular adhesion and growth. (From Tian, B., Liu, J., Dvir, T., Jin, L., Tsui, J.H., Qing, Q., Suo, Z., Langer, R., Kohane, D.S., and Lieber, C.M., *Nature Materials,* 11, 986–994, 2012.)

infections can have devastating results. Other possible signals that can be monitored are hypoxia, hypoglycemia, infection-related changes such as lipopolysaccharide (LPS) levels or pH, growth factor concentrations, cell numbers, apoptosis signals, important cytokine concentrations, and cancerous cells detection. The phagocytosis of a material *in vivo* is largely determined by its shape and size. Thus, nanoscale sensors can also be used in intracellular settings for a more precise measurement of cellular events or direct control of cellular activities.

However, a certain range of problems remains ahead of developments in this area. Possible local or systemic toxicity of the nanostructures together with low biocompatibility might hinder the functionality of the designed systems in physiological conditions. Most of the currently used materials for building nanoelectronic structures are not similar to known biomaterials and the information on their possible side effects is scarce. Another issue is that these materials are generally not biodegradable, which means that they would remain in the implanted area even if they ceased to be functional. One way around this problem would be to develop nanoscale devices from biocompatible materials, but this area has not shown significant advances yet. Another possible problem is biofouling, which readily happens in *in vivo* conditions upon implantation.

Regulatory issues are one of the main obstacles for the incorporation of MEMS into tissue engineering products. TE structures are already combination products that contain biologics, biomaterials, and drugs. The addition of devices into such systems would further complicate their approval for clinical use. Biofouling *in vivo* is another obstacle that might prevent the total functionality of the implanted structure within the product by diminishing the readings via biosensors or clogging the release routes. Devices should be incorporated in such a way as to prevent biofouling or be designed that biofouling would not affect their intended functionality or expected lifetime. This can be prevented by incorporating multiple sensors and shutting down fouled sensors over time. This further proves the necessity for developing nanoscale sensors as there is a limit to the volume of an implant.

Communications technologies and advances in the area of wireless communications provide us with the opportunity to control implanted structures over the long term. Gathering information from the surrounding tissue, and manipulating and turning this information into commands might be possible in the near future.

12.6 CONCLUSIONS

Nanoscale technologies are an integral part of current tissue engineering structures. From the incorporation of nanoscale architectural elements to the development of active scaffolds via bioelectronics, the utilization of cell–material interactions at the nanolevel has an important part in future advances in the field of regenerative medicine. With the current developments in the utilization of microtissues in organ-on-a-chip structures, there is a growing need for higher precision monitoring of engineered tissues through novel visualization techniques and also biosensors. It can be anticipated that these innovations will find their way into TE structures intended for clinical applications, thus significantly improving their success rate and functionality.

ACKNOWLEDGMENTS

This work has been supported by Protip SAS, EuroTransBio BiMot Project (ETB-2012-32), and PMNA (Region Alsace).

REFERENCES

1. Yasuhiko, T. 2001. Recent progress in tissue engineering. *Drug Discovery Today* 6:483–487.
2. Atala, A., S.B. Bauer, S. Soker, J.J. Yoo, and A.B. Retik. 2006. Tissue-engineered autologous bladders for patients needing cystoplasty. *The Lancet* 367 (9518):1241–1246.
3. Khademhosseini, A., J.P. Vacanti, and R. Langer. 2009. Progress in tissue engineering. *Scientific American* 300 (5):64–71.
4. Pibarot, P. and J.G. Dumesnil. 2009. Valvular heart disease: Changing concepts in disease management, prosthetic heart valves selection of the optimal prosthesis and long-term management. *Circulation* 119:1034–1048.

5. Hajicharalambous, C.S., J. Lichter, W.T. Hix, M. Swierczewska, M.F. Rubner, and P. Rajagopalan. 2009. Nano- and sub-micron porous polyelectrolyte multilayer assemblies: Biomimetic surfaces for human corneal epithelial cells. *Biomaterials* 30 (23–24):4029–4036.
6. Khademhosseini, A., G. Eng, J. Yeh, P.A. Kucharczyk, R. Langer, G. Vunjak-Novakovic, and M. Radisic. 2007. Microfluidic patterning for fabrication of contractile cardiac organoids. *Biomed Microdevices* 9 (2):149–157.
7. Lange, P., J.M. Fishman, M.J. Elliott, P. De Coppi, and M.A. Birchall. 2011. What can regenerative medicine offer for infants with laryngotracheal agenesis? *Otolaryngology—Head and Neck Surgery* 145 (4):544–550.
8. Placzek, M.R., I.M. Chung, H.M. Macedo, S. Ismail, T.M. Blanco, M. Lim, J.M. Cha, I. Fauzi, Y. Kang, and D.C.L. Yeo. 2009. Stem cell bioprocessing: Fundamentals and principles. *Journal of the Royal Society Interface* 6 (32):209–232.
9. Badylak, S.F., D.J. Weiss, A. Caplan, and P. Macchiarini. 2012. Engineered whole organs and complex tissues. *The Lancet* 379 (9819):943–952.
10. Freed, L.E., G.C. Engelmayr, J.T. Borenstein, F.T. Moutos, and F. Guilak. 2009. Advanced material strategies for tissue engineering scaffolds. *Advanced Materials* 21 (32–33):3410–3418.
11. LaVan, D.A., T. McGuire, and R. Langer. 2003. Small-scale systems for in vivo drug delivery. *Nature Biotechnology* 21 (10):1184–1191.
12. Basmanav, F.B., G.T. Kose, and V. Hasirci. 2008. Sequential growth factor delivery from complexed microspheres for bone tissue engineering. *Biomaterials* 29 (31):4195–4204.
13. Roberts, S.J., D. Howard, L.D. Buttery, and K.M. Shakesheff. 2008. Clinical applications of musculoskeletal tissue engineering. *British Medical Bulletin* 86 (1):7–22.
14. Nichol, J.W. and A. Khademhosseini. 2009. Modular tissue engineering: Engineering biological tissues from the bottom up. *Soft Matter* 5 (7):1312–1319.
15. Shen, W. 2011. Engineered polypeptides for tissue engineering. In J.A. Burdick and R.L. Mauck (eds), *Biomaterials for Tissue Engineering Applications. A Review of the Past and Future Trends*, pp. 243–277. Springer, Vienna.
16. Kirkpatrick, C.J., S. Fuchs, and R.E. Unger. 2011. Co-culture systems for vascularization—Learning from nature. *Advanced Drug Delivery Reviews* 63 (4–5):291–299.
17. Chung, B.G., L. Kang, and A. Khademhosseini. 2007. Micro- and nanoscale technologies for tissue engineering and drug discovery applications. *Expert Opinion on Drug Discovery* 2 (12):1–16.
18. Taipale, J. and J. Keski-Oja. 1997. Growth factors in the extracellular matrix. *FASEB Journal* 11:51–59.
19. Dong, C.-M., X. Wu, J. Caves, S.S. Rele, B.S. Thomas, and E.L. Chaikof. 2005. Photomediated crosslinking of C6-cinnamate derivatized type I collagen. *Biomaterials* 26 (18):4041–4049.
20. Ruoslahti, E. and M.D. Pierschbacher. 1986. Arg-Gly-Asp: A versatile cell recognition signal. *Cell* 44:517–518.
21. Norman, J.J. and T.A. Desai. 2006. Methods for fabrication of nanoscale topography for tissue engineering scaffolds. *Annals of Biomedical Engineering* 34 (1):89–101.
22. Chen, A., D.K. Lieu, L. Freschauf, V. Lew, H. Sharma, J. Wang, D. Nguyen, et al. 2011. Shrink-film configurable multiscale wrinkles for functional alignment of human embryonic stem cells and their cardiac derivatives. *Advanced Materials* 23 (48):5785–5791.
23. Vautier, D., J. Hemmerle, C. Vodouhe, G. Koenig, L. Richert, C. Picart, J.C. Voegel, C. Debry, J. Chluba, and J. Ogier. 2003. 3-D surface charges modulate protrusive and contractile contacts of chondrosarcoma cells. *Cell Motility and the Cytoskeleton* 56 (3):147–158.

24. Gustafson, T. and L. Wolpert. 1999. Studies on the cellular basis of morphogenesis in the sea urchin embryo. Directed movements of primary mesenchyme cells in normal and vegetalized larvae. *Experimental Cell Research* 253:288–295.
25. Zhang, W., I.P. Ahluwalia, and P.C. Yelick. 2010. Three dimensional dental epithelial-mesenchymal constructs of predetermined size and shape for tooth regeneration. *Biomaterials* 31 (31):7995–8003.
26. Hartgerink, J.D., E. Beniash, and S.I. Stupp. 2001. Self-assembly and mineralization of peptide-amphiphile nanofibers. *Science* 294:1684–1688.
27. Zhang, L., S. Ramsaywack, H. Fenniri, and T.J. Webster. 2008. Enhanced osteoblast adhesion on self-assembled nanostructured hydrogel scaffolds. *Tissue Engineering Part A* 14 (8):1353–1364.
28. Zhang, K.C., M.R. Diehl, and D.A. Tirrell. 2005. Artificial polypeptide scaffold for protein immobilization. *Journal of the American Chemical Society* 127:10136–10137.
29. Lee, S.Y., Y.C. Park, H.S. Cho, K.S. Ra, H.S. Baik, S.-Y. Paik, J.W. Yun, H.S. Park, and J.W. Choi. 2003. Expression of an artificial polypeptide with a repeated tripeptide glutamyl–tryptophanyl–lysine in *Saccharomyces cerevisiae*. *Letters in Applied Microbiology* 36:121–128.
30. Karoly, J., N. Cyrille, M. Francoise, M. Keith, V.-N. Gordana, and F. Gabor. 2010. Tissue engineering by self-assembly and bio-printing of living cells. *Biofabrication* 2 (2):022001.
31. Karp, J.M., J. Yeh, G. Eng, J. Fukuda, J. Blumling, K.Y. Suh, J. Cheng, et al. 2007. Controlling size, shape and homogeneity of embryoid bodies using poly(ethylene glycol) microwells. *Lab on a Chip* 7 (6):786–794.
32. Yeh, J., Y. Ling, J.M. Karp, J. Gantz, A. Chandawarkar, G. Eng, J. Blumling 3rd, R. Langer, and A. Khademhosseini. 2006. Micromolding of shape-controlled, harvestable cell-laden hydrogels. *Biomaterials* 27 (31):5391–5398.
33. L'Heureux, N., T.N. McAllister, and L.M. de la Fuente. 2007. Tissue-engineered blood vessel for adult arterial revascularization. *New England Journal of Medicine* 357 (14):1451–1453.
34. Marga, F., A. Neagu, I. Kosztin, and G. Forgacs. 2007. Developmental biology and tissue engineering. *Birth Defects Research Part C: Embryo Today: Reviews* 81 (4):320–328.
35. Noh, H.K., S.W. Lee, J.M. Kim, J.E. Oh, K.H. Kim, C.P. Chung, S.C. Choi, W.H. Park, and B.M. Min. 2006. Electrospinning of chitin nanofibers: Degradation behavior and cellular response to normal human keratinocytes and fibroblasts. *Biomaterials* 27 (21):3934–3944.
36. Xu, T., K.W. Binder, M.Z. Albanna, D. Dice, W. Zhao, J.J. Yoo, and A. Atala. 2012. Hybrid printing of mechanically and biologically improved constructs for cartilage tissue engineering applications. *Biofabrication* 5 (1):015001.
37. Bettinger, C.J. and J.T. Borenstein. 2010. Biomaterials-based microfluidics for engineered tissue constructs. *Soft Matter* 6 (20):4999–5015.
38. Omori, K., Y. Tada, T. Suzuki, Y. Nomoto, T. Matsuzuka, K. Kobayashi, T. Nakamura, S. Kanemaru, M. Yamashita, and R. Asato. 2008. Clinical application of in situ tissue engineering using a scaffolding technique for reconstruction of the larynx and trachea. *Annals of Otology, Rhinology, and Laryngology* 117 (9):673–678.
39. Petersen, T. and L. Niklason. 2007. Cellular lifespan and regenerative medicine. *Biomaterials* 28 (26):3751–3756.
40. Fuchs, S., X. Jiang, H. Schmidt, E. Dohle, S. Ghanaati, C. Orth, A. Hofmann, A. Motta, C. Migliaresi, and C.J. Kirkpatrick. 2009. Dynamic processes involved in the prevascularization of silk fibroin constructs for bone regeneration using outgrowth endothelial cells. *Biomaterials* 30 (7):1329–1338.

41. Ghanaati, S., R.E. Unger, M.J. Webber, M. Barbeck, C. Orth, J.A. Kirkpatrick, P. Booms, et al. 2011. Scaffold vascularization *in vivo* driven by primary human osteoblasts in concert with host inflammatory cells. *Biomaterials* 32 (32):8150–8160.
42. Tian, L. and S.C. George. 2011. Biomaterials to prevascularize engineered tissues. *Journal of Cardiovascular Translational Research* 4 (5):685–698.
43. Checa, S. and P.J. Prendergast. 2010. Effect of cell seeding and mechanical loading on vascularization and tissue formation inside a scaffold: A mechano-biological model using a lattice approach to simulate cell activity. *Journal of Biomechanics* 43 (5):961–968.
44. Perng, C.K., Y.J. Wang, C.H. Tsi, and H. Ma. 2011. *In vivo* angiogenesis effect of porous collagen scaffold with hyaluronic acid oligosaccharides. *Journal of Surgical Research* 168 (1):9–15.
45. Zhang, H., S. Dai, J.X. Bi, and K.K. Liu. 2011. Biomimetic three-dimensional microenvironment for controlling stem cell fate. *Interface Focus* 1 (5):792–803.
46. Elbjeirami, W.M. and J.L. West. 2006. Angiogenesis-like activity of endothelial cells co-cultured with VEGF-producing smooth muscle cells. *Tissue Engineering* 12 (2):381–390.
47. Chen, Y.-C., R.-Z. Lin, H. Qi, Y. Yang, H. Bae, J.M. Melero-Martin, and A. Khademhosseini. 2012. Functional human vascular network generated in photocrosslinkable gelatin methacrylate hydrogels. *Advanced Functional Materials* 22 (10):2027–2039.
48. McGuigan, A.P., B. Leung, and M.V. Sefton. 2006. Fabrication of cells containing gel modules to assemble modular tissue-engineered constructs. *Nature Protocols* 1 (6):2963–2970.
49. Du, Y., E. Lo, S. Ali, and A. Khademhosseini. 2008. Directed assembly of cell-laden microgels for fabrication of 3D tissue constructs. *Proceedings of the National Academy of Sciences* 105 (28):9522–9527.
50. Yeo, L.Y., H.C. Chang, P.P.Y. Chan, and J.R. Friend. 2011. Microfluidic devices for bioapplications. *Small* 7 (1):12–48.
51. Borenstein, J.T., H. Terai, K.R. King, E.J. Weinberg, M.R. Kaazempur-Mofrad, and J.P. Vacanti. 2002. Microfabrication technology for vascularized tissue engineering. *Biomedical Microdevices* 4 (3):167–175.
52. King, K.R., C.C.J. Wang, M.R. Kaazempur-Mofrad, J.P. Vacanti, and J.T. Borenstein. 2004. Biodegradable microfluidics. *Advanced Materials* 16 (22):2007–2012.
53. Fidkowski, C., M.R. Kaazempur-Mofrad, J. Borenstein, J.P. Vacanti, R. Langer, and Y. Wang. 2005. Endothelialized microvasculature based on a biodegradable elastomer. *Tissue Engineering* 11 (1–2):302–309.
54. Wu, W., A. DeConinck, and J.A. Lewis. 2011. Omnidirectional printing of 3D microvascular networks. *Advanced Materials* 23 (24):H178–H183.
55. Park, S.H., W.Y. Sim, B.H. Min, S.S. Yang, A. Khademhosseini, and D.L. Kaplan. 2012. Chip-based comparison of the osteogenesis of human bone marrow-and adipose tissue-derived mesenchymal stem cells under mechanical stimulation. *PLoS ONE* 7 (9):e46689.
56. Xu, Y., K. Jang, T. Yamashita, Y. Tanaka, K. Mawatari, and T. Kitamori. 2012b. Microchip-based cellular biochemical systems for practical applications and fundamental research: From microfluidics to nanofluidics. *Analytical and Bioanalytical Chemistry* 402 (1):99–107.
57. Eosoly, S., N.E. Vrana, S. Lohfeld, M. Hindie, and L. Looney. 2012. Interaction of cell culture with composition effects on the mechanical properties of polycaprolactone-hydroxyapatite scaffolds fabricated via selective laser sintering (SLS). *Materials Science and Engineering: C* 32 (8):2250–2257.

58. Kocgozlu, L., P. Lavalle, G. Koenig, B. Senger, Y. Haikel, P. Schaaf, J.-C. Voegel, H. Tenenbaum, and D. Vautier. 2010. Selective and uncoupled role of substrate elasticity in the regulation of replication and transcription in epithelial cells. *Journal of Cell Science* 123 (1):29–39.
59. Mijailovich, S.M., M. Kojic, M. Zivkovic, B. Fabry, and J.J. Fredberg. 2002. A finite element model of cell deformation during magnetic bead twisting. *Journal of Applied Physiology* 93 (4):1429–1436.
60. Li, J., M. Dao, C.T. Lim, and S. Suresh. 2005. Spectrin-level modeling of the cytoskeleton and optical tweezers stretching of the erythrocyte. *Biophysical Journal* 88 (5):3707–3719.
61. Remmerbach, T.W., F. Wottawah, J. Dietrich, B. Lincoln, C. Wittekind, and J. Guck. 2009. Oral cancer diagnosis by mechanical phenotyping. *Cancer Research* 69 (5):1728–1732.
62. Overby, D.R., B.D. Matthews, E. Alsberg, and D.E. Ingber. 2005. Novel dynamic rheological behavior of individual focal adhesions measured within single cells using electromagnetic pulling cytometry. *Acta Biomaterialia* 1 (3):295–303.
63. Ward, K.A., W.I. Li, S. Zimmer, and T. Davis. 1991. Viscoelastic properties of transformed cells: Role in tumor cell progression and metastasis formation. *Biorheology* 28 (3–4):301.
64. Guck, J., R. Ananthakrishnan, H. Mahmood, T.J. Moon, C.C. Cunningham, and J. Käs. 2001. The optical stretcher: A novel laser tool to micromanipulate cells. *Biophysical Journal* 81 (2):767–784.
65. Suresh, S., J. Spatz, J.P. Mills, A. Micoulet, M. Dao, C.T. Lim, M. Beil, and T. Seufferlein. 2005. Connections between single-cell biomechanics and human disease states: Gastrointestinal cancer and malaria. *Acta Biomaterialia* 1 (1):15–30.
66. Vaziri, A. and M.R.K. Mofrad. 2007. Mechanics and deformation of the nucleus in micropipette aspiration experiment. *Journal of Biomechanics* 40 (9):2053–2062.
67. Nash, G.B., E. O'Brien, E.C. Gordon-Smith, and J.A. Dormandy. 1989. Abnormalities in the mechanical properties of red blood cells caused by Plasmodium falciparum. *Blood* 74 (2):855–861.
68. Haga, H., M. Nagayama, and K. Kawabata. 2007. Imaging mechanical properties of living cells by scanning probe microscopy. *Current Nanoscience* 3 (1):97–103.
69. Lekka, M., P. Laidler, D. Gil, J. Lekki, Z. Stachura, and A.Z. Hrynkiewicz. 1999. Elasticity of normal and cancerous human bladder cells studied by scanning force microscopy. *European Biophysics Journal* 28 (4):312–316.
70. Lekka, M. and P. Laidler. 2009. Applicability of AFM in cancer detection. *Nature Nanotechnology* 4 (2):72.
71. Le Blanc, K. 2003. Immunomodulatory effects of fetal and adult mesenchymal stem cells. *Cytotherapy* 5 (6):485–489.
72. Yourek, G., M.A. Hussain, and J.J. Mao. 2007. Cytoskeletal changes of mesenchymal stem cells during differentiation. *Asaio Journal* 53 (2):219–228.
73. Worthen, G.S., B. Schwab 3rd, E.L. Elson, and G.P. Downey. 1989. Mechanics of stimulated neutrophils: Cell stiffening induces retention in capillaries. *Science* 245 (4914):183.
74. Fuchs, E. and D.W. Cleveland. 1998. A structural scaffolding of intermediate filaments in health and disease. *Science* 279 (5350):514–519.
75. Bijl, M., P.C. Limburg, and C.G.M. Kallenberg. 2001. New insights into the pathogenesis of systemic lupus erythematosus (SLE): The role of apoptosis. In M. Bilj (ed.), *Apoptosis and Autoantibodies in Systemic Lupus Erythematosus*, pp. 125–138. University of Groningen, Groningen.
76. Lautenschläger, F., S. Paschke, S. Schinkinger, A. Bruel, M. Beil, and J. Guck. 2009. The regulatory role of cell mechanics for migration of differentiating myeloid cells. *Proceedings of the National Academy of Sciences* 106 (37):15696–15701.

77. Davila, J., A. Chassepot, J. Longo, F. Boulmedais, A. Reisch, B. Frisch, F. Meyer, et al. 2012. Cyto-mechanoresponsive polyelectrolyte multilayer films. *Journal of the American Chemical Society* 134 (1):83–86.
78. Vogel, V. and M. Sheetz. 2006. Local force and geometry sensing regulate cell functions. *Nature Reviews Molecular Cell Biology* 7 (4):265–275.
79. Beningo, K.A. and Y.L. Wang. 2002. Flexible substrata for the detection of cellular traction forces. *Trends in Cell Biology* 12 (2):79–84.
80. Fu, J., Y.K. Wang, M.T. Yang, R.A. Desai, X. Yu, Z. Liu, and C.S. Chen. 2010. Mechanical regulation of cell function with geometrically modulated elastomeric substrates. *Nature Methods* 7 (9):733–736.
81. Schoen, I., W. Hu, E. Klotzsch, and V. Vogel. 2010. Probing cellular traction forces by micropillar arrays: Contribution of substrate warping to pillar deflection. *Nano Letters* 10 (5):1823–1830.
82. Lattuada, M. and T.A. Hatton. 2011. Synthesis, properties and applications of Janus nanoparticles. *Nano Today* 6 (3):286–308.
83. Swiston, A.J., J.B. Gilbert, D.J. Irvine, R.E. Cohen, and M.F. Rubner. 2010. Freely suspended cellular "backpacks" lead to cell aggregate self-assembly. *Biomacromolecules* 11 (7):1826–1832.
84. Miranville, A., C. Heeschen, C. Sengenes, C.A. Curat, R. Busse, and A. Bouloumie. 2004. Improvement of postnatal neovascularization by human adipose tissue-derived stem cells. *Circulation* 110 (3):349–355.
85. Corchero, J.L., J. Seras, E. Garcia-Fruitos, E. Vazquez, and A. Villaverde. 2010. Nanoparticle assisted tissue engineering. *Biotech International* 22 (3):13–16.
86. Ito, A., H. Jitsunobu, Y. Kawabe, and M. Kamihira. 2007. Construction of heterotypic cell—Sheets by magnetic force-based 3-D coculture of HepG2 and NIH3T3 cells. *Journal of Bioscience and Bioengineering* 104 (5):371–378.
87. Yang, Q., J. Peng, Q. Guo, J. Huang, L. Zhang, J. Yao, F. Yang, S. Wang, W. Xu, and A. Wang. 2008. A cartilage ECM-derived 3-D porous acellular matrix scaffold for in vivo cartilage tissue engineering with PKH26-labeled chondrogenic bone marrow-derived mesenchymal stem cells. *Biomaterials* 29 (15):2378–2387.
88. Christian, W., T.S. Johnson, and T.J. Gill. 2008. In vitro and in vivo cell tracking of chondrocytes of different origin by fluorescent PKH 26 and CMFDA. *Journal of Biomedical Science and Engineering* 1 (3):163–169.
89. Ruan, G., A. Agrawal, A.I. Marcus, and S. Nie. 2007. Imaging and tracking of tat peptide-conjugated quantum dots in living cells: New insights into nanoparticle uptake, intracellular transport, and vesicle shedding. *Journal of the American Chemical Society* 129 (47):14759–14766.
90. Derfus, A.M., W.C.W. Chan, and S.N. Bhatia. 2004. Intracellular delivery of quantum dots for live cell labeling and organelle tracking. *Advanced Materials* 16 (12):961–966.
91. Probst, C.E., P. Zrazhevskiy, V. Bagalkot, and X. Gao. 2013. Quantum dots as a platform for nanoparticle drug delivery vehicle design. *Advanced Drug Delivery Reviews* 65 (5):703–718.
92. Roda, A., M. Guardigli, E. Michelini, and M. Mirasoli. 2009. Bioluminescence in analytical chemistry and in vivo imaging. *TrAC Trends in Analytical Chemistry* 28 (3):307–322.
93. de Boer, J., C. van Blitterswijk, and C. Löwik. 2006. Bioluminescent imaging: Emerging technology for non-invasive imaging of bone tissue engineering. *Biomaterials* 27 (9):1851–1858.
94. Román, I., M. Vilalta, J. Rodriguez, A.M. Matthies, S. Srouji, E. Livne, J.A. Hubbell, N. Rubio, and J. Blanco. 2007. Analysis of progenitor cell–scaffold combinations by in vivo non-invasive photonic imaging. *Biomaterials* 28 (17):2718–2728.

95. Liu, J., A. Barradas, H. Fernandes, F. Janssen, B. Papenburg, D. Stamatialis, A. Martens, C. van Blitterswijk, and J. de Boer. 2009. In vitro and in vivo bioluminescent imaging of hypoxia in tissue-engineered grafts. *Tissue Engineering Part C: Methods* 16 (3):479–485.
96. Vilalta, M., C. Jorgensen, I.R. Dégano, Y. Chernajovsky, D. Gould, D. Noël, J.A. Andrades, J. Becerra, N. Rubio, and J. Blanco. 2009. Dual luciferase labelling for non-invasive bioluminescence imaging of mesenchymal stromal cell chondrogenic differentiation in demineralized bone matrix scaffolds. *Biomaterials* 30 (28):4986–4995.
97. George, T.C., D.A. Basiji, B.E. Hall, D.H. Lynch, W.E. Ortyn, D.J. Perry, M.J. Seo, C.A. Zimmerman, and P.J. Morrissey. 2004. Distinguishing modes of cell death using the ImageStream® multispectral imaging flow cytometer. *Cytometry Part A* 59 (2):237–245.
98. Basiji, D.A., W.E. Ortyn, L. Liang, V. Venkatachalam, and P. Morrissey. 2007. Cellular image analysis and imaging by flow cytometry. *Clinics in Laboratory Medicine* 27 (3):653–670.
99. Filby, A., E. Perucha, H. Summers, P. Rees, P. Chana, S. Heck, G.M. Lord, and D. Davies. 2011. An imaging flow cytometric method for measuring cell division history and molecular symmetry during mitosis. *Cytometry Part A* 79 (7):496–506.
100. Marangon, I., N. Boggetto, C. Ménard-Moyon, E. Venturelli, M.L. Béoutis, C. Pechoux, N. Luciani, C. Wilhelm, A. Bianco, and F. Gazeau. 2012. Intercellular carbon nanotube translocation assessed by flow cytometry imaging. *Nano Letters* 12 (9):4830–4837.
101. Pampaloni, F., E.G. Reynaud, and E.H.K. Stelzer. 2007. The third dimension bridges the gap between cell culture and live tissue. *Nature Reviews Molecular Cell Biology* 8 (10):839–845.
102. Verveer, P.J., J. Swoger, F. Pampaloni, K. Greger, M. Marcello, and E.H.K. Stelzer. 2007. High-resolution three-dimensional imaging of large specimens with light sheet-based microscopy. *Nature Methods* 4 (4):311–313.
103. Helmchen, F. and W. Denk. 2005. Deep tissue two-photon microscopy. *Nature Methods* 2 (12):932–940.
104. Diaspro, A., P. Bianchini, G. Vicidomini, M. Faretta, P. Ramoino, and C. Usai. 2006. Multi-photon excitation microscopy. *Biomedical Engineering Online* 5:36.
105. Campagnola, P. 2011. Second harmonic generation imaging microscopy: Applications to diseases diagnostics. *Analytical Chemistry* 83 (9):3224.
106. Freund, I., M. Deutsch, and A. Sprecher. 1986. Connective tissue polarity. Optical second-harmonic microscopy, crossed-beam summation, and small-angle scattering in rat-tail tendon. *Biophysical Journal* 50 (4):693–712.
107. Pedersen, J.A. and M.A. Swartz. 2005. Mechanobiology in the third dimension. *Annals of Biomedical Engineering* 33 (11):1469–1490.
108. Bilgin, C.C., A.W. Lund, A. Can, G.E. Plopper, and B. Yener. 2010. Quantification of three-dimensional cell-mediated collagen remodeling using graph theory. *PLoS ONE* 5 (9):e12783.
109. Rice, W.L., D.L. Kaplan, and I. Georgakoudi. 2010. Two-photon microscopy for non-invasive, quantitative monitoring of stem cell differentiation. *PLoS ONE* 5 (4):e10075.
110. Evans, C.L. and X.S. Xie. 2008. Coherent anti-Stokes Raman scattering microscopy: chemical imaging for biology and medicine. *Annual Review of Analytical Chemistry* 1:883–909.
111. Schliwa, M. and G. Woehlke. 2001. Molecular motors: Switching on kinesin. *Nature* 411 (6836):424–425.
112. Liu, P., Y. Ying, Y. Zhao, D.I. Mundy, M. Zhu, and R.G.W. Anderson. 2004. Chinese hamster ovary K2 cell lipid droplets appear to be metabolic organelles involved in membrane traffic. *Journal of Biological Chemistry* 279 (5):3787–3792.
113. Nan, X., W.Y. Yang, and X.S. Xie. 2004. CARS microscopy lights up lipids in living cells. *Biophotonics International* 11 (8):44–47.

114. Brackmann, C., J.O. Dahlberg, N.E. Vrana, C. Lally, P. Gatenholm, and A. Enejder. 2012. Non-linear microscopy of smooth muscle cells in artificial extracellular matrices made of cellulose. *Journal of Biophotonics* 5 (5–6):404–414.
115. Kluge, J.A., G.G. Leisk, R.D. Cardwell, A.P. Fernandes, M. House, A. Ward, A.L. Dorfmann, and D.L. Kaplan. 2011. Bioreactor system using noninvasive imaging and mechanical stretch for biomaterial screening. *Annals of Biomedical Engineering* 39 (5):1390–1402.
116. Niklason, L.E., A.T. Yeh, E.A. Calle, Y. Bai, A. Valentín, and J.D. Humphrey. 2010. Enabling tools for engineering collagenous tissues integrating bioreactors, intravital imaging, and biomechanical modeling. *Proceedings of the National Academy of Sciences* 107 (8):3335–3339.
117. Hofmann, M.C., B.M. Whited, T. Criswell, M.N. Rylander, C.G. Rylander, S. Soker, G. Wang, and Y. Xu. 2012. A fiber-optic-based imaging system for nondestructive assessment of cell-seeded tissue-engineered scaffolds. *Tissue Engineering Part C: Methods* 18 (9):677–687.
118. Greenbaum, A. and A. Ozcan. 2012. Maskless imaging of dense samples using pixel super-resolution based multi-height lensfree on-chip microscopy. *Optics Express* 20 (3):3129–3143.
119. Isikman, S.O., W. Bishara, S. Mavandadi, W.Y. Frank, S. Feng, R. Lau, and A. Ozcan. 2011. Lens-free optical tomographic microscope with a large imaging volume on a chip. *Proceedings of the National Academy of Sciences* 108 (18):7296–7301.
120. Cohen-Karni, T., R. Langer, and D.S. Kohane. 2012. The smartest materials: The future of nanoelectronics in medicine. *ACS Nano* 6 (8):6541–6545.
121. Vrana, N.E., A. Dupret, C. Coraux, D. Vautier, C. Debry, and P. Lavalle. 2011. Hybrid titanium/biodegradable polymer implants with an hierarchical pore structure as a means to control selective cell movement. *PLoS ONE* 6 (5):e20480.
122. Staples, M., K. Daniel, M. Cima, and R. Langer. 2006. Application of micro- and nano-electromechanical devices to drug delivery. *Pharmaceutical Research* 23 (5):847–863.
123. Jin, Q., G. Wei, Z. Lin, J.V. Sugai, S.E. Lynch, P.X. Ma, and W.V. Giannobile. 2008. Nanofibrous scaffolds incorporating PDGF-BB microspheres induce chemokine expression and tissue neogenesis in vivo. *PLoS ONE* 3 (3):e1729.
124. Gao, C., F. Tao, W. Lin, Z. Xu, and Z. Xue. 2008. Ordered arrays of magnetic metal nanotubes and nanowires encapsulated with carbon tubes. *Journal of Nanoscience and Nanotechnology* 8 (9):4494–4499.
125. Hwang, G.L. and K.C. Hwang. 2001. Carbon nanotube reinforced ceramics. *Journal of Materials Chemistry* 11 (6):1722–1725.
126. Supronowicz, P.R., P.M. Ajayan, K.R. Ullmann, B.P. Arulanandam, D.W. Metzger, and R. Bizios. 2001. Novel current-conducting composite substrates for exposing osteoblasts to alternating current stimulation. *Journal of Biomedical Materials Research* 59 (3):499–506.
127. Meng, D., J. Ioannou, and A.R. Boccaccini. 2009. Bioglass®-based scaffolds with carbon nanotube coating for bone tissue engineering. *Journal of Materials Science: Materials in Medicine* 20 (10):2139–2144.
128. Chahine, N.O. and P.-h.G. Chao. 2011. Micro and nanotechnologies for tissue engineering. In J.A. Burdick and R.L. Mauck (eds), *Biomaterials for Tissue Engineering Applications: A Review of the Past and Future Trends*, pp. 139–178. Springer, Berlin.
129. Mattson, M.P., R.C. Haddon, and A.M. Rao. 2003. Molecular functionalization of carbon nanotubes and use as substrates for neuronal growth. *Journal of Molecular Neuroscience* 14: 175–182.
130. Lewitus, D.Y., J. Landers, J.R. Branch, K.L. Smith, G. Callegari, J. Kohn, and A.V. Neimark. 2011. Biohybrid carbon nanotube/agarose fibers for neural tissue engineering. *Advanced Functional Materials* 21 (14):2624–2632.

131. Jan, E. and N.A. Kotov. 2007. Successful differentiation of mouse neural stem cells on layer-by-layer assembled single-walled carbon nanotube composite. *Nano Letters* 7 (5):1123–1128.
132. Fischer, K.E., B.J. Alemán, S.L. Tao, R. Hugh Daniels, E.M. Li, M.D. Bünger, G. Nagaraj, P. Singh, A. Zettl, and T.A. Desai. 2009. Biomimetic nanowire coatings for next generation adhesive drug delivery systems. *Nano Letters* 9 (2):716–720.
133. Tian, B., J. Liu, T. Dvir, L. Jin, J.H. Tsui, Q. Qing, Z. Suo, R. Langer, D.S. Kohane, and C.M. Lieber. 2012. Macroporous nanowire nanoelectronic scaffolds for synthetic tissues. *Nature Materials* 11 (11):986–994.
134. Ainslie, K.M. and T.A. Desai. 2008. Microfabricated implants for applications in therapeutic delivery, tissue engineering, and biosensing. *Lab on a Chip* 8 (11):1864–1878.
135. Heath, J.P. and L.D. Peachey. 1989. Morphology of fibroblasts in collagen gels: A study using 400 keV electron microscopy and computer graphics. *Cell Motility and the Cytoskeleton* 14 (3):382–392.
136. Li, W.J., C.T. Laurencin, E.J. Caterson, R.S. Tuan, and F.K. Ko. 2002. Electrospun nanofibrous structure: A novel scaffold for tissue engineering. *Journal of Biomedical Materials Research* 60 (4):613–621.
137. Davidson, P.M., H. Özçelik, V. Hasirci, G. Reiter, and K. Anselme. 2009. Microstructured surfaces cause severe but non-detrimental deformation of the cell nucleus. *Advanced Materials* 21 (35):3586–3590.

13 Bioinspired Nanomaterials for Bone Regeneration

Esmaiel Jabbari

CONTENTS

13.1 INTRODUCTION

The current clinical methods for treating skeletal defects involve bone transplantation or the use of other materials to restore continuity [1]. Autologous bone grafting has been the gold standard for bone replacement because it provides such essential elements as osteogenic cells, osteoinductive factors, and an osteoconductive matrix for healing. However, the limited supply of autograft bone, donor-site morbidity, and the long recovery time for segmental defects restrict its use in bone repair. Allograft bone, although abundant in supply, has drawbacks that include reduced rates of graft incorporation compared with autograft bone. Furthermore, the long recovery time for segmental defects or partial recovery in the case of nonunions has prompted researchers to look for alternative bone grafts to accelerate the rate of fracture healing.

A titanium cage packed with autologous bone from the iliac crest has been used in interbody spine fusion [2]. The titanium cage provides mechanical stability so an additional fixation plate is not necessary. Titanium cages coated with collagen or poly(D,L-lactide) (PDLA) sponge loaded with recombinant human bone morphogenetic-2 (rhBMP-2) have shown significantly higher bone callus volume (BCV) and mechanical stiffness after 8 and 12 weeks when implanted in the cervical spine [2]. Compared with collagen sponge, a PDLA-coated titanium cage resulted in a higher BCV and a greater progression of interbody callus formation. Although metal cages provide mechanical support at the defect site, they shield stress in the regenerating region during remodeling, resulting in lower bone mass density and less than ideal integration with the host tissue.

Calcium phosphate (CaP) bioactive ceramics such as hydroxyapatite (HA), tricalcium phosphate (TCP), and biphasic calcium phosphate (BCP) have been used as a carrier for rhBMP-2 to induce clinically relevant bone formation [3,4]. Bioactive ceramics have been shown to promote bone ingrowth and certain compositions harden *in situ*. Granules composed of 60% HA and 40% TCP are used as a carrier for rhBMP-2 for posterolateral lumbar spine fusion in human subjects [3]. However, their potential drawbacks are fatigue fracture and their very low strength in shear and tension, which limit their use to fractures that are subject to uniform compressive loading. Furthermore, their initial mechanical strength is less than cancellous bone, which leads to difficulty in maintaining the composite within the defect during surgery.

Stress shielding and particulate wear are potential concerns with the use of nondegradable polymers such as poly(methyl methacrylate) (PMMA). In an attempt to control the degradation profile and improve the stability of the carrier against soft tissue compression, poly(lactide-co-glycolide) (PLGA) sponges or granules have been used to deliver rhBMP-2 for bone repair [5]. rhBMP-2-impregnated porous PLGA particles, implanted intramuscularly in nude mice, induced greater formation of new bone compared with active demineralized bone allograft [6]. Open-cell poly(L-lactide) (PLA) has been used to induce transverse process fusion in a canine model [7]. Recently, low-molecular-weight (LMW) PLGA, as a biodegradable slow delivery system for rhBMP-2, has been tested in the calf muscle of Wistar rats [5]. LMW PLGA accelerated the resorption rate, which in turn released rhBMP-2 at a faster rate, resulting in new bone formation 3 weeks after implantation [5]. To accelerate the resorption rate of the carrier, gelatin sponge coated with PLGA polymer has been used as a delivery system for rhBMP-2 to induce bone formation in the tibial diaphysis of dogs with internal plate fixation [8]. A major drawback of coated gelatin sponge is the need for a second surgical procedure to remove the internal fixation plate.

Bioinspired synthetic composites are an attractive option because material degradation can be designed to coincide with the rate of tissue regeneration for a given application. The flexibility in the design of polymer composites allows the synthesis of a wide range of properties. For instance, their mechanical and degradation properties can be tailored to a particular application by changing the molecular weight to fit a particular application. Degradable polymers based on polyhydroxyalkanoates (PHA) have been used as scaffolds for guided regeneration of skeletal tissues in orthopedics [9]. PHA polymers include poly(L-lactic acid) (PLLA), poly(glycolic acid) (PGA), polycaprolactone (PCL), poly(trimethylene carbonate), poly(butylene terephthalate), poly(hydroxybutyrate), poly(hydroxyvalerate), and poly(dioxanone) (PDS) and their copolymers [10–14]. The development of self-reinforced PGA (SR-PGA) and PLLA (SR-PLLA) has expanded the use of PHA in load-bearing orthopedic applications [15,16]. A hybrid of PLGA and collagen has been used for guided bone regeneration to treat periodontesis [17]. A group of poly(anhydrides) consisting of different ratios of sebacic acid (SA) and 1,6-bis(p-carboxyphenoxy) hexane (CPH) have been developed that degrade by surface erosion while maintaining their bulk mechanical properties [18].

Photopolymerizable RGD-modified poly(ethylene glycol) (PEG) hydrogels have been used to enhance cytoskeletal reorganization of osteoblasts after encapsulation [19]. Alginate hydrogels grafted with adhesion ligands have also been used for encapsulating calvarial osteoblasts [20]. An *in situ* cross-linkable enzymatically degradable poly(lactide-co-glycolide ethylene oxide) macromer has been developed for encapsulating bone marrow stromal (BMS) cells [21]. A novel class of PEG-based gel chain extended with very short hydroxy acid segments (SPEXA macromer) has been developed that exhibits tunable degradation from a few days to many months and a tunable modulus ranging from 1 kPa to 1 MPa [22–24]. The hydroxy acid monomers included the least hydrophobic, glycolide (G), lactide (L), p-dioxanone (D), and the most hydrophobic, ε-caprolactone (C). Chain extension of PEG with short hydroxy acid segments resulted in micelle formation for all hydroxy acid types. The wide range of degradation rates observed for SPEXA gels can be explained by large differences in the equilibrium water content of the micelles for different hydroxy acid monomer types. A biphasic relationship between the hydroxy acid segment length and the gel degradation rate was observed for all monomers, which was related to the transition from surface (controlled by segment length) to bulk (controlled by micelle equilibrium water content) hydrolysis within the micelle phase [22]. The PEG gels chain extended with short hydroxy acid segments are particularly useful for cell delivery in three-dimensional (3-D) matrices because a major fraction of the toxic polymerization photoinitiator is partitioned to the micellar phase, leading to higher viability of the encapsulated cells [22].

Natural [25–27] and synthetic [28–32] polymers reinforced with CaPs exhibit compressive strengths in the range of 2–30 MPa, suitable for trabecular bone replacement. The addition of HA to drawn PLLA not only improved biodegradability and strength but it also increased compatibility with the bone tissue. Cell-based approaches using osteoblasts [33], bone marrow–derived osteoprogenitor cells [34], and stem cells [35] are attractive for bone reconstruction. The rate of extracellular matrix (ECM) deposition and mineralization in cell-based systems depends on the presence of soft tissue for the blood supply and degradation rate, porosity, pore size, and the extent of pore interconnectivity of the scaffold [36,37]. Although progress has been made in developing composites for skeletal tissue regeneration [38,39], these matrices are not applicable for the reconstruction of load-bearing tissues, because they do not mimic the bone's natural micro- and nanostructure.

13.2 BONE STRUCTURE

Bone is a composite material consisting of a collagenous and an apatite phase [40]. The inorganic apatite crystals contribute approximately 65% to the wet weight of the bone while the collagenous phase contributes 20% [40]. Glycoproteins and proteoglycans control the water content of the bone to 15% [41]. The collagenous phase gives bone its form and contributes to its ability to resist bending, while the apatite crystals resist compression [42]. Bone exhibits many orders of hierarchical structures from macroscopic to microscopic, submicroscopic, nanoscale, and subnanostructure length scales [43]. These levels of organization include the cancellous and cortical macrostructures, the osteons (10–500 μm), the lamellae submicrostructures

(1–10 μm), the fibrillar collagen with embedded nanocrystals (100–1000 nm), and the molecular structure of the constituent elements [44]. The bone properties are determined at the nanoscale with CaP nanocrystals embedded in the tough collagen fibrils. Extended sheets of minerals are positioned along the long axis of the collagen molecules and parallel to one another in the neighboring collagen layers. The sheets fuse with one another to form a structurally stable composite. The plate-shaped crystals are 2–3 nm thick and tens of nanometers long and wide. The collagen fibers in the matrix have different levels of organization from 1–4 nm for fibrils to 50–70 nm for fibers and 150–250 nm for bundles.

The collagenous phase plays a central role in the regulation of mineralization; the control of cell division, migration, differentiation, and maturation; the maintenance of matrix integrity; growth factor modulation; and the extent of mineral–collagen interactions [45,46]. Noncollagenous proteins play complex functions during bone formation and remodeling [47]. The collagenous phase facilitates communication with the cellular environment and provides a medium for proteins to maintain their bioactivity [48]. For example, osteonectin (ON) protein is involved in several functions including linking the mineral and collagenous phases and regulating mineralization. An engineered composite should mimic the complexity of the bone ECM with nanoscale structures to control cell–matrix interactions.

13.3 COLLAGEN NANOSTRUCTURE AND ITS EFFECT ON CELL DIFFERENTIATION

BMS cells are a heterogeneous population that gives rise to multiple differentiated connective tissue cells [49,50]. The differentiation of stem cells depends on the growth medium components as well as the type of substrate cells cultured [51,52]. The effect of substrate on the differentiation of progenitor BMS cells was investigated with aligned collagen tubular scaffolds [53]. BMS cells seeded into collagen tubes and cultured in an osteogenic medium were assessed by immunofluorescence for the pattern of specific proteins. The staining patterns for bone-related proteins and mineralization are shown in Figure 13.1. BMS cells displayed cell aggregates positive for osteopontin (OP) (Figure 13.1a), osteocalcin (OC), and ON (not shown) and alkaline phosphatase (ALP; Figure 13.1b). BMS cells formed multilayered mineralized structures positive for alizarin red (Figure 13.1c). The gene expression profile of the BMS cells seeded on the collagen scaffolds showed upregulation of OP (an early-stage marker) and a marked upregulation of OC at Day 12, which was maintained through Day 18. BMS cells showed vessel-like structures positive for platelet endothelial cell adhesion molecule-1 (PECAM-1), Flk-1, α-smooth muscle actin (α-SMA), and tomato lectin (Figure 13.1d). Linear as well as branching tubular structures consisting of endothelial cells surrounded and wrapped by elongated α-SMA-positive cells were observed (Figure 13.1e). In addition, areas of extensive neovascularization were observed (Figure 13.1f).

It is well established that type I collagen promotes osteogenic differentiation of BMS cells in conventional two-dimensional (2-D) monolayer cultures [54,55]. The ECM has profound effects on cell behavior including cell proliferation, migration, and vasculogenesis [69]. Three-dimensional matrices containing collagen have

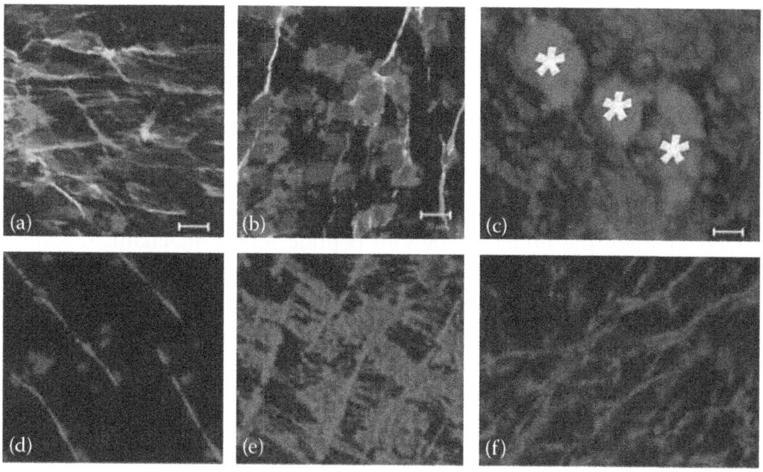

FIGURE 13.1 **(See color insert)** Expression pattern of osteogenic markers osteopontin (a), osteocalcin (b), and alizarin red (c) in collagen tubes; expression pattern of vasculogenic markers α-SMA of capillary-like structures (red, d), parallel sheets (red, e), and tomato lectin in tubular structures (red, f). DAPI stained blue is for nuclei and phalloidin stained green is for actin (scale bars are 20 μm).

been shown to promote the formation of tubular structures resembling capillaries in cultured endothelial cells [70]. These findings demonstrate that the structural organization of the scaffold at the nano- and microscale affects the maturation and differentiation of the BMS cells to multiple pathways including osteogenesis and vasculogenesis.

13.4 PEPTIDE-REINFORCED NANOCOMPOSITES

Hydrogels reinforced with CaP fillers are very attractive for bone tissue regeneration [56,57]. They exhibit the typical mechanical properties of bone and replicate the bone microstructure [58]. Composites with nanoparticles (<100 nm) exhibit a higher stiffness compared with those with microparticles at the same volume fraction [59]. The filler particles do not affect the relaxation time of the polymer chains, as the radius of gyration of the chains in microcomposites is orders of magnitude smaller than the size of the filler particles. However, in nanocomposites the radius of gyration of the chains is of the same order of magnitude as the size of the filler particles, thus increasing the relaxation time of the chains interacting with multiple filler particles. It has been shown that the mechanical performance of the composites is significantly improved by increasing the interfacial adhesion between the polymer and dispersed particle phases [60]. While filled systems with large particles (>1 μm) exhibit linear behavior in a wide range of deformations, nanofilled composites exhibit nonlinearity even at small strains [61,62]. ON is one of the proteins in the bone matrix with a strong affinity for both collagen and HA and is a nucleator of mineralization [63]. The first 17 N-terminal amino acids of ON are responsible for binding to the bone collagen network while a glutamic acid (Glu)-rich sequence binds to HA nanocrystals [64]. ON has the highest affinity for calcium ions among all bone proteins and the apatite formed in the presence of ON has the smallest crystal size [65]. These observations suggest that the switch from OP to ON expression in the ossification front limits the size of the apatite crystals and links the collagen network to the mineral phase.

To test the effect of interactions between the apatite crystals and the collagenous matrix, an acrylate-functionalized Glu peptide derived from ON (Ac-Glu6) was synthesized to study the effect of the Glu peptide on the mechanical response of hydrogel–apatite composites [66]. A similar procedure was used to synthesize acrylate-functionalized neutral glycine and positively charged lysine peptides (Ac-Gly6 and Ac-Lys6). The gelation kinetics of a suspension of Ac-Glu6, Ac-Gly66, or Ac-Lys6 attached to the surface of HA microparticles (diameter, 50 μm) or nanoparticles (diameter, 50 nm) (Figure 13.2a) and the cross-linkable poly(lactide ethylene oxide fumarate) (PLEOF) copolymer were compared with rheometry [67–73]. Nanocomposites (treated and untreated) displayed much higher stiffness compared with microcomposites, at the same volume fraction. The storage modulus of the composites prepared with micron-sized particles was reasonably predicted by the Guth–Smallwood equation [74], while those prepared with nanosized HA could not be explained solely by hydrodynamic effects.

Contrary to the Glu6 peptide, Gly6 and Lys6 had no ionic interactions with HA nanoparticles. HA/Ac-Gly6 and HA/Ac-Lys6 nanocomposites did not show a significant change in shear modulus compared with that without HA surface treatment, as shown in Figure 13.2b. These results demonstrate that the increase in the overall viscoelastic response of the nanocomposite was specific to the Ac-Glu6 peptide and the reinforcement was amplified as the size of the nanoparticles was reduced from 5 μm to 50 nm. The ionic bonds formed by the interaction of Ac-Glu6 with HA had a relatively higher adsorption energy and a longer residence time on the particles. These results demonstrated that one of the factors contributing to bone toughness is the cross-linking of the soft collagenous phase and the hard mineral phase, mediated by specific peptides derived from ON and other soluble proteins in the bone matrix.

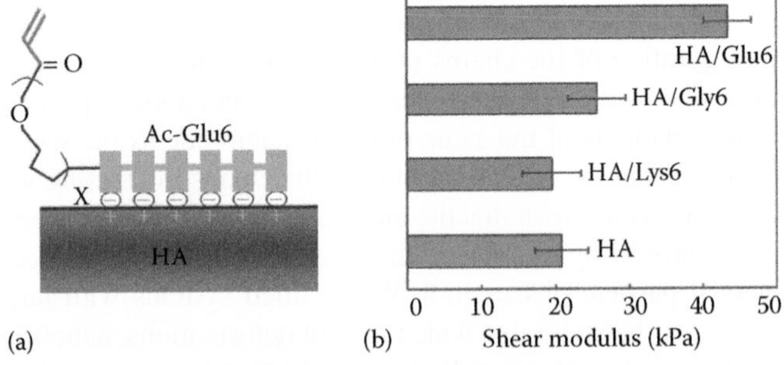

FIGURE 13.2 (a) Schematic diagram showing the binding of the Glu sequence of osteonectin protein to HA nanoparticles. (b) Comparison of the modulus of the nanocomposites treated with the Ac-Glu6 peptide with those treated with Ac-Gly6 and Ac-Lys6.

13.5 BONE-MIMETIC LAMINATED COMPOSITES

Bone is a composite matrix composed of mineralized aligned collagen nanofibers (NFs) [75]. The apatite nanocrystals provide osteoconductivity and stiffness while the collagen fibers provide elasticity and a template for mineralization and maturation of osteoprogenitor cells. Unique factors that contribute to bone toughness are the aligned network of collagen NFs, apatite nanocrystals, and proteins in the bone ECM that link the apatite crystals to the collagen fibers and form a network. In a biomimetic approach, NFs were coated with CaP crystals by incubation in a modified simulated body fluid (SBF) [76,77]. The latter approach could mimic the morphology of the mineralized bone matrix but the approach was limited by the depth of penetration of calcium and phosphate ions to the interior parts of the fiber sheet and crystal nucleation on the fiber surface.

The nucleation, growth, and stabilization of CaP nanocrystals on the collagen fibers in the bone matrix are mediated by ECM noncollagenous proteins such as bone sialoprotein (BSP), ON, OP, and OC [69,78]. Glu sequences ranging from 2 to 10 residues in those proteins regulate the nucleation and growth of CaP crystals on the collagen fibers [78]. The peptide EEGGC, consisting of two Glu residues terminated with a cysteine at one end and two glycine residues at the other end, was synthesized in the solid phase [79]. The peptide was conjugated to an acrylate-terminated LMW polylactide (PLAA) chain by the reaction between the acrylate on the polymer and the cysteine residue on the peptide [79]. The PLA–Glu conjugate was mixed with high molecular weight PLGA and electrospun to form NFs with Glu functionality on the surface (Glu-NF microsheets). CaP nanocrystals were grown on the fibers by incubating the fibers in an SBF with the Glu sequence serving as the chelating agent for calcium and phosphate ions. Figure 13.3a and b show scanning electron microscope images of Glu-NF fibers before and after incubation in SBF, respectively. Figure 13.3b shows the uniform distribution of CaP crystals on NFs within as well as on the surface of the microsheet. The average size of the crystals was below 100 nm. The modulus of the fiber sheet increased fourfold after incubation in SBF and CaP crystal nucleation [79] (Figure 13.3c).

Next, a layer-by-layer (LBL) approach was used to overcome the limited diffusion of calcium and phosphate ions to the interior parts of the microsheets, thus achieving significantly higher CaP to-fiber ratios and thicker sheets to study the osteogenic differentiation of BMS cells on the fiber sheets [79]. BMS cells were seeded on the fiber

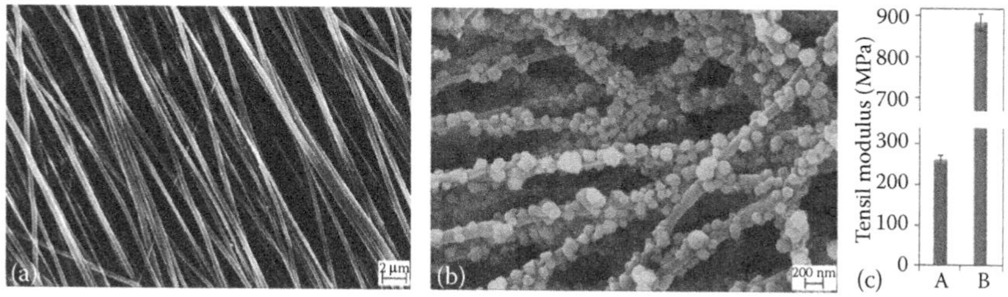

FIGURE 13.3 SEM images of Glu-NF nanofibers before (a) and after (b) incubation in simulated body fluid. (c) Tensile strength of a Glu-NF fiber sheet before and after incubation in SBF. The scale bars in (a) and (b) are 2 μm and 200 nm, respectively.

sheets before and after incubation in SBF for CaP nucleation and they were cultured in osteogenic media for 28 days. The BMS cells spread in the direction of fiber alignment irrespective of their incubation in SBF solution. The ALP activity, calcium content, and collagen type-I expression of BMS cells cultured on the microsheets before and after incubation in SBF are shown in Figure 13.4a–c, respectively. ALP activity increased significantly with incubation in SBF prior to cell seeding. For example, peak ALP activity of the fibers increased from 1600 ± 100 to 5100 ± 400 IU/mg DNA after incubation in SBF. The calcium content of the BMS cells increased from 460 ± 40 to 960 ± 110 mg/mg DNA after incubation in SBF.

When the LBL approach was used to produce CaP-deposited microsheets, there was a dramatic increase in the CaP content to >200% of the fiber weight, indicating that CaP nucleation and growth on the fibers was controlled by the diffusion of calcium/phosphate ions from the SBF medium to the interior part of the microsheets. It should be noted that the CaP content of LBL microsheets was between those of cancellous (160%) and cortical (310%) bone [80], demonstrating that CaP contents as high as that of cortical bone can be achieved by reducing layer thickness. The LBL approach has been used to increase filler loading and the stiffness of composites or fabricate highly ductile oppositely charged multilayered films [81]. Our work demonstrates for the first time that the extent of CaP nucleation and growth on synthetic fibers, and in turn the fiber strength, can be controlled by the incubation time in SBF or by the density of the Glu-containing peptide on the fiber surface.

The tensile modulus and toughness of the microsheets dramatically increased with higher CaP contents of 200% after incubation in SBF, as shown in Figure 13.3c. The fiber size, CaP crystal size, and nucleation and growth of CaP crystals on the fiber surface contributed to the higher modulus and toughness of the microsheets, as shown schematically in Figure 13.5. The Glu peptide initiates nucleation of CaP crystals on the fiber surface (Figure 13.5b). The nucleated CaP crystals continue to

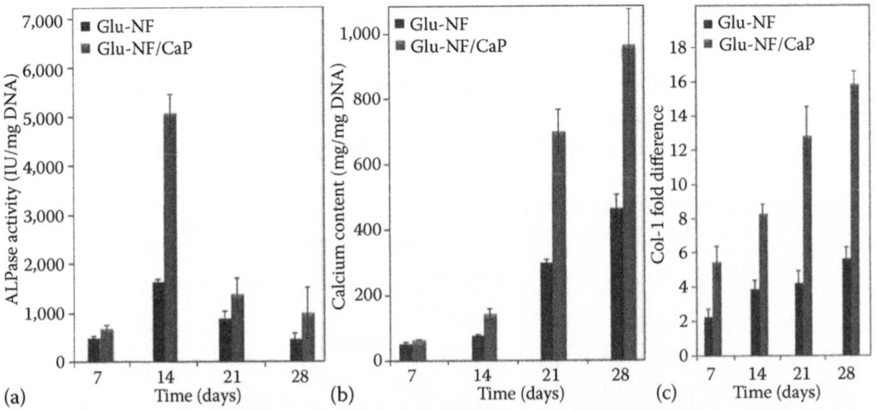

FIGURE 13.4 ALP activity (a), calcium content (b), and collagen type-I mRNA expression (c) of BMS cells seeded on microsheets before (gray) and after (black) incubation in the simulated body fluid (SBF) for CaP deposition and incubated in an osteogenic medium for up to 28 days. Error bars correspond to means ± 1 SD for $n = 3$.

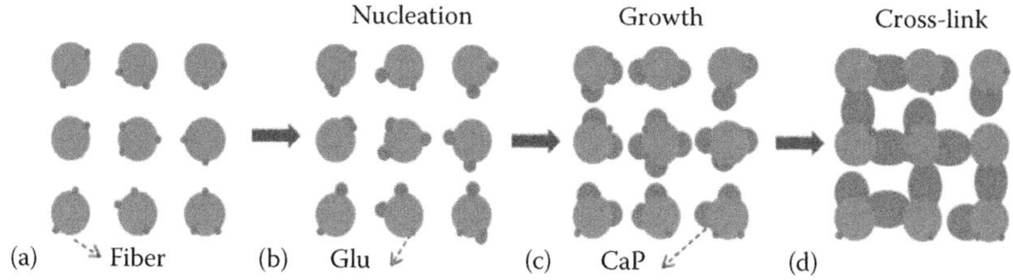

FIGURE 13.5 **(See color insert)** Schematic diagram demonstrating the effect of CaP deposition on Glu nanofibers on the strength of the microsheets. Fiber cross section, Glu peptide, and CaP crystals are represented by brown, red, and gray circles, respectively. The Glu on the fiber surface (a) initiates the nucleation of CaP crystals (b). The CaP crystals continue to grow with the incubation time (c) and start to merge and fuse to form a network of fibers cross-linked by CaP crystals (d).

grow with incubation time (Figure 13.5c) and start to merge and fuse to form a network of fibers cross-linked by the CaP crystals (Figure 13.5d). We speculate that the higher toughness of the microsheets with 200% CaP content is related to the formation of a CaP cross-linked network of fibers.

The extent of osteogenic differentiation and maturation of the BMS cells seeded on the microsheets increased with CaP deposition. This increase was explained by the higher stiffness as well as the higher CaP content and connectivity of the crystals on the microsheets prior to cell seeding, consistent with prior results that suggested matrix stiffness and CaP content enhance osteogenesis of the BMS cells [82]. The findings of this work demonstrate that Glu conjugation coupled with an LBL approach dramatically increases the nucleation and growth of apatite-like nanocrystals on the surface of NFs, leading-to-CaP to fiber ratios >200%, a modulus close to 1 GPa, and higher osteogenic differentiation of the BMS cells. The CaP content of the NFs can be further increased to reach that of cortical bone by reducing the thickness of the individual fiber layers in the LBL approach or by reducing the porosity of the fiber sheets. The CaP-deposited Glu-NF composites are potentially useful as a biomimetic matrix in the regeneration of skeletal tissues.

13.6 CONCLUSIONS

In this chapter, we have summarized recent advances in the development of bioinspired nanomaterials to provide strength and to support complex cell–matrix interactions in bone regeneration. We demonstrated that the surface treatment of apatite nanocrystals with a peptide mimicking the terminal region of the ON glycoprotein of bone resulted in an order of magnitude increase in the modulus of the composite hydrogel, with the reinforcement effect modulated by the size of the crystals. We further demonstrated that osteon–mimetic composites can be generated by the nucleation and growth of CaP crystals on the Glu-modified aligned NFs followed by an LBL lamination approach to dramatically increase CaP-to-fiber ratios, modulus, and osteogenic differentiation of BMS cells.

ACKNOWLEDGMENTS

Preparation of this chapter was supported by research grants to E. Jabbari from the National Science Foundation under grant nos. CBET0756394 and DMR1049381, the National Institutes of Health under grant no. DE19180, and the Arbeitsgemeinschaft fur Osteosynthesefragen (AO) Foundation under grant no. C10-44J.

REFERENCES

1. M. J. Yaszemski, J. B. Oldham, L. Lu, and B. L. Currier, Clinical needs for bone tissue engineering technology, in J. E. Davis (ed.), *Bone Engineering*, pp. 541–547. EM Squared, Toronto, 2000.
2. M. E. Majd, M. Vadhva, and R. T. Holt, Anterior cervical reconstruction using titanium cages with anterior plating, *Spine*, 24, 1604–1610, 1999.
3. T. Ohyama, Y. Kubo, H. Iwata, and W. Taki, β-tricalcium phosphate combined with recombinant human bone morphogenetic protein-2: A substitute for autograft, used for packing interbody fusion cages in canine lumbar spine, *Neurol. Med. Chir.*, 44, 234–241, 2004.
4. T. Akamaru, D. Suh, S. D. Boden, H. S. Kim, A. Minamide, and J. Louis-Ugbo, Simple carrier matrix modifications can enhance delivery of recombinant human bone morphogenetic protein-2 for posterolateral spine fusion, *Spine*, 28, 429–434, 2003.
5. K. Bessho, D. L. Carnes, R. Cavin, and J. L. Ong, Experimental studies on bone induction using low-molecular-weight poly (DL-lactide-co-glycolide) as a carrier for recombinant human bone morphogenetic protein-2, *J. Biomed. Mater. Res.*, 61, 61–65, 2002.
6. B. D. Boyan, C. H. Lohmann, A. Somers, G. G. Niederauer, J. M. Wozney, D. D. Dean, D. L. Carnes, and Z. Schwartz, Potential of porous poly-D,L-lactide-co-glycolide particles as a carrier for recombinant human bone morphogenetic protein-2 during osteoinduction in vivo, *J. Biomed. Mater. Res.*, 46, 51–59, 1999.
7. H. S. Sandhu, L. E. Kanim, J. M. Kabo, J. M. Toth, E. N. Zeegan, D. Liu, L. L. Seeger, and E. G. Dawson, Evaluation of rhBMP-2 with an OPLA carrier in a canine posterolateral spinal fusion model, *Spine*, 20, 2669–2682, 1995.
8. S. Kokubo, M. Mochizuki, S. Fukushima, T. Ito, K. Nozaki, T. Iwai, K. Takahashi, S. Yokota, K. Miyata, and N. Sasaki, Long-term stability of bone tissues induced by an osteoinductive biomaterial, recombinant human bone morphogenetic protein-2 and a biodegradable carrier, *Biomaterials*, 25, 1795–1803, 2004.
9. H. Ueda and Y. Tabata, Polyhydroxyalkanonate derivatives in current clinical applications and trials, *Adv. Drug Deliv. Rev.*, 55, 501–518, 2003.
10. S. H. Rhee, Bone-like apatite-forming ability and mechanical properties of poly(epsilon-caprolactone)/silica hybrid as a function of poly(epsilon-caprolactone) content, *Biomaterials*, 25, 1167–1175, 2004.
11. H. Suh, M. J. Song, M. Ohata, Y. B. Kang, and S. Tsutsumi, Ex vivo mechanical evaluation of carbonate apatite-collagen-grafted porous poly-L-lactic acid membrane in rabbit calvarial bone, *Tissue Eng.*, 9, 635–643, 2003.
12. J. M. Karp, M. S. Shoichet, and J. E. Davies, Bone formation on two-dimensional poly(DL-lactide-co-glycolide) (PLGA) films and three-dimensional PLGA tissue engineering scaffolds in vitro, *J. Biomed. Mater. Res. A*, 64, 388–396, 2003.
13. F. Kandziora, G. Schmidmaier, G. Schollmeier, H. Bail, R. Pflugmacher, T. Gorke, M. Wagner, M. Raschke, T. Mittlmeier, and N. P. Haas, IGF-I and TGF-beta1 application by a poly-(D,L-lactide)-coated cage promotes intervertebral bone matrix formation in the sheep cervical spine, *Spine*, 27, 1710–1723, 2002.

14. C. G. Simon Jr., C. A. Khatri, S. A. Wight, and F. W. Wang, Preliminary report on the biocompatibility of a moldable, resorbable, composite bone graft consisting of calcium phosphate cement and poly(lactide-co-glycolide) microspheres, *J. Orthop. Res.*, 20, 473–482, 2002.
15. S. Leinonen, J. Tiainen, M. Kellomaki, P. Tormala, T. Waris, M. Ninkovic, and N. Ashammakhi, Holding power of bioabsorbable self-reinforced poly-L/DL-lactide 70/30 tacks and miniscrews in human cadaver bone, *J. Craniofac. Surg.*, 14, 171–175, 2003.
16. P. Nordstrom, T. Pohjonen, P. Tormala, and P. Rokkanen, Shear-load carrying capacity of cancellous bone after implantation of self-reinforced polyglycolic acid and poly-L-lactic acid pins: Experimental study on rats, *Biomaterials*, 22, 2557–2561, 2001.
17. G. Chen, T. Sato, T. Ushida, R. Hirochika, and T. Tateishi, Redifferentiation of dedifferentiated bovine chondrocytes when cultured in vitro in a PLGA-collagen hybrid mesh, *FEBS Lett.*, 542, 995–999, 2003.
18. A. K. Poshusta, J. A. Burdick, D. J. Mortisen, R. F. Padera, D. Ruehlman, M. J. Yaszemski, and K. S. Anseth, Histocompatibility of photocrosslinked polyanhydrides: A novel in situ forming orthopaedic biomaterial, *J. Biomed. Mater. Res.*, 64, 62–69, 2003.
19. J. Elisseeff, K. S. Anseth, D. Sims, W. McIntosh, M. Randolph, and R. Langer, Transdermal photopolymerization for minimally invasive implantation, *Proc. Natl. Acad. Sci. USA*, 96, 3104–3107, 1999.
20. E. Alsberg, K. W. Anderson, A. Albeiruti, R. T. Franceschi, and D. J. Mooney, Cell-interactive alginate hydrogels for bone tissue engineering, *J. Dent. Res.*, 80, 2025–2029, 2001.
21. X. He and E. Jabbari, Material properties and cytocompatibility of injectable MMP degradable poly(lactide ethylene oxide fumarate) hydrogel as a carrier for marrow stromal cells, *Biomacromolecules*, 8, 780–792, 2007.
22. S. Moeinzadeh, D. Barati, S. K. Sarvestani, O. Karaman, and E. Jabbari, Nanostructure formation and transition from surface to bulk degradation in polyethylene glycol gels chain-extended with short hydroxy acid segments, *Biomacromolecules*, 14(8), 2917–2928, 2013.
23. S. Moeinzadeh, D. Barati, X. He, and E. Jabbari, Gelation characteristics and osteogenic differentiation of stromal cells in inert hydrolytically degradable micellar polyethylene glycol hydrogels, *Biomacromolecules*, 13, 2073–2086, 2012.
24. S. Moeinzadeh, S. Nouri Khorasani, J. Ma, X. He, and E. Jabbari, Synthesis and gelation characteristics of photo-crosslinkable star poly(ethylene oxide-co-lactide-glycolide acrylate) macromonomers, *Polymer*, 52, 3887–3896, 2011.
25. C. V. Rodrigues, P. Serricella, A. B. Linhares, R. M. Guerdes, R. Borojevic, M. A. Rossi, M. E. Duarte, and M. Farina, Characterization of a bovine collagen-hydroxyapatite composite scaffold for bone tissue engineering, *Biomaterials*, 24, 4987–4997, 2003.
26. J. M. Oliveira, M. T. Rodrigues, S. S. Silva, P. B. Malafaya, M. E. Gomes, C. A. Viegas, I. R. Dias, J. T. Azevedo, J. F. Mano, and R. L. Reis, Novel hydroxyapatite/chitosan bilayered scaffold for osteochondral tissue-engineering applications: Scaffold design and its performance when seeded with goat bone marrow stromal cells, *Biomaterials*, 27, 6123–6137, 2006.
27. T. Furuzono, S. Yasuda, T. Kimura, S. Kyotani, J. Tanaka, and A. Kishida, Nano-scaled hydroxyapatite/polymer composite IV. Fabrication and cell adhesion properties of a three-dimensional scaffold made of composite material with a silk fibroin substrate to develop a percutaneous device, *J. Artif. Organs*, 7, 137–144, 2004.
28. Z. Ajdukovic, S. Najman, L. J. Dordevic, V. Savic, D. Mihailovic, D. Petrovic, N. Ignjatovic, and D. Uskokovic, Repair of bone tissue affected by osteoporosis with hydroxyapatite-poly-L-lactide (HAp-PLLA) with and without blood plasma, *J. Biomater. Appl.*, 20, 179–190, 2005.
29. S. S. Kim, K. M. Ahn, M. S. Park, J. H. Lee, C. Y. Choi, and B. S. Kim, A poly(lactide-co-glycolide)/hydroxyapatite composite scaffold with enhanced osteoconductivity, *J. Biomed. Mater. Res. A*, 80, 206–215, 2007.

30. D. Hakimimehr, D. M. Liu, and T. Troczynski, In-situ preparation of poly(propylene fumarate)/hydroxyapatite composite, *Biomaterials*, 26, 7297–7303, 2005.
31. F. E. Wiria, K. F. Leong, C. K. Chua, and Y. Liu, Poly(ε-caprolactone)/hydroxyapatite for tissue engineering scaffold fabrication via selective laser sintering, *Acta Biomater.*, 3, 1–12, 2007.
32. Y. Hu, C. Zhang, S. Zhang, Z. Xiong, and J. Xu, Development of a porous poly(L-lactic acid)/hydroxyapatite/collagen scaffold as a BMP delivery system and its use in healing canine segmental bone defect, *J. Biomed. Mater. Res. A*, 67, 591–598, 2003.
33. M. C. Kruyt, W. J. Dhert, F. C. Oner, C. A. van Blitterswijk, A. J. Verbout, and J. D. de Bruijn, Analysis of ectopic and orthotopic bone formation in cell-based tissue-engineered constructs in goats, *Biomaterials*, 28, 1798–1805, 2007.
34. S. L. Hall, K. H. Lau, S. T. Chen, J. C. Felt, D. S. Gridley, J. K. Yee, and D. J. Baylink, An improved mouse Sca-1+ cell-based bone marrow transplantation model for use in gene- and cell-based therapeutic studies, *Acta Haematol.*, 117, 24–33, 2007.
35. G. Pelled, K. Tai, D. Sheyn, Y. Zilberman, S. Kumbar, L. S. Nair, C. T. Laurencin, D. Gazit, and C. Ortiz, Structural and nanoindentation studies of stem cell-based tissue-engineered bone, *J. Biomech.*, 40, 399–411, 2007.
36. U. Meyer, H. P. Wiesmann, K. Berr, N. R. Kubler, and J. Handschel, Cell-based bone reconstruction therapies-principles of clinical approaches, *Int. J. Oral Maxillofac. Implants*, 21, 899–906, 2006.
37. J. Handschel, H. P. Wiesmann, R. Depprich, N. R. Kubler, and U. Meyer, Cell-based bone reconstruction therapies: Cell sources, *Int. J. Oral Maxillofac. Implants*, 21(6), 890–898, 2006.
38. E. Jabbari, L. Lu, and M. J. Yaszemski, Synthesis and characterization of injectable and biodegradable composites for orthopedic applications, in S. K. Mallapragada and B. Narasimhan (eds), *Handbook of Biodegradable Polymeric Materials and Their Applications*, vol. 2, pp. 239–270. American Scientific Publishers, Stevenson Ranch, CA, 2004.
39. E. Jabbari, L. Lu, B. L. Currier, A. G. Mikos, and M. J. Yaszemski, Injectable polymers and hydrogels for orthopedic and dental applications, in L. J. Sandell and A. J. Grodzinsky (eds), *Tissue Engineering in Musculoskeletal Clinical Practice*, Chapter 32. American Academy of Orthopaedic Surgeons, Rosemont, IL, 2004.
40. M. J. Glimcher, The nature of the mineral component of bone and the mechanisms of calcification, in F. L. Coe and M. J. Favus (eds), *Disorders of Bone and Mineral Metabolism*, pp. 265–286. Raven Press, New York, 1992.
41. M. A. Fernández-seara, S. L. Wehrli, M. Takahashi, and F. W. Wehrli, Water content measured by proton-deuteron exchange NMR predicts bone mineral density and mechanical properties, *J. Bone Miner. Res.*, 19, 289–296, 2004.
42. M. C. Summitt and K. D. Reisinger, Characterization of the mechanical properties of demineralized bone, *J. Biomed. Mater. Res. A*, 67, 742–750, 2003.
43. W. J. Landis, The strength of a calcified tissue depends in part on the molecular structure and organization of its constituent mineral crystals in their organic matrix, *Bone*, 16, 533–544, 1995.
44. J.-Y. Rho, L. Kuhn-Spearing, and P. Zioupos, Mechanical properties and the hierarchical structure of bone, *Med. Eng. Phys.*, 20, 92–102, 1998.
45. J. A. Buckwalter and R. R. Cooper, Bone biology. Part II: Formation, form, modeling, remodeling, and regulation of cell function, *J. Bone Joint Surg.*, 77A, 1276–1289, 1995.
46. R. Fujisawa, Y. Wada, Y. Nodasaka, and Y. Kuboki, Acidic amino acid-rich sequences as binding sites of osteonectin to hydroxyapatite crystals, *Biochim. Biophys. Acta*, 1292, 53–60, 1996.
47. V. I. Sikavitsas, J. S. Temenoff, and A. G. Mikos, Biomaterials and bone mechanotransduction. *Biomaterials*, 22, 2581–2593, 2001.

48. M. F. Young, Bone matrix proteins: Their function, regulation, and relationship to osteoporosis, *Osteoporos. Int.*, 14(Suppl. 3), S35–S42, 2003.
49. P. Bianco, M. Riminucci, S. Gronthos, and P. G. Robey, Bone marrow stromal stem cells: Nature, biology, and potential applications, *Stem Cells*, 19, 180–192, 2001.
50. D. Woodbury, E. J. Schwarz, D. J. Prockop, and I. B. Black, Adult rat and human bone marrow stromal cells differentiate into neurons, *J. Neurosci. Res.*, 61, 364–370, 2000.
51. P. Carmeliet and A. Luttun, The emerging role of the bone marrow-derived stem cells in (therapeutic) angiogenesis, *Thromb. Haemost.*, 68, 289–297, 2001.
52. C. Maniatopoulos, J. Sodek, and A. H. Melcher, Bone formation in vitro by stromal cells obtained from bone marrow of young adult rats, *Cell Tissue Res.*, 254, 317–330, 1988.
53. J. A. Henderson, X. He, and E. Jabbari, Concurrent differentiation of marrow stromal cells to osteogenic and vasculogenic lineages, *Macromol. Biosci.*, 8, 499–507, 2008.
54. C. W. Lan and Y. J. Wang, Collagen as an immobilization vehicle for bone marrow stromal cells enriched with osteogenic potential, *Artif. Cells Blood Substit. Immobil. Biotechnol.*, 31, 59–68, 2003.
55. M. Taira, S. Toyosawa, N. Ijyuin, J. Takahashi, and Y. Araki, Studies on osteogenic differentiation of rat bone marrow stromal cells cultured in type I collagen gel by RT-PCR analysis, *J. Oral Rehabil.*, 30, 802–807, 2003.
56. S. Ramakrishna, J. Mayer, E. Wintermantel, and K. W. Leong, Biomedical applications of polymer-composite materials: A review, *Comp. Sci. Tech.*, 61, 1189–1224, 2001.
57. J. F. Mano, R. A. Sousa, L. F. Boesel, N. M. Neves, and R. L. Reis, Bioinert, biodegradable and injectable polymeric matrix composites for hard tissue replacement: State of art and recent developments, *Comp. Sci. Tech.*, 64, 789–817, 2004.
58. M. J. Yaszemski, R. G. Payne, W. C. Hayes, R. Langer, and A. G. Mikos, Evolution of bone transplantation: Molecular, cellular and tissue strategies to engineer human bone, *Biomaterials*, 17, 175–185, 1996.
59. S. N. Nazhat, R. Joseph, M. Wang, R. Smith, K. E. Tanner, and W. Bonfield, Dynamic mechanical characterisation of hydroxyapatite reinforced polyethylene: Effect of particle size, *J. Mater. Sci. Mater. Med.*, 11, 621–628, 2000.
60. M. Wang and W. Bonfield, Chemically coupled hydroxyapatite polyethylene composites: Structure and properties, *Biomaterials*, 22, 1311–1320, 2001.
61. X. Wang and C. G. Robertson, Strain-induced nonlinearity of filled rubbers, *Phys. Rev. E*, 72, 031406, 2005.
62. A. S. Sarvestani, X. He, and E. Jabbari, The effect of osteonectin-derived peptide on the viscoelasticity of hydrogel/apatite nanocomposite scaffolds, *Biopolymers*, 85, 370–378, 2007.
63. J. Sodek, B. Zhu, M. H. Huynh, T. J. Brown, and M. Ringuette, Novel functions of the matricellular proteins osteopontin and osteonectin, *Connect. Tissue Res.*, 43, 308–319, 2002.
64. R.-L. Xie and G. L. Long, Elements within the first 17 amino acids of human osteonectin are responsible for binding to type V collagen, *J. Biol. Chem.*, 271, 8121–8125, 1996.
65. A. L. Boskey, D. J. Moore, M. Amling, E. Canalis, and A. M. Delany, Infrared analysis of the mineral and matrix in bones of osteonectin-null mice and their wildtype controls, *J. Bone Miner. Res.*, 8, 1005–1011, 2003.
66. X. He and E. Jabbari, Solid-phase synthesis of reactive peptide crosslinker by selective deprotection, *Prot. Pept. Lett.*, 13, 715–718, 2006.
67. A. S. Sarvestani, X. He, and E. Jabbari, The role of filler-matrix interaction on viscoelastic response of biomimetic nanocomposite hydrogels. In special issue: Nanostructured Materials for Biomedical Applications, *J. Nanomater.*, Article No. 126803, 2008.
68. A. S. Sarvestani and E. Jabbari, A model for the viscoelastic behavior of nanofilled hydrogel composites under oscillatory shear loading, *Polym. Compos.*, 29, 326–336, 2008.

69. A. S. Sarvestani, X. He, and E. Jabbari, Osteonectin-derived peptide increases the modulus of a bone-mimetic nanocomposite, *Eur. Biophys. J. Biophys. Lett.*, 37, 229–234, 2007.
70. A. S. Sarvestani, W. Xu, X. He, and E. Jabbari, Gelation and degradation characteristics of in-situ photo-crosslinked poly(l-lactide-co-ethylene oxide-co-fumarate) hydrogels, *Polymer*, 48, 7113–7120, 2007.
71. A. S. Sarvestani, X. He, and E. Jabbari, Effect of composition on gelation kinetics of unfilled and nanoapatite-filled poly(lactide-ethylene oxide-fumarate) hydrogels, *Mater. Lett.*, 16, 5278–5281, 2007.
72. A. S. Sarvestani, X. He, and E. Jabbari, Viscoelastic characterization and modeling of gelation kinetics of injectable in situ crosslinkable poly(lactide-ethylene oxide-fumarate) hydrogels, *Biomacromolecules*, 8, 406–415, 2007.
73. A. S. Sarvestani and E. Jabbari, Modeling and experimental investigation of rheological properties of injectable poly(lactide ethylene oxide fumarate)/hydroxyapatite nanocomposites, *Biomacromolecules*, 7, 1573–1580, 2006.
74. E. J. Guth, Theory of filler reinforcement, *J. Appl. Phys.*, 16, 20–25, 1945.
75. J. F. Mooney, Disorders of bone and mineral metabolism, *Arth. J. Arth. Rel. Surg.*, 8, 415, 1992.
76. B. Mavis, T. T. Demirtas, M. Gumusderelioglu, G. Gunduz, and U. Colak, Synthesis, characterization and osteoblastic activity of polycaprolactone nanofibers coated with biomimetic calcium phosphate, *Acta Biomater.*, 5, 3098–3111, 2009.
77. W. Y. Liu, Y. C. Yeh, J. Lipner, J. Xie, H.-W. Sung, S. Thomopoulos, and Y. Xia, Enhancing the stiffness of electrospun nanofiber scaffolds with a controlled surface coating and mineralization, *Langmuir*, 27, 9088–9093, 2011.
78. Q. Wang, X. M. Wang, L. L. Tian, Z. J. Cheng, and F. Z. Cui, In situ remineralization of partially demineralized human dentine mediated by a biomimetic non-collagen peptide, *Soft Matter*, 7, 9673–9680, 2011.
79. O. Karaman, A. Kumar, S. Moeinzadeh, X. He, T. Cui, and E. Jabbari, Effect of surface modification of nanofibers with glutamic acid peptide on calcium phosphate nucleation and osteogenic differentiation of marrow stromal cells, *J. Tissue Eng. Regen. Med.* Published electronically July 30, 2013. doi: 10.1002/term.1775.
80. B. H. Li and R. M. Aspden, Composition and mechanical properties of cancellous bone from the femoral head of patients with osteoporosis or osteoarthritis, *J. Bone Miner. Res.*, 12, 641–651, 1997.
81. P. Podsiadlo, M. Qin, M. Cuddihy, J. Zhu, K. Critchley, E. Kheng, A. K. Kaushik, et al., Highly ductile multilayered films by layer-by-layer assembly of oppositely charged polyurethanes for biomedical applications, *Langmuir*, 25, 14093–14099, 2009.
82. K. Chatterjee, S. Lin-Gibson, W. E. Wallace, S. H. Parekh, Y. J. Lee, M. T. Cicerone, M. F. Young, and C. G. Simon, The effect of 3-D hydrogel scaffold modulus on osteoblast differentiation and mineralization revealed by combinatorial screening, *Biomaterials*, 31, 5051–5062, 2010.

14 Nanotechnology for Tissue Engineering and Regenerative Medicine

Şükran Şeker, Y. Emre Arslan, Serap Durkut, A. Eser Elçin, and Y. Murat Elçin

CONTENTS

ABBREVIATIONS

BMP-2	Bone morphogenetic protein-2
DEAEM	Diethylaminoethylmethacrylate
DMAEM	Dimethylaminoethylmethacrylate
ECM	Extracellular matrix

LCST	Lower critical solution temperature
MEMS	Microelectromechanical systems
NEMS	Nanoelectromechanical systems
P(LCL)	Poly(L-lactide-co-ε-caprolactone)
PCL	Poly(ε-caprolactone)
PDMS	Polydimethylsiloxane
PGA	Poly(glycolic acid)
PLA	Poly(lactic acid)
PLGA	Poly(lactic-co-glycolic acid)
PNIPAAm	Poly(N-isopropylacrylamide)
PVCL	Poly(N-vinylcaprolactam)
UCST	Upper critical solution temperature

14.1 INTRODUCTION

Tissue engineering is a rapidly expanding interdisciplinary field involving biology, medicine, and engineering, searching to address the need for the development of biological substitutes [1]. In tissue engineering and regenerative medicine strategies, the damaged tissue is replaced or repaired to restore normal function. Natural and synthetic biodegradable materials capable of supporting three-dimensional (3-D) tissue formation, so-called scaffolds, are used for cell seeding. Scaffolds typically allow attachment, proliferation, and organization of cells. Cells within a scaffold deposit their extracellular matrix (ECM) and degrade the scaffolds, forming an engineered tissue construct that mimics the native tissue. Despite progress in tissue engineering, there are still some shortcomings in current tissue engineering techniques, such as their relative inability to form functional vascularized tissues [2], not providing adequate resources of oxygen, nutrients, gas exchange, and waste removal [3]. Many tissue engineering and regenerative medicine studies based on two-dimensional (2-D) cell-culture models fail to regenerate the *in vivo* cellular microenvironment; the reason for this is that most cells in a culture dish do not maintain their differentiated functions. To overcome these challenges, tissue engineering technologies must evolve to efficiently transport nutrients, growth factors, water, and oxygen to cells and remove metabolic waste products such as lactic acid, carbon dioxide, and hydrogen ions in exchange [4].

Studies on nano/microfiber production techniques in both academic and industrial settings have drawn the attention of investors on a global scale [5]. Various fiber fabrication techniques such as (a) electrospinning, (b) wet spinning, (c) microfluidic spinning, (d) biospinning, (e) interfacial complexation, and (f) melt spinning have been used by researchers to fabricate fibrous materials for various applications, particularly in the fields of tissue engineering and regenerative medicine [6] (Figure 14.1).

Hydrogels are cross-linked networks of polymeric materials, which are not soluble in water, yet they can absorb large quantities of water. Hydrogels containing interactive functional groups in polymeric chains are usually called "stimuli-responsive," "smart," or "intelligent" polymers. The conformation of polymers in a solution can be altered by environmental stimuli such as pH, temperature, ionic

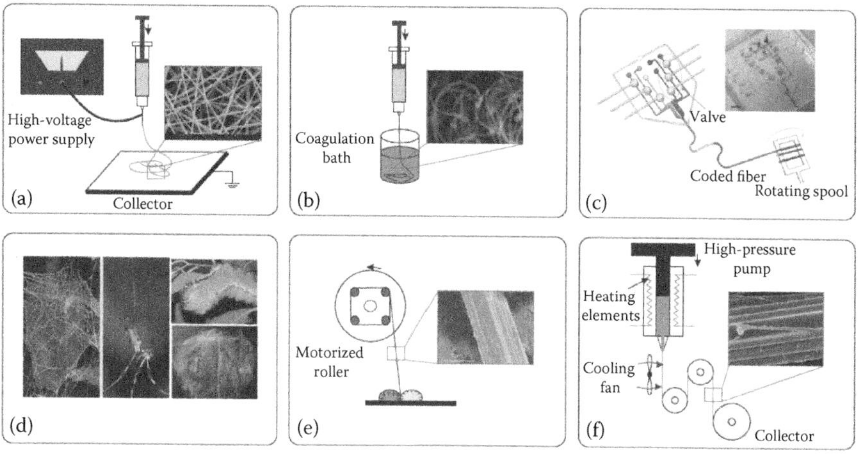

FIGURE 14.1 **(See color insert)** Novel fiber fabrication techniques. (a) Electrospinning: a charged polymer solution is pulled on a collecting plate in the presence of an applied electrostatic force (From Pham, Q.P., Sharma, U., and Mikos, A.G., *Tissue Engineering,* 12, 1197–1211, 2006.) [7]. (b) Wet spinning: a prepolymer solution is injected into a coagulation bath including either a poor solvent or a nonsolvent to form a polymer and cross-link the fibers (From Jalili, R., Aboutalebi, S.H., Esrafilzadeh, D.E. et al., *Advanced Functional Materials,* 23, 5345–5354, 2013.) [8]. (c) Microfluidic spinning: a prepolymer solution is pushed into a microchannel using coaxial flow and the polymer is cross-linked (From Lee, B.R., Lee, K.H., Kang, E., Kim, D.S., and Lee, S.H., *Biomicrofluidics,* 5, 022208, 2011.) [9]. (d) Biospinning: fibers are naturally produced by various insects such as silkworms and spiders (From Mandal, B.B. and Kundu, S.C., *Acta Biomaterialia,* 6, 360–371, 2010.) [10]. (e) Interfacial complexation: fibers are fabricated at the interface of two oppositely charged polyelectrolyte solutions by means of polyion complex formation (From Yim, E.K.F., Liao, I.C., and Leong, K.W., *Tissue Engineering,* 13, 423–433, 2007.) [11]. (f) Melt spinning: a heated polymer melt is extruded through a spinneret to form fiber strands (From Zhmayev, E., Cho, D., and Joo, Y.L., *Polymer,* 51, 4140–4144, 2010.) [12]. (From Tamayol, A., Akbari, M., Annabi, N., Paul, A., Khademhosseini, A., and Juncker, D., *Biotechnology Advances*, 31, 669–687, 2013. With permission.)

strength, electric current, ultrasound velocity, and so on. Along with the onset of changes in environmental conditions, either attractive or repulsive forces direct the physicochemical properties of the polymer, for example, the increase/reduction in both the pore size and permeability, as well as significant changes in the volume and hydrophilic/hydrophobic properties. These changes may be gradual and smooth, or they may be abrupt and discontinuous, depending on the nature of the polymer and the purposes of the procedure [13].

In tissue engineering, micro- and nanofabrication techniques can be used to develop ideal scaffolds that closely mimic, reinforce, or even induce the creation of *de novo* tissue. Tissue engineering scaffolds generated by micro- and nanotechnology have the potential for increased complexity and vascularization [2]. A soft-lithographic technique has allowed for the generation of hydrogel scaffolds that contain a network of fluidic channels [3]. Microfabricated channel networks provide the capabilities for manufacturing microvasculature ready for perfusion. Printed microfluidic channels have been shown to have great potential in allowing the transportation of dissolved oxygen, nutrients, and soluble factors by the perfusion system, and waste can be removed in the same manner from the system [4].

This chapter will provide an overview of some of the recent progress in tissue engineering and regenerative medicine research, using micro- and nanofabrication techniques to overcome challenges through the development and fabrication of micro- and nanostructures. The major topics that will be emphasized are nanofibrous scaffolds, stimuli-responsive systems, and micro/nanofluidic approaches.

14.2 ELECTROSPUN NANOFIBROUS SCAFFOLDS

Over the course of the last decades, the electrostatic fiber formation technique, also known as electrospinning, has received great interest due to its outstanding properties such as ease of operation and versatility, while being inexpensive, scalable, and reliable. The process allows the fabrication of fibers at the nano/microscale from a wide range of synthetic and natural polymers. The well-known technique was first documented in 1897 by Rayleigh and studied by Zeleny in 1914 [14]. Formhals acquired a series of patents in 1934 [15]. Utilizing this material processing method, ultrafine polymer fibers with diameters ranging from 2 nm to several micrometers can be achieved. Nowadays there are a myriad of commercially available electrospun polymers for research and medical use. Electrospinning can be successfully performed using many kinds of polymers. Electrospun polymer meshes are used in a wide range of applications such as filtration devices, textiles, electrical and optical components, sensors, pharmaceuticals, biotechnology, environmental engineering, defense, and security [16]. Another promising field of application is the tissue-engineering approach for regenerative medicine. Owing to their unique properties, such as smaller pores and higher surface area, electrospun fibers have been successfully used as tissue engineering scaffolds [17]. Nano/microfibrous tissue engineering scaffolds can be fabricated from a variety of natural and synthetic polymers, including poly(lactic acid), poly(urethane), silk fibroin, collagen, hyaluronic acid, chitosan/collagen, and cellulose. To increase the yield of nanofiber production, techniques such as multiple spinnerets or nozzle systems, and bottom-up gas electrospinning (Figure 14.2) (known as bubble electrospinning in the literature) have been developed [18].

Over 200 universities and research groups worldwide have used electrospinning to fabricate nano/microfiber-based materials, adding to its popularity. Thus, patents referring to electrospun polymers have grown in recent years. A number of companies, such as eSpin Technologies, NanoTechnics, and KATO Tech, have started to produce electrospun materials for a wide array of applications [19].

14.2.1 Fundamentals of Electrospinning

The fiber formation technique via the electrospinning process differs from the conventional approaches that rely predominantly on mechanical forces and remain within the limits of geometric boundaries. The system involves many parameters such as polymer solubility, electrostatic repulsion between polymer molecules, solution conductivity, applied voltage, distance between nozzle and collector, perfusion

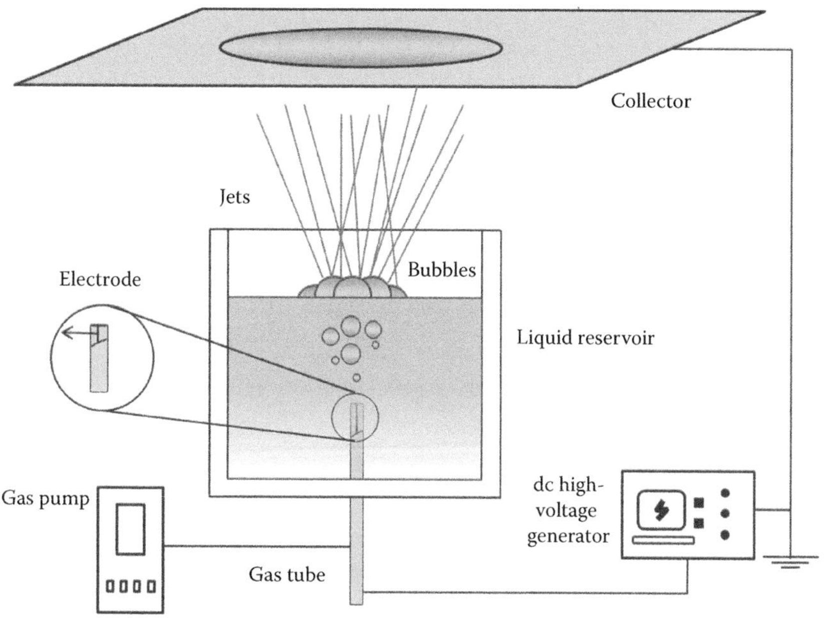

FIGURE 14.2 Schematic illustration of the bubble-electrospinning process. (Adapted from Agarwal, S., Greiner, A., and Wendorff, J.H., *Progress in Polymer Science*, 38, 963–991, 2013.)

rate of polymer solution, and additional ambient parameters such as solution temperature, humidity, and air velocity in the spinning chamber [20].

A simple electrospinning setup includes a direct current (dc) power supply, which generates the essential electrical field between the nozzle and the collector; a metallic needle or spinneret through which a polymer jet is pulled; a perfusion pump that pushes the polymer solution into the nozzle; and finally a grounded collecting plate (usually a metal screen, plate, or rotating mandrel) where nano/microfibers are collected. Two typical types of electrospinning apparatus are shown in Figure 14.3. A high voltage is applied to the polymer solution at the tip of a metallic needle or spinneret. The electrically charged polymer jets toward a grounded metal screen, plate, or rotating mandrel. The polymer jet undergoes whipping motions during the flight from the spinneret to the collector with the effect of the electrical field [21].

Most polymers can be completely dissolved in organic solvents. Although the solvents are evaporated, a minute amount of solvent may be left behind. Since this may limit the usage of electrospun materials for medical and bio-related applications, researchers have now focused on the development of solvent-free electrospinning processes.

14.2.2 Polymers and Processing Parameters in Electrospinning

Using the electrospinning process, it is possible to fabricate fine fibers (nano/micro) for many biomedical applications, including tissue engineering. To that end, synthetic and natural polymers or blends of biomacromolecules, such as proteins,

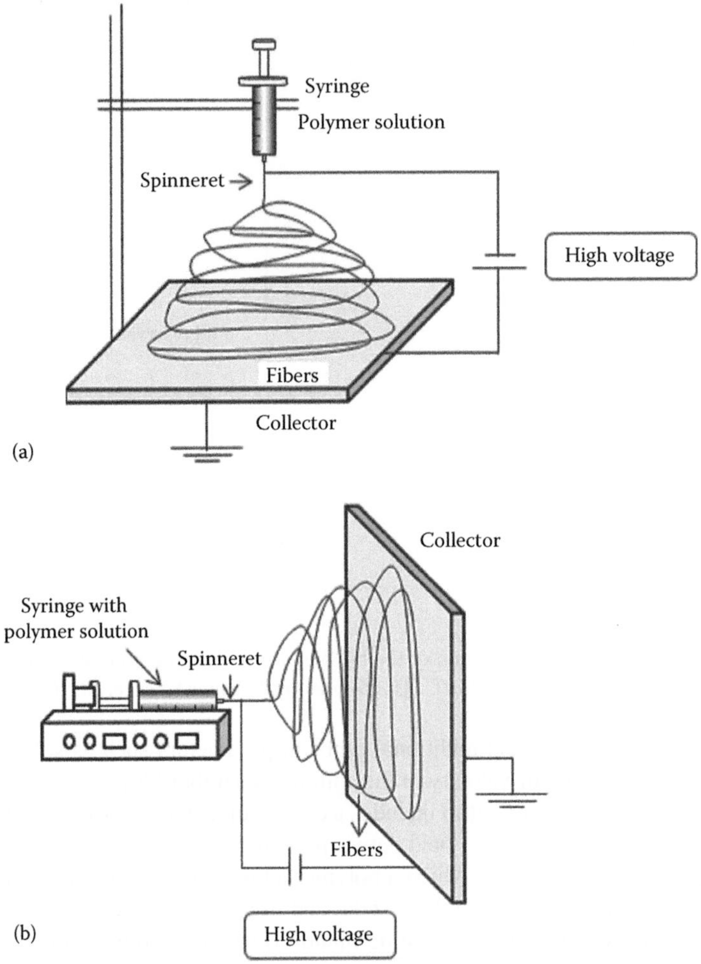

FIGURE 14.3 Schematic diagram of the electrospinning apparatus: (a) a typical vertical setup and (b) a horizontal setup. (Adapted from Bhardwaj, N. and Kundu, S.C., *Biotechnology Advances*, 28, 325–347, 2010.)

nucleic acids, and polysaccharides, can be utilized. Over 200 polymers have been used in electrospinning and have been characterized accordingly within the last few years [22]. Table 14.1 provides a list of the most commonly used polymers for electrospinning. In particular, natural polymers are of great interest for biomedical applications because of their unique properties such as biocompatibility and low immunogenicity, and their natural capacity for binding cells. In recent years, collagen, gelatin, elastin, silk fibroin, and other materials have been processed using electrospinning. Studies suggest that scaffolds fabricated from natural polymers possess better clinical functionality [23]. Additionally, biodegradable synthetic polymers, such as poly(ε-caprolactone) (PCL), poly(lactic acid) (PLA), poly(glycolic acid) (PGA), copolymer poly(lactic-co-glycolic acid) (PLGA), and the

TABLE 14.1
Polymers Used in Various Tissue Engineering Applications

Polymers	Applications	References
Poly(lactide-co-glycolide)	Dental tissue engineering	[25]
Poly(ε-caprolactone)	Bone tissue engineering	[26]
Poly(L-lactide)	Neural tissue engineering	[27]
Polyurethane/carbon nanotubes	Musculoskeletal tissue engineering	[28]
Silk fibroin	Wound dressing	[29]
Chitosan	Cartilage tissue engineering	[30]
Hyaluronic acid	Cell encapsulation and tissue regeneration	[31]
Collagen/elastin	Vascular tissue engineering	[32]

copolymer poly(L-lactide-co-ε-caprolactone) (P(LLA-CL)) have been widely used to fabricate nanofibrous scaffolds for vascular tissue engineering, cardiac tissue engineering, bone tissue engineering, and as wound dressings [24].

To obtain fine fiber–based scaffolds, a number of issues should be considered, such as the solution parameters (including viscosity, conductivity, molecular weight, and surface tension), the process parameters (including applied voltage, distance between tip and collector, and flow rate), and the ambient parameters (including temperature and humidity) [15,21]. A notable parameter is the selection of the solvent to be used in the preparation of polymer solutions, since the first step in the technique involves dissolving polymers in a proper solvent or solvent system. Basically, a solvent has two crucial roles in electrospinning: one is to dissolve the polymer and the other is to carry the dissolved polymer forward onto the collector. To obtain fine fibers from polymers, solvents should possess properties such as good volatility, vapor pressure, and boiling point [22]. The use of chloroform, ethanol, dimethylformamide, trifluoroacetic acid, dichloromethane, and water as solvents has been extensively investigated. Researchers concur that solvents affect the morphology and size of nanofibers [33].

14.2.3 Electrospinning in Tissue Engineering and Regenerative Medicine

The field of tissue engineering is multidisciplinary, combining the principles of engineering and life sciences in order to restore diseased tissues or organs of the human body [34]. The biomaterials used in tissue engineering and regenerative medicine attract great interest because of their crucial roles in providing matrices for cellular growth and proliferation, and supporting new tissue formation in three dimensions [35]. Tissue engineering aims to build biologically functional templates from biodegradable and biocompatible polymers using numerous techniques, including electrospinning to mimic natural ECM formation, which is an important constituent for cell attachment, proliferation, and differentiation. By using the electrospinning technique, nano/microfibrous scaffolds, which mimic natural ECMs, are created for various tissue-engineering applications. Researchers have reported that these nano/microfibrous scaffolds positively influence cell attachment, proliferation, and orientation. Therefore, the technique has become popular with scientists in fabricating nano/microfibrous scaffolds [36].

Polymeric nanofiber-based scaffolds have been constructed for tissue engineering and regenerative medicine applications, such as the nerve [37], heart [38], cartilage [39], bone [40], skin [41], and arterial blood vessels [42]. It has been shown that an electrospun PLGA fiber scaffold supports the attachment and proliferation of fibroblastic cells, such as periodontal ligament fibroblasts, owing to its oriented fiber mesh with high surface area and porosity (>90%).

Incorporating bioactive/signaling molecules or nanoparticles or both can be an effective tool to improve the properties of electrospun polymer scaffolds. For instance, a nanofibrous silk fibroin scaffold was modified with bone morphogenetic protein-2 (BMP-2) and hydroxyapatite nanoparticles via electrospinning [43]. Human bone marrow mesenchymal stem cells were seeded onto this scaffold and the new bone formation was investigated. The findings suggested that this composite could serve as an ideal scaffold for bone tissue engineering [43].

The potential of nanofibrous membranes for use as medical dressings has also been pointed out. The wound-healing properties of electrospun Type I collagen membrane have been evaluated. The findings have shown that cells migrate onto this membrane. The level of wound healing acquired using the nanofibrous collagen membrane was higher than other wound treatment strategies during the early stage of the process [44]. Silk and polyurethane fibers have also been investigated as wound-dressing material [45]. In another study, a nanofibrous polymeric membrane was loaded with an antimicrobial agent, that is, silver ions, to improve the properties of the biomaterial [46].

Ligaments are the bands responsible for joint movement and stability. Athletes usually injure these tissues during heavy sporting activities. Because aligned nanofibers increase cell attachment and proliferation, scientists have examined their potential for ligament tissue engineering. Human ligament fibroblasts seeded onto an aligned polyurethane nanofiber scaffold acquired a spindle-shaped morphology, oriented in the direction of the nanofibers and synthesized ECM proteins, such as collagen [47].

The skeletal muscles are responsible for the movement of the body. Once damaged, it is difficult to regenerate these muscles in adults. Thus, surgical options aside, a tissue engineering approach can be a promising alternative. In this respect, degradable electrospun polyester urethane microfibers were evaluated as a scaffold for skeletal muscle tissue engineering, using human satellite cells, and murine and rat myoblast cell lines [48]. The results demonstrated sufficient mechanical consistency and good cellular responses (such as adhesion and differentiation) for the scaffold, implying its potential for skeletal muscle tissue engineering [48].

Contrary to randomly oriented electrospun nanofibers, tubular aligned nanofiber–based constructs may have the potential for vascular tissue engineering. Smooth muscle cells were seeded onto a biodegradable PLLA-PCL (75:25) nanofibrous tubular scaffold, which was electrospun using a rotating collector wire [49]. The results indicated that the tubular construct mimics the natural ECM of the blood vessels, providing comparable mechanical properties with the human coronary artery, while enabling smooth muscle cell adhesion and proliferation [49].

There are a limited number of clinical therapy options for nerve injuries. As a matter of fact, irreversible loss of function may take place when a nerve injury occurs. Neural tissue engineering may hold promise for such injuries. In this approach, normal or genetically engineered cells and ECM-like scaffolds may be used to repair damaged neural tissue. In one such study, the efficiency of a PLLA-based electrospun construct was demonstrated in terms of neural stem cell adhesion and differentiation toward the neural lineage [50]. In conclusion, the literature in this particular field indicates promise for various tissue engineering applications.

14.3 INTELLIGENT STIMULI-RESPONSIVE HYDROGELS

Stimuli-responsive hydrogels can be classified according to their molecular structure, as (a) linear free chains in solution, (b) covalently cross-linked reversible and physical gels, and (c) chain adsorbed or surface-grafted forms (brushes) (Figure 14.4).

The most studied hydrogels are the pH- and temperature-responsive hydrogels when compared with responsive hydrogels operating with stimuli, such as ionic strength, specific ions or molecules, electric field, magnetic field, and light. Both pH- and temperature-responsive hydrogels have also been preferred for use in micro- and nanoactuation.

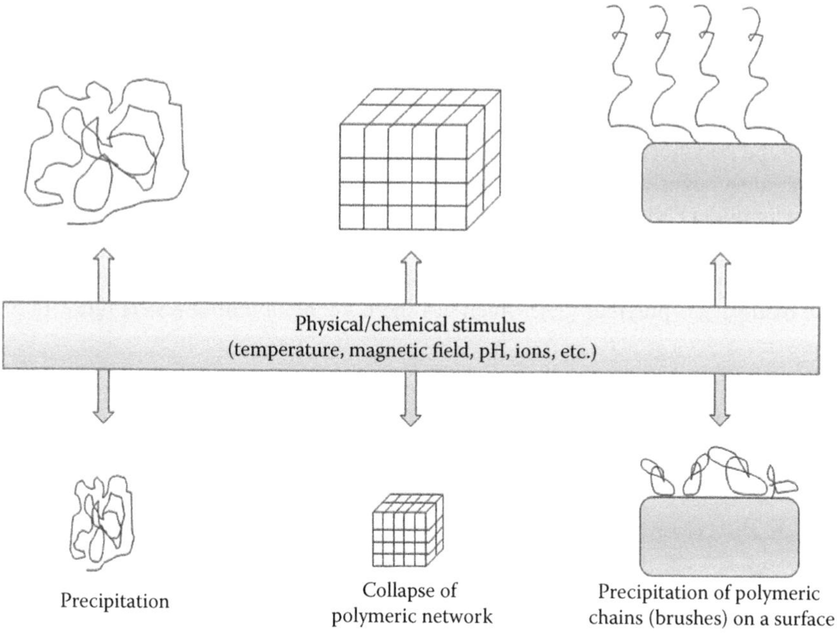

FIGURE 14.4 Classification of stimuli-responsive polymers by their physical forms. (Adapted from Kumar, A., Srivastava, A., Galaev, I.Y., and Mattiasson, B., *Progress in Polymer Science*, 32, 1205–1237, 2007.)

14.3.1 pH-Sensitive Hydrogels

pH-sensitive polymers contain pendant acidic (e.g., carboxylic and sulfonic acids) or basic (e.g., ammonium salts) groups that either accept or release protons in response to changes in the environmental pH [52]. With the changes in the environmental pH, the hydrogel networks change from neutral to charged state, related to the protonation of the amine groups or the deprotonation of the acid groups. Anionic pH-sensitive polymers such as polyacrylic acids [53], cationic hydrogels [54] from cationic monomers such as dimethylaminoethylmethacrylate (DMAEM) and diethylaminoethylmethacrylate (DEAEM), and natural-based polyelectrolytes, especially polysaccharides, have been proposed for the development of pH-responsive hydrogels. Natural biopolymers are especially attractive in biomedicine, including tissue engineering, due to their biocompatibility, biodegradability, and resemblance to the macromolecular environment of the ECM [55].

14.3.2 Temperature-Sensitive (Thermoresponsive) Hydrogels

Temperature-sensitive or thermoresponsive hydrogels are yet another commonly studied class of stimuli-responsive polymeric systems [56]. A major characteristic feature of these polymers is the presence of hydrophobic groups, such as methyl, ethyl, and propyl groups. Thermoresponsive polymers exhibit two types of characteristic behavior in aqueous solution, which are known as the lower critical solution temperature (LCST) and the upper critical solution temperature (UCST). The LCST-type thermoresponsive polymers become insoluble on heating, because the hydrophobic segments of the polymer are strengthened, leading to gel formation. Below the LCST, hydrogen bonding between the hydrophilic segments and water leads to enhanced dissolution. The UCST-type thermoresponsive polymers are formed on cooling the polymer solution [52].

The most significant natural thermoresponsive polymers are poly(N-isopropylacrylamide) (PNIPAAm) with an LCST of 32°C [57], poly(N-vinylcaprolactam) (PNVCL) with an LCST between 32°C and 35°C, depending on the molecular mass of the polymer [58], and poly(ethylene oxide)$_{106}$-poly(propylene oxide)$_{70}$-poly (ethylene oxide)$_{106}$ co-polymer [59], which has the trade name Pluronics® (Figure 14.5).

Poly(N-isopropylacrylamide) (PNIPAAm)	Poly(N-vinylcaprolactam) (PNVCL)	Poly(ethylene oxide)$_{106}$- poly(propylene oxide)$_{70}$- Poly (ethylene oxide)$_{106}$ [Pluronics®]
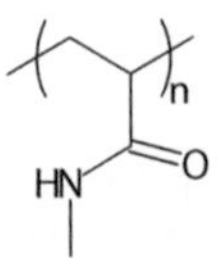	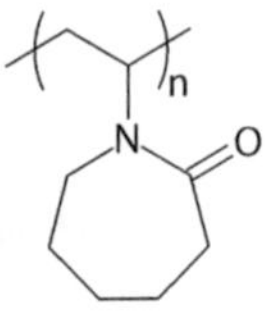	$HO{+}[CH_2{-}CH_2{-}O]_x{+}[CH_2{-}CH(CH_3){-}O]_y{+}[CH_2{-}CH_2{-}O]_x{-}H$

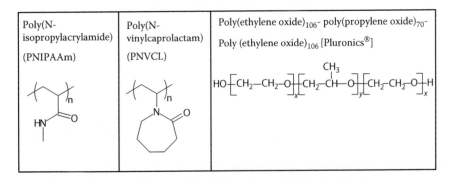

FIGURE 14.5 Chemical structures of some important thermoresponsive polymers.

Most of the thermoresponsive polymers used in biomedical applications display a transition between room and body temperatures.

PNIPAAm is the most widely investigated and used synthetic temperature-responsive polymer. It is soluble below its LCST, but precipitates above its LCST owing to the reversible formation (below LCST) and cleavage (above LCST) of the hydrogen bonds between the –NH and –C=O groups of PNIPAAm chains and the surrounding water molecules [60].

14.3.3 Nanofabrication Techniques and Devices with Responsive Hydrogels

Studies on intelligent hydrogels using nanofabrication techniques provide new opportunities for mimicking natural biosystems because they are more sensitive, effective, and fast analysis systems. Some of their recent developments and applications will be mentioned herein. Stimuli-responsive nanostructured polymer materials and systems in thin films (2-D) and 3-D colloidal nanoparticles, which are excellent candidates for the fabrication of nanostructured stimuli-responsive surfaces and coatings, are summarized in Figure 14.6.

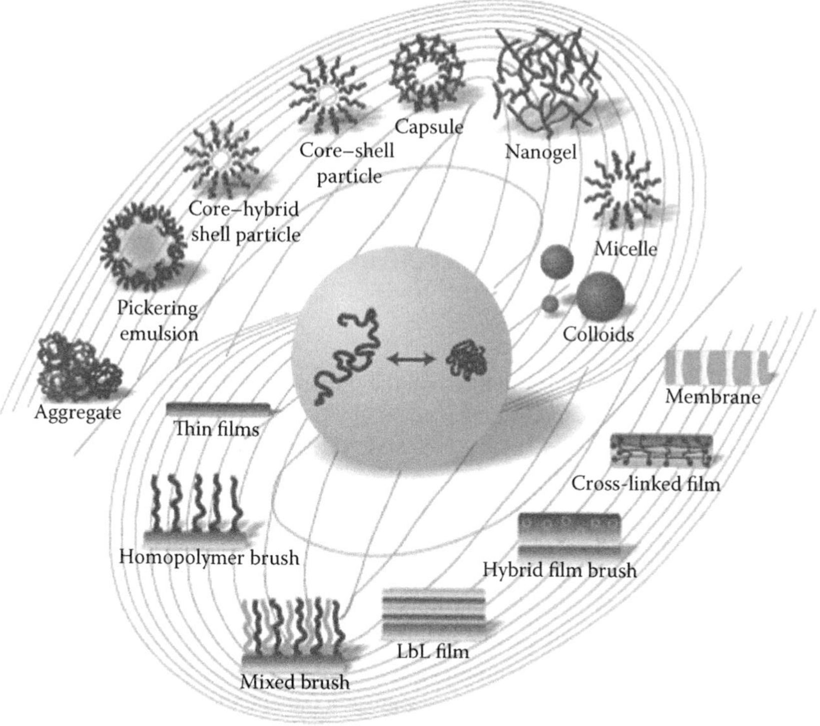

FIGURE 14.6 (**See color insert**) Galaxy of nanostructured stimuli-responsive polymer materials. (From Stuart, M.A.C., Huck, W.T.S., Genzer, J. et al., *Nature Materials*, 9, 101–113, 2010. With permission.)

Nanostructured thin network films (i.e., gel films, in many cases hydrogel films prepared from water-soluble polymers) are materials in which surface confinement brings a range of opportunities for engineering stimuli-responsive properties. An important attribute of gel thin films is their fast kinetics of swelling and shrinking compared with bulk gels. In another system, the swelling of porous bulk gels results in an increase in pore size, providing a unique opportunity for the regulation of transport through the film in a very broad diffusivity range, from the level in solutions down to the level in solids. Thin responsive (hydro)-gel films can be used as freestanding films or on various supports (adhered or covalently grafted). These films can accommodate various chemicals, biomolecules, and nanoparticles [61].

In particular, pH- and temperature-responsive thin polymer films have been used for micro- and nanoactuation. Actuation by means of responsive polymer brushes originates from the variable stretching of grafted macromolecules as a result of the strong steric repulsive interactions between the neighboring chains. Thus, these system are capable of sensing their environment and are often prepared for a broad range of applications such as microanalysis systems, for example, as micropumps, microvalves, microsensors, microfluidics, and also tissue engineering and controlled release systems for drug and gene delivery [62].

Nanoparticles are nanostructured 3-D stimuli-responsive polymer material systems, which can typically be represented by a core–shell architecture formed through the self-assembly of amphiphilic copolymers (polymer micelles or vesicles) or by means of the surface modification of various particles (inorganic or polymeric) with functional polymers. Responsive colloidal particles can be introduced into the interface between two immiscible fluids (i.e., liquid/liquid or liquid/gas), where the particles are strongly pinned owing to their large surface areas (Figure 14.7). External stimuli are used to stimulate the self-assembled structures, which may induce reversible or irreversible disintegration, aggregation, swelling, and adsorption [62].

14.3.4 Applications

14.3.4.1 Delivery Systems

Nanotechnology plays an important role in the therapies of the future as "nanomedicines." The aim of advanced systems is to supply doses of drugs for sustained periods to specific sites in the body and reduce toxic effects [63]. This can be achieved by efficiently loading therapeutic agents into polymeric nanoparticles either dispersed in a polymer matrix or encapsulated in the polymer. To regulate the release of the drug, there are some critical parameters: first is the particle size and the particle size distribution, since a particle size is reduced to the nanometer dimension range to increase its surface area [64]. Another important parameter is the controlled release of drugs by using stimuli-responsive materials that can undergo a physicochemical change in response to environmental triggers.

Micelle-like nanoparticles can easily approach the submicrometer scale. Several types of such micelle-like nanoparticles and vesicles based on block copolymers and polyelectrolyte complexes have been prepared. Pluronic micelles (typically 20–100 nm) for drug delivery have been studied for poorly water-soluble drugs, such as amphotericin B (AmB), propofol, paclitaxel, and photosensitizers, based on the compatibility of

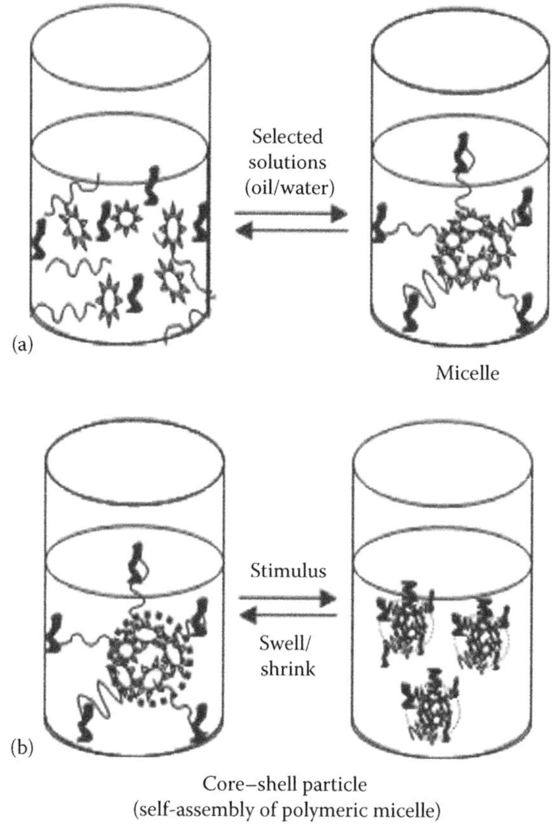

FIGURE 14.7 Scheme demonstrating various configurations of stimuli-responsive nanoparticles: (a) micelles and (b) core–shell particles in suspension and the swelling/shrinking behavior of external stimuli.

a drug and the core of a polymeric micelle [65]. To date, numerous distinctive, intelligent nanoscaled micelles, such as temperature- [13,66], pH- [67], and magnetic field–responsive [68] micelles, have been reported as drug delivery systems.

Polymeric nanoparticles are matrix-type, solid colloidal particles, which generally exhibit greater stability than micelles. Thus, the kinetics of drug release can be adjusted across a very broad range by the conjugation of drugs with macromolecules and the regulation of their transport across the capsule wall. In core–shell particle structure systems, depending on their polarity, drug molecules can be entrapped in (i) the core (nonpolar molecule), (ii) the shell (polar molecule), and (iii) in between the core and the shell (intermediate polarity) [69–71].

14.3.4.2 Tissue Engineering

Stimuli-responsive polymeric biointerfaces have attracted interest in tissue engineering and in bioseparation. Switching and tuning the adhesion between stimuli-responsive materials and proteins/cells have been explored to control cell [72] and protein adhesion [73]. Advanced tissue engineering technology has been significantly developed for the cultivation of cell sheets in a desired structure. A drop in the ambient temperature below the LCST causes a grafted polymer to change its surface properties from hydrophilic to hydrophobic, which causes detachment of the cell stratum. Mammalian cells are normally cultivated on a hydrophobic solid substrate and are detached from the substrate by protease treatment, which often damages the cells by hydrolyzing various membrane-associated protein molecules. A PNIPAAm-grafted surface is hydrophobic at 37°C, since this temperature is above the critical temperature for the grafted polymer and cells grow well on it. A decrease in temperature results in surface transition to the hydrophilic state and the cells can be easily detached from the solid substrate without any damage [74]. Using the cell sheet technology, a cell-dense tissue is formed comprising only cells (without any support material) [75].

Cellular interactions with PNIPAAm-grafted surfaces can be regulated vertically using the thickness of the grafted polymer layers in the nanometer scale range. PNIPAAm-grafted surfaces with 15–20 nm thick layers exhibit temperature-dependent cell adhesion/detachment, while surfaces with layers thicker than 30 nm do not support cell adhesion. The changes in cell adhesion are explained by the limited mobility of the surface-grafted polymer chains as a function of grafting, hydration, and temperature [76]. Lateral regulation of cell adhesion on a smart surface is achieved by nanopatterning of surface chemistry.

Cell sheet engineering favors the formation of scaffold-free, transplantable 3-D tissues from a thermoresponsive cell culture substrate (Figure 14.8). Recovered cocultured cell sheets can be manipulated, moved, and sandwiched between other structures. The double-layered coculture can be achieved by placing a contiguous cell sheet, for example, made up of confluent human aortic endothelial cells recovered from PNIPAAm-grafted surfaces, onto a layer of rat hepatocytes. The double-layered structure of endothelial cells and hepatocytes remains in tight contact during culturing. It has been shown that hepatocytes in the layered coculture system with the sheet of endothelial cells can maintain a differentiated cell shape and albumin expression for over 41 days of culture, whereas the functions disappear within 10 days of culture in control hepatocytes without contact with the endothelial cell sheet [77].

Okano's group has successfully fabricated a microfluidic device using PNIPAAm on the surfaces of tissue culture polystyrene (TCPS) dishes (PNIPAAm-TCPS) using electron beam irradiation polymerization [78]. The flow rates in the channels of the microfluidic device could be controlled in the design. A hydrophobic PNIPAAm surface at 37°C provided convenient cell adhesion, spreading, and growth. By reducing the temperature to 20°C and applying a defined flow, cells were removed from the hydrated PNIPAAm layer. Cells applied with a high shear stress were removed from the substrate faster than those applied with a low shear stress were. The results of this study suggest that the microfluidic device could be a useful cell sheet culture surface for tissue engineering [78].

14.4 MICRO- AND NANOFLUIDIC DEVICES

The design and use of micro- and nanoelectromechanical systems (MEMS and NEMS) have rapidly gained popularity in the biomedical field over the last

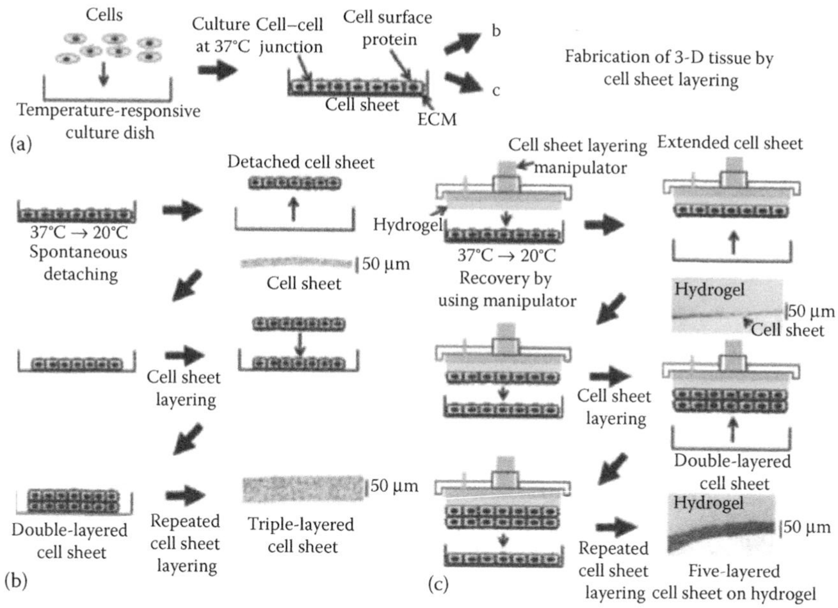

FIGURE 14.8 (**See color insert**) Fabrication of 3-D cell-dense tissues by cell sheet engineering. (a) Cells are cultivated on a temperature-responsive culture dish until reaching confluence. (b, c) The cell sheet is recovered by changing the temperature, and the 3-D tissues are fabricated by the following alternative methods. Using a simple pipetting method, extremely cell-dense 3-D tissue constructs can be fabricated by layering recovered cell sheets. Using the hydrogel-coated, plunger-like manipulator method, the cell sheet is recovered from the dish by adhering to the hydrogel surface within the manipulator. (From Haraguchi, Y., Shimizu, T., Sasagawa, T. et al., *Nature Protocols*, 7, 850–857, 2012. With permission.)

decade [79]. Micro- and nanoscale techniques can be used to produce micro- and nanoscaled constructs (Figure 14.9). Microfabrication approaches such as microfluidics are in widespread use in many fields such as microelectronics, diagnostics, therapeutics, genomics, proteomics, drug discovery, and tissue engineering. In tissue engineering, microfluidic lab-on-a-chip systems are ideal for controlling the spatial and temporal distribution of cellular signaling in cells. Also, microfluidic technologies can regulate the chemical and mechanical environment of a cell culture in ways that are not possible in conventional 2-D culture systems [80]. Microfluidic devices provide an attractive alternative to traditional technologies because of their availability and cost-effectiveness; small requirement for solvents, reagents, and cells; smaller overall size; portability; faster processes; and versatility in design [81].

Fluids in microchannels flow laminarly due to the small channel size (typically <500 μm). There is no turbulent mixing between liquids that flow beside each other within the same hollow microchannel [82]. The ability of fluids to flow laminarly within microchannels is important in order to control the spatial positioning of soluble factors relative to cells in microfluidic devices [83] (Figure 14.10).

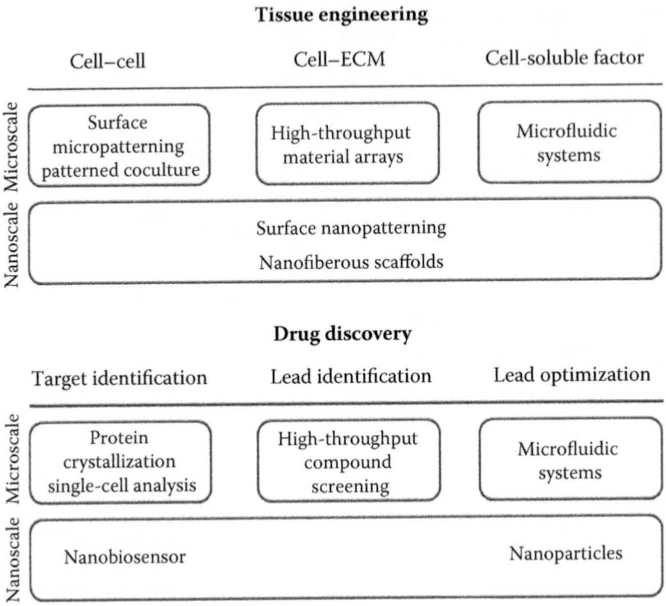

FIGURE 14.9 Micro- and nanoscale approaches used in tissue engineering and drug discovery. (Adapted from Chung, B.G., Kang, L., and Khademhosseini, A., *Expert Opinion on Drug Discovery*, 2, 1–16, 2007.)

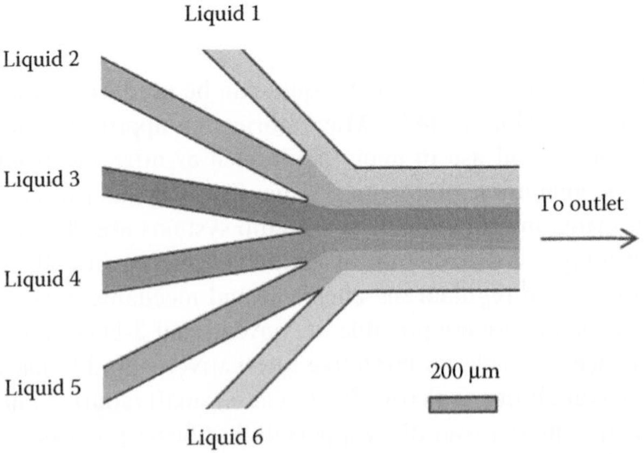

FIGURE 14.10 Six-layered laminar flow in microfluidic channel. (Adapted from Weibel, D.B., Kruithof, M., Potenta, S., Sia, S.K., Lee, A., and Whitesides, G.M., *Analytical Chemistry*, 77, 4726–4733, 2005.)

14.4.1 Fabrication of Micro- and Nanofluidic Devices

Microfluidic devices with many microscopic channels can be fabricated through various surface treatments or lithographic techniques. Currently, soft lithography is one of the most commonly used techniques for the fabrication of micro- and nanofluidic devices. Soft lithography [85] is a set of techniques for microfabrication, which is based on printing and molding using elastomeric stamps fabricated from patterned silicon wafers without the use of expensive "clean rooms" and photolithographic equipment [86] (Figure 14.11). The widely used elastomer material for soft lithography is polydimethylsiloxane (PDMS) due to its advantages of ease of fabrication, physical properties, and economy [87]. In addition, it is nontoxic and nonflammable, and it can be used to culture cells.

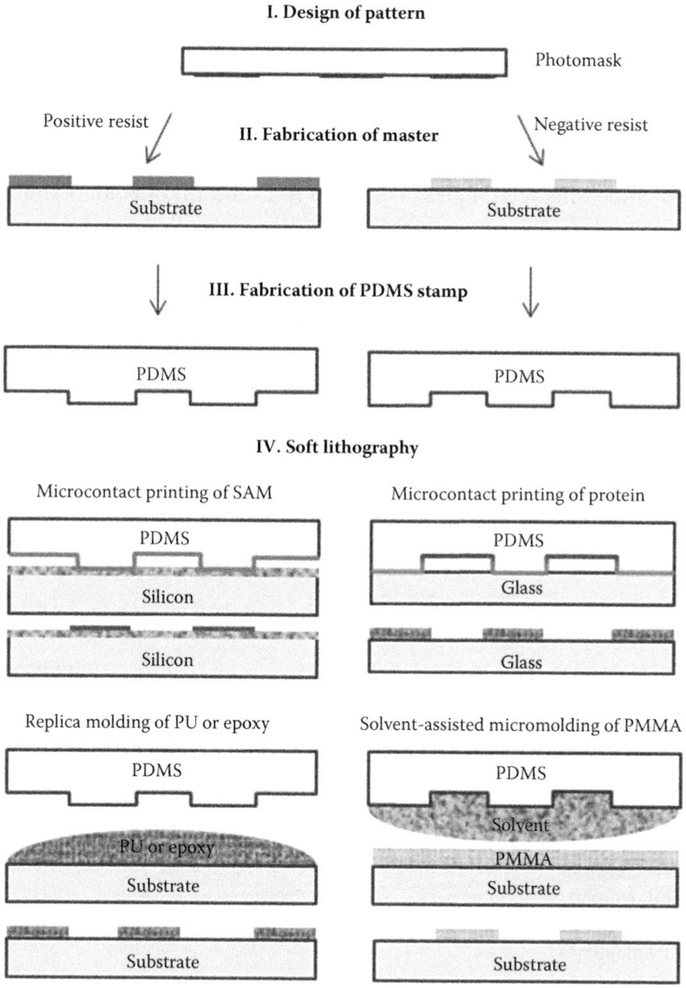

FIGURE 14.11 Schematic illustration of the basic soft lithography techniques. (Adapted from Qin, D., Xia, Y., and Whitesides, G.M., *Nature Protocols*, 5, 491–502, 2010.)

Soft lithography for fabricating microstructures has some advantages over the photolithography technique in biological applications. For example, it offers the ability to manage the molecular structure of surfaces, to pattern the complex molecules relevant to biology, to distribute molecules spatially on a surface, to fabricate channel structures appropriate for microfluidics in a rapid and inexpensive manner, and to pattern and manipulate cells [89].

Soft lithography includes a wide variety of techniques; thus, it can usually be divided into three main categories, each of which can be further divided [90]: replica molding, embossing, and printing.

Replica molding [91] includes the transfer of a pattern from a rigid or elastomeric mold into another material by solidifying a liquid polymer precursor against a topographically patterned mold to fabricate objects with a specific topography. Replica molding can be further divided into subgroups: microtransfer molding, micromolding in capillaries, and ultraviolet (UV) molding. Embossing can be defined as a process that involves imprinting a pattern onto an initially flat surface by pressing a mold into the surface. This technique can generally be divided into two subgroups based on the master type used: nanoimprinting (uses a rigid master) and solvent-assisted micromolding (uses a soft mold as the master). There are differences between the replica molding and embossing techniques. The mold is a solid film in embossing, while in replica molding it is a liquid precursor [90]. Printing includes material transfer from the mold onto the substrate. This technique can be further divided into the subgroups of microcontact printing and nanotransfer printing (Figure 14.12).

14.4.2 Flow Control by Thermoresponsive Polymers

Valves are one of the crucial components of flow control in microfluidic systems. An interesting approach is the use of hydrogels in valves for microfluidics applications.

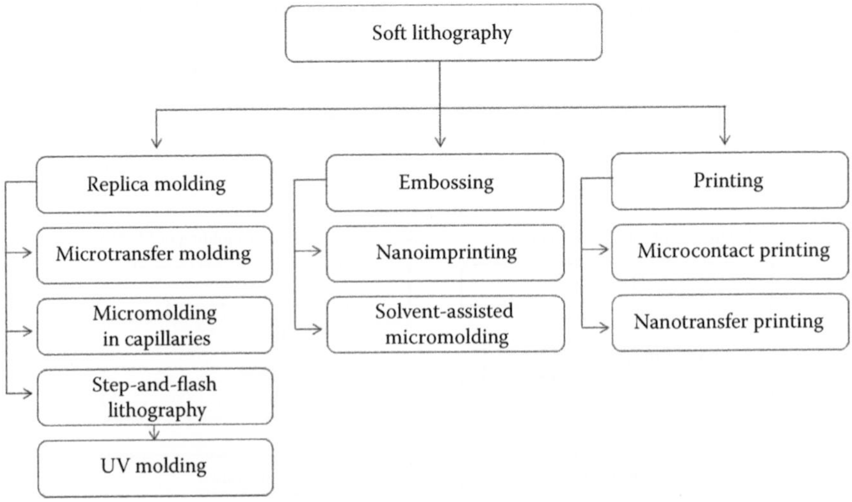

FIGURE 14.12 Soft lithography can usually be divided into three main categories, each of which can be further divided.

The responsive hydrogel-based microfluidics has a significant advantage over conventional microfluidics. Responsive hydrogel-integrated microfluidic devices do not require external power, external control, and complex fabrication schemes to control flow [92]. Here, the valves are formed with responsive hydrogels either by depositing a thin layer on the top of a rigid membrane surface, or by suitable functionalization of the pores, which, in turn, causes shrinking/swelling or on/off in response to external stimuli [93] (Figure 14.13).

Geiger et al. presented a highly functional microfluidic device exhibiting an integrated thermally sensitive hydrogel valve [95]. The valve is normally closed at room temperature. Upon heating above the LCST of 32°C, the polymer valve becomes hydrophobic, shrinks, and forms large pores, thus allowing the solution to flow [95]. Chunder et al. produced a smart surface in a microfluidic channel acting as a thermosensitive valve, which enabled the control of fluid flow temperature change [96]. In this system, when the microvalve is open, the deformation of the membrane reduces the volume of the reservoir chamber and pushes the fluid through the microvalve [96].

Beebe and colleagues have reported an innovative use of responsive hydrogels for microfluidic systems [97]. This method allows pH-responsive hydrogels of different shapes and sizes to be integrated directly into microfluidic systems. The system is based on poly(hydroxyethyl methacrylate)-co-acrylic acid (polyHEMA-co-AA). The approach combines lithography, photopolymerization, and microfluidics to create functional components (valves) within microchannels for local flow control by

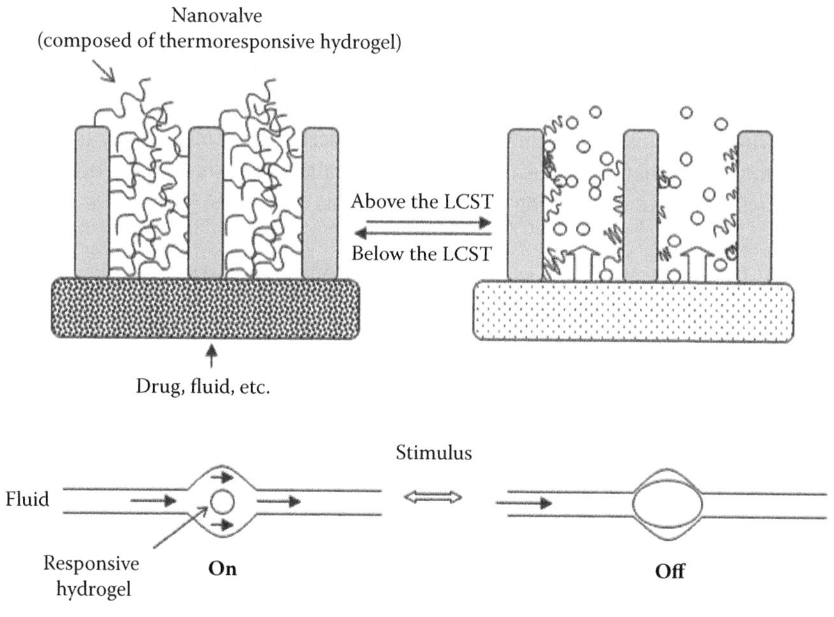

FIGURE 14.13 Schematic overview of two types of membrane valves with stimuli-responsive hydrogel. (Adapted from Yang, Q., Adrus, N., Tomicki, F., and Ulbricht, M., *Journal of Materials Chemistry*, 21, 2783–2811, 2010.)

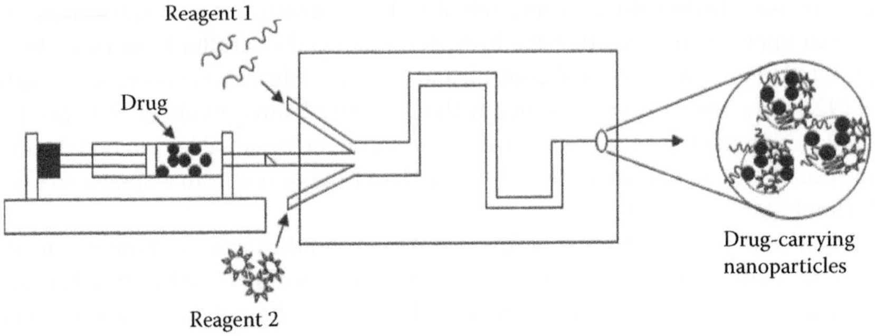

FIGURE 14.14 Microfluidic system for the synthesis of nanoparticles. (Adapted from Tsui, J.H., Lee, W., Pun, S.H., Kim, J., and Kim, D.H., *Advanced Drug Delivery Reviews*, 65, 1575–1588, 2013.)

reversible expansion and contraction of the hydrogels depending on the pH of the surrounding environment [97].

A microfluidic platform can also be used for the reproducible synthesis of nanoparticles in highly controlled steps (Figure 14.14). The ability of microfluidic systems to rapidly mix reagents and to provide homogeneous reaction environments, continuously varying reaction conditions, rapid temperature control, and addition of reagents at precise time intervals during the progress of a reaction are some of the key features that make microfluidic systems very useful for the synthesis of nanoparticles. Microfluidics succeeded in synthesizing smaller and more homogeneous PLGA-PEG nanoparticles compared with bulk synthesis [98], which enabled control over the rate of mixing in conjunction with control of the precursor composition. Thus, this system may be used to tune nanoparticles' size, homogeneity, and drug loading and release. Another advantage of microfluidics technology is its ability to manipulate small volumes of fluid, enabling the synthesis of large libraries of nanoparticles without consuming excessive amounts of expensive reagents.

14.4.3 Tissue Engineering and Drug Screening Applications

14.4.3.1 Organ-on-a-Chip

There are fundamental differences between the cellular microenvironment under *in vivo* and *in vitro* conditions. The cells in the body exist in a controlled microenvironment that is firmly regulated with respect to interactions with surrounding cells, soluble factors, and ECM molecules, while cells in traditional culture dishes exist in a 2-D environment [83]. Understanding basic biological processes such as mitosis, translation, gene regulation, cell communications, and cell interactions involves examining molecular and cellular changes at the molecular level. Micro- and nanoscale technologies provide a suitable tool to investigate cell–microenvironment interactions (i.e., cell–cell, cell–matrix, and cell–soluble factor interactions) *in vitro* and can be used to control culture conditions and carry out high-throughput experimentation through their size and topography. Therefore, the extracellular signals that regulate a cell's fate can be investigated using these systems [100]. Miniaturized cell culture platforms provide detailed information about cellular behavior at the scale found in living systems. Microfluidic systems incorporating cells with physiological activity have been developed as an alternative to *in vivo* experiments.

Organ-on-a-chip systems are being used for tissue engineering and drug development studies. Huh et al. developed an actual lung-on-a-chip microfluidic device that reproduces the critical structural, functional, and mechanical properties of the human alveolar–capillary interface, which is the basic functional unit of the living lung [101]. In this study, two of the channels fabricated by soft lithography were separated by a thin (10 μm), porous, flexible membrane made of PDMS. This membrane was coated with ECM proteins (fibronectin or collagen). Then, human alveolar epithelial cells and human pulmonary microvascular endothelial cells were cultured on opposite sides of the ECM-coated membrane (Figure 14.15). The cellular response to a pulmonary infection of bacterial origin and to silica nanoparticles was demonstrated using this system. The results indicated that the developed lung-on-a-chip microfluidic device could reconstitute multiple physiological functions observed in the whole living lung. These findings have shown that this bioinspired microdevice may enhance the capabilities of cell culture models and allow low-cost alternatives to animal and clinical studies for drug screening and toxicology studies [101].

Hepatotoxicity is a medical term that is used to describe liver damage, particularly damage caused by the use of drugs. Recently, liver-on-a-chip systems have been evaluated for use in hepatotoxicity studies. Toh et al. have developed a microfluidic

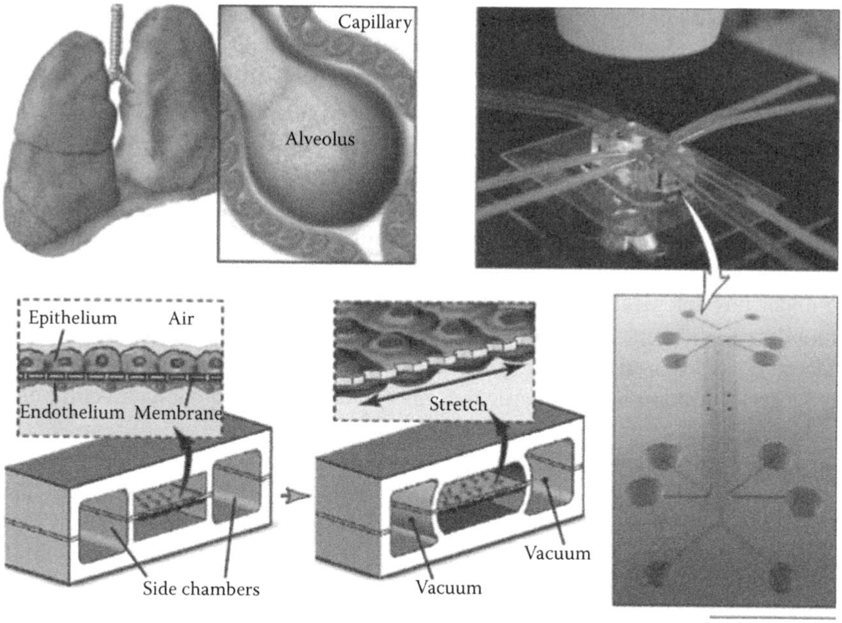

FIGURE 14.15 Schematic representation of the lung-on-a-chip microfluidic device. Epithelial cells and endothelial cells are cultured on opposite sides of the ECM-coated membrane. (From Huh, D., Hamilton, G.A., and Ingber, D.E., *Trends in Cell Biology*, 21, 745–754, 2011. With permission.)

3-D hepatocyte chip (3D HepaTox Chip), each channel engineered to maintain the synthetic and metabolic functions of hepatocytes, to test *in vitro* drug toxicity [102]. The results indicate that *in vitro* toxicity data can be correlated to *in vivo* toxicity data and the 3D HepaTox Chip (liver-on-a-chip) has potential utility in predicting *in vivo* toxicity [102].

The intestine is also a very important organ of the body for drug metabolism. Kimura et al. have developed an integrated microfluidic system for the long-term perfusion culturing and online monitoring of intestinal tissue models [103]. This microfluidic system consists of two independent channels separated by a semipermeable membrane on which Caco-2 cells are seeded and cultured. Caco-2 cells were successfully cultured for over 30 days and the polarized transporter activity of the cells was measured with an on-chip optical detection system [103].

Park et al. have developed a circular microfluidic compartmentalized coculture platform that can be used for central nervous system axon myelination research [104]. The brain-on-a-chip system consists of two compartments: a soma compartment and an axon/glia compartment linked through arrays of axon-guiding microfluidic channels. The results showed that the microfluidic platform can be used as an excellent *in vitro* coculture system platform for studying localized axon–glia interaction and signaling [104].

A recent approach to microfluidic systems has been to determine the cell migration rate. For example, van der Meer et al. have developed a microfluidic wound-healing assay to measure the endothelial cell migration rate [105]. Briefly, a confluent endothelial cell monolayer inside a microfluidic channel was exposed to the protease trypsin by flushing the channel with three parallel fluid streams to generate a wound model. Cell viability, monolayer integrity, and cell migration rate results were compared with the conventional wound-healing assay. The results indicated that the microfluidic system is a useful research tool for the wound-healing assay [105].

High blood pressure is a major risk factor for many cardiovascular diseases. Understanding the molecular basis of the manifestations of cardiovascular diseases requires micro- and nanoscale approaches. Günther et al. have developed a microfluidic system for the routine determination of the resistance artery structure and function under physiological conditions [106]. In the study, small arteries were loaded, immobilized, and kept intact in microenvironments on a microfluidic device. The results indicated that the physiological response of a small blood vessel exposed to varying concentrations of biochemical drugs could be investigated via this microfluidics system [106].

In another study, Plouffe et al. have developed a unique microfluidic device capable of capturing circulating endothelial progenitor cells (EPCs) by understanding their surface chemistries and adhesion properties [107]. Six different antibodies (anti-CD34, anti-CD31, antivascular endothelial growth factor receptor-2 [VEGFR-2], anti-CD146, anti-CD45, and anti-von Willebrand factor [vWF]), designed to match the surface antigens on ovine peripheral blood-derived EPCs, were immobilized on the microfluidic device surface. Cell suspensions flowed through the microfluidic device. As a result, the heart-on-a-chip microfluidic device may provide a platform for isolating progenitor cells for tissue engineering applications and cell-based revascularization strategies, and also for addressing challenges in cardiovascular disease [107].

14.4.3.2 Body-on-a-Chip

Another new biomimetic microfluidic device is a more sophisticated organ-on-a-chip microdevice, called body-on-a-chip, animal-on-a-chip, human-on-a-chip, or living-on-a-chip systems [108]. For years, the pharmaceutical industry has searched for experimental model systems to accelerate the drug-development process and to develop drugs that are safer and more effective in humans at a lower cost. Traditional animal testing approaches are expensive, and they often fail to predict human responses [101]. In the human body, all organs are functionally integrated with each other. The integrated "human-on-a-chip" models have been developed in recent studies (Figure 14.16). Sung et al. have developed a microfluidic network consisting of Matrigel-encapsulated colon cancer cells (HCT-116) and hepatoma cells (HepG2/C3A) cultured in separate chambers representing the liver, tumor, and marrow, to test the cytotoxicity of anticancer drugs [109]. The results indicate that the 3-D hydrogel cell cultures have the potential to test the metabolism-dependent toxicity of anticancer drugs in a more physiologically realistic environment.

Zhang et al. have developed a multichannel 3-D microfluidic cell culture system, culturing four human cell types (C3A, A549, HK-2, and HPA) to represent the liver, lung, kidney, and adipose tissue for potential application in human drug screening [110]. The results show that more sophisticated organ-on-a-chip microdevices could potentially be used to supplement or even replace animal models in drug screening studies. These studies are promising for conducting an investigation of the

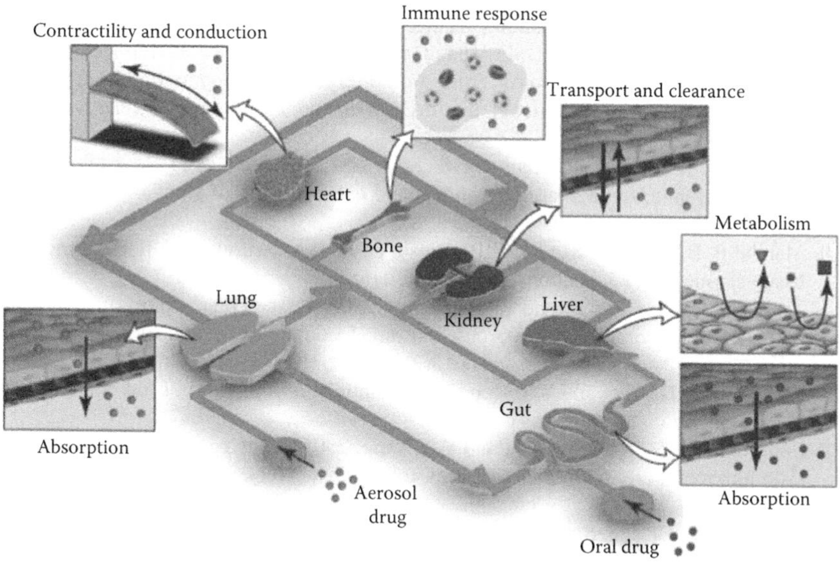

FIGURE 14.16 The human-on-a-chip model. In this system, interconnected compartments contain cells specific to different organs, linked through a microfluidic channel system. (From Huh, D., Hamilton, G.A., and Ingber, D.E., *Trends in Cell Biology*, 21, 745–754, 2011. With permission.)

human physiology in an organ-specific context and drug development and toxicology responses [110].

ACKNOWLEDGMENT

YME acknowledges the support of the Turkish Academy of Sciences, TÜBA (Ankara, Turkey).

REFERENCES

1. Elçin Y. M. 2003. *Tissue Engineering, Stem Cells and Gene Therapies*, AEMB Series, vol. 534, New York, Kluwer Academic/Plenum.
2. Chung B. G., L. Kang, and A. Khademhosseini. 2007. Micro- and nanoscale technologies for tissue engineering and drug discovery applications. *Expert Opinion on Drug Discovery* 2(12): 1–16.
3. Lee W., V. Lee, S. Polio, et al. 2010. On-demand three-dimensional freeform fabrication of multi-layered hydrogel scaffold with fluidic channels. *Biotechnology and Bioengineering* 105: 1178–1186.
4. Zhang Y., Y. Yu, H. Chen, and I. T. Ozbolat. 2013. Characterization of printable cellular micro-fluidic channels for tissue engineering. *Biofabrication* 5: 1–11.
5. Wang H. S., G. D. Fu, and X. S. Li. 2009. Functional polymeric nanofibers from electrospinning. *Recent Patents on Nanotechnology* 3: 21–31.
6. Tamayol A., M. Akbari, N. Annabi, A. Paul, A. Khademhosseini, and D. Juncker. 2013. Fiber-based tissue engineering: Progress, challenges, and opportunities. *Biotechnology Advances* 31: 669–687.
7. Pham Q. P., U. Sharma, and A. G. Mikos. 2006. Electrospinning of polymeric nanofibers for tissue engineering applications: A review. *Tissue Engineering* 12(5): 1197–1211.
8. Jalili R., S. H. Aboutalebi, D. E. lzadeh, et al. 2013. Scalable one-step wet-spinning of graphene fibers and yarns from liquid crystalline dispersions of graphene oxide: Towards multifunctional textiles. *Advanced Functional Materials* 23(43): 5345–5354.
9. Lee B. R., K. H. Lee, E. Kang, D. S. Kim, and S. H. Lee. 2011. Microfluidic wet spinning of chitosan-alginate microfibers and encapsulation of HepG2 cells in fibers. *Biomicrofluidics* 5: 022208.
10. Mandal B. B. and S. C. Kundu. 2010. Biospinning by silkworms: Silk fiber matrices for tissue engineering applications. *Acta Biomaterialia* 6(2): 360–371.
11. Yim E. K. F., I. C. Liao, and K. W. Leong. 2007. Tissue compatibility of interfacial polyelectrolyte complexation fibrous scaffold: Evaluation of blood compatibility and biocompatibility. *Tissue Engineering* 13(2): 423–433.
12. Zhmayev E., D. Cho, and Y. L. Joo. 2010. Nanofibers from gas-assisted polymer melt electrospinning. *Polymer* 51: 4140–4144.
13. Soppimath K. S., D. C. W. Tan, and Y. Y. Yang. 2005. pH-triggered thermally responsive polymer core–shell nanoparticles for drug delivery. *Advanced Materials* 17: 318–323.
14. Rutledge G. C. and S. V. Fridrikh. 2007. Formation of fibers by electrospinning. *Advanced Drug Delivery Reviews* 59: 1384–1391.
15. Huang Z. M., Y. Z. Zhang, M. Kotaki, and S. Ramakrishna. 2003. A review on polymer nanofibers by electrospinning and their applications in nanocomposites. *Composites Science and Technology* 63: 2223–2253.
16. Bhardwaj N. and S. C. Kundu. 2010. Electrospinning: A fascinating fiber fabrication technique. *Biotechnology Advances* 28: 325–347.

17. Jang J. H., O. Castano, and H. W. Kim. 2009. Electrospun materials as potential platforms for bone tissue engineering. *Advanced Drug Delivery Reviews* 61: 1065–1083.
18. Agarwal S., A. Greiner, and J. H. Wendorff. 2013. Functional materials by electrospinning of polymers. *Progress in Polymer Science* 38: 963–991.
19. Ramakrishna S., K. Fujihara, W. E. Teo, T. Yong, Z. Ma, and R. Ramaseshan. 2006. Electrospun nanofibers: Solving global issues. *Materials Today* 9(3): 40–50.
20. Teo W. E. and S. Ramakrishna. 2006. A review on electrospinning design and nanofibre assemblies. *Nanotechnology* 17: R89–R106.
21. Schiffman J. D., and C. L. Schauer. 2008. A review: Electrospinning of biopolymer nanofibers and their applications. *Polymer Reviews* 48: 317–352.
22. Khadka D. B. and D. T. Haynie. 2012. Protein- and peptide-based electrospun nanofibers in medical biomaterials. *Nanomedicine: Nanotechnology, Biology, and Medicine* 8: 1242–1262.
23. Sell S. A., P. S. Wolfe, K. Garg, J. M. McCoo, I. A. Rodriguez, and G. L. Bowlin. 2010. The use of natural polymers in tissue engineering: A focus on electrospun extracellular matrix analogues. *Polymers* 2: 522–553.
24. Li M., M. J. Mondrinos, M. R. Gandhi, F. K. Ko, A. S. Weiss, and P. I. Lelkes. 2005. Electrospun protein fibers as matrices for tissue engineering. *Biomaterials* 26: 5999–6008.
25. Inanç B., Y. E. Arslan, Ş. Şeker, A. E. Elçin, and Y. M. Elçin. 2009. Periodontal ligament cellular structures engineered with electrospun poly (DL-lactide-co-glycolide) nanofibrous membrane scaffolds. *Journal of Biomedical Materials Research Part A* 90(1): 186–195.
26. Yoshimoto H., Y. M. Shin, H. Terai, and J. P. Vacanti. 2003. A biodegradable nanofiber scaffold by electrospinning and its potential for bone tissue engineering. *Biomaterials* 24: 2077–2082.
27. Yang F., R. Murugan, S. Wang, and S. Ramakrishna. 2005. Electrospinning of nano/micro scale poly(L-lactic acid) aligned fibers and their potential in neural tissue engineering. *Biomaterials* 26: 2603–2610.
28. Sirivisoot S. and B. S. Harrison. 2011. Skeletal myotube formation enhanced by electrospun polyurethane carbon nanotube scaffolds. *International Journal of Nanomedicine* 6: 2483–2497.
29. Min B. M., G. Lee, S. H. Kim, Y. S. Nam, T. S. Lee, and W. H. Park. 2004. Electrospinning of silk fibroin nanofibers and its effect on the adhesion and spreading of normal human keratinocytes and fibroblasts in vitro. *Biomaterials* 25: 1289–1297.
30. Nettles D. L., S. H. Elder, and J. A. Gilbert. 2006. Potential use of chitosan as a cell scaffold material for cartilage tissue engineering. *Tissue Engineering* 8(6): 1009–1016.
31. Ji Y., K. Ghosh, X. Z. Shu, et al. 2006. Electrospun three-dimensional hyaluronic acid nanofibrous scaffolds. *Biomaterials* 27: 3782–3792.
32. Buttafoco L., N. G. Kolkman, P. E. Buijtenhuijs, et al. 2006. Electrospinning of collagen and elastin for tissue engineering applications. *Biomaterials* 27: 724–734.
33. Wannatong L., A. Sirivat, and P. Supaphol. 2004. Effects of solvents on electrospun polymeric fibers: Preliminary study on polystyrene. *Polymer International* 53: 1851–1859.
34. Hong J. K. and S. V. Madihally. 2011. Next generation of electrosprayed fibers for tissue regeneration. *Tissue Engineering: Part B, Reviews* 17(2): 125–142.
35. Liu W., S. Thomopoulos, and Y. Xia. 2012. Electrospun nanofibers for regenerative medicine. *Advanced Healthcare Materials* 1(1): 10–25.
36. Kai D., G. Jin, M. P. Prabhakaran, and S. Ramakrishna. 2013. Electrospun synthetic and natural nanofibers for regenerative medicine and stem cells. *Biotechnology Journal* 8: 59–72.
37. Xie J., M. R. MacEwan, A. G. Schwartz, and Y. Xia. 2010. Electrospun nanofibers for neural tissue engineering. *Nanoscale* 2: 35–44.

38. Prabhakaran M. P., D. Kai, L. G. Mobarakeh, and S. Ramakrishna. 2011. Electrospun biocomposite nanofibrous patch for cardiac tissue engineering. *Biomedical Materials* 6: 055001.
39. Chen J. P. and C. H. Su. 2011. Surface modification of electrospun PLLA nanofibers by plasma treatment and cationized gelatin immobilization for cartilage tissue engineering. *Acta Biomaterialia* 7: 234–243.
40. Ramachandran K. and P. I. Gouma. 2008. Electrospinning for bone tissue engineering. *Recent Patents on Nanotechnology* 2: 1–7.
41. Zhong S. P., Y. Z. Zhang, and C. T. Lim. 2010. Tissue scaffolds for skin wound healing and dermal reconstruction. *WIREs Nanomed Nanobiotechnology* 2: 510–525.
42. Zhu Y., Y. Cao, J. Pan, and Y. Liu. 2010. Macro-alignment of electrospun fibers for vascular tissue engineering. *Journal of Biomedical Materials Research Part B: Applied Biomaterials* 92(2): 508–516.
43. Li C., C. Vepari, H. J. Jin, H. J. Kim, and D. L. Kaplan. 2006. Electrospun silk-BMP-2 scaffolds for bone tissue engineering. *Biomaterials* 27(16): 3115–3124.
44. Rho K. S., L. Jeong, G. Lee, et al. 2006. Electrospinning of collagen nanofibers: Effects on the behavior of normal human keratinocytes and early-stage wound healing. *Biomaterials* 27: 1452–1461.
45. Khil M. S., D. I. Cha, H. Y. Kim, I. S. Kim, and N. Bhattarai. 2003. Electrospun nanofibrous polyurethane membrane as wound dressing. *Journal of Biomedical Materials Research Part B: Applied Biomaterials* 67(2): 675–679.
46. Jia J., Y. Y. Duan, S. H. Wang, S. F. Zhang, and Z. Y. Wang. 2007. Preparation and characterization of antibacterial silver-containing nanofibers for wound dressing applications. *Journal of US-China Medical Science* 4: 52–54.
47. Chainani A., K. J. Hippensteel, A. Kishan, et al. 2013. Multi-layered electrospun scaffolds for tendon tissue engineering. *Tissue Engineering Part A* 19(23–24): 2594–2604.
48. Vasita R. and D. S. Katti. 2006. Nanofibers and their applications in tissue engineering. *International Journal of Nanomedicine* 1(1): 15–30.
49. Mo X. M., C. Y. Xu, M. Kotaki, and S. Ramakrishna. 2004. Electrospun P(LLA-CL) nanofiber: A biomimetic extracellular matrix for smooth muscle cell and endothelial cell proliferation. *Biomaterials* 25: 1883–1890.
50. Yang F., R. Murugan, S. Ramakrishna, X. Wang, Y. X. Ma, and S. Wang. 2004. Fabrication of nano-structured porous PLLA scaffold intended for nerve tissue engineering. *Biomaterials* 25: 1891–1900.
51. Kumar A., A. Srivastava, I. Y. Galaev, and B. Mattiasson. 2007. Smart polymers: Physical forms and bioengineering applications. *Progress in Polymer Science* 32: 1205–1237.
52. Qiu Y. and K. Park. 2001. Environment-sensitive hydrogels for drug delivery. *Advanced Drug Delivery Reviews* 53: 321–339.
53. Peppas L. B. and N. A. Peppas. 1989. Solute and penetrant diffusion in swellable polymer: IX. The mechanism of drug release from PH-sensitive swelling-controlled systems. *Journal of Controlled Release* 8: 267–274.
54. Siegel R. A., I. Johannes, C. A. Hunt, and B. A. Firestone. 1992. Buffer effects on swelling kinetics in polybasic gels. *Pharmaceutical Research* 9: 76–81.
55. Mano J. F., G. A. Silva, H. S. Azevedo, et al. 2007. Natural origin biodegradable systems in tissue engineering and regenerative medicine: Present status and some moving trends. *Journal of the Royal Society Interface* 4: 999–1030.
56. Gill E. S. and S. M. Hudson. 2004. Stimuli-responsive polymers and their bioconjugates. *Progress in Polymer Science* 29: 1173–1222.
57. Heskins M. and J. E. Guillet. 1968. Solution properties of poly (N-isopropylacrylamide). *Journal of Macromolecular Science, Part A: Pure and Applied Chemistry* 2(8): 1441–1455.

58. Suwa K., K. Morishita, A. Kishida, and M. Akashi. 1997. Synthesis and functionalities of poly(N-vinylalkylamide). V. Control of a lower critical solution temperature of poly (N-vinylalkylamide). *Journal of Polymer Science Part A: Polymer Chemistry* 35: 3087–3094.
59. Wanka G., H. Hoffmann, and W. Ulbricht. 1994. Phase diagrams and aggregation behavior of poly (oxyethylene)-poly (oxypropylene)– poly(oxyethylene) triblock copolymers in aqueous solutions. *Macromolecules* 27: 4145–4159.
60. Feil H., Y. H. Bae, J. Feijen, and S. W. Kim. 1993. Effect of comonomer hydrophilicity and ionization on the lower critical solution temperature of N-isopropylacrylamide copolymer. *Macromolecules* 26: 2496–2500.
61. Tokarev I. and S. Minko. 2009. Stimuli-responsive hydrogel thin films. *Soft Matter* 5: 511–524.
62. Stuart M. A. C., W. T. S. Huck, J. Genzer, et al. 2010. Emerging applications of stimuli-responsive polymer materials. *Nature Materials* 9: 101–113.
63. Koo O. M., I. Rubinstein, and H. Onyuksel. 2005. Role of nanotechnology in targeted drug delivery and imaging: A concise review. *Nanomedicine: Nanotechnology, Biology, and Medicine* 1: 193–212.
64. Panyam J. and V. Labhasetwar. 2003. Biodegradable nanoparticles for drug and gene delivery to cells and tissue. *Advanced Drug Delivery Reviews* 55(3): 329–347.
65. Kwon G. S. 2003. Polymeric micelles for delivery of poorly water-soluble compounds. *Critical Reviews in Therapeutic Drug Carrier Systems* 20: 357–403.
66. Wei H., X. Z. Zhang, Y. Zhou, S. X. Cheng, and R. X. Zhuo. 2006. Self-assembled thermoresponsive micelles of poly(N-isopropylacrylamide-b-methyl methacrylate). *Biomaterials* 27: 2028–2034.
67. Lee E. S., K. Na, and Y. H. Bae. 2005. Super pH-sensitive multifunctional polymeric micelle. *Nano Letters* 5: 325–329.
68. Park J. H., G. Maltzahn, E. Ruoslahti, S. N. Bhatia, and M. J. Sailor. 2008. Micellar hybrid nanoparticles for simultaneous magnetofluorescent imaging and drug delivery. *Angewandte Chemie International Edition* 47(38): 7284–7288.
69. Kataoka K., A. Harada, and Y. Nagasaki. 2001. Block copolymer micelles for drug delivery: Design, characterization and biological significance. *Advanced Drug Delivery Reviews* 47: 113–131.
70. Guo X., S. Yuan, S. Yang, K. Lv, and S. K. Yuan. 2011. Mesoscale simulation on patterned core-shell nanosphere model for amphiphilic block copolymer. *Colloids and Surfaces A: Physicochemical and Engineering Aspects* 384: 212–218.
71. Kaditi E., G. Mountrichas, and S. Pispas. 2011. Amphiphilic block copolymers by a combination of anionic polymerization and selective post-polymerization functionalization. *European Polymer Journal* 47: 415–434.
72. Lutolf M. P., J. L. Lauer-Fields, H. G. Schmoekel, et al. 2003. Synthetic matrix metalloproteinase-sensitive hydrogels for the conduction of tissue regeneration: Engineering cell-invasion characteristics. *Proceedings of the National Academy of Sciences* 100: 5413–5418.
73. Alarcon C. D. H., T. Farhan, V. L. Osborne, W. T. S. Huck, and C. Alexander. 2005. Bioadhesion at micro-patterned stimuli-responsive polymer brushes. *Journal of Materials Chemistry* 15: 2089–2094.
74. Hatakeyama H., A. Kikuchi, M. Yamato, and T. Okano. 2006. Bio-functionalized thermoresponsive interfaces facilitating cell adhesion and proliferation. *Biomaterials* 27: 5069–5078.
75. Haraguchi Y., T. Shimizu, T. Sasagawa, et al. 2012. Fabrication of functional three-dimensional tissues by stacking cell sheets in vitro. *Nature Protocols* 7(5): 850–857.
76. Kikuchi A. and T. Okano. 2005. Nanostructured designs of biomedical materials: Applications of cell sheet engineering to functional regenerative tissues and organs. *Journal of Controlled Release* 101: 69–84.

77. Tsuda Y., A. Kikuchi, M. Yamato, A. Nakao, Y. Sakurai, and M. Umezu. 2004. The use of patterned dual thermoresponsive surfaces for the collective recovery as co-cultured cell sheets. *Biomaterials* 26: 1885–1893.
78. Tang Z., Y. Akiyama, K. Itoga, J. Kobayashi, M. Yamato, and T. Okano. 2012. Shear stress-dependent cell detachment from temperature-responsive cell culture surfaces in a microfluidic device. *Biomaterials* 33: 7405–7411.
79. Ashraf M. W., S. Tayyaba, and N. Afzulpurkar. 2011. Micro electromechanical systems (MEMS) based microfluidic devices for biomedical applications. *International Journal of Molecular Sciences* 12: 3648–3704.
80. Villa-Diaz L. G., Y. Torisawa, T. Uchida, et al. 2009. Microfluidic culture of single human embryonic stem cell colonies. *Lab on a Chip* 9: 1749–1755.
81. Sia S. K. and G. M. Whitesides. 2003. Microfluidic devices fabricated in poly(dimethylsiloxane) for biological studies. *Electrophoresis* 24: 3563–3576.
82. Huh D., G. A. Hamilton, and D. E. Ingber. 2011. From 3D cell culture to organs-on-chips. *Trends in Cell Biology* 21: 745–754.
83. Khademhosseini A., R. Langer, J. Borenstein, and J. P. Vacanti. 2006. Microscale technologies for tissue engineering and biology. *Proceedings of the National Academy of Sciences* 103: 2480–2487.
84. Weibel D. B., M. Kruithof, S. Potenta, S. K. Sia, A. Lee, and G. M. Whitesides. 2005. Torque-actuated valves for microfluidics. *Analytical Chemistry* 77: 4726–4733.
85. Xia Y. and G. M. Whitesides. 1998. Soft lithography. *Annual Review of Materials* 28: 153–184.
86. Zhao X. M., Y. Xia, and G. M. Whitesides. 1997. Soft lithographic methods for nanofabrication. *Journal of Materials Chemistry* 7(7): 1069–1074.
87. Friend J. and L. Yeo. 2010. Fabrication of microfluidic devices using polydimethylsiloxane. *Biomicrofluidics* 4: 026502.
88. Qin D., Y. Xia, and G. M. Whitesides. 2010. Soft lithography for micro- and nanoscale patterning. *Nature Protocols* 5: 491–502.
89. Whitesides G. M., E. Ostuni, S. Takayama, X. Y. Jiang, and D. E. Ingber. 2001. Soft lithography in biology and biochemistry. *Annual Review of Biomedical Engineering* 3: 335–373.
90. Gates B. D., Q. Xu, J. C. Love, D. B. Wolfe, and G. M. Whitesides. 2004. Unconventional nanofabrication. *Annual Review of Materials Research* 34: 339–372.
91. Xia Y. N., E. Kim, X. M. Zhao, J. A. Rogers, M. Prentiss, and G. M. Whitesides. 1996. Complex optical surfaces formed by replica molding against elastomeric masters. *Science* 273: 347–349.
92. Kumar A., A. Srivastava, I. Y. Galaev, and B. Mattiasson. 2007. Smart polymers: Physical forms and bioengineering applications. *Progress in Polymer Science* 32: 1205–1237.
93. Argentiere S., G. Gigli, M. Mortato, I. Gerges, and L. Blasi. 2012. Smart microfluidics: The role of stimuli-responsive polymers in microfluidic devices. Chapter 6, in: R. T. Kelly (ed.), *Advances in Microfluidics*, InTech Corporation, Rijeka, pp. 128–154.
94. Yang Q., N. Adrus, F. Tomicki, and M. Ulbricht. 2010. Composites of functional polymeric hydrogels and porous membranes. *Journal of Materials Chemistry* 21(9): 2783–2811.
95. Geiger E. J., A. P. Pisano, and F. Svec. 2010. A polymer-based microfluidic platform featuring on-chip actuated hydrogel valves for disposable applications. *Journal of Microelectromechanical Systems* 19(4): 944–950.
96. Chunder A., K. Etcheverry, G. Londe, H. J. Cho, and L. Zhai. 2009. Conformal switchable superhydrophobic/hydrophilic surfaces for microscale flow control. *Colloids and Surfaces A: Physicochemical and Engineering Aspects* 333(1–3): 187–193.
97. Beebe D. J., J. S. Moore, J. M. Bauer, et al. 2000. Functional hydrogel structures for autonomous flow control inside microfluidic channels. *Nature* 404(6778): 588–590.

98. Karnik R., F. Gu, P. Basto, et al. 2008. Microfluidic platform for controlled synthesis of polymeric nanoparticles. *Nano Letters* 8(9): 2906–2912.
99. Tsui J. H., W. Lee, S. H. Pun, J. Kim, and D. H. Kim. 2013. Microfluidics-assisted in vitro drug screening and carrier production. *Advanced Drug Delivery Reviews.* 65(11–12): 1575–1588.
100. Young E. W. K. and D. J. Beebe. 2010. Fundamentals of microfluidic cell culture in controlled microenvironments. *Chemical Society Reviews* 39(3): 1036–1048.
101. Huh D., B. D. Matthews, A. Mammoto, M. Montoya-Zavala, H. Y. Hsin, and D. E. Ingber. 2010. Reconstituting organ-level lung functions on a chip. *Science* 328: 1662–1668.
102. Toh Y. C., T. C. Lim, D. Tai, G. Xiao, D. van Noort, and H. Yu. 2009. A microfluidic 3D hepatocyte chip for drug toxicity testing. *Lab on a Chip* 9: 2026–2035.
103. Kimura H., T. Yamamoto, H. Sakai, Y. Sakai, and T. Fujii. 2008. An integrated microfluidic system for long-term perfusion culture and on-line monitoring of intestinal tissue models. *Lab on a Chip* 8: 741–746.
104. Park J., H. Koito, J. Li, and A. Han. 2009. Microfluidic compartmentalized co-culture platform for CNS axon myelination research. *Biomedical Microdevices* 11(6): 1145–1153.
105. van der Meer A. D., K. Vermeul, A. A. Poot, J. Feijen, and I. Vermes. 2010. A microfluidic wound-healing assay for quantifying endothelial cell migration. *American Journal of Physiology: Heart and Circulatory Physiology* 298: H719–H725.
106. Günther A., S. Yasotharan, A. Vagaon, et al. 2010. A microfluidic platform for probing small artery structure and function. *Lab on a Chip* 10: 2341–2349.
107. Plouffe B. D., T. Kniazeva, J. E. Mayer Jr., S. K. Murthy, and V. L. Sales. 2009. Development of microfluidics as endothelial progenitor cell capture technology for cardiovascular tissue engineering and diagnostic medicine. *FASEB Journal* 23: 3309–3314.
108. Baker M. 2011. Tissue models: A living system on a chip. *Nature* 471: 661–665.
109. Sung J. H. and M. L. Shuler. 2009. A micro cell culture analog (mCCA) with 3-D hydrogel culture of multiple cell lines to assess metabolism-dependent cytotoxicity of anti-cancer drugs. *Lab on a Chip* 9: 1385–1394.
110. Zhang C., Z. Zhao, N. A. Abdul Rahim, D. van Noort, and H. Yu. 2009. Towards a human-on-chip: Culturing multiple cell types on a chip with compartmentalized microenvironments. *Lab on a Chip* 9: 3185–3192.

Index

E

F

O

P

Q

T

U

V

W

X

Y

Printed and bound by CPI Group (UK) Ltd, Croydon, CR0 4YY
18/10/2024
01776208-0011